T0279672

This collection of essays examines the ways in which disputes and controversies about the application of scientific knowledge are resolved. Four concrete examples of public controversy are considered in detail: the efficacy of Laetrile, the classification of homosexuality as a disease, the setting of safety standards in the workplace, and the utility of nuclear energy as a source of power. The essays in this volume show that debates about these cases are not confined to matters of empirical fact. Rather, as is seen with most scientific and technical controversies, they focus on and are structured by complex ethical, economic, and political interests.

Drs. Engelhardt and Caplan have brought together a distinguished group of scholars from the sciences and humanities, who sketch a theory of scientific controversy and attempt to provide recommendations about the ways in which both scientists and the public ought to seek more informed resolutions of highly contentious issues in science and technology. *Scientific controversies* is offered as a contribution to the better understanding of the roles of both science and nonscientific interests in disputes and controversies pertaining to science and technology.

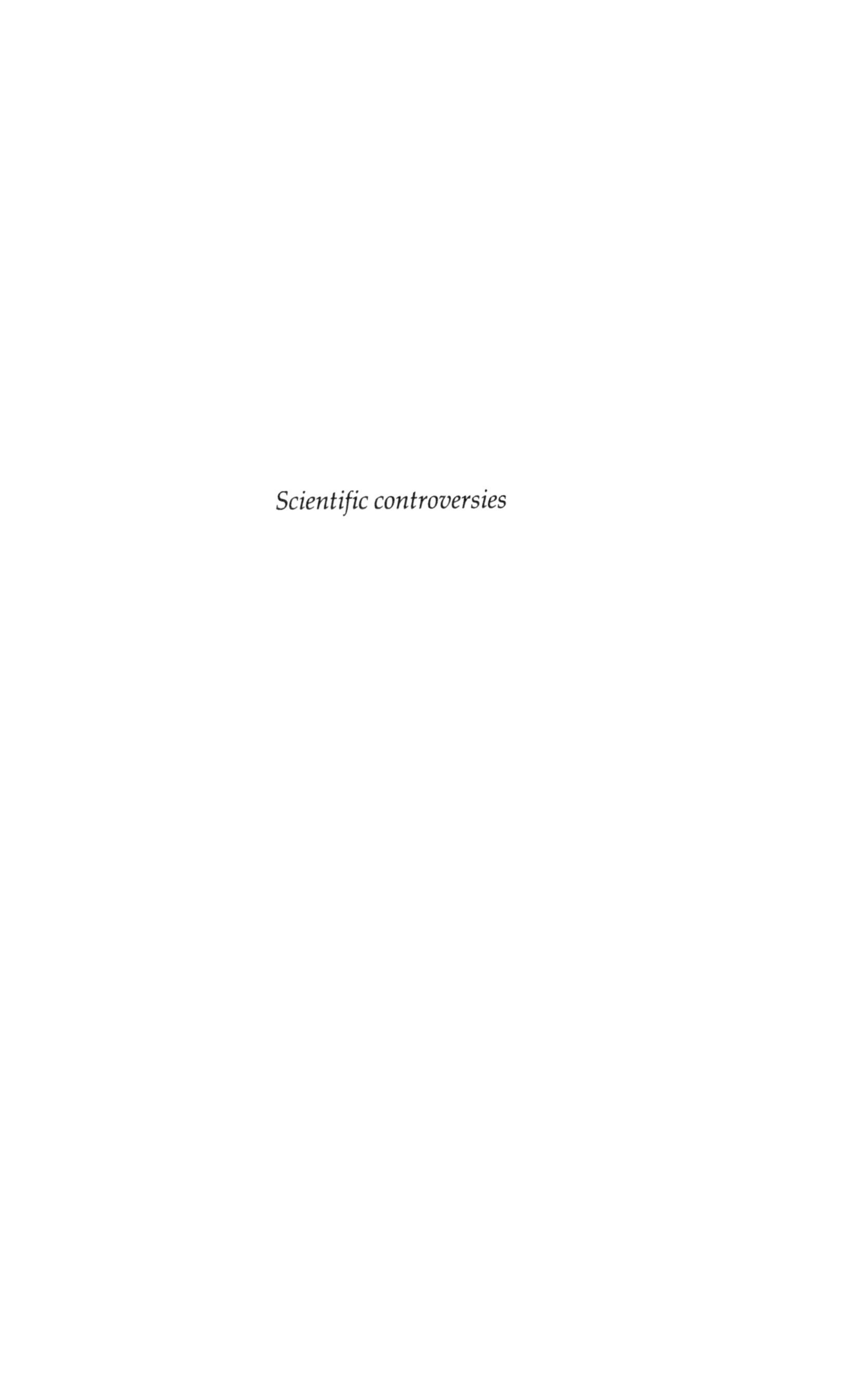

Scientific controversies

Scientific controversies

Case studies in the resolution and closure of disputes in science and technology

EDITED BY

H. TRISTRAM ENGELHARDT, JR.

AND

ARTHUR L. CAPLAN

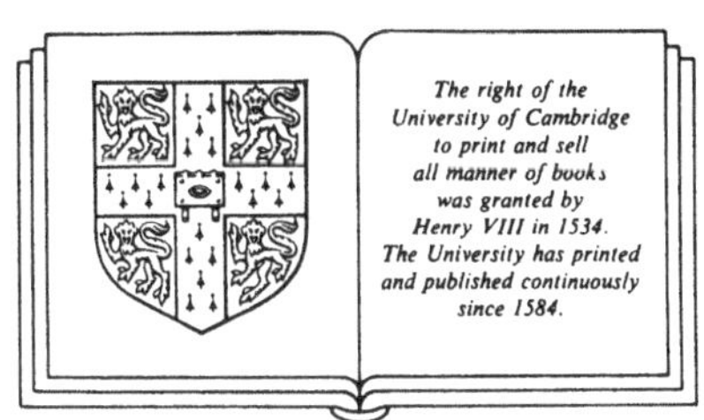

CAMBRIDGE UNIVERSITY PRESS

Cambridge

New York Port Chester

Melbourne Sydney

Published by the Press Syndicate of the University of Cambridge
The Pitt Building, Trumpington Street, Cambridge CB2 1RP
40 West 20th Street, New York, NY 10011, USA
10 Stamford Road, Oakleigh, Melbourne 3166, Australia

First published 1987
Reprinted 1989

Library of Congress Cataloging in Publication Data

Main entry under title:

Scientific controversies.

Includes index.

1. Science – Social aspects – Case studies.
2. Technology – Social aspects – Case studies.
3. Science – Philosophy – Case studies.
4. Technology – Philosophy – Case studies.
I. Engelhardt, H. Tristram (Hugo Tristram), 1941– . II. Caplan, Arthur L.
Q181.S3826 1987 303.4'83 84-28480
ISBN 0 521 25565 1 hard covers
ISBN 0 521 27560 1 paperback

British Library Cataloguing in Publication Applied for

Transferred to digital printing 2003

CONTENTS

Preface vii *List of contributors* ix

Introduction
Patterns of controversy and closure: the interplay of knowledge, values, and political forces
H. Tristram Engelhardt, Jr., and Arthur L. Caplan *page* 1

PART I. THEORETICAL PERSPECTIVES

1. Ethical theory and the problem of closure
   Tom L. Beauchamp 27
2. Scientific controversy and its termination
   Ernan McMullin 49
3. The political anatomy of controversy in the sciences
   Everett Mendelsohn 93
4. Controversies involving science and technology: a theoretical perspective *Ronald N. Giere* 125
5. Politics, public policy-making, and the process of reaching closure *Robert F. Rich* 151
6. The role of experts in scientific controversy
   Garland E. Allen 169
7. The continental drift debate *Henry Frankel* 203
8. How history and politics affect closure in biomedical discussions: the example of the Soviet Union
   Loren R. Graham 249
9. Scientific disputes over policy *Allan Mazur* 265
10. Controversies and the authority of science
    Dorothy Nelkin 283
11. Post-Skinner and post-Freud: philosophical causes of scientific disagreements *Alasdair MacIntyre* 295

PART II. CONTEMPORARY CASE STUDIES

Laetrile

12. Resolution of the Laetrile controversy: past attempts and future prospects *Gerald E. Markle and James C. Petersen* 315
13. Federal regulation of Laetrile *Robert S. K. Young* 333
14. Quasi libertarianism and the Laetrile controversy *Baruch A. Brody* 343
15. Judicial deflection of scientific questions: pushing the Laetrile controversy toward medical closure *Robert L. Schwartz* 355

Homosexuality

16. Politics, science, and the problem of psychiatric nomenclature: a case study of the American Psychiatric Association referendum on homosexuality *Ronald Bayer* 381
17. The diagnostic status of homosexuality in *DSM-III*: a reformulation of the issues *Robert L. Spitzer* 401
18. On arriving at the American Psychiatric Association decision on homosexuality *Irving Bieber* 417

Safety in the workplace

19. Values in the debate over workplace safety and health: the rancorous rhetoric about regulation *Gilbert S. Omenn* 437
20. The successful experiment that failed *Marvin S. Legator* 465
21. The power of efficiency: balancing benefits and costs in regulating occupational exposure to toxic substances *Taylor R. Durham* 487
22. Closure in occupational safety and health: the benzene and cotton dust decisions *Mark MacCarthy* 505

Nuclear power

23. Nuclear fear: a history and an experiment *Spencer R. Weart* 529
24. Closure and controversy: Three Mile Island *Cora Bagley Marrett* 551
25. Understanding the nuclear power controversy *Douglas MacLean* 567

PART III. CONTROVERSY, CLOSURE, AND THE PUBLIC

26. The role of the mass media in scientific controversy *Rae Goodell* 585
27. The National Commission on Human Experimentation: procedures and outcomes *Stephen E. Toulmin* 599
28. The forms and norms of closure *Ruth Macklin* 615

Author index 625 *Subject index* 632

PREFACE

Contemporary societies take science seriously. They presume that science can resolve disputes over factual issues such as the question of whether Laetrile should be used to treat cancer, whether homosexuality is a disease, whether nuclear power is safe, or whether a particular concentration of benzene in the workplace is tolerable. Such disputes are, however, not purely "scientific," as the essays in this volume show. They are formed around and structured by complex ethical and political interests. In fact, more than purely knowledge-directed concerns shape scientific controversies generally, even those concerning the theory of continental drift. A better understanding of the interplay between knowledge-directed and value- or politically directed forces in scientific disputes is required for an appreciation of the very character of culture. Ours is a culture that looks to scientific investigations, panels, and commissions for the determination of facts and the resolution of scientific disputes. This volume is offered as a contribution to the better understanding of the role of science and of the place of nonscientific interests in what may appear, at first glance, to be purely scientific undertakings.

The original conferences and research that produced this volume were supported by a grant from the National Endowment for the Humanities (No. AV-30691-78-200). This project, which addressed the character of scientific disputes with a heavy ethical or political overlay, spanned the years 1978–82. We wish to underscore our debt of thanks to the Endowment and, in particular, to Richard Hedrick, who served as the program officer for the grant. His guidance contributed in many ways to the successful completion of the project. The conference and research group meetings also benefited from the contributions and participation of a great number of individuals to whom the editors of this volume acknowledge their gratitude. These include Garland E. Allen, Jane L. Backlund, Thomas Bartosiewicz,

Ronald Bayer, Tom Beauchamp, Irving Bieber, Baruch A. Brody, Vern L. Bullough, Eric J. Cassell, Daryl Chubin, Robert Coulson, Ruth Cowan, Paul De Forest, John Duffy, Gerald Dworkin, Noel Farley, Gerald Feinberg, Henry Frankel, Sylvia D. Fries, Mary Ann Gardell, Barbara Gastel, Ronald N. Giere, Rae Goodell, Stephen Jay Gould, Loren Graham, Richard Green, Mary Gualandi, John T. Kalberer, Roger E. Kasperson, Robert W. Kates, Peter Klopfer, Jane C. Kronick, Larry Laudan, Marvin Legator, Bernd Löwe, Mark MacCarthy, Alasdair MacIntyre, Douglas MacLean, Ernan McMullin, Gerald E. Markle, Cora Bagley Marrett, Allan Mazur, Everett Mendelsohn, James C. Miller, Murdo Morrison, Dorothy Nelkin, Gilbert Omenn, Beverly Paigen, James C. Petersen, Robert F. Rich, Randy J. Rydell, Robert L. Schwartz, Margery W. Shaw, Peter Singer, Robert L. Spitzer, Frederick Suppe, Stephen E. Toulmin, Paul A. Walker, Spencer R. Weart, and Robert S. K. Young, among others. The contributions of the Hastings Center and its director, Daniel Callahan, were also central to the success of the project. He, with H. Tristram Engelhardt, Jr., were the codirectors of the research groups sponsored by the project.

From 1982 to early 1984, the materials developed from the project have been worked, reworked, and transformed into the essays of this volume. This has involved integrating the results of a number of different studies on the nature of scientific controversies: (1) case studies of controversy, such as the use of Laetrile in cancer treatment, the classification of homosexuality as a disease, the safety of nuclear power, and the assessment of risks in the workplace; (2) the role of journalism in articulating scientific disputes; (3) the place of arbitration in resolving scientific disputes; and (4) the general intellectual assessment of the nature of scientific disputes that possess a heavy ethical and political overlay. This has also involved controversies among participants in the project regarding the nature of controversies. This volume is unlikely to quiet controversies among its readers, but we hope that it will enhance their appreciation of the subject.

Hastings-on-Hudson, New York
Houston, Texas
March 6, 1986

Arthur L. Caplan
H. Tristram Engelhardt, Jr.

CONTRIBUTORS

Garland E. Allen Department of Biology, Box 1137, Washington University, St. Louis, Missouri 63130

Ronald Bayer The Hastings Center, 360 Broadway, Hastings-on-Hudson, New York 10706

Tom L. Beauchamp Joseph and Rose Kennedy Institute of Ethics, Department of Philosophy, Georgetown University, Washington, D.C. 20057

Irving Bieber 132 East 72nd Street, New York, New York 10021

Baruch A. Brody Center for Ethics, Medicine, and Public Issues, Baylor College of Medicine, Texas Medical Center, Houston, Texas 77030

Arthur L. Caplan The Hastings Center, 360 Broadway, Hastings-on-Hudson, New York 10706

Taylor R. Durham Department of Business Administration/Economics, North Adams State College, North Adams, Massachusetts 01247

H. Tristram Engelhardt, Jr. Center for Ethics, Medicine, and Public Issues, Baylor College of Medicine, Texas Medical Center, Houston, Texas 77030

Henry Frankel Department of Philosophy, University of Missouri – Kansas City, 5100 Rock Hill, Kansas City, Missouri 64110

Ronald N. Giere Department of History and Philosophy of Science, 130 Goodbody Hall, Indiana University, Bloomington, Indiana 47405

Rae Goodell Science Writing Program, Department of Humanities, Massachusetts Institute of Technology, Cambridge, Massachusetts 02139

Loren R. Graham Program in Science Technology and Society, Massachusetts Institute of Technology, Cambridge, Massachusetts 02139

Marvin S. Legator Division of Genetic Toxicology, Department of Preventive Medicine and Community Health, University of Texas Medical Branch, Galveston, Texas 77550

Mark MacCarthy Committee on Energy and Commerce, U.S. House of Representatives, Rayburn Office Building, Washington, D.C. 20515

Alasdair MacIntyre, Department of Philosophy, Vanderbilt University, Nashville, Tennessee 37235

Ruth Macklin Department of Community Health, Albert Einstein College of Medicine, 1300 Morris Park Avenue, Bronx, New York 10461

DOUGLAS MACLEAN Center for Philosophy and Public Policy, Woods Hall, University of Maryland, College Park, Maryland 20742

ERNAN MCMULLIN Program in the History and Philosophy of Science, University of Notre Dame, Box 556, Notre Dame, Indiana 46556

GERALD E. MARKLE Department of Sociology, Western Michigan University, Kalamazoo, Michigan 49008

CORA BAGLEY MARRETT Department of Sociology, University of Wisconsin, Madison, Wisconsin 53706

ALLAN MAZUR Department of Sociology, 500 University Place, Syracuse University, Syracuse, New York 13210

EVERETT MENDELSOHN Department of the History of Science, Science Center 235, Harvard University, Cambridge, Massachusetts 02138

DOROTHY NELKIN Program in Science Technology and Society, Cornell University, 622 Clark Hall, Ithaca, New York 14853

GILBERT S. OMENN Department of Environmental Health, School of Public Health and Community Medicine, University of Washington, Seattle, Washington 98195

JAMES C. PETERSEN Department of Sociology, Western Michigan University, Kalamazoo, Michigan 49008

ROBERT F. RICH School of Urban and Public Affairs, Carnegie-Mellon University, Schenley Park, Pittsburgh, Pennsylvania 15213

ROBERT L. SCHWARTZ University of New Mexico School of Law, 117 Stanford, N.E., Albuquerque, New Mexico 87131

ROBERT L. SPITZER New York State Psychiatric Institute, 722 West 168th Street, New York, New York 10032

STEPHEN E. TOULMIN Department of Philosophy, University of Chicago, 1126 East 59th Street, Chicago, Illinois 60637

SPENCER R. WEART Center for the History of Physics, American Institute of Physics, 335 East 45th Street, New York, New York 10017

ROBERT S. K. YOUNG Division of Oncology, Food and Drug Administration, 5600 Fishers Lane, Rockville, Maryland 20857

INTRODUCTION

Patterns of controversy and closure: the interplay of knowledge, values, and political forces

H. TRISTRAM ENGELHARDT, JR., AND ARTHUR L. CAPLAN

For some time the analysis of science by philosophers, sociologists, historians, and others has been dominated by discussions about theory change and development. Metascientific studies have not progressed far beyond bitter wrangles as to the adequacy of such concepts as "paradigms," "research traditions," "themata," and "theories" for adequately describing developments over time in various fields of scientific inquiry. In many respects these disputes have shed a great deal of light on the manner in which scientific ideas change and evolve, but they have also obscured important questions that need to be asked about conceptual development in science in order to understand this process fully and to facilitate the formulation of public policy with respect to science and technology.

This volume takes as its focus the notion of controversy. Controversies characterize intellectual change and developments within and about science. Disputes among scientists and between scientists and laymen are easy to locate in the history of science. Moreover, controversies play a central role in the way those who do science and those who depend upon its findings approach the subject. By analyzing controversies in and about science, this volume seeks to advance the analysis of scientific change beyond the sterile and confining boundaries of current disputes concerning the role of internal and external factors in fueling scientific change.

Science, ethics, and politics are marked by controversies. As one would expect, the ways in which controversies in each area arise as matters of dispute and approach solution appear to be somewhat divergent. Scientific controversies, for example, are usually seen to be the sorts of disputes that are to be resolved by appeal to facts and to rigorous reasoning concerning facts. Political controversies are held, by contrast, to involve issues properly amenable to resolution by negotiation. Ethical issues fall somewhere in between. Some of them are viewed as issues resolvable by appeal to the nature of reason or morality; others, by appeal to the facts of the situation. In addition, they are regarded, at least at times, as properly resolved through common agreement. In a number of instances, all three genres of controversy are interwined. Many scientific controversies have heavy political and ethical overlays.

This volume is concerned with controversies of the latter character. Disputes concerning the effectiveness of Laetrile as a cancer drug, the question of whether homosexuality is a disease, the nature of the risks associated with the use of nuclear power, and the character of hazards presented by workplace pollutants have been taken as case studies. General themes have been sought by comparing such studies with less political examples, such as the debates in geology concerning continental drift. The contributors include philosophers, historians, sociologists, physicians, and a lawyer, as well as natural scientists. The result is a geography of controversies, fashioned by exploring case studies in order to determine under which circumstances different forms of rationality are pursued with success or failure, as well as where, for extra-epistemic considerations (i.e., concerns other than establishing an empirical truth), particular controversies come to a close. In outlining the interplay among these concerns, this volume maps the relations among scientific, moral, and political interests. It focuses on the ways in which scientific debates with heavy political or ethical overlays come to be regarded as having been resolved.

In this volume we have used the term *closure* to indicate the conclusion, ending, or resolution of a controversy. The term trades upon ambiguities. We use it in the eighth sense given by the *Oxford English Dictionary*: "a bringing to a conclusion; end, close" (q.v.). This sense is noncommittal in that it does not specify the means by which closure is achieved. However, the dictionary also lists more specific meanings pertinent to the analyses in this volume. The sixth sense defines closure as "an agreeing upon terms, a coming to an arrangement *with*; agreement, union, unity" (q.v.). The ninth sense is relevant as well: "the closing of a debate in a legislative assembly by vote of the house or by other competent authority" (q.v.). One of the tasks of this volume is to suggest distinctions among the various ways in which controversies are brought to a conclusion: by negotiation, by political procedure, or by the means supposedly more usual in scientific controversies, namely, the appeal to facts and observations.

It has been widely assumed that scientific controversies, even those with a heavy political and ethical overlay, are amenable to closure by appeal simply to facts and the rational consideration of facts. Indeed there has recently been a resurgence of objectivist views of ethics supporting the hope that ethical controversies can be resolved on the basis of reasons that can bind generally.[1] Also, much of Western public policy has presupposed that scientific controversies are resolvable by rational analysis and the investigation of the facts. Science has been presumed to be objective.[2] Yet it has become ever clearer that cultural assumptions influence scientific reasoning and findings, even apart from controversies that have explicitly interwoven scientific, political, and ethical issues.[3] As a result, the nature of scientific controversies with heavy political and ethical overlays has been thrown into question – a matter of some importance, since understanding this problem is central to many issues of public policy-making. The essays in this volume reassess aspirations to rationality in such controversies.

These essays spring from a three-year series of meetings held at the Hastings Center. These meetings focused on better appreciating the ways in which the inner logic of scientific investigation, predicated upon a concern with evidence and with reasoning on the basis of evidence, involves values and political concerns. In great measure this involved approaching once more the traditional externalist – internalist debates of the history and philosophy of science. It also required examining the extent to which the movement of science toward more ample and secure knowledge claims is shaped, directed, or impeded by forces external to science, such as the values and political forces of the scientist's cultural milieu. Here the focus of

1 John Rawls, for example, presumes that his device of the original position and the use of reflective equilibrium can disclose generally rationally defensible understandings of justice. It is true that Rawls qualifies this claim in some passages. However, his theory of justice has supported a major revival of views concerning the general rational defensibility of ethical claims regarding justice. See John Rawls, *A Theory of Justice* (Cambridge, Mass.: Harvard University Press [Belknap Press], 1971). A number of analogous accounts have developed. See, for example, Bruce A. Ackerman, *Social Justice in the Liberal State* (New Haven: Yale University Press, 1980), and Norman Daniels, ed., *Reading Rawls* (New York: Basic Books, 1981).

2 Although many qualifications are in order, Karl Popper has been taken to defend a view somewhat similar to this. See his *Logic of Scientific Discovery* (New York: Basic Books, 1959).

3 H. Tristram Engelhardt, Jr., and Daniel Callahan, eds., *The Foundation of Ethics and Its Relationship to Science*, 4 vols. (Hastings-on-Hudson, N.Y.: Hastings Center, 1976–8, 1980). See also Paul Feyerabend, *Against Method* (New York: Schocken Books, 1975); Norwood R. Hansen, *Patterns of Discovery* (Cambridge: Cambridge University Press, 1961); Thomas S. Kuhn, *The Structure of Scientific Revolutions*, 2d ed. (Chicago: University of Chicago Press, 1970); Larry Laudan, *Progress and Its Problems* (Berkeley: University of California Press, 1977).

concern was upon those scientific debates that have explicit and heavy ethical and political overlay, a condition that characterizes a great proportion of the scientific disputes that gain general public attention. The goal of the project was to achieve a better understanding of the character of scientific reasoning in areas where its controversies are of direct importance to both public policymakers and to the public in general, and in so doing to gain a better understanding of an important area of science. Since the conclusion of these meetings, the essays have been refashioned to permit further exploration of the interplay of scientific, ethical, and political concerns.

Scientific controversies with a heavy political and ethical overlay

One of the first problems in the analysis of a scientific controversy with a strong political or ethical dimension is to determine the actual number and nature of the controversies involved. There is the problem of the individuation of controversies. What appears at first glance to be only one controversy may under closer examination be resolvable into two or more controversies. One might think, for example, of the public dispute concerning Laetrile, which in part concerned the efficacy of Laetrile as a treatment for cancer, a point explored in this volume by Robert Young. In part this dispute also concerned the additional issue of the right to choose freely one's own course of treatment, an issue explored in this volume by both Robert Schwartz and Baruch Brody. Thus at least two quite different controversies were intermingled in this dispute, one involving rules for establishing a scientific claim and the other involving political rules regarding access to particular medications. These two elements are, in principle, both distinguishable and separable. They are intermingled in one debate out of historical accident, owing to the drug control laws of the United States.

This is as one would expect when scientific debates are drawn into or blended with political disputes. When it originates in a libertarian view of the right of individuals to self-medication, an appeal to the efficacy of Laetrile may be only a rhetorical ploy. Given a more statist view, the issues of efficacy and risk become more germane. As a consequence, the political discussion has its own complex history, drawing in part from libertarian concerns, in part from the suspicions of Americans regarding established elites (in this case, the U.S. Food and Drug Administration), and in part from the idea that the state should function to protect the best interests of its subjects. The controversy can only with qualification be understood as an argument focused on what should count as scientific evidence and how one ought to reason on the basis of such evidence.

When one is involved in a conflict of political forces, one is not immediately concerned with finding *the* correct answer. One is interested, rather, in theories of the state and in conflict management, in the sense of finding the most prudent and efficient balancing of forces. A scientific controversy with a heavy political and ethical overlay is not, then, one controversy but a scientific controversy (or controversies) plus a controversy (or controversies) concerning social and political theories and viewpoints. This becomes quite evident, for example, in the history of the Laetrile controversy as sketched in this volume by Gerald Markle and John Peterson.

We need therefore at the outset to frame some preliminary distinctions among the various ways in which controversies can reach closure. This will allow us to make better distinctions among controversies, as well as among the roles of various knowledge and value claims, in characterizing disputes. Tom Beauchamp (Chap. 1, this volume) characterizes five modes through which controversies end.

1. *Sound argument closure:* occurs if, and only if, a correct position has been reached in a context of controversy, thereby rendering opposition views incorrect.

2. *Consensus closure:* occurs if, and only if, a consensus has been reached, in a context of controversy through a means other than sound argument closure or some form of procedure or negotiation, that some position is best and that opposition views are incorrect.

3. *Procedural closure:* occurs if, and only if, a controversy is terminated by formal, procedurally governed efforts to end the sustained discussion that characterizes the controversy.

4. *Natural death closure:* occurs if, and only if, a controversy has come to an end through a gradual natural death, as by fading away because of waning interest.

5. *Negotiation closure:* occurs if, and only if, a controversy is settled through an intentionally arranged and morally unobjectionable resolution acceptable to the principals in the controversy.

A similar classification is offered by Ernan McMullin. McMullin stresses the difference between the roles played by factors that the participants in the controversy regard as epistemic (knowledge-based) factors, and therefore internal to the very understanding of science, and those that are recognized as nonepistemic factors, and therefore external to the undertaking of science as an endeavor of rational, empirical investigation. It should be noted that what may appear as nonepistemic factors to some may be viewed by others as epistemic factors. The concept of "epistemic" is context-dependent. For example, for those who credited the epistemic claims of late medieval theology, the appeal to Church authority in the effort to resolve the controversies raised by Galileo counted as an appeal to epistemic, not nonepistemic factors. McMullin offers a three-part classification of the termination of disputes concerning science and technology:

1. *Resolution:* A controversy is resolved when an agreement is reached on the merits of the case in terms of what the participants take to be standard epistemic factors.
2. *Closure:* A controversy reaches closure when it is terminated on the basis of nonepistemic factors, such as the authority of the state; the pride, ambition, or laziness of a controversialist; or the withdrawal of publication facilities.
3. *Abandonment:* Controversies may terminate through participants losing interest.

Beauchamp's and McMullin's classifications, although they differ, are helpful because they underscore the different roles played in controversies by epistemic and nonepistemic factors.

Timeless truth, historical realities, and the context-bound nature of controversy

One view of scientific controversies presupposes them to be purely rational debates in which those disputing agree on (1) how to acquire evidence relevant to the dispute and (2) how to reason with the evidence in order to reach a rationally defensible conclusion that will resolve the controversy. Such controversies present the possibility of a sound argument closure, in Beauchamp's taxonomy, or one of resolution, in McMullin's classification. In the case of the Laetrile dispute, one might think of the Federal Drug Administration (FDA) and the scientific establishment as constituting a single community of controversy that reached a resolution of the question about the efficacy of Laetrile on the basis of commonly shared rules of evidence and inference. Such controversies are commonly alleged to be apolitical and impersonal.

In many reconstructions of disputes, it is as if the various political, economic, social, and psychological differences among the persons participating in the controversy did not matter. Instead, the participants become placeholders for anonymous reasoners. The paradigm is that of a grand syllogism reasoned through by the scientific community, viewed as the epistemic equivalent of the impartial, rational, fully informed observer posited in ethical theory. In terms of such a view, and given a faith in the knowability of reality, a final and true answer is sought, which should in principle be available to terminate the controversy.

The rules for acquiring evidence and drawing conclusions, however, change in science over time. One must, as a consequence, qualify scientific controversies with a sociohistorical subscript to identify a scientific controversy with a particular scientific community, its rules for selecting evidence relevant to a debate, and its rules for reasoning on the basis of such evidence. That is, a scientific controversy becomes identified with a particular scientific community, understood as a group of stakeholders in a scientific debate who at a particular point in history share common rules of evidence and inference.[4] A scientific controversy in this sense of a rational dispute is bound to a particular group that defines the dispute through the group's understanding of the possibilities for solving the controversy. In this process, as Everett Mendelsohn shows, controversies are subjected not only to external but also to internal political forces. Scientific and professional authorities and their conflicts influence the character of scientific debates and their forms of resolution, so that even when scientific controversies are somewhat shielded from external sociopolitical forces, there are still social and political forces internal to communities of scientists that must be taken into account.

One sees the range of such influences when one compares, for example, the case studies in this volume of the geologists' debate concerning the theory of continental drift, sketched by Henry Frankel, and the biomedical discussions in the Soviet Union analyzed by Loren Graham. Moreover, many of the controversies turn on conflicting metaphysical assumptions and philosophical viewpoints that are themselves historically conditioned. Here – at least with regard to scientific disagreements between behaviorists and Freudians – Alasdair MacIntyre is in agreement with Mendelsohn's portrayal of the interplay among scientific goals and philosophical and metaphysical presuppositions. The final resolution of a scientific controversy, if such is to be available at all, often requires the resolution of philosophical problems, and these themselves are embedded in a history of controversy. Further, as Garland Allen emphasizes, economic and other class interests shape the character of scientific controversies, in part by influencing the views of participants, in part by influencing patterns of research funding.

[4] By *rules of inference* we mean established warrants for drawing conclusions from evidence so as to bring closure to a dispute concerning the significance of the evidence.

These forces bind the sociology and history of scientific controversies with the epistemologies of such controversies, as is shown in this volume by Dorothy Nelkin, Allan Mazur, and Robert Rich. One must also, as Ruth Macklin, Tom Beauchamp, and Ernan McMullin warn, distinguish among the various kinds of interests at stake in scientific controversies. In fact, because of the diversity of interests of the participants, one must decide whose controversy it is. This marks a departure from examining scientific controversies *sub specie veritatis*, (i.e., from "the" supposed point of view of the "truth"), where truth is seen as univocal, ahistorical, and the property of "the" scientific community, to viewing truth as defined by a particular community through its particular sociohistorically conditioned rules of evidence and inference. One might think of the way in which Charles S. Peirce identified "the" truth with the ultimate conclusions of the ultimate scientific community.[5] One appeals to that community as an intellectual possibility in criticizing actual communities and their rules of evidence and scientific reference. The scientific community best able to make true knowledge claims is thus an intellectual standpoint from which one both justifies and criticizes the deliverances of any particular community. Indeed it is for this reason that a scientific controversy that ends with closure in McMullin's terms (i.e., on the basis of social forces, not on rational grounds), perpetually invites the reopening of the controversy on rational grounds (at least, if the termination of the controversy was not also justifiable on rational grounds). Of course, there is the additional problem that over the course of time the grounds for rational resolution themselves change.

The prescriptive and descriptive senses of closure

Since science frequently makes (or at least many individual scientists make) universalistic claims of delivering or seeking to deliver "the" truth about empirical reality, and since in many scientific undertakings there is great cross-cultural and transnational agreement among scientists concerning which rules of evidence and inference ought to be employed, one may overlook the need to identify the particular community of scientists engaged in a dispute. One may act as though the controversy involved "the" scientific community rather than *a* scientific community. Since science seeks a univocal account of empirical reality, this notion of "the" scientific community can

[5] Charles S. Peirce, "Pragmatism and Pragmaticism," in *Collected Papers* (Cambridge, Mass.: Harvard University Press [Belknap Press], 1965), 5.311, 5.316.

function as a regulative ideal, as a goal toward which the effort of scientists is directed, even though the goal itself is unattainable.[6] Yet one will need to recognize that the study of the theories of knowledge of particular scientific communities must involve history and sociology, since the rules of inference and evidence are historically and socially conditioned. As a consequence, one can distinguish two senses of sound argument closure of scientific controversies on the basis of the kinds of sound arguments:

1. *Sound argument closure in the strict sense,* which is the ideal. It would occur on the basis of rules of evidence and inference unconditioned by the history or culture of the participants. This is a regulative ideal, never actually achieved, but to which one can appeal in indicating how particular "sound" arguments depart from this goal.

2. *Sound argument closure in the broad sense,* in which the rules of evidence and rules of inference are historically, socially, and culturally conditioned. Actual communities resolve scientific controversies by appealing to rules of evidence and inference that are, as far as can be determined by the participants, correct and undistorted by the presuppositions of the participants. Historians and philosophers of science will, however, be able to disclose distortions.

The first is an ideal, a goal used in attempting to frame an atemporal view of scientific reasoning, of how reasoning ought to occur apart from the influences of particular social and historical circumstances. The second involves the actual sociohistorically conditioned reasonings that exemplify any particular controversy (e.g., how one would have reasoned had one been a physician in the eighteenth century accurately following the views of Thomas Sydenham and François Boissier de Sauvages). When one compares groups of scientists participating in disputes, some will appear to be closer and others farther from sound arguments in the strict sense.

In the case of the Laetrile debate, for example, disparate sets of arguments have been invoked by at least two communities of scientists: (1) those agreeing with the scientific establishment's rules for identifying and assaying evidence in terms of that establishment's statistically disciplined modes of data gathering; and (2) those relying on more clinical methods that appear to support the possible effectiveness of Laetrile. And, as has already been noted, differences in modes of identifying evidence and of arguing from evidence were

[6] Alasdair MacIntyre, "A Crisis in Moral Philosophy: Why Is the Search for the Foundations of Ethics So Frustrating?," in *Knowing and Valuing,* H. Tristram Engelhardt, Jr., and Daniel Callahan, eds. (Hastings-on-Hudson, N.Y.: Hastings Center, 1980), pp. 18–35.

intensified by the fact that many members of the nonorthodox scientific community were also members of political groups with special civil libertarian concerns regarding the right to self-medication. The existence of differing political interests not only marks such a debate with a special complexity; it tends to skew the evidence for each of the competing groups.

Economic interests play a similar role, as has been shown by the history of the economic competition between orthodox and nonorthodox physicians or even among various proponents of orthodox medicine. Consider, for example, the nature of the controversy that would be engendered by a full-scale attempt to determine the efficacy of chiropractic. That chiropractic still thrives, often providing its own diagnoses and therapies, is only in part a result of differing views about what counts as science. Differences in scientific methodology are intensified by differences in economic interest that influence the psychology of discovery so that data are seen differently or different data are given to the observers.[7]

This point can be put even more generally. Scientists do not engage in science as disembodied knowers, a point developed at length by Stephen Toulmin.[8] Differences in views of proper scientific method are often conjoined with membership in conflicting interest groups. The interest groups may be political, economic, or structured around special concerns of prestige and special personal investments. Consider the debates regarding the comparative efficacy of coronary bypass surgery and medical management of coronary artery disease.[9]

[7] One should not forget how difficult it is to free science from the distortions of cultural, moral, and other expectations. A good example is the nineteenth-century assumption that masturbation is a disease. Reputable physicians were convinced of the data showing that this habit was a serious danger to health. See H. T. Engelhardt, Jr. *The Foundations of Bioethics* (New York: Oxford, 1985), pp. 157–64.

[8] Stephen Toulmin, "How Can We Reconnect the Sciences with the Foundation of Ethics?," in Engelhardt and Callahan, *Knowing and Valuing*, pp. 18–35.

[9] See Glenda K. Barnes et al., "Changes in Working Status of Patients Following Coronary Bypass Surgery," *Journal of the American Medical Association* 238 (1977): 1259–62; Marvin L. Murphy et al., "Treatment of Chronic Stable Angina," *New England Journal of Medicine* 297 (1977): 621–7; K. Detry et al., "Veterans Administration Cooperative Study of Surgery for Coronary Arterial Occlusive Disease: III. Methods and Baseline Characteristics, Including Experience with Medical Treatment," *American Journal of Cardiology* 40 (1977): 665–71; G. M. Lawrie et al., "A Debate on Coronary Bypass," *New England Journal of Medicine* 297 (1977): 1464; G. M. Lawrie et al., "Improved Survival beyond Five Years in 1,108 Patients Undergoing Coronary Bypass," *American Journal of Cardiology* 41 (1978): 355; R. C. Read et al., "Survival of Men Treated for Chronic Stable Angina Pectoris," *Journal of Thoracic Cardiovascular Surgery* 75 (1978): 1–16; M. E. DeBakey and G. M. Lawrie, "Aortocoronary-artery Bypass: Assessment after 13

Beyond conflicting economic interests, surgeons and internists are mutually suspicious of each other's judgments concerning therapeutic results. A concrete scientific controversy usually cannot be reduced simply to a controversy regarding facts and the proper warrants for drawing conclusions from evidence.

The scope and character of the domain of scientific facts can also be at issue, as has been the case with regard to the debate about whether homosexuality should be classified as a mental disorder. In part, there has been the issue of whether psychiatry can disclose that certain activities, or propensities to activities, are abnormal in the sense of giving a prima facie warrant in favor of treatment. One finds this point put poignantly by Robert Spitzer, who in his essay accepts the fact that homosexuality should count as a disorder only if it is ego-dystonic, although he is reluctant to grant such grace to the satisfied shoe fetishist. Irving Bieber, however, argues that homosexuality is pathological. In this dispute the debate between scientific communities cuts more deeply than the Laetrile debate with regard to the soundness of types of particular studies and the reliability of anecdotal evidence of cures. One can, however, recognize similarities. As Ronald Bayer persuasively argues, a strong interplay of interest in knowledge, values regarding proper conduct, and political forces is manifest in this controversy.

Controversies among communities of thought

One can, on the basis of the essays in this volume, advance at least some preliminary conclusions regarding the character of controversies. To begin with, scientific disputes resist closure or resolution when the stakeholders in the debate belong to

1. Different scientific communities with different appreciations of the evidence at stake: For example, one might compare the members of the scientific establishment who are convinced, on the basis of FDA-recognized data, of the inefficacy of Laetrile with the physicians who are convinced of its efficacy on the basis of their clinical experience with its use.

2. Competing social groups with different views of social control. Here one might think of those who would regulate self-medication for the good of citizens (e.g., the FDA as a paternalistic regulator)

Years," *Journal of the American Medical Association* 239 (1978): 837–9; G. M. Lawrie and G. C. Morris, "Factors Influencing Late Survival after Coronary Bypass Operation," *Annals of Surgery* 187 (1978): 665–76; H. N. Hultgren et al., "Aortocoronary-artery Bypass Assessment after 13 Years," *Journal of the American Medical Association* 240 (1978): 1353–4.

versus those who would preserve certain liberties for self-medication (e.g., libertarian supporters of the right to use Laetrile).

The greater the public involvement in a dispute, the greater the probable complexity of the debate, for one is likely to draw to the debate people who may not agree to the same rules for acquiring scientific evidence and for reasoning on its basis. In fact, there may be ways in which public involvement can distort the debate because, as Rae Goodell notes, of the difficulty of presenting through the mass media a sufficiently comprehensive account of what is at stake in a scientific controversy. There is a tendency to polarize scientific debates artificially and to construe them as having two sides, whereas in fact they are often multisided.

As a result, participants are likely to talk past each other regarding the significance of the scientific evidence and to be confused by the ambiguity introduced by the political or social elements of the dispute. The result is a form of distracting intellectual cacophony, owing to the multiple contexts in which the dispute is pursued, the different languages used, and the different goals pursued by the disputants.

Our study of scientific controversies with heavy political and ethical overlays has, in short, revealed the complexity of such disputes and the difficulty of individuating a controversy – that is, of identifying a single controversy. If one takes as one's investigative focus the possible grounds for the closure or the failure to achieve closure of such controversies, one is brought to dissecting what appears at first sight to be one controversy into a number of controversies, where "a" scientific controversy is defined by the existence of "a" community of disputants who share common rules of evidence and of reasoning with evidence. If such rules are not shared, then the dispute is not a single scientific controversy. If the disputants do not share common rules for fair negotiating, then there is not "a" political controversy (i.e., "a" controversy in principle amenable to resolution on the basis of epistemic considerations or on the basis of rules for fair procedure or negotiation). Each community defines the controversy in a different fashion, and initially, at least, they do not share means for resolving the controversy.

Thus, in addition to the involvement of political and ethical concerns defining different communities of controversy, communities of scientists may or may not share a sufficiently common understanding of good science so as to be able to settle a dispute, as a scientific controversy, by sound argument. Indeed, the term *scientific* is itself ambiguous. If the term is used broadly to include any area in which sound argument closure is possible, one uses it in a way similar to the German *wissenschaftlich*. Further distinctions (e.g., "empirical scientific," "mathematical," "mid-twentieth century physical scientific") would depend upon the character of the rules of evidence, and perhaps of inference. Distinctions should be multiplied insofar as it is useful towards identifying particular controversies amenable to a particular mode of closure. One would wish to identify controversies in this sense in order to decide when it is appropriate to seek further evidence or to seek grounds for negotiating an end to a dispute. Also, a comparative epistemology is helpful here in determining why certain information is a clincher in an argument for closure for one community of controversy but not for another. A mapping of the relationships among the various communities and their assumptions can provide a helpful geography of the relations among the controversies involved in a debate.

One will need to note, as Ernan McMullin argues in this volume, that scientific disputes or controversies can occur, apart from ethical and scientific issues, at three levels. There can be controversies of fact, where scientists agree concerning basic assumptions but disagree regarding results of observations. Protracted controversies of this sort are not frequent. They can, however, hinge on ways of construing data that support alternative theories. Here one finds a second level, that of controversies about theory, which are the most prominent. They occur when two or more competing theories are available to account for the data at hand. The most difficult level is that involving controversies of principle, which may be either methodological or ontological. One needs to determine the level at which a scientific debate is joined by its various participants, as is shown by Mendelsohn, Rich, and Mazur. The mapping of controversies requires a number of axes to portray all of the dimensions involved.

Closures: counting the ways

What one sees is the result of the varying capacities of reason to resolve controversies. These various roles of epistemic and nonepistemic considerations in the conclusion of controversies suggest that Ernan McMullin's and Tom Beauchamp's classifications can be combined to indicate where the mode of closure is determined by an unwillingness to appeal to reason; a failure to appeal to reason; an incapacity of reason; or the capacity of reason to resolve a controversy through disclosing a resolution or at least fair means for fashioning a solution.

1. *Closure through loss of interest:* Here one should place Beauchamp's *natural death closure* and what McMullin terms *abandonment*.

If the participants lose interest, the controversy ends not on rational grounds but because the participants have ceased to appeal to rational grounds for resolution either through sound argument or through fair principles of negotiation.

2. *Closure through force:* This is the mode of ending a controversy that McMullin terms *closure.* From an intellectual point of view, it is unsatisfactory in that there is no rational basis for resolution. But here, unlike closure through loss of interest, there is a positive commitment to set aside the presumptions of fair negotiation and procedure.

3. *Closure through consensus:* As Beauchamp indicates, a controversy may terminate when the participants embrace a particular viewpoint, not through general rational arguments, negotiation or established procedures but through nonepistemic influences that lead to a community of belief. One may take here as an example the conversion of individuals to a particular religious viewpoint that commits them to hold a creationist view of the origins of the world. Members of political groups may similarly possess such a consensus regarding, for example, their mission or destiny.

4. *Closure through sound argument:* This is the mode of ending a controversy termed *resolution* by Ernan McMullin. It identifies a state of affairs in which the controversy has been resolved so that the inquiring mind can now see that a particular solution is the appropriate solution. It contrasts with both resolution by force and resolution by natural death in that there are reasons for – not simply causes for – the controversy's ceasing. It is distinguishable from consensus closure, for a generally justifiable system of arguments is used to deliver a solution. Although consensus closure can involve a restricted rationality, the consensus does not offer reasons for its conclusions. Unlike sound argument closure, it cannot deliver its resolution in epistemic terms alone.

One needs to distinguish further, as has been indicated earlier, between

a. *sound argument closure in the strict sense,* where rules of evidence and inference would be true, valid, and undistorted by the history and the culture of the participants.

b. *sound argument closure in the broad sense,* in which a controversy is resolved in terms of rules of evidence and of inference, recognized by the participants as rationally justified rules, although the rules are distorted by the historically and culturally conditioned presuppositions of those participants.

One must also introduce two further distinctions to classify the possible division of the disputants into more than one scientific community.

i. *sound argument closure – broad sense, simple mode.* This occurs when scientists form a community of investigators with common rules of evidence and inference, and when sufficient data can be secured to resolve the controversy. The controversy is not deep. Much is agreed to by all stakeholders. As an example, one might think of the participants in the debate about continental drift.

ii. *sound argument closure – broad sense, complex mode.* This occurs when scientists form two or more communities with different appreciations of the rules of evidence or inference with regard to the issues at stake. For example, one might think of the debate in the Laetrile dispute between the scientific establishment and nonorthodox practitioners. Sound argument closure is in principle possible if sufficient appeals can be made to general background views of science that are held in common.

5. *Closure through negotiation.* Unfortunately, many controversies appear to be amenable neither to consensus closure nor to sound argument closure. Insofar as the participants do not lose interest in such debates, and insofar as the use of force is barred on moral grounds, not to mention intellectual grounds, one is forced to find a fifth way. This must be created through the negotiation of the participants in the controversy by appeals to various considerations and through various peaceable manipulations. A fabric of agreement will need to be fashioned. Insofar as this form of closure involves immediate negotiation, Beauchamp terms it *negotiation closure.* Insofar as it involves an appeal to a procedure that has been accepted for the resolution of such controversies (presumably by prior negotiation to establish the relevance and authority of the procedure), Beauchamp terms this *procedural closure.* In contrast with sound argument or consensus closure, which discover or disclose a resolution, negotiation closure creates one.

Here, as in the case of sound argument closure, one could distinguish between negotiation closure in the strict sense, which would follow ideal rules of fair negotiation, and negotiation closure in the broad sense, which follows historically and socially conditioned rules of fair negotiation. One could distinguish as well between a simple and complex mode of negotiation closure, in order to indicate when one has a dispute involving a single community or multiple communities of understanding regarding rules of fair negotiation.

In actual cases of dispute, two or more of these modes of closure may be involved. As a consequence, there will be different senses in which the different elements of the dispute will have come to a termination. The first, second, and third modes of closure produce conclusions that can be termed neither fair nor correct. Closure in such cases is achieved neither by reliance upon sound arguments nor by reliance upon fair principles of negotiation. If by accident the agreement achieved by force or consensus is in fact correct, this is a fortuitous outcome that does not change the character of the process, which either does not appeal to reason or in fact rejects reason. However, the termination of a controversy by consensus is "fair" in the sense of respecting the views of those involved. There is not, though, an explicit reliance on principles of procedural fairness such as occurs in negotiations that do not rely on force. The fourth and fifth senses of closure offer an interesting contrast between closures that are correct and those that are fair. Means of closure in terms of sound argument aim at proceeding in a way that ensures correctness. Means of resolving controversies through negotiation aim at fairness, as Ronald Giere and Robert Rich suggest. In a peaceable, secular, pluralistic society such as ours, where closure through force is to be avoided on moral principle and where closure through consensus is unlikely, given the heterogeneity of the groups involved, the focus will be upon correctness and fairness. Since scientific disputes with a heavy political and ethical overlay can only in part be resolved through sound argument closure, such disputes will likely reach conclusions that are in part correct and in part fair.

Creating policy in a context of controversy

Communities of scientists must create policy in order to work together despite areas of disagreement. One might think, for example, of the dispute concerning the classification of homosexuality as a mental disorder, as outlined in Ronald Bayer's essay in this volume. Disease classifications not only indicate clusters of phenomena regularly occurring together but, in addition, indicate disapproval of those clusters (i.e., as disorders, as pathological circumstances, as failures to achieve a minimal level of physiological or psychological excellence), as well as establishing a basis for warranting treatment of those conditions. Consequently, the classification of homosexuality as a mental disorder invited both scientific controversy and political dispute.[10] On the one hand, many homosexuals did not want homosexuality to be considered a disorder. Yet on the other hand, many psychiatrists wanted to enable homosexuals to receive treatment, especially third-party reimbursable treatment, should they wish it. A

[10] Michael Ruse, "Are Homosexuals Sick?", in A. Caplan, H. T. Engelhardt, Jr., and J. J. McCartney, eds., *Concepts of Health and Disease* (Reading, Mass.: Addison-Wesley, 1981), pp. 693–724.

compromise was the category "ego-dystonic homosexuality."[11] The Task Force on Nomenclature and Statistics, of the American Psychiatric Association, thus was not simply engaged in seeking to discover through sound arguments whether homosexuality is a disorder. Scientific and methodological arguments concerning the capacity of psychiatry to discover the nature of sexual deviancy played a role in the final conclusions. However, since social considerations did as well, the final conclusion was also in part a negotiated conclusion, a social creation.

Within such bodies as the Task Force on Nomenclature and Statistics, the Assembly of the American Psychiatric Association, and the Council on Research and Development, the use of voting to approve a taxonomy is unavoidable when there is not sound argument recognized by the participants that will resolve the controversy. It may appear odd to vote on a "scientific" matter. A form of sound argument closure should, one might expect, simply deliver an answer. However, one must establish by negotiation formal procedures to bring closure to a scientific dispute when more than one community of scientists exists (as in the case in psychiatry) or when a conclusion has not yet been reached by sound argument and one intends to engage in common activities or undertakings. Thus, even if the taxonomy in question is not a nosology with immediate ethical and political implications but simply a taxonomy adopted for the systematic collection of data, a procedure must be available to allow a community of scientists to engage in systematic data collection despite the lack of full theoretical agreement on the part of all the scientists.

Those not endorsing but acquiescing in the majority view already constitute their own scientific community on the point of the correctness of the particular nomenclature. One thus fashions *one* political community, although in a strict sense there is not *one* scientific community. As long as such disparate scientific communities remain a part of a "political" association of scientists, they may find it worth their while to agree to procedures for common activities. Their continuing as members will depend on a judgment of the balance of advantages versus disadvantages to be drawn from association. Which is to say, negotiation closure enters into the life of science as soon as science becomes a large-scale social undertaking. Negotiation closure and its creation, procedural closure, are vehicles for resolving disputes when sound argument closure is not (yet) successful but where common action is worth the costs of ignoring particular areas

[11] American Psychiatric Association, *Diagnostic and Statistical Manual of Mental Disorders*, 3d ed. (Washington, D.C.: American Psychiatric Association, 1980).

of disagreement. Thus, a social community is formed even where a scientific community is not fully realized by common understandings of rules of evidence and inference.

Negotiating the "facts" for public policy

The more science is enlisted in framing social decisions, the more it is necessary to have common social understandings of the significance of scientific claims or of the consequences of applying science. Science pure, unapplied, and unappealed to in framing social policies is under little pressure to develop a common understanding regarding accounts of reality that have little bearing on scientific theory. It is enough to be in agreement regarding rules of evidence, rules of reasoning with evidence, and general theoretical understandings. Within those limits, it may be accepted by most scientists that probabilities of complex states of affairs will vary from study to study and that different studies to assess the correctness of those probabilities will themselves vary in assigning a probability to the correctness of the probabilities advanced by the first studies.

However, social policy usually requires acting as if *a* particular account is correct. One must often base social policy upon particular understandings of empirical facts (e.g., the risks of a certain sort of nuclear reactor). The choice of any particular scientific account entails particular social policy costs. Thus, deciding that one or ten parts of benzene per billion is safe or unsafe will depend, among other things, upon relying on some rather than other studies. Choices must be made concerning which group of studies to rely on, in part on the basis of the type of arguments that play a role in sound argument closure. Some studies will be accepted because they are more rationally convincing in better following accepted rules of evidence and inference than others.

However, choices will also be made, as Ronald Giere notes, on the basis of the costs of being wrong. The problem is how to move from a gray world of scientific findings and uncertainties to a more black-and-white world of public policy, and of how to resolve the controversies that such passages involve.

An excellent example of the role of interests in the interpretation of the significance of data is offered by the essays of Gilbert Omenn, Marvin Legator, Mark MacCarthy, and T. R. Durham in their study of controversies concerning workplace safety. As Gilbert Omenn shows in his review of the role of the Occupational Safety and Health Administration (OSHA), different notions of where the burden of proof lies lead to different construals of the data at hand. For example, management often quite understandably takes the attitude that manufacturing procedures are innocent until proven dangerous, whereas labor, on the other hand, argues that it is best to err on the side of safety. As a consequence, the discussions regarding occupational safety and the role of OSHA have been politicized by the competing factions who have a vested interest in particular interpretations of data and in the character of regulations. Moreover, as Legator suggests, the possibility of future litigation may make management cautious in developing data that could in the future be part of large civil actions. More fundamentally, the question concerns what amount of assurance about what level of risk is worth what cost. Here there is no correct answer to be discovered but rather an answer to be fairly negotiated by the parties involved.

Such considerations routinely play a role, for example, in medical diagnosis and choice of treatment. In much of diagnosis there is a considerable risk of making the wrong diagnosis and therefore instituting the wrong treatment. To decide on the costs of various mistakes, one must weigh the costs of false positives (which include the risks of treatment for the wrong disease as well as the failure to treat the patient in a timely fashion for the disease the patient does in fact have), versus the cost of false negatives, of not treating the patient.[12] In the case of setting standards for the use of nuclear power, similar considerations obtain. If one has adopted standards that turn out to be too lax, one will pay the cost in the loss of lives, health, and property. However, if one sets the standards too strictly, one will unduly increase the monetary costs of electric power and perhaps encourage the use of forms of generating power that will in fact be more costly in terms of lives, health, and property than nuclear power. As a consequence, one must decide upon a prudent balancing of the risks of over- versus under-diagnosis, of over- versus under-regulation, of over- versus under-assessment of costly consequences.

Such assessments require a prior assessment of the comparative significance of the various possible benefits and harms involved in the various possible choices. There is unlikely, though, to be initial agreement with respect to such rank orderings. The result is often a scientific dispute with a heavy political and ethical overlay, such as that regarding proper standards for exposure to benzene in the workplace. Such disputes are complex, turning on competing views regarding (1) the actual merit of various studies to determine probabi-

[12] H. Tristram Engelhardt, Jr., *The Foundations of Bioethics*, pp. 185–89.

lities of risks (i.e., empirical, scientific, "sound argument" considerations); (2) the likely costs of various errors of assessment and of various possible policies (i.e., again, empirical scientific "sound argument" considerations); and (3) the ranking of the various possible benefits and harms (an endeavor in value theory, not empirical science). The points of disagreement are likely to be multiple and intricate. They are, moreover, likely to be intractable if approached from the point of view of sound argument closure. To begin with, the disputants qua scientists may differ as to their understanding regarding how the evidence should be interpreted. But most divisive will be the differences in rankings of possible benefits and harms, and in views of what harms should be exchanged for what benefits.

There will, for example, be differing views on the extent to which it is or is not proper to compare, say, a 95-percent chance of ten persons dying each year, owing to one form of acquiring energy, to a .095-percent chance of ten thousand persons dying in a catastrophe, owing to the use of nuclear power (to pick two fictitious figures). Or one might try to assess the propriety of trading lower gasoline prices for a slightly higher risk of workers and others dying owing to exposure to benzene. These disputes will not be open to a definitive rational resolution. It is not possible to appeal to a disinterested moral observer or group of disinterested observers in order to acquire an answer. A particular answer depends upon having a particular moral sense, and that cannot be imputed to a disinterested value observer or group of "rational" contractors in order to resolve the question, without begging the question. One would need to be able to identify a proper concrete value sense to impute, and that identification itself would require an appeal to a value sense.

Although one will not be able to discover or disclose by sound moral reasoning the authoritative moral sense to which one ought to appeal, there is still the possibility of negotiating the values by which to decide which facts to weigh and how. In assessing the possible harms and benefits of nuclear power, one is not simply describing the possibilities. In embarking on public policy, one is forced to choose among possible interpretations and to act on some but not others and eventually to pay the consequence of possible misjudgments.

Spencer Weart, for example, offers a sketch of how different assumptions cast nuclear power in a context of offering future promise or future threat. "Facts" regarding safety, after all, do not exist nakedly but within frameworks of expectations and general cultural understandings. Moreover, the facts themselves are far from clear. For example, Cora Marrett's analysis of the efforts by the President's Commission on the Accident at Three Mile Island to control and focus the debate of the commission shows how the scope of examination was circumscribed to facilitate the pace of decision-making and to enhance the public acceptance of the commission's report. This maneuver, however, left many fundamental questions regarding nuclear power untouched. As Douglas MacLean shows, the results have consequently been less than satisfactory.

The controversy regarding nuclear power, as a result, remains open and less tractable to negotiation than the issues of workplace safety. Interest groups have not been able, at least to date, to fashion a satisfactory preliminary compromise through negotiation. In part this is due to the circumstances of the situation, as Giere argues, for calculations yield a range of expected results divergent enough to support significantly different views of likely outcomes. As a result, no party to the controversy appears to be compelled by sound argument to accept a particular answer. Instead one is faced with an attempt to negotiate, though the stakeholders possess disparate views of the risks and benefits at stake.

Bargaining as a means of negotiating scientific disputes with heavy political overlays

Negotiations regarding how to weigh different interpretations of data are not simply guided by background views regarding what ways of exchanging benefits and risks are morally proper or prudent for the community as a whole. It may also be to the economic, political, or social advantage of one party to bring a second party to agree to a particular exchange of harms for benefits. The first party may in fact agree with the second party about what a reasonable and fair exchange might be, *ceteris paribus*. However, all things are usually not equal, and the first party may therefore seek to bring the second to agree to a different exchange, given some special inducements in order to make the exchange acceptable (e.g., increased pay for hazardous work).

The seeming controversy is then in fact really a haggling in the market between differently advantaged traders. Appeals to scientific data and differing value theories may only be the rhetorical ploys of buyers and sellers. This haggling can be seen as the economic, or at times political (i.e., when it involves a bargaining for political powers and privileges), dimension of a dispute, in contrast to the scientific and value theoretical elements of a dispute. In the marketplace, for example, closure is reached when the bid and ask prices coincide.

Some negotiations can then be understood as marketplace transactions in which individuals accede to a point, given an agreed-upon balancing of the interests of the various parties. The metaphor of the market is useful in suggesting the nondeducibility of the outcomes, as Ronald Giere suggests in his chapter. One must here again recall the distinction between resolutions through sound argument, which offer "correct" solutions, and resolutions through negotiations or through appeals to a procedure, which seek to be "fair." The market metaphor underscores this difference between a goal of negotiation closure (i.e., fairness) and the goal of sound argument closure (i.e., correctness).

One must be careful, though, to avoid overextending the scope of this point. Negotiations of scientific disputes presume a sense of correct outcomes, outcomes true as far as can be determined with the facts at hand. In order to respond to the uncertainty of outcomes, a prudent balancing of possible harms and benefits is sought. This may be established in a fashion that is tantamount to a market exchange. One might, for example, imagine the bargaining between management and labor with regard to the establishment of particular standards for workplace safety. In other cases the bargaining and exchange is less explicit. However, there may still be a negotiation over exchanges of possible harms for benefits. When there is no single appropriate answer, answers will be chosen because they are more useful in avoiding certain risks and securing certain benefits. Still, the facts do set constraints and some limits of reasonableness.

Reaching closure

As the foregoing suggests, there is not a single answer, or even a single genre of answers, to the question of how closure can be reached or why it fails to be achieved. Scientific disputes with a heavy political and ethical overlay are numerous controversies blended together. Each controversy has its own mode of resolution. An account of the closure, or of the failure of the closure, of a scientific dispute with a heavy political and ethical overlay is possible only through accounts of the various controversies that constitute the dispute. A geography of scientific controversies and of the numerous pitfalls to closure must, in short, be multidimensional.

As this volume shows, scientific controversies are as complex as the weave of human interests in truth, fairness, and individual advantage. The understanding of controversies, and of their resolutions, requires what the humanities always require: a careful attention to the complexities of the human condition. In the essays that follow, the reader finds a number of approaches to delineating the character of scientific disputes, to determining the ways in which such disputes are constituted out of a number of overlapping controversies, and to showing how they can come to conclusion or fail to come to conclusion correctly and fairly. Insight in these areas is central not only to reasonable public policy planning but to undertanding our complex society, which proceeds by various approaches to resolving scientific debates with heavy political and ethical overlay.

Our society has both a practical and intellectual need to be clearer regarding how one can reasonably bring conclusions to such debates. By understanding their complexity, one can better appreciate why such disputes are at times confounding and confusing debates, as in the instance of the debates concerning Laetrile or the classification of homosexuality. By appreciating the interplay of the factual and evaluational dimensions of the debates concerning workplace safety and nuclear power, we can perhaps better acknowledge what elements of such debates can be resolved through the discovery of new facts and what elements can only be resolved through developing means to create arbitrary standards of safety that will fairly balance the interests of the stakeholders. There is an advantage in knowing where we can hope to know more truly versus where we can hope to negotiate solutions more fairly.

PART I. THEORETICAL PERSPECTIVES

1

Ethical theory and the problem of closure

TOM L. BEAUCHAMP

Definitions of closure

Can ethical theory be employed to close or settle controversial moral problems? Could there be a policy or method for achieving closure on these problems? Answers to these questions depend somewhat on the sense of the term *closure* under consideration. Accordingly, I shall first discuss possible meanings of this term. I shall not, however, confine the analysis to the ordinary English meanings of the term, which often incorporate too many senses and fail to make distinctions important to philosophical argument. The term *closure* fails to distinguish, for example, between termination of controversy by final resolution and termination that stops short of either final resolution or truth. Accordingly, an analysis of the complex English usages of this word would in all likelihood prove fruitless.[1]

My procedure will be the following: I first examine five distinct senses of *closure*. I then concentrate on the fifth definition, which is stipulative. I make no pretense that this fifth sense is either descriptive of ordinary English, or reforming (uncovering some deeper meaning), or the only important sense of the term. Stipulative definitions do not constitute profound philosophy, but both stipula-

[1] It has been thought that if analyses of concepts that eventuate in the logically necessary and sufficient conditions of a term's correct application cannot be achieved by an examination of ordinary language, then perhaps a translation into some other language – e.g., an empiricist or a phenomenalistic language – would turn the trick. We would then allegedly have achieved a deeper meaning of *knowledge, material object, causation, closure,* or whatever term is under consideration. No one, however, has devised an unobjectionable method for achieving synonymy by translation, and problems of the epistemic status of appeals to deeper synonymy are at present unresolved. I believe that such a quest would certainly fail in the case of *closure,* even though I agree that any univocal definition must in some respects be reforming.

tion and comparison to alternative definitions can introduce clarity and reduce misunderstanding. These are my initial purposes. I shall then take the fifth stipulated understanding of closure and use it to develop a policy for achieving closure.

In all cases, closure is an outcome, not a process, for closure is the termination of a controversy. However, we cannot escape asking what processes lead to this outcome and, correspondingly, what the different senses of closure are. In what follows, then, I shall provide both a taxonomy of types of processes leading to closure and the senses of closure attached to each process. I shall even label each sense of closure by reference to the type of process involved. My five categories thus express ways of ending controversies, as well as senses of the term *closure*.

Sound argument closure

The first sense of closure, which I shall refer to as *sound argument closure*, is the following: Closure occurs if, and only if, a correct position has been reached in a context of controversy, thereby rendering opposition views incorrect. Philosophers interested in problems of justification (as distinct from purely conceptual or purely meta-ethical investigations) have been interested in finding either a general method for the resolution of problems or a particular resolution of a single problem. In either case, closure is achieved if the correct answer to the central question(s) is provided. The controversial issue is then truly decided, even if social or professional controversy continues thereafter. Accordingly, closure in this sense can occur even if agreement has not been reached in the context on the merit of any given position and even if there is a pervasive conviction of nonresolution.

Closure in this sense is not confined, of course, to closure on moral problems. Early modern theories of a heliocentric universe decided, and so resolved, a controversy in astronomy long before some principals in the discussion regarded the controversy as closed. Similarly, Newtonians ended or settled certain disputes with Cartesians well before it was recognized in France that the controversy was closed. Such disputes are decided by the weight of evidence and argument. The epistemological problem of how to determine that facts are objective, evidence good, and premises correct in contexts of controversy is, of course, a major and significant one for any account of criteria of soundness. In particular, there are problems about whether moral premises are ever true or correct and related problems about whether fairness and goodness, rather than correctness, are at stake in moral controversies and about whether judgments of soundness and fairness are relative to systems of thought. Loren Graham has cogently argued elsewhere in this volume (chap. 8) that moral problems are regarded in the Soviet Union as subject to correct, "scientific resolution." Far from being seen as rare, sound argument closure is pervasively accepted as the only meritorious standard. Most moral systems in recent Western philosophy clearly focus on fairness, obligation, and goodness, rather than on soundness of arguments, which are generally associated with outcomes of truth rather than outcomes of fairness. These problems about the concept of sound argument closure cannot be resolved short of treating the most abiding controversies in epistemology and ethical theory about relativism and objectivity, a task that can scarcely be undertaken here. However, it should be observed that in this first sense an argument must be sound (whatever the precise criteria of soundness), and not merely thought to be sound. Many clearly unacceptable arguments about both truth and fairness have been thought to be sound, even by whole cultures.

I suggest that in the strict sense of "sound argument," this form of closure is almost never achieved in the arena of controversial moral or public policy debates where dilemmas are confronted. The nature of a "moral dilemma" will be analyzed later, but it should be observed now that I am referring to a genuine dilemma created by acceptable but conflicting general principles; I do not simply mean a controversy merely *regarded* by principals as dilemmatic. Slavery and genocide, for example, do not present moral dilemmas, even if some plantation owners or Nazis have thought they do. Where genuine dilemmas are present, some positions may have better arguments supporting them than do other positions, but the better arguments cannot be said to render opponents' arguments incorrect. For example, there are powerful and perhaps even airtight arguments showing that under very general conditions civil disobedience is justified in constitutional democracies. Rawls has delineated these general conditions, and I take his conclusions to be well defended.[2] His arguments are far superior to those of even able opponents – such as the confused arguments advanced by former Justice Fortas in a book on the subject.[3] Nonetheless, there remains room for legitimate disagreement concerning the precise conditions under which civil disobedience is justified and as to whether particular acts, such as those of Daniel Ellsberg, actually fall under the justifying conditions.

[2] John Rawls, *A Theory of Justice* (Cambridge, Mass.: Harvard University Press, 1971), Chap. 6.

[3] Abe Fortas, *Concerning Dissent and Civil Disobedience* (New York: New American Library, 1968).

Consensus closure

A second sense of closure is connected to the first:[4] Closure occurs if, and only if, in a context of controversy a consensus has been reached that some position is correct or fair and opposition views incorrect or unfair. Here it does not matter whether a correct or fair position has been reached. It does not even matter whether, as a matter of justification and method, some point of view is well defended. Nor need principals believe that a permanent solution has been found, or even a definitive one. It only matters that there is a consensus agreement that the force of one position has overwhelmed others. Issues central to a controversy have in that context been resolved, and it is therefore routinely treated as no longer a matter of controversy. Settlement can come by any route whatever, and the settlement may be appropriate or inappropriate. Though it is unlikely, it is even possible that the controversy could be entirely decided in terms that Ernan McMullin has characterized as nonepistemic factors (see chap. 2, this volume). That is, the weight of evidence might play no role at all in bringing about the consensus. It follows that it is not a logically necessary condition of closure in this second sense that there be argument or theory, and the premises and conclusions of arguments can be irrelevant or false. It also does not matter whether the controversy later reemerges in substantially the same form. Consensus *at the time*, in the context, is sufficient.

Closure in this sense is comparatively frequent. Again, Loren Graham has offered elsewhere in this volume a useful set of examples of this type of closure in the Soviet Union, where a prevailing ideology virtually requires closure in a relatively short period of time (chap. 8). But even where such an ideology is absent, this form of closure is not rare. In North American society, a general consensus has now been reached, or is about to be reached, on the recombinant DNA issue and has for some while been reached on the issue of punishment for attempted suicide. Such a consensus was long ago reached on laws against bigamy.

Procedural closure

A third sense of closure is more familiar: Closure occurs if, and only if, an issue is ended by formal, procedurally governed efforts to terminate the sustained discussion that characterizes controversy. I call this sense of closure *procedural closure*. Here a correct resolution, or "sound argument closure," need not be present (even if it is a

[4] My formulation of this second sense is indebted to Ernan McMullin for several criticisms and suggested reformulations. His preferred term is *resolution*.

superior resolution), for – as in the second sense – settlement could rest on irrelevant, inappropriate, or even false grounds. Considerable disagreement among disputing parties may also be sustained indefinitely. This sense of closure is conceptually connected to "cloture." A legislature, in invoking cloture, seals off debate; there remains only the matter of voting aye or nay, and the controversy is then formally closed – or at least temporarily so. Social institutions employ many such methods for authoritatively arbitrating disputes and getting on with business. Courts, for example, function as arbitrating media formally charged to end controversies between disputing parties, and in cases involving such moral problems as abortion and capital punishment, procedural closure may be achieved in the courts through the votes of justices. Similarly, controversies over the Alaska pipeline and the dam at Tocks Island were ended when duly constituted bodies authoritatively determined to build the pipeline and not to build the dam.

It is doubtful that this third form of closure works effectively to end moral controversies more frequently than does the first form. That is, procedural closure usually works only temporarily, for the documents, arguments, or votes that it generates seldom if ever reach to the deepest layers of controversy. I would not maintain that procedural closure almost never ends debate, as I did in the case of sound argument closure, but only rarely is moral controversy closed off from further discussion, even where procedures for closure are institutionalized. Maneuvers are typically made in legislatures, for example, to reintroduce debates on abortion and capital punishment, even though some method had been found in previous sessions to achieve procedural closure. Similarly, new cases before the courts that differ from previous precedent cases may reintroduce a controversy thought to have been resolved by previous court decisions. It may be that a few moral issues, such as that of the morality of having more than one spouse, have been "finally" closed in some societies at some time by established procedures, but the occurrence is uncommon. Similarly in science, this form of closure is rare. The Lysenko affair, as discussed by Ernan McMullin (chap. 2, this volume), fits this sense rather than the second sense, because authoritative decision, and not the consensus of disputants, brought about closure. (Consensus closure was achieved in the late 1960s, however.) Such events are rare in the history of science.

Closure by natural death

A fourth sense of closure is the following: Closure occurs if, and only if, a controversy has come to an end through a gradual natural death.

I shall refer to this form of closure as *natural death closure*. Although controversial moral problems do not frequently find closure in this sense, such closure is not rare. Issues are sometimes closed out even though there has been no definite intention to close them out, no procedure for closure employed, and no sound argument resolution known to have been provided. Interest in the issues simply disappears, and no new arguments are found to advance them. They vanish. The controversy over Bentham's felicific calculus now seems closed in this sense, despite ongoing controversies over his general utilitarian philosophy, and many points of controversy in Thomas Aquinas's *Summa theologica* occupy this status – for example, his concern about whether heretics ought to be burned at the stake. An example of a naturally dead moral controversy evident in recent scientific and technological advances is found in writings on positive eugenics, in which breeding patterns for marriage and human reproduction were proposed. Despite recent attempts to revive this controversy,[5] it is at the moment effectively dead.

Sometimes natural death closure occurs when an issue has not actually died but has crested as a topic of controversy and so has lost its status as an ongoing issue of importance. It may never die or be closed in the aforementioned senses; but its standing as an issue has been crippled, perhaps because combatants have tired, and it is no longer the focus of sustained controversy. Natural death closure of this latter type has occurred in philosophical, theological, and scientific circles in which the teleological argument for the existence of God was once much discussed. After Hume's *Dialogues* and Kant's first *Critique* were read and unemotionally examined, the issue crested as a controversy, except in introductory courses in philosophy, enlightened divinity school seminars, and occasional scholarly books such as F. R. Tennant's attempt to revive the issue in the late 1920s and R. G. Swinburne's more recent efforts.

Ronald Giere has pointed out to me that this concept of natural death closure perhaps unjustifiably stretches the general notion of closure beyond acceptable conceptual boundaries because in this case controversy simply disappears or is abandoned, rather than achieving resolution. That is, controversy is overtaken by a dynamic flow of events and either is never resolved or is recast in the form of some related or more circumscribed controversy. Although Giere's criticism has its appeal, I think it unobjectionable to speak of a controversy's being closed, in much the same way as a case is

[5] See Joseph Fletcher, *The Ethics of Genetic Control* (Garden City, N.Y.: Doubleday, 1974), esp. pp. 29f, 125f, 153, 158f, and 180.

sometimes closed by a police department or a court – namely, in those cases where disputing parties simply stop disputing. It is irrelevant whether controversial points were closed in either the first, second, or third senses above, for the case is closed, even if no such resolution occurred and irrespective of the reason for closure. Moreover, it is also true of at least the first and third senses of closure that central issues in a dispute sometimes continue to be discussed and may ultimately turn into some related but distinguishable controversy.

Negotiation closure

A fifth sense of closure – the sense in which I use the term almost exclusively hereafter – incorporates some elements of earlier definitions but excludes other elements: Closure in this fifth sense occurs if, and only if, a controversy is settled through an intentionally arranged and morally unobjectionable resolution acceptable to the principals in the controversy, even if they regard the resolution as compromising their ideal solution. This fifth sense suggests the reaching of conclusions that are well defended by argument, perhaps argument from ethical theory. However, it is not required – as sound argument closure requires – that there be only one correct, best, or fairest answer to the central question(s) constituting the controversy. Closure can be achieved in this sense by compromise – in contrast to sound argument closure, where such compromise is impermissible. This fifth sense does not require that an agreed-upon resolution of the controversy be fully satisfactory and so noncontroversial in terms of the preferred moral standards of each principal. It requires only that the resolution achieved be minimally satisfactory. Thus, as in the case of three previous senses of closure, this fifth sense paradoxically permits a controversy to be closed when disputing parties still hold unclosed (in one of the other senses of closure) differences of opinion on the moral issue generating the controversy. I shall refer to this sense of closure as *negotiation closure*: The negotiation is a process, and compromise is the outcome. (A negotiation of course need not eventuate in a compromise. There may even be a settlement without compromise.)

I recognize that this fifth sense of closure is somewhat counterintuitive. Because closure may not have been achieved in any of the four previously mentioned senses – not even the second, for no consensus in the relevant sense may have been achieved – it seems odd to speak of the *controversy* as having been settled. But this is true of three of the other senses as well. Closure will often not be achieved in *all* of these senses merely because it has been achieved in some *one* of them.

Moreover, it will not do to object (as, again, Giere has done) that negotiations commonly occur through a procedure, and hence negotiation closure is merely a form of procedural closure. The process of negotiation entails no formal procedure paralleling closure. Although it is true that formalized, structured forms of mediation and arbitration are procedural, not all forms of negotiation conform to these models. For example, a student who negotiates a grade with a professor need use no formal procedures such as those outlined in grievance codes. Similarly, out-of-court settlements and changes in scientific design in order to protect human subjects are often negotiated in the absence of formal (or even implicit) rules and procedures.

Some may say, however, that negotiation closure is unacceptable as a definition of the term because it goes beyond the broadest boundaries of the accepted meanings of closure. But, as I mentioned earlier, my definition is stipulative, and such definitions are neither true nor false. They are only useful or useless, appropriate or inappropriate. If closure in this fifth sense *can* be beneficially brought about, this definition would obviously be appropriate. By contrast, if sound argument closure is never achievable in moral controversy, and if (as I think) the first and third senses of closure present no challenge to ethical theory, then a discussion of *policies* for closure in these senses is useless. This partially accounts for my concentration on negotiation closure.

I realize, of course, that a negotiated compromise may also constitute a consensus conviction, and that the latter is generally more desirable than the former. It may therefore seem somewhat odd that I focus on the fifth sense rather than the second. I choose this route for the following reason: I am interested in a policy to achieve morally unobjectionable resolutions, whether or not actual consensus is achieved. If consensus does occur, so much the better. But since negotiated compromise is sufficient, consensus resolution is not necessary. (It is also worth noting that negotiation closure can in principle eventuate in sound argument closure.)

In concluding this section, some further differences between sound argument closure and negotiation closure deserve attention. The distinction between justifying a position believed to be ideal and justifying the taking of a position different from the one believed to be ideal (in a context of controversy) is important. Suppose that two persons, P_1 and P_2, are members of a government advisory panel that serves the President. Each respectively holds morally justified positions M_1 and M_2, where M_1 and M_2 are both well-reasoned and morally defensible positions. "Morally defensible" here means "based on arguments correctly grounded in moral principles themselves defensible by appeal to well-reasoned moral theories." Suppose further that P_1 recognizes about P_2, and P_2 about P_1, that each holds well-supported positions (M_1 and M_2) from which they cannot be budged, and thus they both recognize a state of ultimate disagreement over an ideal solution. From the perspective of each, it would be ideal if closure could be achieved exclusively by further argument favorable to the preferred position, but this outcome is recognized to be impossible. In those cases when an end to the controversy is more important than a continuation of argument in defense of one's preferred views, justifying the *taking or acceptance* of a satisfactory position different from the one believed to be ideal emerges as the central problem.

Methods of achieving negotiation closure

Several ways of achieving negotiation closure have been employed in the past to resolve or reduce moral controversy. None has proved to be a consistently reliable means to end controversy, but each has proved a constructive method and deserves consideration prior to any attempt to delineate a policy for achieving closure. I shall now examine five such methods. (These means may also be ways to closure in one or more of the other senses, and some of my examples may fit these senses better than negotiation closure. Nonetheless, in all cases the examples can be reconstructed as instances of negotiation closure.)

Obtaining factual information

First, closure has been achieved by obtaining relevant factual information on which the controversy turns. It is sometimes presumed that moral disputes are (by definition of "moral") purely moral and not in any significant sense scientific, conceptual, or metaphysical. This presumption is too simplistic, for evaluative disputes – that is, disputes over what ought or ought not to be done – often have nonevaluative elements as central ingredients. For example, debates about the proper allocation of resources for research, prevention of disease and accidents, and education have erupted into controversies over the acceptable level of expenditures when there has been uncertainty about the facts – that is, uncertainty as to whether the proposed measures can actually prevent illness or accidents, improve treatments, or promote health. Or, to take a second and more concrete example, when the National Commission for the Protection of Human Subjects of Biomedical and Behavioral Research began to debate issues surrounding psychosurgery, controversy centered on the moral permissibility of using amygdalotomy and other risky surgical procedures (even with consenting subjects). As soon as factual studies were completed that pointed both to the relatively low risk of these procedures in the hands of qualified surgeons and to widespread patient satisfaction with the surgery, agreement crystallized among the commissioners regarding the conditions under which it would be morally justified to use the technique (even though in the minds of some commissioners these conditions were not ideal and even though some of the factual information itself proved controversial).[6] Similarly, new scientific information about recombinant DNA research, particularly data pertaining to the K-12 strain of *Escherichia coli*, turned this controversy in an unanticipated direction. The data presumably indicated that this strain is probably not pathogenic, even when genes for known toxins are introduced. Because the greatest fears about recombinant DNA research had been that the research might create an organism of pathogenic capability against which known antibodies would be inefficacious and that might produce widespread contagion, this new information had a dramatic effect on the moral and political controversy.[7]

As these cases indicate, disagreement often focuses on risks or on the causation of harm. Controversies about saccharin, Laetrile, toxic substances in the workplace, IQ research, fluoridation, radioactive waste disposal, and flu vaccines have similarly been bedeviled by factual uncertainty. While these have also often been value disputes about the proper bounds of liberty, attention should not be deflected from the role of factual information. New information may have only a limited bearing on some issues, but it has an overpowering influence on others. The problem, of course, is that rarely if ever is all the information obtained that would be sufficient to settle the factual sources of disagreement. (I here set aside controversies over criteria of the relevance of factual findings.)

Although it is logically possible that disputants will continue to dispute when they have reached agreement about all relevant facts (and have not overlooked hidden factual disagreements), I myself

[6] See U.S. Department of Health, Education and Welfare, *Report and Recommendations: Psychosurgery*, Publication No. (OS) 77–0001 (Washington, D.C.: U.S. Government Printing Office, 1977), esp. Chaps. 4 and 6.

[7] See "Lobbying Derails Bills on DNA Regulation," *Science and Government Report* 7 (October 15, 1977): 1–2, and Nicholas Wade's 1977 articles: "Gene-Splicing: At Grass-Roots Level a Hundred Flowers Bloom," *Science* 195: 558–60; "Recombinant DNA: NIH Rules Broken in Insulin Gene Project," *Science* 197: 1342–5; "Confusion Breaks Out over Gene Splice Law," *Science* 198: 176.

have never encountered such a case. Of course most controversies are so complex that they turn not merely on causal matters such as safety and efficacy but also on methodological problems regarding the best strategy for achieving a goal. For example, in the ongoing controversy over use of the synthetic narcotic methadone, a central question is whether supplementary drug programs or alternative strategies not involving drugs are better ways of curing heroin addiction. But even in the case of this question of strategy, factual data about withdrawal, the resale of methadone, rehabilitative outcomes, duration of addiction, and analgesic effects have a direct bearing on the complex set of issues defining the controversy.

A hypothesis worthy of serious consideration, in my view, is that if all relevant facts (and definitions critical to conceptualization of the facts) about such matters as risks to safety and health, likely long-range consequences, numbers of people affected, previous historical abuses, and similar matters were available, it would be unlikely that most moral disagreements would remain. I believe, for example, that if all of the relevant genetic facts were available many current issues concerning genetic screening legislation would disappear. Obviously no empirical probability can be assigned to this outcome. Indeed to my knowledge there is no empirical study of the closure of controversies that tends either to confirm or to disconfirm this hypothesis, and thus at present it will perhaps remain as controversial as the controversies at which it is directed. Nonetheless, as I have suggested, at least some anecdotal evidence tends to confirm rather than disconfirm it, and it is without doubt, as Charles Stevenson once put it, "a useful heuristic maxim."[8]

Conceptual analysis

Controversies have also been settled by conceptual analyses that eventuate in conceptual agreement. In some cases a *formal* conceptual analysis may not be necessary to achieve agreement, because stipulating a definition or clearly explaining a term's meaning may prove sufficient. Controversies over the morality of euthanasia, for example, are often needlessly entangled because different senses of the term are employed. If one party equates euthanasia with mercy killing, a second equates it with assisted suicide, and another equates it with voluntarily elected natural death, a "controversy" is blind.

[8] Charles L. Stevenson, "The Nature of Ethical Disagreement," *Sigma* (1948); as reprinted in *Facts and Values: Studies in Ethical Analysis* (New Haven: Yale University Press, 1963), pp. 7–8.

One may even have a noncontroversy, for the parties may be addressing entirely separate issues through their conceptual assumptions. Reaching conceptual agreement is no guarantee that a dispute will be settled, but the parties may at least shift to another level. For example, much controversy could be deflected in present controversies over the use of institutional review boards to scrutinize social science research if disputing parties could come to agreement, perhaps a negotiated agreement, over the meanings of the terms *experimentation, research, human subject, social science, minimal risk,* and *behavioral science.* Similarly, in the controversy over whether homosexuality is a disease, controversy could have been quieted at many turns through careful attention to the meanings of the central terms *homosexuality* and *disease.*

In the opening line of perhaps the most widely studied treatise on ethics written in the twentieth century, G. E. Moore offered the following hypothesis:

> It appears to me that in Ethics, as in all other philosophical studies, the difficulties and disagreements, of which its history are full, are mainly due to a very simple cause: namely to the attempt to answer questions, without first discovering precisely *what* question it is which you desire to answer ... The work of analysis and distinction is often very difficult ... But I am inclined to think that ... if only this attempt were made, many of the most glaring difficulties and disagreements in philosophy would disappear.[9]

To this flawless observation it need only be added that this problem is scarcely confined to philosophical controversy.

Adopting a common framework of moral principles

Progress toward negotiation closure can also be achieved by the mutual adoption of a common framework of moral principles. If this method involves a complete shift from one radically different framework to another, closure is probably doomed to be rarely achieved. Differences that divide people at the level of their most cherished principles and presuppositions are deep divisions, probably more impenetrable than any others. On the other hand, closure has been achieved through various forms of moral and legal negotiations that involve the adoption of a new or changed moral framework. James Childress and I recently argued, in our *Principles of*

[9] G. E. Moore, *Principia Ethica* (Cambridge: Cambridge University Press, 1903), Preface, p. vii.

Biomedical Ethics, for an approach that presents such a moral framework. After we came to agreement over the framework, the path to the resolution of some of our own moral disagreements was eased. Similarly, the National Commission for the Protection of Human Subjects of Biomedical and Behavioral Research began its deliberations by unanimously adopting a common framework of moral principles that provided the general background for deliberations about particular problems. The commissioners developed a framework of three moral principles: respect for persons, beneficence, and justice. These principles were analyzed in the light of contemporary philosophical ethics and then applied to a range of controversial problems that confronted the commission.[10] There is ample evidence in the transcripts of this body's deliberations to indicate that the common framework of moral principles facilitated discussion of these controversies and led to many agreements that might otherwise have been impossible.

Using examples and counterexamples

Fourth, closure has been achieved, or at least facilitated, by the dialectical use of example and opposed counterexample in testing a general principle or recommendation. If, for example, one wishes to ascertain the conditions (if any) under which deception is justified in psychological research, a way to begin is to examine examples of both paradigmatically justified and paradigmatically unjustified deception, to compare these examples with similar examples that differ only in minor respects, and to attempt to determine which variables tend to count against deception, which in its favor, and why. One could then construct plausible principles of justified deception and attack these principles by counterexamples. Finally, the principles could be modified and subjected to further counterexamples until a "most acceptable" principle is discovered. Such use of example, counterexample, and reformulation serves as a format for weighing the strength of conflicting considerations.

An attempt to reach closure by the weighing of example, counterexample, and reconstruction occurred when the commission mentioned earlier came to consider the level of risk that could justifiably be permitted in nontherapeutic scientific research involving children as subjects. Based on principles of acceptable risk used in its own

[10] These principles and their analysis by the commission have been published by the U.S. Department of Health, Education and Welfare as *The Belmont Report: Ethical Principles and Guidelines for the Protection of Human Subjects of Research* (Washington, D.C.: U.S. Government Printing Office, 1978).

previous deliberations, the commissioners were inclined to accept the view that only procedures involving minimal risk could be justified (where "minimal risk" is analogically determined by the level of risk present in standard medical examinations of patients). However, a series of examples from the history of medicine was cited that revealed how certain diagnostic, therapeutic, and preventive advances would have been unlikely, or at least retarded, unless procedures presenting more than minimal risk had been employed. Counterexamples of overzealous researchers who placed children at too much risk were then thrown up against these examples, and the debate continued for several months. Eventually a majority of the commissioners abandoned their original principle that nontherapeutic research presenting more than minimal risk was unjustified. They came to accept the view that a higher level of risk can be justified by the benefits provided to other children. Once a negotiated compromise was arranged, the controversy about the involvement of children as research subjects ended (although a small minority of commissioners never agreed).[11] Some regarded the final resolution as less than ideal but as nonetheless justified – a clear example of negotiation closure.

Exposing inadequacies and unexpected consequences

Finally, an important method of philosophical inquiry in general is that of exposing the inadequacies and unexpected consequences of an argument. This method of inquiry has often been brought to bear on moral disagreements and has led to significant shifts in position. Besides the obvious case of pointing to an inconsistency, there are many ways of attacking an argument. For example, a number of writers have discussed the nature of "persons" when treating problems of abortion, fetal rights, and the definition of "death." Some of these writers have not appreciated that their arguments about persons were so broad that they carried important and unacceptable implications for both infants and animals, as well as for human fetuses. Their arguments provided reasons that they had not noticed for denying to infants rights that are granted to adults, or for granting (or denying) to fetuses rights that are granted to infants, and in some cases for granting (or denying) to animals rights that are granted to infants. It may of course be correct to hold that infants have fewer

[11] U.S. Department of Health, Education and Welfare, *Report and Recommendations: Research Involving Children*, Publication No. (OS) 77–0004 (Washington, D.C.: U.S. Government Printing Office, 1977), Chap. 9 and "Recommendations."

rights than adults, or that fetuses and animals should be granted the same rights as infants. The present point is only that if an argument leads to conclusions that a proponent is not prepared to defend and did not previously anticipate, then part of the argument will have to be changed, and the distance between those who disagree will presumably be reduced by this process. This style of argument is often supplemented by one or more of the other ways of reducing disagreement that I have mentioned.

Much moral disagreement may not be resolvable by any of the five means that we have now discussed. "Moral expertise" is a matter of skill in using these methods, not a method of providing definitive answers. No contention has been made in this section that moral disagreements can always be successfully negotiated by use of these methods, or even that every rational person must accept the same method for approaching such problems. There is always a possibility of ultimate disagreement, especially over hard cases. However, *if* something is to be done in contexts of controversy, resolution is most likely by use of the methods outlined in this section.

Nonetheless, no matter how instructive these ways of finding closure may be, merely explicating them does not reach my final objective, which is to outline a policy for negotiation closure. This is to be a policy for how closure should be attempted, as distinct from how it has been successfully achieved.

A policy for negotiation closure

What, then, is the problem of a closure policy? In part, it is little more than a common philosophical problem about justification. Philosophers interested in science, knowledge, and ethics have traditionally explored two different kinds of problems about justification. First, many human beliefs surpass the actual evidence available to support them; how to justify or criticize these beliefs is a matter of philosophical curiosity. For example, we hold beliefs about the existence of material objects where the only evidence for doing so is certain ways of being appeared to; and we believe that there are laws of nature, even though the only evidence for this belief consists in constancies of correlation among similar events. Most problems in epistemology and the philosophy of science center on how to bridge a gap between evidence and a broader belief based on the evidence. Often this gap is a logical gap, where conclusions are not deducible from supporting premises.

In philosophical ethics there are related problems about supporting or justifying conclusions, especially in supporting general (universalized) principles. There are gaps, for example, between wide cultural and philosophical acceptance of basic moral principles, such as respect for persons and the formal principle of justice, and the further conclusion that these principles are universally valid moral truths. It is sometimes even said that the entire institutions of factual inference and moral inference stand in need of justification. However that may be, just as we seek ways of bridging gaps between factually well-supported premises and broader conclusions in order to justify what we believe as a matter of fact to be true, so we seek ways of bridging gaps between morally well-supported premises and broader conclusions in order to justify what we believe as a matter of morals to be correct. The problem of induction is perhaps the most familiar problem of this sort in epistemology, and the problem of intrinsic value is an equally familiar problem in ethics.

Second, in both epistemology and ethics a recurrent problem is that different people accept the same evidence, or at least the same descriptive premises, and nonetheless reach significantly different conclusions. For example, some people believe in miracles and divine revelation, whereas others do not, yet they may agree on descriptions of all relevant historical occurrences. Similarly, some believe that science has a distinct method or logic, whereas others do not, yet they do not hold different scientific beliefs.

Some of these problems of justification may not be resolvable by any means whatsoever. I am inclined to think that this is so for the ontological, legal, and moral problems of fetal status that plague the debate over abortion. Some problems of theological belief, particularly epistemic problems concerning faith, also seem to me unresolvable. However, if something is to be done about these problems of justification in contexts of controversy, a resolution in any context – moral, epistemic, or otherwise – is most likely to occur by obtaining agreement on basic premises. For this reason Descartes clung to foundationalism (sound argument closure, that is) as a method for achieving closure in epistemology. My presumption is that if something is to be done in ethics to achieve negotiation closure (or sound argument closure, for that matter), something analogous to foundationalism must be pursued: There must be a policy for constructing a position from the foundations up, or at least from as near the foundations as possible. This would be a policy that fair-minded individuals could pursue to achieve as much closure – or conflict resolution, anyway – as possible; that is, it would be a policy created to decide what is to be done to bring about closure on issues where there are apparently irreconcilable differences of opinion.

This policy must be composed of a set of directives for achieving negotiation closure. Such directives are prescriptive rules for the acceptance, rejection, and modification of beliefs that are held by one or more principals in a controversy. If parties enmeshed in controversies could be persuaded to accept the strictures of the policy, they would be bound to the acceptance of the set of beliefs that most closely conforms to the policy's directives and to the rejection of beliefs the policy requires to be rejected. Negotiation closure would be achieved, then, when the principals conformed to the directives and thereby settled their differences.

What directives should constitute this policy for negotiation closure? I suggest five directives as the policy's components.

1. The policy should require two forms of consistency. Clearly, it should require that beliefs be internally consistent; otherwise the resolution would not yield the same results when used by different parties in the same circumstances – a disaster for a closure policy. More significantly, the policy should require that beliefs be consistent with the set of moral propositions that no one can rationally deny. This is not the place to engage in a defense of the nature and commitments of rational belief in morals, but most of us believe that criteria can be adduced for the reasonableness and unreasonableness of moral beliefs, just as they can for factual beliefs. In both contexts we correctly refer to arguments and resolutions as either reasonable or unreasonable. To say that no one can rationally deny certain moral and factual beliefs is minimally to say that these beliefs are less subject to skeptical doubt than are the beliefs required to create skeptical doubt. This is not to say that such beliefs are certain but only that they are heavily protected and are at the heart of our systems of morality and factual knowledge. Aristotle's formal principle of justice and Mill's harm principle are cases in point. To consider only the latter, the harm principle says that coercive interference with a person's liberty is justified if through his or her actions the person produces significant harm to other people or to public institutions. I would maintain that the harm principle itself is less subject to critical attack than any moral proposition that might be employed to attack it. Any beliefs introduced in an attempt at closure must be consistent with the set of such rational principles or beliefs.

2. The policy should require that obtainable factual information having a significant bearing on the outcome of the controversy be assembled. As indicated in the previous section, many controversies turn more on incomplete factual information pertaining, for example, to risks of harm than on differences of normative principle.

3. The policy should prefer and should specify a method for distinguishing between universalizable moral principles and those

nonuniversalizable claims rooted in self-interest, obstructive forms of ideology, religious bias, and so forth. The policy also should be able to distinguish between relevant and irrelevant universalizable principles. I do not deny that this requirement will be difficult to put into practice and will never be so exacting as to produce a Kantian categorical imperative that faithfully distinguishes valid from invalid generalizations. Still, this component in a policy for closure seems to me of the highest importance. Many moral problems where conflict should be reduced are rooted in ideological, selfish, or interest-group preferences thought to capture the sole correct moral position. These may be major impediments to closure, whereas appeals to universalizability might usefully break down these barriers. Use of the Golden Rule or of questions such as "What would happen if everyone acted like that?" are examples of such appeals to consistency. They may ultimately prove inefficacious in many settings, but this is no more a mark of their invalidity than would consistency appeals in scientific arguments that went unheeded by certain parties be a mark of their invalidity.

4. The policy should require that beliefs be favored that are most coherent with the total set of beliefs, including the moral and factual beliefs mentioned in components 1 through 3, the sole qualification being that the dialectical adjustment of principles and particular judgments (as discussed in the section entitled "Methods of Achieving Negotiation Closure") be encouraged rather than inhibited by the attempt to achieve coherence. This requirement will seem unduly conservative to some, for it is not calculated to promote bold revisionary thinking in morals. Still, this requirement seems to me morally unobjectionable as a component in a policy for closure, even if it would be objectionable as a sole general requirement for moral thinking. Again, a parallel to epistemology may prove instructive: Coherence theories in epistemology are objectionable because they contain no provisions for distinguishing between true and false propositions and because they make it difficult to argue against highly peculiar but internally coherent systems of thought. Nevertheless, I cannot see that a coherence requirement is objectionable as a component part of a foundationalist program in epistemology, and I similarly cannot see that it is objectionable as a component part of a policy for closure.

5. The policy should require that alternative resolutions be given priority rankings based on (1) the *comprehensiveness* of the resolution as a means of dealing with the range of problematic situations and (2) the *simplicity* of the resolution. The resolution should be comprehensive, but at the same time it should not incorporate more rules and

concepts than are necessary, and certainly no more than affected persons are able to remember and apply without confusion.

The policy constituted by these five components proposes that it is morally justified for any person, *P*, at a time, *t*, in a situation of moral controversy to accept resolution r_1 (by following the directives of the policy) rather than resolution r_2, whatever r_2 may be. This idea of moral justification may seem peculiar, especially since r_1 is a negotiated position potentially regarded by all parties as morally less satisfactory than their ideal. I think, however, that the sense of "morally justified" involved is not peculiar or difficult to explain. It may be analyzed as follows: Moral justification occurs by following a general moral policy, the acceptance of which is best suited to achieve the general ends of all moral policies (the fair adjustment of pro and con interests through rules that limit conflict and promote welfare). It is no more peculiar to accept this view of moral justification than it would be to accept an analogous account of epistemic justification in circumstances of epistemological controversy. We could say, following R. B. Brandt, that the statement "it is epistemically justified ['rational'] for person *S* at time *t* in his circumstances to place more confidence in belief *p* than in *q*" may be analyzed epistemologically as follows: "There is a general epistemic policy, the acceptance of which is our best bet for achieving the ends of the cognitive enterprise, and this policy directs, for the situation of *S* at *t*, that *S* place more confidence in *p* than in *q*."

It would be useful to have both ethical and epistemic policies that function to resolve controversies. Although most moral and epistemological thinkers would no doubt prefer a substantive moral theory or epistemic theory composed of the one and only set of correct moral and epistemic principles, fair-minded people will agree that it is quite undecided what constitute the full sets of correct moral and correct epistemic principles. It is this unsettled situation that gives rise to the need for a negotiation policy in morals, especially when there exists a special need for closure of a controversy.

However, we need to consider further the distinction between doing the best one can to achieve a morally justified position supporting closure and doing the best one can to support a "best" moral argument supporting a point of view. I am presupposing a meta-ethical thesis not argued for in this paper, viz., that moral principles are abstract generalizations that guide moral thinking but cannot be applied with the specificity and precision needed to resolve many moral problems. Principles simply do not function in many cases to dictate particular outcomes to the exclusion of other outcomes. The most recalcitrant controversies, where closure is most difficult, are almost always dilemmatic. They are dilemmatic because parties on all sides of the controversy can correctly marshal moral principles on behalf of their substantially different conclusions. I earlier promised to explore this point further, and I shall do so now.

Dilemmas occur if good reasons for mutually exclusive alternatives can be cited by all the principals, and yet if any one set of reasons is acted upon the resulting actions will be desirable in some respects but undesirable in others. Consider, for example, the issue of whether we should allow prisoners to become involved as voluntary subjects of nontherapeutic research. Principles of autonomy dictate that prisoners not be denied the opportunity to participate, yet principles of nonmaleficence require that prisoners be protected from harm that might be caused by the research itself or through the coercive actions of those in control of prisons. One moral resolution has it that prisoners should be free to volunteer under some prison conditions (e.g., with sanitary facilities, minimum security provisions, consent referees, and so on) and be prohibited from "volunteering" under other prison conditions (e.g., where there exist powerful parole boards, unmonitored punitive possibilities, undue influence through payments by drug companies, and so on). This ideal solution unfortunately leaves the dilemma untouched. Under current prison conditions a mixture of these conditions is present and, because of various social and political forces, simply will not be eliminated. Most moral dilemmas similarly present a need to balance competing ideal claims in the hope that negotiation closure may be found.

The policy proposed here is intended to handle this sort of dilemmatic situation. However, one can only do one's best. Even when the policy is carefully followed, the best efforts may result in failure. Ultimate disagreement over hard problems is to be expected; it is one reason why we have procedural methods of closure. Indeed, it is one reason why we have politics as well as ethics. However, such "failures" exhibit the fruitlessness of further attempts to convince an opponent; they do not reveal a deficiency in the policy itself. Deficiencies in policy must finally be gauged by comparison with alternative policies intended to achieve closure, or so I shall now argue.

Justifying a closure policy

Because philosophical problems of justification have been emphasized in the preceding section of this essay, the problem of justifying the recommended policy deserves consideration. Alternatives to the recommended policy are clearly possible, and the skeptical view that it is impossible to justify any closure policy might also be defended. This skeptical claim seems to me indefensible, but it might be modified in a weakened form to the claim that it is impossible to show that a policy will lead or will probably lead to acceptable negotiated beliefs, just as it is impossible in epistemology to show that a policy will lead or will probably lead to true beliefs. David Hume argued that it could not be shown either with certainty or with measurable probability that any given policy would eventuate in true beliefs.[12] Hume is correct, both in matters epistemological and matters moral, but this provides no reason to suppose that a policy cannot be justified on grounds that it is the best policy one could hope to obtain in order to reach the goals for which it is intended.

This point can be clarified by one final analogy to problems of justification in epistemology, in this case an analogy to the problem of induction. According to orthodox interpretations of his philosophy, Hume argued an especially significant thesis among those just mentioned: He argued that there could not be a noncircular justification of any inductive policy for reaching true factual beliefs.[13] The arguments that Hume allegedly used need not be outlined here; it is sufficient to note that according to this skeptical thesis there could be no justification of ultimate principles of inductive inference, and so no justification of any particular policy of induction from among the range of possible policies for reaching true beliefs (including a counterinductive policy). Herbert Feigl and Wesley Salmon have correctly argued that this thesis fails to consider pragmatic justifications that function to vindicate one policy as preferable to others. Here a "policy" is a set of validating principles that serve to warrant individual inferences (deductive, inductive, or moral, as the case may be). To vindicate such a policy is to justify a set of ultimate principles of inference. Vindication occurs when it is shown that a particular policy or set of validating principles is best adapted for reaching the objectives intended to be served by the policy. On the hypothesis that there is an order in nature, Salmon argues, induction is the method best equipped for reaching correct predictions (and so true beliefs), and the best policy of induction would be the one most suitable for achieving correct predictions among the possible inductive policies aimed at this result.[14]

[12] *A Treatise of Human Nature*, Bk. 1.

[13] See Hume's *Enquiry Concerning Human Understanding*, secs. IV–V. Tom Mappes and I have argued that Hume never held this skeptical thesis; see *American Philosophical Quarterly* 12 (1975): 119–29.

[14] Herbert Feigl, "De Principiis Non Disputandum . . . ?", in *Philosophical Analysis*,

The same line of argument may be used to justify a closure policy. The policy's justification finally rests on its suitability as a process of reasoning leading to morally satisfactory negotiation closure. If there is a policy best adapted for achieving this form of closure, then it and no other is justified. Of course justification depends significantly on the particular sense of closure at stake – one reason why different senses of the term were distinguished earlier. The policy best suited for achieving procedural closure, for example, may be a very unsuitable policy for negotiation closure. Also, if there is no policy that leads better than any other to some form of closure, then bringing about closure in that sense is all a matter of guesswork, and there can be no policy to bring it about – just as, if there is no order in nature, then no inductive policy can be justified, and inductive inferences are no better than guesses. Whether some policy is more successful than another as a means to an end is largely an empirical question. Even if it is never put to the test, it is an empirical question whether the policy that I have supported will lead more consistently to negotiation closure than might some alternative policy. The answer can be provided only by scientific investigation, not by further philosophical inquiry.

Finally, two important questions have not been addressed here, and I offer no answers to them. First, under what conditions should the recommended policy (or any alternative) be invoked? It may be that the policy should be invoked only rarely and that controversies should often flourish without attempts at closure, especially in a pluralistic and open society. Nothing that I have said here asserts otherwise, and clearly closure is sometimes prematurely attempted. I thus would not maintain that closure is always or even generally better than nonclosure. Second, it is reasonable to ask what procedures would have to be involved in order to implement the policy. This question relates the problem of negotiation closure to problems of procedural closure. This problem cannot be addressed here, but further study of these two important questions is surely in order.[15]

Footnote 14 (*cont.*)
ed. Max Black (Ithaca, N.Y.: Cornell University Press, 1950), pp. 119–56; W. Salmon, "Should We Attempt to Justify Induction?", *Philosophical Studies* (1957): 38–42, and "Symposium on Inductive Evidence," *American Philosophical Quarterly* 2 (1965): 265ff. Hans Reichenbach provided the original arguments leading to these vindicationist views. Cf. his *Experience and Prediction* (Chicago: University of Chicago Press, 1938), pars. 38–9.

15 The comparisons between ethics and epistemology that I have made, and a number of substantive points, have been influenced by an unpublished paper of Richard Brandt's entitled "The Concept of Rational Belief," from which I learned a great deal.

2

Scientific controversy and its termination

ERNAN McMULLIN

Writing in 1921, Norman Campbell answered the question posed in the title of his book *What Is Science?* with the succinct definition: "Science is the study of those judgements concerning which universal agreement can be obtained".[1] Two possible objections occurred to him. The first was that "it is notorious that men of science differ among themselves" just as acrimoniously as philosophers and linguists do. He replies:

> I do not say that all the propositions of science are universally accepted – nothing is further from my meaning; what I say is that the judgements which science studies and on which its final propositions are based are universally accepted. Difference of opinion enters not with the subject-matter, but with the conclusions that are based on them.[2]

A second objection is that if universal agreement is a necessary condition for the subject matter of science, "a single cantankerous person ... could overthrow the whole fabric of science." His response to this is that such a challenge could come only if the person were lying or suffering from hallucination or sensory defect. After a discussion of these different possibilities, he concludes with some satisfaction:

The first draft of this paper was written while I was enjoying the hospitality, as a Mellon-Scaife Fellow, of the Center for the Philosophy of Science at the University of Pittsburgh in 1979. I would like to express my appreciation to the staff of the center, as well as to the members of the Department of History and Philosophy of Science with whom I was able to discuss many of the issues treated in this paper. I would also like to thank Phillip Sloan, Alasdair MacIntyre, and Art Caplan for suggestions on specific topics.

[1] *What Is Science?* (1921; New York: Dover, 1952), p. 27.

[2] Ibid., p. 30.

> It is a very remarkable fact that [in such cases] we can always find a test which . . . enables us to restore universal agreement. It is this fact, which could not be anticipated, which makes science possible, and gives its great importance to the test of universal agreement.[3]

My reason for beginning with Campbell is not to make his view of science a focal point of discussion in this essay. Rather, it is to illustrate how much the climate has changed, not only in philosophy of science but in science itself, over the last sixty years. Although there would still be some who would agree with Campbell that science must be based on an agreed-upon set of fundamental propositions, there are few indeed who would suppose that universal agreement can be attained as easily as he assumes. More to the point, such agreement would no longer be thought necessary; indeed, some would even regard it as undesirable. The classical theories of science, whether of Aristotle, of Descartes, of Kant, or of the positivists, all took for granted two theses: foundationalism (that science must be built on a foundation of propositions, themselves unproblematically true), and logicism (that science possesses a logical method that will allow one to determine which of two theories is the better one in any given case). As these theses came more and more under challenge, the roles played by disagreement and controversy in science came, in consequence, to be viewed quite differently.

The challenge came from many quarters. It came from those like Popper, Quine, and Hanson, who showed that the observation statements from which experimental science begins are not as unproblematic as they seem. Then, there were many who noted that the criteria for theory acceptance are not nearly as cut-and-dried as the notion of a "logic of science" would lead one to expect. Further, the new emphasis on science as the product of a quite special sort of community, which is itself part of a broader community, led to a new appreciation of the importance of psychological, social, and political factors to an understanding of the history or of the characteristic structures of science. No one thinks of science today in terms of definitive truth, of "final propositions," to use Campbell's terms. Disagreement and the withholding of complete assent are no longer regarded as extraneous to science; their presence is understood to be continual and essential.

Different philosophers of science have attributed very different weights to the "disagreement" factor. Kuhn tends to stress agreement; the "normal" situation is one in which a single paradigm is in

[3] Ibid., p. 34.

possession and disagreement is peripheral. There are mechanisms in the training of young scientists and in the life of the scientific community generally that tend to bring about consensus. Toulmin's evolutionary metaphor of scientific change leads him in the same direction. Popper and the neo-Popperians, on the other hand, insist that the clash between rival theories is a permanent part of the scientific landscape and that a successful theory must never be allowed to attain the unchallenged status that would prevent it from making further progress.

In what follows, I propose to examine not the general theme of disagreement in science but the narrower one of scientific controversy. The essay will be divided into two parts. The first will deal in a general way with the nature of controversy in science. The term *controversy* has the imprecision usual in the terminology of metascience, the imprecision that scientists tolerate in terms such as *observe* or *explain*, though not in terms like *energy* or *selection*. But it can be sharpened by looking at common features of episodes in the life of science that are generally taken to qualify as "controversies." Having done this, we can go on in the second part of the paper to ask how scientific controversies end. What are the characteristic ways in which controversies in science terminate?

Controversy in science

As a rough initial definition, let us say that a controversy is a publicly and persistently maintained dispute. A *scientific* controversy is concerned with a matter of belief. Each side argues that the other is wrong and that they themselves are right, or at least have the better case. The difference between them is one of belief, of knowledge claim. Further, the issue is held to be determinable by scientific means – that is, each side in the controversy claims the authority of "science" for their view. There may also be disagreement in regard to an action to be taken (e.g., the siting of a nuclear power plant). But insofar as such a controversy is a scientific one, it rests on a difference of assessment of the *reasons* for action – that is, on a difference of belief.

To count as controversy, the disagreement must be a *continuing* one. There is argument and counterargument. And the exchange is *public*; it is expressed on both sides in writing or orally, so that others can come to judge the merits of the case. It is this last condition that is of most significance to us as we ask how controversy functions in science. A disagreement, no matter how profound, between two scientists is not enough to constitute a controversy until the terms of their disagreement have been made known to the scientific community generally. Others are thus enabled to take part. In principle, a controversy is open to everyone who is qualified to understand the issue at stake. It is thus a *community* activity, even though it may begin by involving only two people. The protagonists appeal to others as though to judges; the outcome of the controversy will depend, not just on the actions and arguments of the protagonists but on the response to these of the scientific community generally, or at least of that part of the community that concerns itself with the disputed topic.

When a disagreement is public, does that suffice to constitute it as "controversy"? Not quite. We would not be inclined to say that the shape of the earth is still a matter of controversy, despite the fact that there are, reputedly, still a few who defend, as publicly as circumstances permit, a flat-earth view. Nor would scientists allow that the equivalence of inertial reference frameworks (the theme of Einstein's special theory of relativity) is any longer a matter of controversy, despite the occasional voice raised in support of older absolutist views. Indeed, most physicists in the 1950s would have held that the indeterminist implications of quantum theory were no longer in controversy, even though no less an authority than Einstein continued to reject them.

For a disagreement to count as "controversy," then, it must seem to the community to be worth taking seriously. Challenge from someone perceived (rightly or wrongly) as a crank or as incompetent does not suffice to create controversy. Even when the challenge comes from an Einstein, it will depend on whether the community considers a response worthwhile. The community can, of course, be mistaken in this judgment. But if no defense of the "orthodox" position against the challenger seems necessary, no real controversy will occur.

Does it depend, then, on the community's perception of the merits of the case? In a way, yes. But more has to be said about who constitutes the "community" here. Supposing scientists are united in their judgment that a particular drug is worthless, or that a particular theory such as neo-Darwinian evolutionary theory is adequately substantiated: Would this mean that controversy about these claims is impossible? Obviously not. If a sizable number of nonscientists reject the scientists' claim, for whatever reason, then a controversy is likely to occur. Scientists may see no merit whatever in the opposing case. But they are forced to take this case seriously because they form part of a wider community where many defend it. They must respond, because their authority to settle the case is questioned. Such controversies (and we have seen more and more of them lately) are genuine controversies, although they would not be called "scientific" in the ordinary, restricted sense of that term.

A scientific controversy can be said to exist, then, only when substantial parts of the scientific community see some merit on both sides of a public disagreement. This may help us to decide when a controversy may be said to begin or (more to our purposes here) to end. When Wegener proposed his hypothesis of continental drift in 1915, it was almost entirely ignored (see Frankel, Chap. 7, this volume). Only ten years later did his work have enough support for geologists to consider it even worth discussing. When masses of new evidence, especially from the ocean beds, and a theoretical explanation of how continental movement might occur became available in the 1960s, virtually all geologists went over to the plate-tectonic model, as the new view was now called. There were still a few holdouts, as there always are when a fundamental theory change occurs. Could their presence warrant our speaking of the controversy as still continuing?

This becomes a matter of definition, and it would be both artificial and unnecessary to try to stipulate a definition sharp enough to allow us to answer such questions as this unambiguously. Suffice it to say that if one side of a scientific controversy is widely believed to have lost all force, the controversy is to all intents and purposes over. This is still quite vague: "Widely believed." *How* widely? "Lost all force": Must it (can it) lose *all* force? "To all intents and purposes": Is it, then, not *quite* over? But it is precise enough for our purposes.

The emphasis above on the role of community in the determination of controversy may serve to introduce one further point that must be made regarding the nature of scientific controversy. A controversy is a historical event; it has a date and a place. It is not just an abstract relationship of evidence and hypothesis. One may, of course, take a controversy to be crystallized in a set of documents, as is customarily done with the Leibniz–Clarke controversy, for instance. This is perfectly legitimate; one may weigh the merits of the arguments advanced on either side, leaving aside matters of personality and of historical contingency. But the controversy itself did involve personality and contingency. In particular, if one is interested in how controversies end, there is no alternative but to take them in their full historical reality. A controversy can end because someone has not read a particular book he or she should have or because a government agency cuts off funds from a research project. Controversy is at least as various in its causalities as any other complex human action is.

The understanding of controversy belongs in the first place to the historian, then, not to the logician. It is a topic for the historian's peculiar skills; it requires the sort of patient unraveling of reason and motive that distinguishes historical from other sorts of analysis. When someone ask, "How do controversies occur?" or "How do they end?", it requires an immediate reference to the historical record. If any generalizations are to be made, they will depend in the first instance on the value of the case histories that are adduced in evidence. In this respect, controversy differs from other topics of concern to the philosopher, such as explanation or justification. Not that these latter need no reference to history or practice. They do. But controversy is a human act, a social episode, whereas explanation and justification are, to a first approximation at least, expressible as propositional relationships.

In this respect, the topic of controversy resembles that of *discovery*, about which so much has been written of late in philosophy of science. Discovery also is a human act, involving the same sort of confusion of personal and social factors as controversy does. It was for this reason that fifty years ago the logical positivists abandoned scientific discovery to the psychologists and the historians in order to focus on the more manageable themes of justification, falsification, and explanation. Discovery was too messy, in their view, to yield "philosophical" (i.e., logical) insights. Presumably they would have said the same of controversy. Both precede science, if science be taken as demonstration or as confirmation. And both involve factors other than propositional ones that are capable of logical relationship. This affinity between the topics is worth stressing because the second-order questions that face us in regard to scientific controversy have already been extensively discussed in the context of the debates about discovery.[4]

Controversy in ethics and in law

What about other sorts of controversy? Two types are particularly worth noting because of the contrast they provide with controversy in science. Moral issues are the source of some of the most intractable controversies in contemporary Western society. The kind of consensus that characterized earlier more homogeneous, more tradition-oriented societies has largely vanished. It is not just that there is widespread disagreement on fundamental ethical matters; there is

[4] See, for example, the papers from the 1978 Reno Conference on Scientific Discovery collected in two volumes by T. Nickles: *Scientific Discovery, Logic and Rationality*, and *Scientific Discovery: Case Studies* (Dordrecht: Reidel, 1980).

little agreement as to how such controversy itself should be adjudicated. Here is the main difference between controversies in ethics and in science – in the natural sciences, at least. There would be today a fairly high degree of consensus about the proper means of resolving controversies in science. So that even where two parties totally disagree on a matter of scientific judgment, they are likely to agree (at least in broad outline) on what would count as relevant considerations on either side of the case.

In his influential book *Ethics and Language*, Charles Stevenson distinguished between two types of disagreement, one in belief and one in attitude.[5] A disagreement in attitude expresses itself by approval and disapproval; it is a matter of "values" (as opposed to the "facts" of the scientists, which decide disagreements of belief). Stevenson took ethical disagreement to involve such differences of attitude in an essential way. One could use persuasive techniques to try to bring the opposing party to share one's values. Indeed, since differences of belief need not be involved, such techniques, rather than strictly cognitive considerations, are the best hope for achieving some sort of "resolution." Thus, for the positivist tradition that Stevenson represented, controversy in science and controversy in ethics are altogether different in nature and in mode of settlement.

On the other hand, for those who believe that ethical judgements can be justified (and they are in the majority among contemporary moral philosophers), the two types of controversy are not to be so sharply contrasted. Ethical controversy can be decided by the application of norms, whether of the utilitarian or the deontologist or the natural law philosopher. Thus, if the controversy is between two parties who agree in normative theory, the procedure for resolution can be relatively straightforward. But if such agreement is lacking, then the controversy at the ethical level rapidly becomes intractable; witness the current debate in the U.S. on abortion. What appears to be an ethical controversy about the morality of a particular type of action disguises a much more fundamental disagreement about the proper means of resolving ethical debates in the first place. Whereas natural science can appeal to some such criterion as accuracy in prediction at the meta-level when disagreements in regard to the proper mode of controversy resolution arise, ethics has no such norm available to it as a long-range means of achieving meta-ethical consensus. Thus, although controversy involving meta-ethical dis-

[5] New Haven: Yale University Press, 1944. The distinction was later popularized by I. Copi in his best-selling *Introduction to Logic* (New York: Macmillan, 1953), Chap. 2.

agreement or disagreement in normative theory need not (and should not) reduce to opposing declarations of "value" or "attitude," the strategies available for its resolution in epistemic terms are oblique and not especially effective, as a rule.

A second context for controversy resolution in which the contrast with science is even more apparent is that of law.[6] Article III of the Constitution of the United States declares that the judicial power of the federal courts shall extend to all "cases" in law or equity arising under the Constitution and all "controversies" between states or citizens. "The Supreme Court has derived from these two words an entire body of doctrine describing the circumstances in which federal courts may or may not exercise their subject-matter jurisdiction."[7] For a federal court to claim jurisdiction, there must be a "controversy,"[8] and so the determination of what constitutes a "controversy" becomes a crucial matter, although one on which the Constitution itself has nothing explicitly to say.

The parties to a controversy may be citizens, states, or the federal government itself. The issue ordinarily involves an alleged infringement of a right or an alleged violation of a law. Or it may involve an alleged conflict within the law itself – between state laws and the Constitution, for example. What has to be decided is whether a right has been violated, a crime committed, or whether there is a conflict of law. A preliminary decision may be needed as to whether in fact a "controversy" exists, although this is ordinarily a straightforward affair. In criminal cases, for instance, there would rarely be any doubt about the presence of a controversy. The main function of the court is to close the controversy, insofar as this is possible.

The doctrine of "controversy" developed by the Supreme Court is complicated and not altogether coherent. It may best be understood as resting on three linked concerns, all of them broadly constitutional in nature. First is concern for the separation of powers: executive, legislative, and judicial. The boundary lines between these are not always clear, and so the federal courts (and the Supreme Court, in particular) have to exercise care that the judicial power does not

[6] Vanessa Merton drew my attention to the importance of the technical notion of controversy in U.S. constitutional law. I am indebted to my colleagues Fernand Dutile and John Robinson for their aid in following it up.

[7] Lawrence Tribe, *American Constitutional Law* (Mineola, N.Y.: Foundation, 1978), pp. 52–3. Under the heading "Limits on the Exercise of Federal Judicial Power: The 'Case or Controversy' Requirement," Tribe gives an extensive analysis of the often conflicting precedents, pp. 52–82.

[8] In the end, "case" and "controversy" tend to reduce to one another, so that it becomes correct to say that either can be used as the measure for the limits of judicial power.

intrude upon the prerogatives of the executive or the legislative branches of government. To recognize a particular dispute as a "controversy" (or to call it "justiciable," a term that has recently been introduced as a synonym for the controversy requirement) is in part to say that making it subject to review in federal court respects the proper balance between the judicial branch and the other two powers.

The second concern is with proper formulation of the dispute. It has to be sufficiently "ripe" to be capable of decision; the matter must have crystallized to the point that it is justiciable. The context must be a genuinely adversarial one, in order to "assure that concrete adverseness which sharpens the presentation of issues upon which the court so largely depends for illumination of difficult constitutional questions."[9] Thus collusive or feigned suits can be dismissed. Hypothetical cases are not to be considered. The court's role is judicial, not advisory. Were it to be advisory, the issues might not be posed in the sharpest light possible, and thus the best decision might not be reached.[10]

Finally, there must be harm or the prospect of harm to one of the litigants, so that there would be an actual hardship if the case were not to be heard. The legal phrase is "injury in fact," and the court has presumably an obligation to provide redress where an alleged injury has happened or will happen if the court does not act.[11] If the original issue has become moot, if the circumstances have changed so that the litigant no longer faces the originally anticipated harm, then it has ceased to qualify as a justiciable controversy.

[9] *Baker* v. *Carr, United States Reports* 369 (1962): 186, 204, quoted in *The Constitution of the United States of America: Analysis and Interpretation*, L. S. Jayson, J. Killian, S. Beckey, et al., eds. (Washington, D.C.: U.S. Government Printing Office, 1973), p. 636.

[10] The constitutional bar to the court's serving in an advisory capacity (an issue that came up as early as 1793) has been explained differently at different times: "The earliest cases of judicial refusal to issue advisory opinions were justified primarily in terms of the need to maintain the constitutional separation of powers. The refusal to give advice has more recently been explained in terms of the character of the judicial function." Tribe, *American Constitutional Law*, p. 57. The courts have come closer to the advisory role in the "declaratory judgments" recognized first in the Declaratory Judgement Act, 1934 (28 U.S.C.A., §2201), and recently invoked frequently in issues involving civil rights. But the Act still insists that there must be an "actual controversy" even in such cases.

[11] The doctrine of "legal standing" is a rather confused one. It is rooted both in this requirement that there be an "injury in fact" on one side and also in the earlier requirement that there be a genuinely adversarial situation so that the issue will be presented in its sharpest form. For there to be a justiciable controversy, each party must have "standing."

Besides these constitutional concerns, the Supreme Court also seems to be influenced by more pragmatic considerations to deny the status of controversy, on occasion, to suits about whose justiciability a substantial doubt has been raised. The ever-pressing backlog of cases and the desire of the court to avoid thorny political challenge undoubtedly affect some of the decisions reached and help to make this doctrine of "controversy" the thicket it is for students of constitutional law.

The contrast between controversy in law and controversy in science could hardly be greater. In law, what matters first is whether there is an actual controversy or not, and the conditions for this are laid down. The controversy is most often concerned with alleged violations of rights, not with matters of belief. There are agreed-upon procedures for evaluating and eventually for terminating controversy; once the appeals are exhausted, the controversy is closed. And once it is closed, it cannot ordinarily be reopened. The analysis of controversy in science will turn us to very different issues.

The analysis of controversy in science

The historian who sets out to analyze complex human interactions is alert to many kinds of clues: individual beliefs, feelings, desires; the influence of social institutions; the circumstances that cast people together and shape their relations; and so on. When the object of the historian's interest is a scientific controversy, the focus often narrows to cognitive factors such as beliefs held, methods followed, arguments constructed. Broader issues of personality and institution may be played down or left aside altogether. This is because a controversy is a disagreement about an issue, and what matters most, as a rule, to an understanding of it is the issue itself and the considerations advanced on each side. Instead of speaking of beliefs and decisions, however, the historian may choose to speak in the abstract of reasons and proofs. These are propositional entities, of themselves without causal force. What matters to the controversy as a real historical event is, of course, not the reason or the argument but someone's perception of that reason *as* a reason, someone's acceptance of an argument *as* an argument of a certain strength. We need not delay over the much-debated issue as to whether reasons (understand: the perception of specific propositions *as* reasons) function as causes or not, or the allied question as to the extent to which the historian's analysis is a causal one. What will turn out to be of some importance to us, however, is that the historian has a choice between, broadly, the history of ideas and historical explanation of a more concrete sort.

Where the topic is the career of a particular theory or the evolution of a particular concept, the former approach is obviously indicated. But controversy is much more strongly marked by historical contingency. It is not an abstract disagreement that can be adequately set down in propositional form. For completeness, one must know the people involved, their backgrounds, skills and beliefs, often their feelings and desires. If the topic is how a particular controversy ended, this is even more true. No matter how well we understand the issues in the Leibniz–Clarke controversy, it will not tell us how the controversy actually ended.

Are there any useful distinctions that might help us to categorize the sorts of factors that tend to play a role in controversy? The customary distinction between "internal" and "external" history of science has occasioned so much debate and has been understood in such a variety of ways that it must somehow be improved upon. In order to focus the issue, it may be helpful to begin by simply listing, as illustration, some of the factors mentioned in a recent analysis of a major scientific controversy.

The introduction of the quantum theory of matter in the mid-1920s occasioned a vigorous and long-lasting disagreement concerning its indeterministic implications and their effect on the overall acceptability of the theory. Here are some of the considerations cited by Forman, for example: (1) Born succeeded in interpreting the new wave function in terms of probability distributions. (2) Heisenberg was influenced by the philosophical critique of classical causal notions popular in the German intellectual milieu of the day. (3) The indeterminism of the new theory was seized upon by many German physicists in order to achieve for their science a more favorable image in the eyes of a public disillusioned by war and its aftermath. (4) Fowler, because of his contacts with many central European physicists, was one of the few British scientists to take an early interest in the revolutionary implications of the new theory. (5) Although Jeans probably did not hold strong views on determinism one way or the other, it would have been out of character for him not to take the opportunity of his 1926 presidential address to the Royal Astronomical Society to outdo Eddington on this new philosophical issue, as he had done on others.[12]

Epistemic and nonepistemic factors

A scientific controversy may be viewed as argument, as a quest for well-founded knowledge (*epistēmē*). The factors that are believed by

[12] Paul Forman, "The Reception of Acausal Quantum Mechanics in Germany and Britain," in *The Reception of Unconventional Science*, S. Mauskopf, ed. (Boulder, Colo.: Glenview Press, 1978), pp. 11–50.

the protagonists to bear on this may be called "epistemic." They may be found first in published work and will include observation reports, hypotheses, interpretations, assumptions, criticisms and responses, considered not only in logical interrelation but in temporal sequence. Beyond the scientific considerations made explicit in the documents of the controversy, there will be such implicit factors as principles of method. There may be philosophical disagreements, as in the Bohr–Einstein debate of the 1930s. These are epistemic because they are part of the knowledge structure that the protagonists are setting at risk. What counts is what *they* proposed, believed, assumed to be relevant to the merits of the case that they were debating.

But not all of the influences on the course of controversy are epistemic. Among the more obvious sorts of nonepistemic factors are personality traits (e.g., ambition, laziness, sensitivity to criticism, carelessness); institutional pressures (e.g., loss of research funding, changing university needs); political influences (e.g., tipping a controversy in favor of the side with stronger political backing); hostility between scientists, directed along national lines; "chance" events (e.g., an unplanned meeting at a railroad station, a longtime friendship between two of the protagonists; the fact that someone trained at Cambridge rather than at Berkeley; death or illness of a major protagonist). Such factors obviously affect the outcome of controversy; no account of the historical event would be complete without invoking them.

Might one say that epistemic factors involve knowledge in some essential way, whereas the others are noncognitive and operate as causes, not as reasons? This will not do. The German scientists who (according to Forman) supported the new and unintuitive quantum mechanics because it enhanced the public image of science were acting for a *reason*. They could, presumably, have specified this as one reason why they came to the support of the Heisenberg-Bohr side of the controversy, although undoubtedly they would have been unwilling to admit such a motivation publicly. But motives such as these are nonepistemic (although possibly quite rational and certainly cognitive). That is, they do not form part of the *truth* argument; they would not be cited, by those influenced by them, as relevant to the merits of the case under debate.

How broadly should the label "epistemic" be taken? During the Copernican debate, many (like Tycho Brahe) were influenced by the apparent tension between a realist Copernican astronomy and certain biblical passages regarding the motion of the sun. Was this belief epistemic or not? Ought the influence of theology, in cases such as this one, be called an "epistemic" factor? Clearly it should, since theology was believed by the protagonists in the Copernican debate to carry weight in its own right. What about Church authority in the case of Galileo and Descartes? This raises a set of difficult and controversial historical issues. Galileo certainly believed the Church to have the right to declare the truth on issues bearing on salvation. Was the inerrancy of the literal word of Scripture one of these? His *Letter to the Grand Duchess Christina* (1615) argues against this, though not in an entirely consistent way. Was the declaration of the Congregation of the Index in 1616 a nonepistemic factor for Galileo, then, something to be obeyed on the grounds of political expediency but not because of any possible epistemic weight that it might carry?

It should be obvious that the decision as to whether these factors were regarded as epistemic or nonepistemic is a crucial one for anyone writing about the Galileo case or about Descartes's decision not to enter the lists. It should also be clear that the application of the distinction in such cases, involving as it does historical insight into the sources of belief and action, can be extremely delicate. More broadly, when someone engaged in controversy is known to have made a move for which there would have been, within the world view of the time, a plausible epistemic reason, ought the historian to assume that it was for this reason that the controversialist did what he or she did, or ought one to investigate the further possibility that nonepistemic factors may have played the decisive role?

Questions like this one, involving the proper balance between epistemic and nonepistemic factors in the explanation of episodes in the history of thought, have deeply divided historians of ideas and historical sociologists for a long time past.[13] They have been one of the chief preoccupations of the sociology of knowledge. And they have recently led to vigorous disagreements between philosophers of science and sociologists of science. The issue is clearly crucial to the analysis of controversy. Before embarking on it, another distinction must first be introduced.

Standard versus nonstandard factors

When I described something as an "epistemic" factor just now, no position was taken as to whether it *ought* to have played the part it did in the debate. But the historian may also choose to *evaluate* the arguments advanced and the research strategies followed in a scientific

[13] A much fuller discussion of this distinction and the coordinate one between standard and nonstandard factors that follows may be found in my essay "The Rational and the Social in the History of Science," in *Scientific Rationality: The Sociological Turn*, J. R. Brown, ed. (Dordrecht: Reidel, 1984), pp. 127–63.

controversy. In order to determine the structure of the controversy as science, the historian may concentrate on those factors that in his or her view constituted the disagreement as a properly scientific one. These "standard" factors must be set off from the "nonstandard" ones that also influenced the course of the controversy. They are called "standard" because their designation involved the choice and application of some standard of what counts as science.

Why not use the more familiar term *rational*, then? First, because rationality extends well outside the boundaries of science, as we now define it. One might in retrospect call neo-Platonism a nonstandard factor in seventeenth-century scientific debates; it would be less appropriate to call it nonrational. Was it nonrational for Leibniz and Newton to call on theology in their debates over the nature of space, matter, and force? In the least contrived sense of that term, certainly not. If we today decide that theology is irrelevant to mechanics, the term *nonstandard* conveys that it is we who make that proposal, and we, therefore, who have to justify it. Besides, *rational* is an extraordinarily ambiguous term, and the degree of its ambiguity is not apparent to the casual eye.

The decision as to whether a particular factor is standard or not thus depends on the notion of science employed by the person using the distinction. The boundary between standard and nonstandard will vary, then, as notions of scientific rationality vary. It is an evaluative distinction, either imposed on the grounds of present usage or proposed as a reconstruction of what it is that made the history of science develop in precisely the way it did. The influence of the standard factors is presumed to account for the progress of science, for its success in fulfilling its aims. And these aims are, in their broadest statement at least, usually believed to be more or less invariant.

Combining the distinctions between epistemic and nonepistemic, and between standard and nonstandard, give us three kinds of factors: standard epistemic, nonstandard epistemic, and nonepistemic. (Nonepistemic factors are automatically nonstandard.) This gives the first clue as to why the single internal–external distinction was inadequate. If the metaphysics and theology that helped to shape the categories of seventeenth-century science are described as "external" factors, it obscures the way in which they operated. If they are taken as "internal," it seems to put them on a par with the other, conventionally "scientific" factors, which again may obscure a point that needs to be made. The "internal–external" way of distinguishing between the factors that influence such features of the life of science as controversy, although it may provide a useful first approximation for some purposes, tends to lend itself too easily to a positivistic exclusion of such nonstandard epistemic factors as philosophy or world view from the bounds of the rational and to suggest that their presence immediately indicates that "external" factors are intruding into the normal operation of science.

A closer look at the category of "external" reveals an ambiguity that is the second clue. Recall Merton's finding that Puritanism fostered an emphasis on the importance of works that correlated with the new empirical-instrumental attitude of mid-seventeenth-century English science. This has been taken by many to be the very paradigm of an "external" factor in the explanation of how this attitude arose. Yet the Puritans might have said that they had a perfectly good *reason* for an empiricist approach to nature. Social factors, so-called, are often – indeed usually – mediated by reasons. Whether these reasons would be accepted outside the group in question is another matter. But from the mere fact that a belief is characteristic of a particular group one surely cannot infer that it is "external" to the proper procedures of science in this case. It can assuredly play an epistemic role. Whether it is to be described as a standard factor or not depends, of course, on the standpoint of the speaker and on a careful analysis of how the belief was arrived at and what part it, and beliefs like it, have historically played in scientific work.

Because controversy is where the consensus of the community breaks down, the distinctions outlined above are immediately relevant to its analysis. A special sensitivity to the "style" of science characteristic of a particular epoch, school, or person, is needed in order to recognize those failures of style that would be especially indicative of the influence of nonepistemic factors in the scientific process.[14] But such factors almost inevitably must play a role in controversy, and it would be unlikely in practice that one could explain the course of a particular controversy without invoking them. Nonstandard factors are also likely to be found among the differences

[14] Karl Mannheim proposed a criterion of "immanence" in this connection. Immanent beliefs are those that find a coherent place in the system of thought of the person or of the society. Nonimmanent beliefs are ones that do not "fit in," that are not supported by the rest of the system and therefore need some special explanation for their presence. His *Ideology and Utopia* (1927–9) gives a very sensitive treatment of the notion of the "style" of a particular work or thinker. His starting point was an idealist and historicist one in which every product of thought was ultimately traced to a nonrational matrix in the Weltanschauung of the day. He never took seriously the possibility that intellectual traditions could come to have a genuine (even if partial) autonomy. His attention was focused on philosophical and political thought; he never attempted to extend his classic analysis of the sources of agreement and disagreement to the natural sciences.

separating controversialists, especially at the most basic explanatory level of such sciences as physics or biology. Consensus is usually harder to achieve in the nonstandard domain; a difference in regard to standard epistemic factors can be more easily led towards agreement. But nonstandard epistemic considerations can be discussed and weighed; they are not immune to demands for justificatory argument. In that respect, they differ fundamentally from nonepistemic factors in regard to the role they can play in controversy.[15]

To account for the way in which one party to a scientific controversy might explain the design chosen for an experiment by saying that he or she was trained at Berkeley might answer a historian's or a sociologist's question. But it would not enable one to decide what epistemic authority this design should carry. And it is this latter, quite special, sort of question that in the end distinguishes the history of science from that of other social institutions. It is the sort of question to which an answer must be available if one is to decide whether a controversy has really been "resolved" or not. That will be the concluding topic of this essay. But before we can get to it, some discussion is needed of the variety of forms that controversy may take and of the characteristic ways in which it tends to terminate.

The anatomy of controversy

A rather old-fashioned distinction between fact, theory, and principle may help us to a rough classification of the main forms that controversy may take. One of the most popular clichés in recent philosophy of science is that facts are theory-laden, that the categories of fact and theory are not as sharply separate as classical empiricism supposed. Nonetheless, for our purposes here, one can readily distinguish between assertions of supposed fact (theory-laden though they be) and proposed explanations of fact (theories). Principles and theories are not sharply distinct either, but, as we shall see, there are levels of generality that can easily enough be separated.

Controversies of fact

These have to do with regularities that are asserted to exist in the world. When scientists disagree publicly about the observational

[15] I am clearly not accepting the "strong thesis" proposed by the Edinburgh school of sociology of science, according to which there ought be no asymmetry between the explanation of "rational" and "nonrational" episodes in the history of science. See my essay, "The Rational and the Social in the History of Science," sec. 5.

basis of a claim made, although agreeing (as far as one can tell) about the assumptions that go into making the observation, we shall call this a "controversy of fact." Such controversies are not as frequent in natural science as they used to be, in part because the modes of experimental control have become more accurate. The scientist insists, where possible, on *reproducibility*; one of the two main reasons for this is to ensure that one has a genuine fact. The irrepressible *Journal of Irreproducible Results* reminds us that sometimes this demand may exclude something interesting. But in general scientists require a result to be reproducible according to criteria that are generally accepted within the community. If it is not and its original proponent still insists on the factual character of the claim, a controversy may well ensue.

One recalls such cases as the "canals of Mars" controversy, in which the American astronomer Percival Lowell made a series of observations of Mars in the 1890s in which he claimed to observe straight-line traces on the surface of Mars that he called "canals" and interpreted as the product of a Martian civilization. Others, working with similar telescopes, could not see the canals and were particularly skeptical of Lowell's claim that the canals are often paired: two parallel lines running for hundreds of miles. Photographs did not show canals, but since the human eye has powers of integration that the camera lacks, this did not settle the controversy, which went on for years. No one supposes today that Lowell was dishonest. Some note the influence of expectation; others point out that the eye, in perceiving patterns, tends to "read" streaks as lines.

Controversies of this type often hinge on a factual claim that if corroborated would tend to undermine an attractive theory. According to the simpler "falsificationist" views of scientific method, a theory should be abandoned if an observational claim incompatible with it is established. But of course if the theory has been a highly successful one up to this point, its defenders may well argue that it is much more likely that some error has been made in reporting or interpreting the observation. In a series of experiments from 1921 to 1925, D. C. Miller set out to test the null results of the Michelson-Morley experiment of 1881, the results that were so important in gaining wide acceptance for Einstein's special theory of relativity. Unlike his predecessors, Miller found (as indeed they had expected to find) a positive value for the velocity of the earth through the ether. This result was incompatible with Einstein's theory, but so strongly entrenched had the theory become in the meantime that Miller's result was assumed by most to be erroneous, although no one at the time could show where the error lay. There were some left – Miller himself among them – who found the new theory unacceptable on philosophical grounds. They urged that it had now been unequivocally refuted.

The subsequent controversy was never more than a minor one. The prestige of the new theory (despite its unintuitive aspects), and the many possible sources of error in the complex Miller experiment, were enough between them to prevent the acceptance of Miller's claim. Thirty years later, Shankland showed how temperature effects in the massive equipment might have been responsible for Miller's results. But the controversy had died out long before, even though no one had tried to test Miller's measurements by attempting to reproduce them; the main reason, presumably, was that the experiment was an extremely elaborate and costly one.

Occasionally controversies of fact turn out to involve fraud, as in the famous "case of the midwife toad," immortalized in Koestler's book of the same name, or the Piltdown Man discovery, which did not fit into the best-supported schemas of human ancestry of the time. Factual claims in historical sciences like paleontology or archaeology pose a particular challenge because the criterion of reproducibility cannot be directly applied to them. Another especially intractable domain is that of statistical data. The compilation and interpretation of such data (think of parapsychology, for instance) leave so much ground for challenge that the category of "fact" sometimes becomes quite strained in their regard.

The Laetrile controversy might seem to be in part, at least, one of fact. Is there or is there not a significant correlation between the administration of Laetrile to cancer patients and subsequent cure of the cancer? Unfortunately, the vagueness of the term *cure*, the nonstandardization of the substance "Laetrile," and the theoretical problems about what should count as a significant correlation combine to reduce the factual component of the controversy so far that resolution becomes very difficult.

Controversies of fact are most often handled, then, by attempting to reproduce the disputed results. Where this is not an appropriate demand, indirect means must be found, hidden assumptions must be made explicit, vague terms clarified. The difficulties in the way of this remind us once more how impregnated with assumption and theory the "facts" of science really are.

Controversies of theory

The commonest source of controversy in science is theoretical difference. Two or more theories are put forward to account for the same problem situation. Popper insists that a plurality of theories ought to be regarded as the normal situation in science. This would mean that controversy between the defenders of different views is at the heart of scientific advance; it is from such controversy that progress (however it be construed) derives. Of course, scientists may entertain a plurality of alternative hypotheses without necessarily becoming "defenders" of any one of them. In recent nuclear physics, a variety of different nuclear models has been proposed. Physicists have felt free to make use of each in turn, depending on the type of problem being treated. The same has been true in elementary-particle physics, where, although the quark model and S-matrix theory each has its enthusiastic advocates, many physicists seem to feel little partisanship and move from one theory to the other as context suggests. In these parts of physics, the notion that irreducibly different models may simultaneously be entertained has gained reluctant acceptance. But elsewhere in science, if theories are seen to be mutually incompatible every effort is made (very often with vigorous controversy ensuing) to eliminate one or the other.

The criteria for theory assessment are not automatic nor clear-cut in their operation. Nor is there full agreement as to how they should be defined nor what relative weight to assign them. There is no logic, therefore, that will allow one in all cases to decide on the merits of a controversy of theory. It is, in general, only when one of the rival theories accumulates a significantly better record over the course of time that resolution becomes possible. In the meantime, nonepistemic factors can play a major, sometimes even a decisive, part.

Take, for instance, the phlogiston–oxygen debate. In 1770, the phlogiston theory was the dominant one in chemistry. But Lavoisier's oxygen hypothesis, gradually formulated in the years 1772–7, explained the calx (oxide) to metal and metal to calx transitions that had forced the supporters of phlogiston to invoke the unsatisfactory notion of negative weight. It furnished a basis for understanding most interactions involving acids. It explained combustion as oxidation. Cavendish's discovery of the composition of water in 1783 allowed a whole series of predictions to be made regarding the generation of water in various chemical interactions, on the assumption that one of the constituents in water is oxygen. These predictions were verified. The publication in 1789 of Lavoisier's *Traité elémentaire de chimie* organized all of this with the aid of a new system of nomenclature that brought order and coherence and a certain degree of predictive power to chemistry for the first time.

The defenders of phlogiston had in the meantime attempted to keep pace. But they had been forced into vagueness and occasional inconsistency. Above all, they gave the impression of adjusting after the fact, instead of anticipating the facts as their rivals were doing. As Lavoisier drily put it (1783), phlogiston, by not being strictly defined, had become a "veritable Proteus" that could "accommodate itself to every explanation into which it is pressed."[16]

Within five years of the appearance of the *Traité*, nearly all chemists had gone over to the new "French" system. One of the greatest of them, Joseph Black, then 63, wrote to Lavoisier in 1791 to say that despite the fact that he had been a defender of phlogiston for thirty years, he had overcome the "mere conservatism" of his earlier hostility to the new theory; he had been "conquered by the clarity of your demonstrations and the soundness of your scheme."[17] Kirwan, one of the most articulate advocates of phlogiston, wrote to Berthollet in 1792, "I lay down my arms and abandon the cause of phlogiston."[18] Neither the deepening political differences between revolutionary France and its neighbors nor generational differences seem to have made much difference in this case. Joseph Priestley, it is true, remained unconvinced. In 1796, he wrote a new defense, *The Doctrine of Phlogiston Established*, even while admitting that he was probably the only spokesman for phlogiston left. The controversy at this point was, to all intents and purposes, over.[19]

Theoretical conflict in this case affected the entire field of chemistry. An instance of a more limited debate is the controversy (1838–61) about the geological origins of the parallel markings on the sides of some mountains in remote Lochaber, in Scotland – the Roads of Glen Roy, as they were called. Martin Rudwick has given us a detailed and

16 "Reflections on Phlogiston," in *Oeuvres* (Paris: 1862), 2: 640; quoted by Alan Musgrave, "Why Did Oxygen Supplant Phlogiston?", in *Method and Appraisal in the Physical Sciences*, C. Howson, ed. (Cambridge: Cambridge University Press, 1976), pp. 181–209. Musgrave argues that this controversy provides a "textbook case" for Lakatos's methodology of scientific research programs. Be this as it may, it certainly does illustrate the operation of epistemic factors in bringing about resolution. I am indebted to Musgrave for the references to this debate below. See also J. B. Conant, *Harvard Case Histories in Experimental Science* (Cambridge, Mass.: Harvard University Press, 1957), vol. 1, where selections from some of the primary sources are given.

17 Cited in J. H. White, *The History of Phlogiston Theory* (London: 1932), p. 133.

18 Cited in H. Hartley, *Studies in the History of Chemistry* (London: 1971), p. 48.

19 Priestley defended the phlogiston theory until his death in 1804, but he was well aware that it was regarded by his colleagues as a lost cause. After his death, one Robert Harrington took up the cudgels. In 1819, he offered a prize to anyone who could refute the defense of phlogiston that he had published that year. See *An Elucidation and Extension of the Harringtonian System of Chemistry, Explaining All the Phenomena Without One Single Anomaly* (London: 1819), and White, *History*, p. 175.

elegant analysis of this debate.[20] In 1839, in his first major scientific paper, Charles Darwin argued that the Roads were ancient sea beaches, a testimony to the gradual subsidence of the Scottish land mass and an instance of a worldwide phenomenon of elevation and subsidence caused by movements in the fluid magma under the earth's crust. These movements were also responsible, he suggested, for earthquakes, volcanoes, and the gradual subsidence of coral reefs. Darwin had visited Glen Roy and made a careful study of the Roads and of the surrounding landscape. His reconstruction of the formation of that landscape as resulting from a retreating sea, pausing at various levels to form the beaches, was detailed and ingenious. The only rival hypothesis was that of McCullough (1816) and Lauder (1821), which construed the Roads as lake beaches. Darwin dismissed this theory on the grounds that there were no traces of the barriers that would have been needed to create lakes at the requisite levels.

A couple of years later (1842), Agassiz proposed a new "lake" hypothesis: that the Roads were the remains of lake beaches, but the lakes were glacial ones. The barriers lacking in the original lake hypothesis were now provided by glaciers, whose melting had, of course, left little trace. This was a possibility that Darwin had not thought of; "I forgot the glacier theory," he wrote to Lyell. His perceptions of Glen Roy had been influenced by his South American experiences of fossil sea beaches, whereas Agassiz's were formed by his detailed knowledge of the glaciers of the Alps. But Darwin was not convinced, even though the main objection to the original lake hypothesis had now been removed. Glen Roy fitted in too well with his global tectonic theory for it to be easily abandoned to the glacialists. Nor were the other leading English geologists persuaded. Darwin's paper had been a masterpiece of argument; Agassiz's was little more than a brief sketch of an hypothesis. And Agassiz's global "Ice Age" theory (for he too saw Glen Roy in the light of a global theory) was suspect to Lyell and the uniformitarian geologists of Britain.

In 1847, Milne published an effective critique of Darwin's argu-

[20] "Darwin and Glen Roy: A 'Great Failure' in Scientific Method?", *Studies in the History and Philosophy of Science* 5 (1974): 97–185. The details of this story given here are all derived from Rudwick's paper. One may be tempted to describe it as a "debate" rather than a "controversy," since the public perception of the disagreement was rather limited. Many of the exchanges between the protagonists took the form of correspondence rather than published articles. There was not the sort of protracted and vigorous public disagreement between Darwin and Agassiz that would qualify the episode as "controversy" in the fullest sense.

ments, but instead of opting for the glacial alternative he chose to return to the original lake hypothesis. Darwin was shaken, but he described the ad hoc hypothesis of barriers that had somehow vanished as "monstrous." It was only in 1861 that Jamieson, in part encouraged by Darwin, reviewed the history of the controversy, made a new study of the Roads, and decided that Darwin's marine hypothesis was untenable. It could not explain the precisely horizontal character of the Roads, nor the fact that similar "roads" had not been found elsewhere in Scotland, nor the apparently recent character of the beaches, nor the association of each beach with an overflow channel of exactly the sort that the glacial hypothesis would demand. The objection that glaciers far thicker than those of the Alps would be required in order to provide the needed barriers was met by urging that the proper analogue was not the Alps but the great glaciers of the north polar region.

The controversy was over. Darwin conceded, in a characteristically emotional way, "Now and for evermore I give up and abominate Glen Roy and all its belongings."[21] Many years later, he was to recall: "I gave up the ghost with more sighs and groans than on almost any other occasion in my life."[22] Rudwick remarks that Jamieson's paper "made a clear, persuasive, and virtually watertight case for the glacial lake hypothesis," one that was taken by Lyell and the geological community generally to "settle the problem finally."[23] Darwin's tenacity in maintaining his view as long as he did is explicable, Rudwick suggests, not so much by such nonepistemic factors as personal pride[24] as by a variety of epistemic factors: the structural parallel between the argument of the Glen Roy paper and that of his earliest work on the origin of species; the consonance of the Glen Roy hypothesis with his tectonic theory; and his reliance on methodological themes from Lyell, Whewell, and Herschel. Contingencies such as the sketchiness of the Agassiz paper by contrast with the thoroughness of the Darwin one also played a part in prolonging the controversy. But its resolution was about as "standardly epistemic" as anyone could wish.

One final instance of a controversy of theory is the highly publicized debate in the 1950s between advocates of the rival expanding-universe and steady state cosmologies. The discovery of the systematic shift of the light of galactic nebulae towards the red end of the

[21] Darwin to Lyell, 1861, in Rudwick, "Darwin and Glen Roy," p. 152.

[22] Darwin to Prestwich, 1880, in ibid., p. 153.

[23] Rudwick, "Darwin and Glen Roy," p. 153.

[24] He notes, for instance, how open-minded Darwin was in taking the opposing arguments at their face value and even sometimes strengthening them.

spectrum had been the main support of various "big bang" theories from the 1930s on. From the point of view of classical physics, the apparent singularity some ten to twenty billion years ago from which the expansion began was anomalous. Cosmologists themselves had always preferred "Copernican" solutions, ones that made large regions of space-time as homogeneous as possible. In 1948, Bondi, Gold, and Hoyle carried the "perfect cosmological principle" as far as it could be carried and postulated a steady state cosmic model in which the large-scale distribution of matter in space-time remains constant. To achieve this, they had to postulate the sudden appearance of matter throughout space-time in the form of hydrogen atoms. Although the required rate of appearance was very slow, this "continuous creation," as they called it, violated conservation laws, the most cherished, perhaps, of all the basic principles of physics.

Each of the models appeared to be able to account for the basic observational data; what decided allegiance seemed to be the individual physicist's intuitions about conservation and beginning-in-time.[25] Which was the more shocking: violation of conservation, or an initial singularity in space-time? Attempts were made to isolate predicted in advance the existence of an isotopic radiation of this sort might be tried. It was not difficult to find such differences, but the outcome of observational trial was for a time inconclusive. The theories were so complex and involved so many assumptions that it was always possible to make a small modification here or there to account for whatever result was obtained. It was more often the steady state model that had to be adjusted in this way; still, this was not, for a time at least, taken to be a decisive argument against it.[26]

Finally in 1965 the discovery of the 3°K radiation permeating all of space gave an advantage to the expanding universe model that had predicted in advance the existence of an isotopic radiation of this sort as a relic of the "big bang" itself. Defenders of the steady state view could find no plausible modification to handle the new finding. And so they reluctantly conceded defeat. What had proved decisive was the way in which one of the models had been more successful, over the course of time, in anticipating new results.

Nonetheless, Hoyle could not bring himself to accept a model that incorporated a "big bang." He had made it clear from the beginning that his repugnance for this notion was in part due to its affinity with the Judeo-Christian concept of a cosmic creation in time. A theory

[25] McMullin, "Is Philosophy Relevant to Cosmology?", *American Philosophical Quarterly* 18 (1981): 177–89.

[26] See J. D. North, *The Measure of the Universe* (Oxford University Press, 1965), especially the final chapter.

that implies a past time singularity beyond which the history of the universe cannot be traced simply cannot (he insisted) be a good theory.[27] And so he has continued to seek alternative solutions, proposing, for example, a model that explains galactic red shift by a steady increase in the masses of elementary particles over time while galactic distances remain constant. The fact that his model has a singularity in the past when all particle masses were zero, as well as cosmic states of negative mass prior to that, unintuitive though these consequences would seem to most physicists, does not dismay him. In the scale of likelihood, it seems that almost anything would be preferable, in his view, to a "big bang."

What made this controversy so intractable was not so much the complexity of the theories involved and the consequent difficulty of using the standard predictive methods of deciding between them as it was the differences of principle and the influence of such nonstandard factors as theological (or antitheological) belief.[28] This brings us, then, to our third type of controversy.

Controversy of principle

Where the question at issue is not simply that of deciding between two theories but, in addition, of deciding about a more general matter of principle that will also affect other parts of science, controversy becomes much more difficult to resolve. Such principles may be

[27] Hoyle argues that a "sensible physical principle" leads to the steady state model, but the 3°K microwave background radiation appears to vitiate this model. "The dilemma is, then, that normal physical and mathematical reasoning seems to lead us to a contradiction between theory and observation. Many people are happy to accept this position. They accept [the big bang model] without looking for any physical explanation of the beginning of the particles. The abrupt beginning is deliberately regarded as *meta*physical – that is, *outside* physics . . . To many people, the thought process seems highly satisfactory because a 'something' outside physics can then be introduced at T = 0. By a semantic maneuver, the word 'something' is replaced by 'god' except that the first letter becomes a capital, God, in order to warn us that we must not carry the enquiry any further." F. Hoyle and J. Narlikar, *The Physics-Astronomy Frontier* (San Francisco: Freeman, 1980), p. 428. Ought we, then, to return to the steady state model? "Possibly so, but at this stage, after ten years or more of attempting to 'tough it out,' we prefer to try to break out of the restraining cycle of the argument just described." He then goes on to speculate about a model involving negative as well as positive masses and claims that it can be made to yield the required 3°K prediction. See also his *Astronomy and Cosmology* (San Francisco: Freeman, 1975), p. 684.

[28] McMullin, "How Should Cosmology Relate to Theology?", in *The Sciences and Theology in the Twentieth Century*, A. R. Peacocke, ed. (Notre Dame, Ind.: University of Notre Dame Press, 1981), pp. 17–57; see sec. 3.

either *methodological*, bearing on the procedures and criteria of science, or *ontological*, having to do with the basic categories and forms in terms of which we schematize the natural world. The cosmological debate of the 1950s involved a difference of ontological principle. There was no disagreement about the procedures of proof and explanation proper to natural science. But there was a deep disagreement about the notions of time, of beginning, of conservation.

The controversy between Newton and his critics about the adequacy of the notion of force in the *Principia* was almost entirely one of principle. It was not a matter of arbitrating between two theories of mechanics, each with its own predictive consequences. Rather, the central metaphor of action at a distance, for which Newton vainly kept trying to find an acceptable philosophical matrix, was anathema, on a variety of different philosophical grounds, to the Cartesians, to Leibniz, to Berkeley, and to many other critics who were perfectly prepared to allow the practical value of the *Principia* for predictive purposes.[29]

After Newton's death, the controversy died away, in part because the grounds for his emphasis on the inherent passivity of matter (which were both philosophical and theological) no longer seemed cogent, and in part because the notion of forces acting in space had lost its strangeness. What was required for the resolution of the controversy was an adequate natural philosophy relating matter, spirit, force, power, and space. Boscovich, Priestley, Kant, in turn tried to provide this. It would be a matter of disagreement as to whether a satisfactory resolution of the philosophical issue has even yet been reached.[30] Mach resurrected the controversy for a time, and it could be said to lie not far from the surface of many of the debates surrounding the theory of relativity. Here, then, is a controversy that even after three centuries is not quite over, although it can hardly be said to have been troubling the consciousness of physicists very much in the last few decades.

The vigorous debates surrounding the quantum theory of matter were of a similar character. There was no question about the predictive successes of the new theory. Nor was there a viable alternative. But critics such as Einstein, and later Bohm, Vigier, and others urged that because of its departure from the deterministic pattern of classical mechanics, it could at best be regarded only as a stopgap. It could not, in their view, be reconciled with a larger

[29] For an account of this controversy, see E. McMullin, *Newton on Matter and Activity* (Notre Dame: University of Notre Dame Press, 1978).

[30] See, for instance, L. Sklar, *Space, Time, and Spacetime* (Berkeley: University of California Press, 1974), Chap. 3.

philosophical view of nature, whose sources are wider than physics alone.

Such controversies (as Einstein himself noted) are obviously not resolved by the predictive record of the theory over time. Fundamental philosophical analyses have to be made, long-term conceptual implications investigated, possible changes in world view carefully weighed. The continuing predictive success of the theory and the lack of working alternatives counts heavily in such a debate (as the example of Newtonian physics shows). But it is not quite decisive. It is, indeed, important that controversy not be too readily abandoned in such a case; the consequences of premature closure are much too grave.

Where the disagreement concerns a straightforwardly methodological principle, it can, in the short term at least, be quite intractable. If the protagonists cannot agree as to what would be the proper means of adjudicating between them, there may be no third-level consideration to which common assent is given and to which an appeal can then be made. Controversies of this sort have not occurred often in the history of science, but their presence has constituted the major support for Feyerabend's assessment of basic theory change as being incapable of rational arbitration. Lacking agreed-upon criteria for achieving resolution, the scientist may have to fall back on propaganda. Nonepistemic factors are then likely to carry the day.

The paradigm case of such controversy has always been taken to be the Aristotelian–Galilean debate of the seventeenth century. It is asserted that what was mainly at issue there was the nature of science itself – of proof, of explanation, of observation. Yet when we look more closely, the discontinuities at this level do not seem quite so great. In his mechanics, Galileo maintained the Aristotelian aim of a demonstrative science. But he argued that the only way to carry through such an aim was to follow mathematical rather than syllogistic form. Again, he noted Aristotle's respect for observation and went on to argue that the results of telescopic observation should suffice to convince anyone empirically inclined that, for example, Jupiter is the center of motion for its moons and the sun for Venus. He assumed that anyone desirous of practicing science would abhor incoherence, and he argued persuasively that there were important incoherences in Aristotelian physics.[31]

His primary aim was to show that Aristotle's account failed in its *own* terms. He could not show that his own account was going to

[31] McMullin, "The Conception of Science in Galileo's Work," in *New Perspectives on Galileo*, R. Butts and J. Pitt, eds. (Dordrecht: Reidel, 1978), pp. 209–57.

succeed, but at least he had some successes to point to. He was as yet uncomfortable with the kind of hypothetical argument to which the new science of stars and of the constituent corpuscles of bodies seemed to be limited. Although he avoided causal argument in mechanics, he left it open to others to try that way again. He could claim some limited predictive successes, even in topics like the nature of the lunar surface. And prediction, it was beginning to seem, could plausibly be argued to be a necessary (though obviously not a sufficient) condition for a theory's being called a genuine form of understanding.

Arguments like these were clearly not coercive. They took time to weigh fully. But there was enough continuity between the Galilean undertaking and that of the "mixed sciences" of the Aristotelian tradition to make a dialogue possible in the meantime. It was not as though there were no common criteria to which an appeal could be made. What made agreement difficult to obtain was the impossibility of assigning relative weight to the losses and gains inherent in adopting one or other of the rival accounts. In the circumstances, nonstandard and nonepistemic factors affected the discussions in important ways.

Among natural scientists, the active period of debate did not last long. The Galilean research program was a dynamic one, suggesting new lines of inquiry. It had scored striking successes in handling age-old problems in optics, astronomy, and mechanics. These successes gradually came to validate the shifts in method and aim. It was not, then, a direct resolution of methodological issues on philosophical grounds. Rather, the practice of the new science developed gradually, and as a result the new science slowly acquired philosophical defenses. This is a complex process, but it is basically an epistemic one. Weber and Merton have shown how nonstandard factors influenced the formation of the new ideal of science. But by 1800, a defense of this ideal would not have needed to rely on anything other than the achievements of the new science itself.[32] The resolution of controversies of principle, in short, is slow, and oblique, and practice-oriented.

Mixed controversies

A controversy that involves science in some way and also an issue of moral or political principle on which the community is divided will be

[32] This is, of course, a very large claim, in the face of Feyerabend's arguments to the contrary and the real difficulty in assessing the influence of sociopsychological factors on the adoption or rejection of an entire world view.

called "mixed." There must be disagreement on both the scientific and the nonscientific aspects of the case for the controversy to count as "mixed." Thus, the debate about the use of life-prolonging techniques for the terminally ill would be regarded as a purely moral, and not as a mixed, controversy. Mixed controversies usually involve an application of science to some human purpose – that is, a technology of some sort. It is from this application that the moral or political issue derives. Nuclear science of itself raises no moral issues. But the application of nuclear science to electrical power generation or to armaments very quickly does make for moral and political dilemmas. What is notable about these debates is that the presence of moral disagreement makes it far more difficult to arrive at a consensus on the scientific issues (e.g., the hazards of low-level radiation or of nitrites in bacon). The influence of nonepistemic factors is thus also likely to be greater in such controversies.

A further difference between debates of this kind and the other sorts of "pure" scientific controversy discussed earlier is that a mixed controversy, because of its moral component, is usually concerned with the wisdom of a particular *action*. It involves not just belief, as the other types of controversy do, but action as well. Thus, decisions are called for, and these decisions have to be made in the absence of a consensus. One may not be able to wait for resolution of the issues, as one can in the case of a controversy of fact or theory.

Such controversies generally hinge on decisions with regard to the relative value of different human goods, such as health or security. Technology, the shaping of the human environment by the human intelligence, is as old as humanity and has always had profound implications for the quality of human living. It has only been within the last century and a half that advances in theoretical science have begun to quicken the pace and broaden the scope of technological advance. It is technological change, above all else, that sparks ethico-scientific controversy. Frequently the long-term effects of the change (e.g., on the environment, as with the internal combustion engine or pesticides; or on the human person, as with contraceptive pills or meat from hormone-treated animals) are hard to assess. Scientists may be induced to give probability estimates, when in fact their forecasts are little more than guesses. The influence of nonepistemic factors is all too evident in such cases.

The moral issue posed by such changes is often a new one, or at least one for which there are no precise precedents. Nuclear armaments pose a threat to the entire human race of a quite novel sort. Recombinant DNA techniques offer the possibility of reshaping human nature itself. What is to guide one in an enterprise such as this? Moral wisdom needs time to form; people of sensitivity often have to live with the new situation and assess the alternatives and the consequences over a long period of time before the moral response that is most appropriate within their value commitments, taken as a whole, becomes clear to them.

Controversies of this kind clearly cannot be settled by means of the techniques used in the resolution of controversies of fact, theory, or even principle. Theory and physical principle are tested by the long-run fertility of the schema they provide. The one giving the best fit to the way the world is, is the one that, given time and assiduous trial, will prevail. Methodological principle is tested by the quality of the knowledge that it enables one to acquire. But moral principle is tested by the quality of the life that it enables one to lead. And the criteria for this assessment are themselves in debate. Different theories of ethics, different practical moral orientations, will lead to different criteria here.

In Part II of this volume, numerous case studies are presented of such mixed controversies as those over the supposed cancer drug Laetrile, over the classification of homosexuality as a disease, over the use of nuclear energy for electrical power generation, and over the enforcement of safety standards in the workplace. Although quite diverse in character, they illustrate the interaction of two sorts of controversy, scientific and moral-political, upon one another. And they effectively convey just how intractable this interaction can lead the joint "mixed" controversy to be.

The end of controversy

Our concern here is with the ways in which scientific controversies typically terminate. The preceding discussions in this essay suggest three main possibilities, which I shall term *resolution*, *closure*, and *abandonment*.

Resolution

The controversy may be *resolved*: that is, agreement may be reached on the merits of the case. The participants themselves and the scientific community of their time are the judges of whether resolution has occurred. The factors involved in satisfactory resolution are necessarily epistemic ones. And they will be *standard* epistemic ones in the eyes of the participants themselves. Later commentators may adjudge them nonstandard and may be led to describe the resolution as an inadequate or merely apparent one. For a later chronicler to describe the termination of some historical controversy as an "inadequate resolution" would be to say that although to those involved the issues may have appeared to be satisfactorily resolved, the factors influencing this judgment included nonstandard or nonepistemic ones that, in the view of the critic, ought not to have been given the weight they were.

The outcome of controversy resolutions is that one or other of the contested views (or perhaps a modified "middle" view) is accepted by both sides. The force of the counterarguments has in one way or another been eliminated or at least sufficiently diminished. "Acceptance" in this context (as in science generally) is, of course, provisional. It does not imply complete conviction of truth or assurance that the view arrived at is a definitive one.

Closure

As we have seen, however, not all scientific controversies end in resolution. Nonepistemic factors may play an important part in terminating a controversy: such factors as the authority of the state, the loss of a research grant, or the laziness of a controversialist. The list of possibilities is virtually endless, since the termination of controversy involves human decision in an essential way, and the variety of nonepistemic factors that may affect that decision is boundless.

The term *closure* may be used for controversy termination in which such factors play a significant role and in which resolution is, in consequence, not achieved. This is a somewhat broader use of the term than the usual one. *Closure* suggests one particular sort of nonepistemic factor, namely the employment of external authority to declare a controversy ended. A controversy that is "closed" is (in the ordinary understanding of this term) not resolved; thus resolution is not a species of closure. Because the term *closure* conveys this suggestion, and because external authority is one of the commonest nonepistemic factors likely to influence the outcome of a controversy, the term seems an appropriate one to cover nonepistemic forms of active termination generally.

One other term that might be used here is *procedural closure*; other authors in this volume have preferred it. I avoid it because the employment of procedures is not peculiar to closure; procedures (like the experimental testing of hypotheses) are also used in resolving a controversy. Of course there are different sorts of procedure, an epistemic sort as distinct from the nonepistemic procedure of deciding by vote or by the interposition of a judge. The term *procedural* is not precise enough, since one would have to distinguish between different sorts of procedure, if the intention is to contrast "procedural" termination with resolution. Further, many nonepistemic factors (the dishonesty of one of the controversialists, for example, or the influence of party affiliation) do not operate here as procedures, strictly speaking. So this label will not do.

When a controversy ends by closure, the original disagreement still to some extent persists. The issues have not been resolved in a manner that the community of the day would regard as adequately epistemic, were they to know the facts of the case. Of course, they may *not* know the facts of the case – the fact, for instance, that one of the controversialists has been ordered or bribed to withdraw. In this case, what was really closure may appear to the community to be resolution. Even the participants themselves may be deluded in this regard, although this is less likely. Once again, nonepistemic factors are by definition irrelevant to the merits of the case at issue; they may effectively terminate the public expression of disagreement that is a necessary feature of controversy. But they do not bring about resolution.

We saw above that in the law court the normal way of terminating the "controversy" between plaintiff and defendant is one of closure. An authority intervenes to terminate debate after the issues have been tested in adversarial fashion. Evidence is called, but epistemically relevant evidence is often disallowed on nonepistemic grounds, and since a verdict must be reached, various principles (e.g., that the defendant is innocent until proved guilty) and procedures (e.g., the giving of a verdict by a jury) are introduced to facilitate a decision in the absence of sufficient evidence to resolve the original controversy. Indeed, it might be said that the sort of disagreement that constitutes a controversy in constitutional terms is not one that would as a rule lend itself to resolution in the epistemic sense.

Cases of pure closure are relatively rare in science, at the level of major controversy at least. (One supposes it to be more frequent where minor controversies are concerned: a department head, a dissertation director, a funding agency, and a journal editor all have certain limited powers of closure.) In most cases where nonepistemic factors intervene to bring about results that the epistemic factors operative in the controversy would not of themselves have sufficed to bring about, a combination of influences is at work, epistemic factors working towards resolution and nonepistemic factors causing closure. Because of weariness or increasing age, a participant may not scrutinize the new arguments of an opponent carefully enough and may give up too easily in consequence. In such cases, one can say that there are elements of closure present sufficient to prevent its being described as resolution proper.

As an instance of almost pure closure, the Lysenko case is without parallel in the entire history of science, perhaps.[33] In the mid-1930s the Soviet agronomist T. D. Lysenko led an attack on classical genetics, specifically on the gene concept and the theory of natural selection, proposing instead a vague Lamarckian notion of the inheritance of acquired characteristics, based supposedly on his own experiences with plant breeding and on the general principles of dialectical materialism. The debate spread into the press. Not only were geneticists drawn into it; so were other scientists, party members, and ultimately the entire reading public in the USSR.

Stalin supported Lysenko, and pressures were put on geneticists to renounce the principles of "Morganist" genetics. Some refused and were imprisoned. In 1940, Nicolai Vavilov, the leading geneticist of the USSR and a fearless critic of Lysenko's views, was arrested and charged with sabotage. Sentence of death in 1941 was commuted to a prison term; he died a year later in prison. For the next five or six years controversy ceased; no one dared to defend the genetic theories that were being taught everywhere else in the world.

But a "closed" controversy does not stay closed. The lamentable effects of Lysenko's theories on Soviet agriculture were plain for all to see. Criticism mounted, and Lysenko turned once again to Stalin. A new and much more thorough campaign was mounted against "bourgeois theories of chromosomal heredity." Hundreds of scientists were dismissed, laboratories were closed, university departments reorganized, textbooks discarded, and stocks of drosophila destroyed. It was closure on the largest scale ever attempted. In 1952, there was a brief challenge to Lysenko. In 1958, there was a further wave of repression, since news of the stunning new achievements in genetics in the West kept seeping in.

In 1962, criticism began again, and Lysenko turned this time to Khrushchev, who once again gave him the support he needed. The fall of Khrushchev in 1964 brought an end of Lysenko's thirty-year dominance. Theoretical genetics was reintroduced, the odd mélange of plantbreeders' maxims and dialectical thought that Lysenko had imposed was repudiated. The controversy this time was resolved, not closed.

How "pure" had the earlier closure been? Medvedev, in his powerful and moving account of the controversy, in the last stages of which he was himself a participant, describes it as a struggle between

[33] See David Joravsky, *The Lysenko Affair* (Cambridge, Mass.: Harvard University Press, 1970).

science and pseudoscience and lays out the nonepistemic factors that he believes to have been responsible.[34] Joravsky notes that Lysenko did use some arguments of a vaguely Lamarckian kind; he also advanced a more or less instrumentalist critique of the gene as a fictional entity. But Joravsky attributes his success not to argument but to political power, to his uncanny ability to mold an army of people who would follow him without question.[35]

To a believer in the transformative role of the Communist party in Soviet society, could it have seemed that the declarations of 1936 and 1948 constituted evidence for the truth of the Lysenko doctrine? Could this have been a case where a normally nonepistemic factor such as the exercise of political power became an epistemic factor for the believer? Joravsky mentions the behavior of the geneticist Alikhanian at the climactic Academy session of the Lenin Academy of Agricultural Sciences in August 1948.[36] Early in the session he defended genetics against Lysenko's attacks. But when the support of the Central Committee of the Communist party for Lysenko was announced, Alikhanian immediately declared that for a fervent communist this must be considered a valid reason to regard the Lysenko position as true. He went on to renounce the genetics he had formerly taught. Joravsky believes that this was a sincere recantation but adds that such were, in his view, very rare. For the vast majority of those involved, it was (and was perceived as) closure.

Abandonment

There is a further way in which controversies in science sometimes end. They may just wither away. Some of the main protagonists may die or grow old. No way of resolving the issues may be found, despite continued efforts, so the scientists lose interest. Such controversies cannot be said to have ended because they were resolved. Yet they are not closed, either. They are just not around any more. As time passes, an abandoned controversy may come to be implicitly resolved as new evidence becomes available or new theories are formulated. But later resolution does not take away from the fact that the termination of the public controversy had occurred earlier.

Stephen Gould has recently drawn attention to a striking case of abandonment. Haeckel's principle that ontogeny recapitulates phylogeny was generally accepted among embryologists in the 1870s and the 1880s; indeed, it was regarded as the most basic explanatory

[34] Zhores Medvedev, *The Rise and Fall of T. D. Lysenko*, trans. I. M. Lerner (New York: Columbia University Press, 1969).

[35] *Lysenko Affair*, p. 209.

[36] Ibid., p. 222.

principle available in that science. But in the early 1890s the new, more experimental approach of Roux and Driesch directed attention to a different set of issues in embryology, and they proposed a different method of handling them. They did not attack the biogenetic law directly; they simply ignored it as being of very limited applicability: "Recapitulation was not 'disproved'; it could not be, for too many well-established cases filled expectations. It was, instead, abandoned as a universal proposition and displayed as but one possible result of a more general process."[37]

This shift of attention to the earliest stages of development also redirected the study of heredity and was one of the factors leading to the Mendelian synthesis and the new science of genetics.[38] In the early days of this science, the same sort of "benign neglect" of the biogenetic law continued. At that point, interest in the formerly dominant principle had almost disappeared. It was only later that geneticists realized that their new science made the biogenetic law untenable. At that point, the issue was resolved. But it had, to all intents and purposes, been abandoned earlier.[39]

A more clear-cut example, already mentioned above, was the abandonment in the mid-eighteenth century of the controversy over whether or not matter could be regarded as an active principle. It collapsed after Newton's death; the success of his own gravitational mechanics had made his principle of the passivity of matter seem less and less plausible, and his reasons for defending the principle had lost their authority. This was resolution of a sort. But in fact there was no final joining of the issues; the controversy ceased not because one side had been shown wrong but because hardly anyone was interested in defending it.

A controversy about controversy

Much of this essay has been concerned with taxonomies of a relatively uncontroversial kind. It would be edifying if we could end

[37] Stephen Gould, *Ontogeny and Phylogeny* (Cambridge, Mass.: Harvard University Press, 1977), p. 206.

[38] Ibid., p. 207.

[39] This is not a clear-cut instance of controversy abandonment. There was no real controversy about Haeckel's law in the 1870s and 1880s. Darwin leaned rather more toward von Baer's law, which had the embryo develop from the more generalized to the more specialized, and this position was ultimately incompatible with Haeckel's law. But there was not much discussion of this tension. It comes closer, therefore, to the abrupt abandonment of a theory without controversy. The abandonment was not complete; de Beer thought the law worth a book-length attack forty years later.

a paper on controversy without ever engaging in controversy ourselves. Is there, in fact, a controversy about controversy in science? One has only to read the contributions to this volume to realize that there is, indeed, a quite fundamental disagreement among the contributors. It is of a philosophical sort, and it is one on which neutrality is impossible. In ending this essay, I will try to characterize this controversy by criticizing (very briefly, because of lack of space) the arguments given by Everett Mendelsohn (Chap. 3, this volume), who represents one side in the debate, and by suggesting the line of argument I would use in support of the other side of the debate.

It is widely agreed today, for the reasons sketched earlier, that controversy in science is a complex human action in which sociopsychological factors inevitably play a part. It would further be agreed that one reason for that is that scientific knowledge is theory-laden, even at the observation level, and that theory itself is ordinarily underdetermined by the evidence available in its support. Where the disagreement comes is in regard to the role played by sociopsychological factors in supplying for this lacuna.

I see an interesting contrast between Kuhn and Mendelsohn in this regard.[40] Kuhn holds that in the "mature" sciences, such as physics and biology, theory appraisal is governed by such epistemic values as predictive accuracy, coherence, and fertility.[41] These values do not function as logical rules and thus do not always lead to the resolution of disagreement. How they are understood, and the relative weight given each of them, may vary from one scientist to another. Much depends on the exemplars with which the individual scientist was trained. These values are not just contingent social conventions. They have been found to serve the basic end of science, which is problem solving. They are thus epistemic in general character, although one can, of course, ask sociological questions about the manner in which they are learned and the sort of consensus that underlies them.

What makes controversy inevitable in science is that such values do not facilitate ready resolution of theoretical conflict. It is not that sociopolitical factors intervene to prevent it. Kuhn, in fact, argues that "the practitioners of a mature science are effectively insulated from

[40] Mendelsohn, who quotes Kuhn frequently in his support, does not convey to the reader how fundamental the disagreement is between his construal of "social interests" and that of Kuhn.

[41] "Objectivity, Value Judgement and Theory Choice," in *The Essential Tension* (Chicago: University of Chicago Press, 1977), pp. 320–9. For more detail, see my paper, "Values in Science," in *PSA 1982*, P. Asquith and T. Nickles, eds. (East Lansing, Mich.: Philosophy of Science Association, 1983), pp. 3–25.

the cultural milieu in which they lead their extraprofessional lives."[42] Thus although social analysis can help us to understand how such sciences operate, the social factors that bear on controversies that occur within such sciences nearly all originate within the scientific community itself. These factors thus have an ultimately epistemic bearing.[43]

Mendelsohn, on the other hand, argues that what usually lies at the root of controversy in science is a difference in social interests, social interests deriving in large part from sources outside the scientific community itself. Among these "social interests" he places religion and philosophy, which he views as external to science in the epistemic sense. (I argued in an earlier section that this approach simply will not do, if a historical understanding of science is sought.) We can, he says, understand the prevalence of controversy in science only if we realize that normal scientific action and choice is "structured and organized by contingent goal orientations."[44] In terms of the distinction that I made earlier, he thus views the termination of controversy in science as predominantly a matter of closure rather than resolution.

Mendelsohn appeals to the historical record, with an occasional allusion to neo-Marxist political analysis. In an earlier essay, he analyzed the growth of new epistemologies of science in the seventeenth century as a response to the political realities of the day, an attempt to "gain social space" at a time when science was weak.[45] There is an obvious affinity between his approach and that of the Edinburgh group in the sociology of science. However, it might be argued that the so-called strong program goes further than Mendelsohn does in arguing, on philosophical grounds, a "principle of unrestricted sociality"[46] that would make social analysis the ultimate explainer in all aspects of the history of science, excluding any kind of autonomy purported to derive from "rationality, validity,

[42] Ibid., p. 119.

[43] Of course, since Kuhn rejects scientific realism and the convergence view of theory growth that usually goes with it, the label "epistemic" that I have attached here would have to be correspondingly reinterpreted.

[44] This phrase is quoted (with approval) from Donald MacKenzie and Barry Barnes, "Scientific Judgement: The Biometry–Mendelism Controversy," in *Natural Order*, Barry Barnes and Steven Shapin, eds. (London: Sage, 1979), p. 201.

[45] "The Social Construction of Scientific Knowledge," in *The Social Production of Scientific Knowledge*, E. Mendelsohn, S. Weingart, R. Whitley, eds. (Dordrecht: Reidel, 1977), 1: 3–26; see p. 17.

[46] See my essay, "The Rational and the Social in the History of Science," sec. 5.

truth or objectivity."[47] Mendelsohn would not, I think, support so strong a claim.

Since he is basing his approach on the history of science, it is important to decide first of all whether the sample that he provides is truly representative of scientific controversy generally. The three cases that he discusses are the Harvey–Descartes debate, the Pasteur–Pouchet controversy, and the nineteenth-century controversy concerning vivisection. In addition, he briefly mentions eugenics and Darwinism.

Two general points may be made. The first is that this is not a representative sample. It appears to be chosen in order to make the very point that Mendelsohn is attempting to establish. Were the cases to serve simply as illustration, this would be acceptable, but if they are to serve as inductive warrant, objection must be raised. Two of the five examples are mixed controversies bearing much more on ethical than on scientific issues (vivisection and eugenics). A third, Darwinism, is surely unique in that it involves issues regarding the nature, origin, and destiny of man that even today still engage scientists and nonscientists in sharp debate.

This leaves two cases. And in regard to them, Mendelsohn's main point seems to be that differences of a metaphysical, methodological, or religious kind separated the debaters and made the controversy between them incapable of proper resolution. The differences themselves he describes as differences in "interests." Harvey and Descartes had different "metaphysical interests." Descartes, for example, had an "interest . . . in using the Harveian concept of circulation to support the mechanical world view for which he was battling." But this is an epistemic factor, and a fairly standard one at that, since coherence between a particular scientific theory and a broader scheme of explanation can serve as warrant in either direction. To call it an "interest," with the sociopolitical overtones that Mendelsohn attaches to this term, could be misleading. If he merely means that because broader metaphysical issues were involved the controversy was more difficult to resolve than it otherwise would have been, the point can be readily conceded. But when he says that the controversy remained unresolved "because the contestants were using the knowledge gained, each to serve an end guided by interests cognitive, metaphysical, and social," one gets the impression that these "interests" were extrascientific factors that somehow prevented proper empirical resolution of the differences. But this way of cutting

[47] D. Bloor, *Knowledge and Social Imagery* (London: Routledge & Kegan Paul, 1976), p. 1.

between the scientific and the extrascientific harkens back to a positivism that Mendelsohn would assuredly disavow.

In Descartes's last reference to the controversy, he rejects Harvey's notion of a "contractile faculty" moving the heart. Such a faculty, he exclaims, would be "much more difficult to conceive than anything which is claimed to be explained by it."[48] The criterion of conceivability, which Descartes relied on so heavily in his scientific work, rested upon elaborate philosophical argumentation. Harvey rejected Descartes's explanation of the heart's pumping motion as caused by an innate heat because he could find no empirical evidence for this claim. This, for him, was the more important criterion; it worried him not at all that he had not himself really explained the heart's motion. The demand for explanatory completeness with which Descartes challenged Harvey was the same as that by which he challenged Galileo in mechanics. The issue was an epistemic one; it could be (and was) defended by methodological argument on both sides. Once again, that a difference of this second-level kind would make a controversy harder to resolve is obviously correct. But that in pointing to it we have somehow disclosed the "political anatomy," in Mendelsohn's words, of controversy in the sciences does not seem to follow, if by this he means that the factors separating the two sides are to be characterized as political in their basis.[49]

The Pasteur–Pouchet debate was unusual in two respects. It concerned not a theory but a question of fact: Did spontaneous generation occur or not? And it was to be "decided" by a commission set up by the Académie des Sciences. Mendelsohn himself notes that "judiciary proceedings" of this sort were unknown in Britain and Germany. The controversy and its closure are, therefore, of a kind rare in the history of science, of a kind, furthermore, that might lend itself more directly to political analysis than would ordinarily be the case. We clearly cannot accept such a debate as a basis for inductive argument concerning the nature of scientific controversy generally.

48 "Description du corps humain," first published with *Traité de l'homme* in 1664; in *Oeuvres*, C. Adam and P. Tannery, eds. (Paris: Vrin, 1964), 10: 241.

49 Harvey's analogy between the heart and the king as sources of power in their respective domains is the sole evidence that Mendelsohn advances for the claim that the controversy reflects political interests. But the analogy that Harvey relies on in his argument is that between the heart and the *sun*. The reference to the king's role in the Commonwealth is a parenthetical one, indicating only that a similar analogy holds between king and sun. Admittedly, the microcosm-macrocosm argument is a nonstandard one from *our* point of view. But it was common in Harvey's day and need not be interpreted as a disguise for an underlying political interest, though some "political interest" may surely have been present.

Yet even apart from this, the Pasteur–Pouchet debate is not quite as political as Farley and Geison[50] (on whom Mendelsohn draws) try to show. It is true that Pasteur's use of empirical evidence was not as methodologically sound as it could have been. It is also likely that motives of professional advancement played some part. It is possible that broader philosophical and religious considerations influenced the contestants. It should again be emphasized, however, that these latter must not automatically be regarded as nonepistemic social factors. And Mendelsohn himself admits that the supposed tie in France between spontaneous generation and materialism was viewed in exactly the opposite way in England, where a similar debate took place ten years later.

A better case would have to be made, in any case, for the claim that political or religious factors played a significant role in the rejection of Pouchet's arguments. The commission had requested "precise and rigorous experiments" from which unequivocal results could immediately be drawn.[51] They had no difficulty in deciding that Pouchet had not satisfied this demand. If there were philosophical factors at work, they were to be found not so much in materialism as in positivism, the philosophy of science embraced by Flourens and Bernard (two members of the commission)[52] as well as by many others in the Académie at that time. That the Académie remained so skeptical of Darwinism can be understood best, perhaps, by recalling Bernard's famous denial (in his *Introduction to the Study of Experimental Medicine*) of scientific interest to theories that attempt to go beyond experimentally determined phenomena. That methodology affects the outcome of controversy is hardly to be doubted, but once more it does not warrant the strongly political cast that Mendelsohn gives to the Pasteur–Pouchet debate. One is led to conclude, then, that the case histories brought forward by Mendelsohn are not nearly sufficient to warrant the general claim that he wishes to make about the prevalence of sociopolitical forms of controversy closure in science.

I have addressed Mendelsohn's argument here not only because of his eloquent advocacy of the sociopolitical in the discussions

[50] John Farley and Gerald Geison, "Science, Politics, and Spontaneous Generation in 19th Century France: The Pasteur–Pouchet Debate," *Bulletin of the History of Medicine* 48 (1974): 161–98.

[51] "Prix Alhumbert pour l'année 1862," *Comptes rendus des séances de l'Académie des Sciences* 55 (1862): 977–9.

[52] Mendelsohn suggests that what the members of the commission had in common was the Catholic religion. Positivism seems, however, a better clue to the outcome of the debate. Bernard, who was secretary of the commission and the moving force in the rejection of Pouchet's claims, was not in fact a Catholic.

from which the papers in this volume originate but also because he recalls a popular trend in recent historiography of science: the debunking of classical internal history of science. What I have argued against here is the use of carefully selected cases studies like those concerning craniology or eugenics to suggest a broadly inductive argument for the pervasive and decisive presence of sociopolitical factors in all aspects of scientific work, specifically in the termination of controversy.

The resolution of controversy

Throughout this essay, scientific controversy has been presented as a complex human action in which social, political, and psychological factors play an important part but in which epistemic factors are more likely to be determinative. What are the grounds, finally, for my confidence in this latter claim? Does it rest on a historical induction? And if it does, can it avoid the sort of objection I have just brought against Mendelsohn? Is the protagonist not likely, once again, to choose a sample that will nicely support the desired thesis?

It would be easy to produce an extensive list of case studies (beginning with Frankel's thorough analysis of the continental drift debate; Chap. 7, this volume) in support of the claim that controversies in science ordinarily end in resolution.[53] But inductive argument, in a domain where representative sampling is so difficult to certify, is not likely to carry conviction, although it is surely worth trying. Instead of developing a historical-sociological thesis about what "normally" happens, I shall recall some elements from the account of controversy termination given earlier that may warrant a modest philosophical thesis, one that is in part grounded in the historical record of science.

What is crucial in scientific controversy, as we have seen, is the role of the scientific community. In a debate between two groups of people, each with a considerable interest in prevailing, the dangers of distortion are evident. But the controversialists must make their arguments public; they must allow them to be subject to scrutiny. It is that scrutiny and the discipline that it imposes on the controversialists that offers the best hope of proper epistemic resolution. This

[53] The list would include such essays as Jarrett Leplin's "The Concept of an *Ad Hoc* Hypothesis," *Studies in the History and Philosophy of Science* 5 (1975): 309–45; Mary Jo Nye's "The Nineteenth-century Atomic Debates and the Dilemma of an 'Indifferent Hypothesis," same journal 7 (1976): 245–68; David Papineau's "The *Vis Viva* controversy: Do Meanings Matter?", same journal 8 (1977): 111–42.

system is obviously not foolproof. The community may be less exigent in its demands for cogency of argument than it should be; it may tend to be dazzled by the famous and underestimate the arguments of the unknown. It may exhibit a systematic socially produced bias. And so on. Thus the controversy may terminate in closure or in a resolution that is only apparent (according to some standard set of norms applied from the outside). Consider the function of the community once again. When a controversy ends, a certain result is taken to have been established, provisionally at least. This is not just entered in a book and left there. It is there to be built on. Whether it be a fact, a theory, or a principle, it is set to work. There will be cross-checking, testing, interrelating, all of the usual tedious but necessary day-to-day work of the scientist. Discrepancies of fact, weaknesses of theory, and inadequacies of principle are likely in the long run to show. That is precisely what the complex communal structure of scientific inquiry is supposed to ensure.

The aftermath of the Pasteur–Pouchet debate affords a good illustration of this process. Pasteur's conclusions were challenged by an English pathologist, H. C. Bastian, who claimed to show (as had Pouchet) that spontaneous generation can occur in many boiled infusions. "These experiments are moreover so simple, and may be so easily repeated, that the evidence which they are capable of supplying lies within the research of all."[54] Among the experimenters who took up the challenge was the redoubtable John Tyndall, who set himself to redesigning the experimental basis of the case against spontaneous generation, as well as to scrutinizing the assumptions Pasteur had made in his argument.

In a series of elegant experiments, he undermined the claims made by Bastian while avoiding the presuppositions that had left Pasteur's work open to challenge.[55] The fact was that the decision of the Paris commission in 1862 had not resolved the controversy, though it did come near to closing it for a time. It was indeed the specific questions raised by Pasteur's critics that enabled Tyndall to rethink the issue

[54] "Note on the Origin of Bacteria," *Proceedings of the Royal Society* 21 (1873): 129–31; see p. 130.

[55] Pasteur and Pouchet assumed that boiling water would kill all organisms and that sugared yeast water (Pasteur's medium) and hay infusion (Pouchet's) were equivalent as test media. Both assumptions proved erroneous. This was what made it possible for the experimental results of each to be basically correct without either set having the force their designer attributed to them. See the case history on spontaneous generation in James B. Conant's *Harvard Case Histories in Experimental Science* (Cambridge, Mass: Harvard University Press, 1957), 2: 489–539, where the relevant documents are reproduced.

and produce an experimental design that proved convincing to nearly everyone.[56]

A controversy suffering premature closure is, however, unlikely to be opened in the same terms again, as happened in the Pasteur case. In the meantime, the alternatives and the challenges are likely to have changed. Nor will it necessarily be recognized *post factum* that the controversy ought not to have been terminated in the way it was. What is more likely to happen is that the side that benefited from the closure will be held up to fresh criticism and will be forced to defend itself against new challenges. Since controversies are the frail human affairs they are, closure, rather than resolution, is bound to happen some of the time. Nevertheless, this does not license one to draw a subjectivist implication regarding the direction of science generally.

A successful theory in the natural sciences furnishes the basis for a long-term research program. The "fertility" of the theory (in the technical sense attached to that term by natural scientists) is an essential criterion of the epistemic quality of the program.[57] The history of such sciences as chemistry, geology, and cytology exhibits a pattern of progressive disclosure of structure.[58] Once an element or relationship finds a place in that structure and guides researchers over a period of time to new complications of structure, it is rarely discarded. Occasionally, with the discovery of new instrumental techniques, it becomes possible to test the theoretical entities in this structure in a more direct way (e.g., the atomic-level discontinuities revealed in point emission microscopy). It is hard to avoid the conclusion that the structures presented to us in the models of contemporary biochemistry or geophysics are in fact a reasonably reliable guide to the structures of the world.[59]

[56] Pasteur himself had remarked that his claim could never be definitively proved, since a critic could always claim that cases of spontaneous generation occur though none have yet been substantiated. Conant concludes: "Only a slow development – a gradual elimination of one alleged case of spontaneous generation after another, an ever-widening demonstration of the fruitfulness of Pasteur's hypothesis, an intensive study of the propagation of pure cultures – has led to the general conviction of the truth of this hypothesis, on which all modern microbiology is based" (ibid., p. 538).

[57] See E. McMullin, "The Criterion of Fertility and the Unit for Appraisal in Science," in *Essays in Memory of Imre Lakatos*, R. S. Cohen, P. K. Feyerabend, and M. W. Wartofsky, eds., Boston Studies in the Philosophy of Science No. 39 (Dordrecht: Reidel, 1976), pp. 395–432.

[58] See, for example, Michael Simon, *The Matter of Life* (New Haven: Yale University Press, 1971); E. McMullin, "Structural Explanation," *American Philosophical Quarterly* 15 (1978): 139–47.

[59] See especially the essays by J. Leplin, R. Boyd, and E. McMullin in *Scientific Realism*, Jarrett Leplin, ed. (Berkeley: University of California Press, 1984).

The insights into the structure of matter that the last two centuries of natural science seem to warrant give reason for confidence that the epistemic factors are working there in the way they should and that the sociopsychological setting of inquiry in the national sciences is on the whole conducive to the discovery of genuine explanation.[60] This is, admittedly, a much larger topic than the end of controversy. But it is important to notice the intimate relationship between the two.

[60] The reader will have noted that this argument bears on the natural sciences only. The structure of the debate is much more complex in the social sciences, and resolution is correspondingly more difficult to attain there. The role of the explanatory theory is crucial to the resolution of most controversy in the natural sciences. Such theories are rare in the social sciences. Furthermore, political and social value differences are often integral to controversy in economics or sociology, whereas they must be counted as nonepistemic factors where the natural sciences are concerned. For these and other reasons, the means of resolution of deep-rooted controversies in the social sciences are often not very effective. My treatment of scientific controversy in this essay has had to be restricted to the simpler context of the natural sciences.

3

The political anatomy of controversy in the sciences

EVERETT MENDELSOHN

Setting the problem

> Controversies are an integral part of the collective production of knowledge; disagreements on concepts, methods, interpretations and applications are the very lifeblood of science and one of the most productive factors in scientific development.[1]

This sociologist's claim signals a shift from what had been the more traditional understandings of the nature of science and its procedures. It is a shift that has been some time in the making and draws attention to the site of a new controversy itself in which a restructured sociology of knowledge of the natural sciences has taken on some of the strong claims for special characteristics of science. It is certainly distinct from the type of claim that the philosopher Norman Campbell made at the turn of the century when he defined science in terms of the areas of agreement that could be achieved.[2] It is also distinct from the views thoughtfully developed by Robert K. Merton over several decades in the sociology of science in which he viewed almost all forms of conflict as representing intrusions of ethnocentrism and caste mores into the critically important norm of universalism, which he saw as one of the focal norms for scientific activity.[3] (The one area of controversy that escaped this criticism in Merton's system was that over priority of scientific discovery. This kind of dispute, in this view, fruitfully falls well within the operative social system of science.)

[1] Helga Nowotny, "Controversies in Science: Remarks on the Different Modes of Production of Knowledge and Their Use," *Zeitschrift für Sociologie* 4 (1975): 37.

[2] Norman Campbell, *What Is Science?* (1921); New York: Dover, 1952.

[3] Robert K. Merton, *The Sociology of Science: Theoretical and Empirical Investigations* (Chicago: University of Chicago Press, 1973), collects the most important papers. See esp. "Social Conflict over Styles of Sociological Work" (1961), pp. 47–69.

The claim that controversies are fundamental to the production of knowledge in the sciences also runs afoul of the thoughtful, historical-sociological interpretation of the institutionalization of science of Joseph Ben David. He proposed the idea that the seventeenth-century institutionalization of the sciences, with the attendant formation of organizations, the establishment of norms, and the development of methods, had created a secure method for resolving disputes. This distinguished science from the other intellectual pursuits and arguments of the seventeenth century (particularly among the scholastics) that were incapable of resolving disagreements or controversies.[4] Rules were established that would govern the procedures of science. The new methods developed within the sciences were likened to juridical norms, a "court of reason," and thereby provided a secure means of producing consensus among scientists. The corollary notion of the existence of a scientific community with shared norms and established procedures through the use of the language of "community" gave the illusion of a unified group within which real disagreement would be foreign.[5]

Thomas Kuhn introduced a sharp break with an image of science that saw conflict as an unwanted externality and instead, in the course of his own retreat from positivism, gave credence to the idea that conflict was a natural outcome of the structures of the scientific enterprise. One can also see an implicit conflict model in the theories of the evolution of knowledge put forward by Stephen Toulmin, among others, in which there is thought to be a natural selection of concepts analagous to the biological process. This evolutionary model strongly implies that a mechanism for resolution of debate was built into the system, whereas in Kuhn's image of science debate could well continue, although the enterprise would be guided by the growing consensus on a new paradigm.[6]

Even if the notion of dispute in the sciences is naturalized, the key question remains. We must ask not why disagreements or disputes arise but why these disagreements become long-standing debates or controversies of great depth and broad constituencies. Why, in many

[4] Joseph Ben David's views are developed in *The Scientist's Role in Society: A Comparative Study* (Englewood Cliffs, N.J.: Prentice-Hall, 1971).

[5] Warren Hagstrom, *The Scientific Community* (New York: Basic Books, 1965) summarizes and develops the "community" concept that had been implicit in the works of others.

[6] Thomas Kuhn, *The Structure of Scientific Revolutions* (Chicago: University of Chicago Press, 1970), 2d ed.; Stephen Toulmin, *Human Understanding*, Vol. 1 (Princeton: Princeton University Press, 1972).

instances, do figures from outside the scientific specialty that was the initial home of the disagreement become involved? Why do these disputes "go public" and come to involve even parties from outside the sciences themselves?

That debates of this sort are found in the history of science has long been known, but only recently, as the historiography of the sciences has caught up with the shifting understandings of the scientific enterprise, did these debates become recognized as important to an understanding of science itself and in turn become the focus of serious study. Previously they had been passed over as aberrations or accidents in an otherwise smooth progress of scientific development. But let me make it clear that debates and controversies of this sort are still seen by some as representing "intrusions" into the orderly processes of scientific activity. There are also deep problems in the historiography of debates and controversies, and even deeper problems concerning the issues of the resolution or closure of debates. For depending upon the implicit or explicit model of the sciences that one holds, the failure to achieve a recognized end to a controversy might well create profound tensions. It is the establishing of a new consensus, the recognition of a new paradigm that is seen as the hallmark of progress, even in those theories of science in which controversy has become a natural element. If debate does not close, what may we say about the intellectual organization of the sciences? One answer, of course, proposed by Paul Feyerabend, is the somewhat quixotic "anything goes," a kind of anarchistic model.[7]

Other questions arise. Who recognizes the emergence of a debate? Who recognizes closure, if and when it is achieved, and gives it legitimacy? Only historians looking backward? Or the parties in the dispute or controversy themselves, or onlooking third parties contemporaneous with the conflict? Further, which individuals and parties should be examined in the study of debates and controversies, and what are the appropriate topics to be examined? Here the historian is often guided by his or her sense of what is legitimate to the debate, and consequently by his or her implicit model of the sciences themselves. Philosophers seem most often to find the "formal" factors of the sciences worthy of understanding, and the "idealized" structures, with their focus on the intellectual components, worthy of attention. Indeed, I often have the sense that philosophers in their discussions talk about the grounds on which the

[7] Paul Feyerabend, *Against Method: Outline of an Anarchistic Theory of Knowledge* (London: New Left Books, 1975).

debate *should* have been fought, and the bases on which closure or resolution *could* have occurred, had the parties only behaved as their idealized model would have them behave.

By contrast, my question is this: Can we engage in a more historical or naturalistic study of debates, avoiding prejudice by too narrow or prior construction and too preliminary a drawing of the boundaries?

My own examinations of the history of science lead me directly to the contention that there is a social nature to controversies in the sciences, and I am brought to agree with the further claim of Nowotny that "debates are fought for reasons which ultimately are to be found outside the realm in which they purport to be fought and for reasons that are extraneous to the intellectual content of the debate."[8] But even this statement might imply a prejudgment of such boundaries as "inside and outside," as "internal and external," and give to them a sense of greater stability or impermeability than I believe they have. The elements often identified as extraneous are placed there by way of convention, rather than as the result of historical research. Surely religion was central, not extraneous, to the debates between the Newtonians and the Leibnizians in seventeenth-century cosmology and physics.[9]

But we can rightly inquire at this point what connection those "reasons" outside the intellectual focus have to that focus itself, or how the exchange mediated between those elements that fall on opposite sides of the boundaries that we traditionally talk of. Kuhn caught something of the sense of the problem and provides an element for beginning its answer: "Like the choice between competing political institutions, that between paradigms proves to be a choice between incompatible modes of community life."[10] The key, of course, is the concept "community life," and Barnes makes what is implied explicit. The modes of community life are incompatible because they are goal-oriented, related to purposes, aims, goals, objectives, and interests. Both thought and action are goal-oriented in this manner, and therefore the elements in a scientific controversy that are judged against each other are judged in relation to "context-situated instrumental interests" or, as Habermas (from whom Barnes has adapted the notions) puts it, "interests in prediction and con-

[8] Nowotny, "Controversies," p. 35.

[9] The basis of the debate is laid out in H. G. Alexander, ed., *The Leibniz–Clarke Correspondence* (New York: Philosophical Library, 1956).

[10] Kuhn, *Structure of Scientific Revolutions*, p. 94. Cited by Barry Barnes and Donald MacKenzie, "On the Role of Interests in Scientific Change," in Roy Wallis, ed., *On the Margins of Science: The Social Construction of Rejected Knowledge*, Sociological Review Monograph No. 27 (Keele: University of Keele, 1979), p. 51.

trol." The contention here is that concepts or paradigms are not abstract theories but rather are resources or instruments that can be put to use and, as such, fall within the category of goal-oriented knowledge.[11]

The activities that scientists carry out are socially structured activities and are then related to social interests. These interests cover a broad spectrum. At one extreme is the narrow professional or specialist concerns of the sort that Barnes would call "esoteric social interests,"[12] as, for example, the need to maintain or strengthen the given scientific discipline or specialty in which one is involved against another field. This may be considered an "internal" social interest. Or, they might be narrow national interests or styles reflecting the concern to defend or promote a given school or tradition in the sciences. An interest of this sort would help account for the different national styles or modes of action we have so often found in the sciences.[13] Farther along this spectrum are the broader direct social influences and interests, such as religion, social class, or political position. That these kinds of interests are sometimes made explicit in scientific controversy will become apparent in several of the cases examined below.

It is in the range of social interests, from the professional to the broadly social, that we can find part of the explanation for why controversies have the trajectory and duration that they have. Barnes and MacKenzie also make this point: "Opposed paradigms and hence opposed evaluations may be sustained and probably are in general sustained by divergent sets of instrumental interests, usually related in turn to divergent social interests."[14] Thus we should expect a mix of esoteric and the broad social interests to become part of the actual substance of a debate and to be reflected in the language and conceptual structure of the debate.

This kind of explanation of controversies is not one easily accepted within the scientific enterprise, which in the course of its own history has structured its ideological defenses in such a way as to avoid any implication of links between the intellectual content and procedures

[11] Barnes and MacKenzie, "Interests in Scientific Change," p. 52. Cf. Jürgen Habermas, *Knowledge and Human Interests*, trans. Jeremy J. Shapiro (Boston: Beacon Press, 1971), esp. pp. 301–17.

[12] Barnes's own development of the concept of "interests" appears in Barry Barnes, *Interests and the Growth of Knowledge* (London: Routledge and Kegan 1977).

[13] See John T. Merz, *History of European Thought in the Nineteenth Century*, 4th ed., Vol. 1 (Edinburgh: Blackwood, 1923).

[14] Barnes and MacKenzie, "Interests in Scientific Change," p. 54.

of science, on the one hand, and its goals, objectives, and interests on the other. In an earlier exploration of the social formation of the epistemologies of science in the seventeenth century, I analyzed just this attempt to establish boundaries as a means of gaining social space at a time when science was weak. This demarcation serves to give science the ability to claim a "higher truth value" for its activities and statements, and thus a greater legitimacy as a way of knowing. That this claim is found throughout the history of science can be seen in Pasteur's explicit disclaimer (discussed later in this chapter) of any role in his own science for religion, philosophy, and politics, even though he recognized that they were served in his debate against spontaneous generation and hence against materialism. Did he protest too much? His contemporaries were not so wary, and Tyndall, for example, was quite willing to celebrate his participation in science's battle against religion and idealism and for materialism. There appears to be an element of real self-interest in attempts to claim a divorce between science and normative or political thought. Habermas recognized the contradiction in this position, and in developing a theme earlier raised by Marcuse he put the issue very bluntly: "Because science must secure the objectivity of its statements against the pressure and seduction of particular interests, it deludes itself about the fundamental interest to which it owes not only its impetus, but the conditions of possible objectivity themselves."[15]

The point to draw from this discussion is that science is a socially organized and institutionalized mode of the production of knowledge in which structures of social organization exist at every level of the institution and its activities. This is true from the most esoteric actions of the scientist in the laboratory,[16] to the broader disciplinary and professional structures, to the very location of science and scientists in the fuller social context of numerous political and social interests. Thus we are not concerned with an occasional biased intrusion into an otherwise unsullied set of rational scientific procedures when examining controversies but rather with normal scientific action and choice that is "structured and organized by contingent goal

[15] *Knowledge and Human Interests*, p. 311. See also Herbert Marcuse, *One Dimensional Man: Studies in the Ideology of Advanced Industrial Society* (Boston: Beacon Press, 1964). See my examination of boundary establishment, E. Mendelsohn, "The Social Construction of Scientific Knowledge," in E. Mendelsohn, P. Weingart, and R. Whitley, eds., *The Social Production of Scientific Knowledge* (Dordrecht: D. Reidel, 1977), pp. 3–26.

[16] See the recent study by Bruno Latour and Steve Woolgar, *Laboratory Life: The Social Construction of Scientific Fact* (Beverly Hills, Calif: Sage, 1979). They apply ethnomethodology to the laboratory scene.

orientations."[17] Only by recognizing this can we understand why controversies last as long as they do. For here the very essential parts of science – experimental procedures and judgments, technical choices and commitments – are influenced and structured by the interests of the participants. And when there are potential consequences to the outcomes, as there are in any goal-oriented activity, then differences really matter. A mere disagreement in this context becomes a sharp controversy.

Because it is in controversies that interests become most sharply drawn and the structures and processes of science most clearly tested, they provide a particularly strategic site for examining the complex set of relationships between the social and the conceptual in science.[18] The conflicting claims of the contesting groups in a debate magnify (and potentially distort) the more commonplace daily procedures of scientific activity and thus give them greater visibility.

In studying the histories of many debates, we find that their authors have reduced them to their abstract intellectual contents with at most some reference to the human actors involved but more often focused wholly on the cognitive structures. What has often been overlooked are the elements of the subtler and more complex model developed earlier, and thus only a partial account is rendered. Indeed, the student of debates and controversies must be willing to see the social interest and the links between the cognitive and the social in order even to recognize the evidence that is present and not dismiss it as mere intrusion. Scientific papers themselves are often misleading, in that their authors have adopted patterns of "retroactive falsification" and attempt to meet the formal requirements of objectivist science and consciously ignore both the actual pattern of scientific investigation and the specific pattern of evaluation and choice. The public addresses of scientists, however, often tell more and unmask the activity to reveal some of the natural content of scientific endeavor. As we shall see, Pasteur's "victory speech" gives a clear indication of a scientist knowledgeable about the social context and aware of interests that knowledge might serve. Similarly, John Tyndall, in his public lectures, reveals something of the goals and

[17] Donald MacKenzie and Barry Barnes, "Scientific Judgment: The Biometry–Mendelism Controversy," in Barry Barnes and Steven Shapin, eds., *Natural Order: Historical Studies of Scientific Culture* (Beverly Hills, Calif.: Sage, 1979), p. 201.

[18] See, for example, the claim made by A. J. Webster, "Scientific Controversy and Socio-Cognitive Metonymy: The Case of Acupuncture," in Barnes and Shapin, *Natural Order*, p. 121.

interests that lay behind his science and its link to his crusade for materialism.

The use context of knowledge or action also gives clues to the patterns of social interest and social choice.[19] For example, the attempts to build a new professional discipline in physiology in Britain in the middle decades of the nineteenth century loom large in the battle for live animal experimentation taken on by Michael Foster and John Burdon-Sanderson. Professional interest guided their actions, whereas social mores and class sensibilities directed the role of the opponents of vivisection in that stunning controversy. The construction of statistical techniques by the biometricians in turn-of-the-century Britain cannot be understood without reference to the use that these techniques were to be put to in providing a basis for proper choice and policy in eugenics and its move to create a better social order. The extent of Karl Pearson's involvement in controversies gives a fuller sense of the politicized role that the new statistics were playing on the British scene, and it shows also that the moves to establish the disciplinary legitimacy of statistics should be viewed as cloaked not only in narrow professional interests but in broadly political and social ones.[20]

The fact that fields of knowledge, with a fairly narrow and esoteric intellectual base, can in their controversies involve parties from outside the discipline and even from outside science itself provides the strongest clue that interests beyond the narrow intellectual ones are at stake. In the case of the debate over the use of live animals in medical experiments, it was clear that public interest, public mores, and public norms were being defended by members of the public against the more restricted vision of experts in animal experimentation about what was involved in the controversy.

[19] The historical case studies of Barnes, MacKenzie, Shapin (see, for example, their papers in Barnes and Shapin, *Natural Order*), Lyndsay Farrall ("Controversy and Conflict in Science: A Case Study – The English Biometric School and Mendel's Laws," *Soc. Stud. Sci.* 5 [1975]: 264–301), Peter Buck, and others bring out this point explicitly. See also my earlier studies on professionalization in the science and the sociology of knowledge context of reductionism in nineteenth-century German biology. E. Mendelsohn, "The Emergence of Science as a Profession in Nineteenth-century Europe," in Carl Hill, ed., *The Management of Scientists* (Boston: Beacon Press, 1964), pp. 1–36; "Revolution and Reduction: The Sociology of Methodological and Philosophical Controversy in Nineteenth-century Biology," in Y. Elkang, ed., *The Interaction between Science and Philosophy* (Atlantic Highlands, N.J.: Humanities Press, 1974), pp. 407–26.

[20] Donald A. MacKenzie has made a strong case in his recent study, *Statistics in Britain, 1865–1930: Social Construction of Scientific Knowledge* (Edinburgh: Edinburgh University Press, 1981).

The social and political procedures by which debates and controversies are carried on are of interest. The use of scientific and professional authority, and the limits of this authority when a disagreement erupts into a controversy, provide an interesting research site. The organized social bodies that provide the forums for debates and controversies are worth noting – the academies and scientific societies and the various committees and juries established by them. The courts of law were called upon when the boundaries of the debate were broadened to include issues of interests to the polity as well as to the organized sciences. Similarly, legislative bodies or parliaments became both a focus for debate and, at times, for the resolution or closure of debate. Voluntary organizations and social movements, sometimes narrowly constructed of the practitioners of science and at other times spilling broadly into the lay public, have also become involved in debates or controversies – movements favoring mesmerism, evolutionism, and antivivisection are examples of this.

But if controversies occur and are a natural part of the scientific landscape, how do they end? "Resolution" of a controversy represents a coming together of the conflicting parties and the emergence of a consensus. The term "closure," on the other hand, might be used to mean a more formal structure for ending a controversy or debate that permits a partial resolution, but not necessarily a dissolution, of the disagreement. Are closure or resolution ever achieved? If so, how? And if not, does it make a difference for science?

The several cases examined below seem to show that resolution is by no means always found. Indeed, on looking at the way these debates drew to a close, we are led to ask how often resolution is ever achieved in real debates on the part of the participants themselves. On the contrary, it appears that scientific work is able to continue and that the knowledge and explanatory modes used by both parties of a conflict become a part of the broad body of knowledge and technique in the sciences and can be, and often are, used even while disagreement persists.

Closure, on the other hand, has been achieved. Sometimes it comes with the judgment of a commission, which gives greater credence to one party as opposed to another. At other times legislative acts establish procedures by which a contested area is monitored and regulated, and the sting is thereby taken out of a debate. In other cases those who are weaker, sometimes in intellectual argument, sometimes in political strength, are driven from the scene, and although they maintain their intellectual position, they no longer have the ability or means to pursue open confrontation. It is clear that often a party maintains its point of view even as its opposite has gained greater recognition and legitimation within the scientific community. Often the only real closure comes with the death of a participant in the debate. After all, if interests do inform judgment and action, we cannot expect a resolution or a consensus to emerge unless those interests themselves undergo change. What we can expect is that the more powerful (a combination of both intellectual elements and social interests) will come to prevail.

Let us turn now to several cases and examine these points in the detail of real history rather than abstract discourse.

Examining cases

Harvey versus Descartes

In 1628, William Harvey published his famous treatise *De motu cordis et sanguinis*, in which he not only proposed a theory of the circulation of blood throughout the whole body but also propounded a theory of the motion of the heart. We know that Harvey's slim but forceful treatise was not very well received. And although later generations of physiologists (not to mention historians) celebrated it as one of the great publications in the history of science, his contemporaries received the book and the theories therein with notable coolness.[21] Harvey himself is remembered as having commented some years after publication that no one over fifty years of age ever gave a positive response to his theories. We know that many of his fellow members of the Royal College of Physicians, certainly having heard Harvey lecture on these ideas for some years prior to publication and assuredly having had access to the book itself, either ignored it in their own works on human anatomy and physiology or alluded to it in somewhat disparaging fashion.

It remained for René Descartes, then a young French philosopher, to be the first man of science to respond favorably to Harvey's notions of circulation. He did so initially in several letters to colleagues, then in his ill-fated *Traité du monde* (1632), a book that was withheld from publication following Galileo's condemnation, and finally in his famous *Discours* of 1637, where Descartes warmly embraced the concept of the circulation of the blood but rejected Harvey's ideas of

[21] There are two important points of entry to Harvey and his reception – the standard biography by Geoffrey Keynes, *The Life of William Harvey* (Oxford: Clarendon Press, 1966), and the new study by Robert G. Frank, Jr., *Harvey and the Oxford Physiologists* (Berkeley: University of California Press, 1980).

the motion of the heart. In its place he propounded a theory of his own.[22]

Historians and physiologists looking back at Descartes's move have tended to offer a fairly simple lament. "Poor Descartes," they say, "No training in biology. Didn't understand Harvey's system, and consequently misrepresented it." Not at all. Descartes fully understood Harvey's concept and rejected it because it ran counter to the metaphysical position that he had adopted and the interest he had in using the Harveian concept of circulation to support the mechanical world view for which he was battling. By replacing Harvey's idea, that the heart contracted because of its muscular nature with his own view that the heart expanded vigorously because of the rapid increase of heat within it, Descartes had coopted the Harveian system to serve a metaphysical interest.[23] He needed concepts that were derivable from mechanics, and not those that carried any taint of "faculties" or "powers." "In supposing the heart to move in the way Harvey describes," he claimed, "we must imagine some *faculty* or *inherent power* that causes movement."[24] Descartes, instead, wanted to consider the motion of the heart as being based on heat and dilation, for "then whatever is of scientific interest in its operation is a simple exemplification of broad mechanical principles." This, of course, is exactly where Descartes wanted to take the motion of the heart, this fundamental physiological activity. He had embraced circulation of the blood because it seemed to him a way of undermining the Aristotelian principles of the special nature of circular motion. Instead he was able to locate circularity in so proximate and earthly an activity as the movement of blood in the body. In this way the myth of circular motion would be undercut, and then by adding to it a motion of the heart based on heat and dilation, the living body itself would seem to be an embodiment of fundamental mechanical causes.

But this was not William Harvey's intent, and indeed he responded vigorously to Descartes's proposal. The heart, he argued, is a contractual muscle, and its motion is based solely on that fact. He took aim at his philosophical opponent and put forward his perspec-

[22] I have discussed this in an earlier study: Everett Mendelsohn, "The Changing Nature of Physiological Explanation in the Seventeenth Century," in *L'aventure de la science. Mélanges Alexandre Koyré*, vol. 1., I. B. Cohen and René Taton, eds. (Paris: Herman, 1964), pp. 367–86.

[23] John Passmore, "William Harvey and the Philosophy of Science," *Australasian Journal of Philosophy* 36 (1958):85–94 provides a provocative study of these points.

[24] See Mendelsohn, "Physiological Explanation," pp. 375–6 and René Descartes, "*La description de corps humain*," in *Oeuvres*, C. Adam and P. Tannery, eds. (Paris, 1909), 11: 234–4.

tive quite clearly: "I have not endeavored from causes and probably principles to demonstrate my propositions, but, as of higher authority, to establish them by appeals to sense and experiment after the manner of the anatomists."[25] Harvey could offer no alternative causal explanation for the motion of the heart and instead was defending in this instance a metaphysics, making a claim based on observation and experiment but quite willing to accept the notion of an innate contractility of muscle rather than seek alternative cause. It should be noted in parallel that Descartes was willing to admit to the existence of an innate heat or "fire without flame" that burned in the heart, a heat capable of causing the dilation that was the basis of the heart's motion, but of course it was a fire or a heat that he had to accept solely by inference. For him, however, it was an acceptable inference because it was based on mechanical principles and explanations rather than on "faculties or powers."

This debate between Harvey and Descartes, which lasted over a period of years, was carried out prior to the formation of the scientific societies and academies that came later in the seventeenth century and seemed to serve as a means of focusing controversy and seeking closure or resolution to debate. Instead, the controversy between Harvey and Descartes was carried out in print (as well as, we know, in letters to friends and disciples), and in this sense was clearly a public debate aimed at gaining adherents and encouraging disciples.

At first glance it would seem that Harvey was arguing a set of principles based on method and evidence, as compared to Descartes, who had established a consciously enunciated metaphysical task. Looking somewhat more deeply, however, it is clear that Harvey also was defending a set of metaphysical interests, one guiding theme of which was to be antiatomistic and antimechanical.[26] A second theme is reflected in Harvey's very special view of the heart itself, a view that seems clearly to reflect his political interests and social position as a staunch defender of the monarchy and physician to the crown. Harvey is anxious to keep the heart distant from what he would consider such base causes as heat, or other mechanical influences: "The Heart of creatures is the foundation of life, the Prince of all, the Sun of their Microcosm, on which all vegetation does depend, from whence all vigor and strength does flow. Likewise the King is the foundation of his Kingdoms, and the Sun of his Microcosm, the *Heart*

[25] William Harvey, *A Second Disquisition to John Riolan, Jun. in Which Many Objections to the Circulation of the Blood are Refuted* (1679), in Harvey, *Works*, trans. Robert Willis (London, 1847), p. 134.

[26] See Walter Pagel, *William Harvey's Biological Ideas* (Basel: Karger, 1967).

of his Commonwealth from whence all power and mercy proceed."[27] Indeed, if this concept of centrality of power is to be given any meaning, the heart itself could not be subject to other, lesser causes.

In the conduct of their debate, neither Harvey nor Descartes sought any adjudication or resolution by third parties. Their interest was not only in convincing their disciples but in influencing the future. Neither, it should be noted, changed his point of view during the course of his own life, and indeed each repeated variants of it in subsequent works. The debate itself was carried out primarily within the group that could be called natural philosophers and physicians, although it now seems clear that the controversy was only partly about science itself. Finally, there was no resolution of the debate in the lifetime of either of the contestants. Instead, death took the debaters off the scene, and in time new generations of disciples turned to new sets of problems.[28] And even these problems often fall within one or another of the basic explanatory schemes of Harvey and Descartes.

At one level, it would seem that further examination and fuller evidence should have resolved this debate. But that was not its character (nor, I would suggest, that of most deep controversies), because the contestants were using the knowledge gained, each to serve an end guided by interests, cognitive, metaphysical, and social.

Spontaneous generation

As a concept in biology, spontaneous generation suffered decisive defeats at several different points in history, beginning as early as the seventeenth century, only to reappear in alternate form at later times.[29] But as the traditional historiography of the subject would have it, nowhere was the defeat of spontaneous generation more decisive than in the confrontation between Louis Pasteur and Félix Pouchet in the 1860s in France. In a recent reexamination of this confrontation, two historians of science have gone far to identify several elements that demand a reassessment of the traditional historical account, with its focus so completely fixed on experi-

[27] Quoted by Christopher Hill, "William Harvey and the Idea of Monarchy," in Charles Webster, ed., *The Intellectual Revolution of the Seventeenth Century* (London: Routledge & Kegan Paul, 1974), p. 160, from Harvey's dedication to Charles I in *De motu cordis*.

[28] See Frank, *Harvey and the Oxford Physiologists*.

[29] See John Farley, *The Spontaneous Generation Controversy from Descartes to Oparin* (Baltimore: Johns Hopkins University Press, 1977).

mentation.[30] The story, as generally related, would have it that Pasteur won his victory over Pouchet and really brought the issue of spontaneous generation to resolution by the superiority of his method and the achievement of his skill. Indeed, we are led to believe that here was a case of science, in the form of experimental method, triumphing over belief, in the guise of an a priori metaphysic. Interestingly enough, the ink on the Farley-Geison paper was hardly dry when a challenge was brought by another student of the debate, whose own study substantiated the traditional historiography.[31] Indeed, I wonder how often it is that the historiography of deep controversies is itself the theme of continued or "meta" controversies. It might suggest that the stakes in such a controversy are high enough so that they are carried to the second order and that the interests being defended are strong enough to propel the argument forward from the scientific focus to the historical one. That the Pasteur–Pouchet debate itself was not resolved or brought to real closure is seen in the fact that it had to be repeated only a few years later in England when Huxley and Tyndall took on Bastian in a British version of the more famous French confrontation.[32]

The debate itself has fairly familiar elements. It began in 1859 when Félix Pouchet published his monograph on what he called *Hetérogenie* (this, incidentally, was the same year in which Charles Darwin published his *Origin of Species*, a move that I will contend later in this essay is closely linked to the controversy itself). Pouchet was at the time 60 years old, a biologist who had worked for a good deal of his scientific life on problems of animal generation.[33] His challenger, Louis Pasteur, was 37, a chemist who had only recently turned to biological problems.[34] Within several months of the publication of Pouchet's volume, Pasteur had carried out his own series of experiments that seemed to contradict the conclusions drawn by Pouchet. Pasteur was contemptuous in tone of the work of his elder compatriot and substituted a series of experiments that by any standards appear

[30] John Farley and Gerald Geison, "Science, Politics and Spontaneous Generation in Nineteenth-century France: The Pasteur–Pouchet Debate," *Bulletin of the History of Medicine* 48 (1974): 161–98.

[31] Nils Roll-Hansen, "Experimental Method and Spontaneous Generation: The Controversy between Pasteur and Pouchet, 1859–1864," *Journal of the History of Medicine* 34 (1979): 273–92.

[32] See Farley, *Spontaneous Generation*, pp. 121–41.

[33] Felix-Archimède Pouchet, *Hetérogenie ou traité de la generation spontanée base sur des nouvelles experiences* (Paris: 1859). See the very brief biographical sketch in *DSB*.

[34] See Gerald L. Geison's long and very thoughtful biographical-scientific appraisal in *DSB*, Vol. 10.

to be innovative and brilliant. So far, however, what we have is a disagreement over experimental results and interpretation of the results. How, then, did the disagreement become a controversy and a major debate?

In 1862 a step was taken that formalized the status of the debate, provided focus and attention to it, and even established the means by which resolution might be achieved. The Académie des Sciences announced a prize that would "be given to him who by well-conducted experiments throws new light on the question of so-called spontaneous generation."[35] In order to award the prize, the Académie appointed a commission to survey the submissions and judge a winner. Why, one might ask, did the Académie single out spontaneous generation as the object of such special attention? Although the Académie did have a tradition of offering such prizes from time to time, it certainly did not do so for every issue and every scientific disagreement or dispute. Those in authority in the Académie obviously felt that this was an issue of some special importance. Certainly the intrinsic merits of the dispute would not at this time have seemed to recommend it for such notoriety. Nor were there many in the Académie with the scientific background or competence to handle the technical elements of the discussion competently. It seems clear that the Académie, in bringing its authority to bear on this issue, had its eyes on other elements than merely spontaneous generation. It is worth pointing out in this context that by making the move that it did, the Académie established a formal structure for adjudicating controversy and indeed cast it in terms very similar to those of a judicial proceeding. Historically, scientific academies might well be viewed as serving the function of adjudication of disagreements and controversies. Not all of them, however, moved so directly to foreshorten debate and to reach closure as did the Paris Académie in this instance, even if they have often served to organize the areas of agreement and consensus and to provide social structures within which disagreement can be resolved.[36]

[35] See Farley and Geison, "Science, Politics, and Spontaneous Generation," p. 181. Details of the debate are given in Georges Pennetier, *Un débat scientifique, Pouchet et Pasteur (1858–1868)* (Paris: 1907).

[36] This role of scientific societies has not received much direct attention, although elements of it can be recognized in the more traditional studies of the histories of the various societies and academics; see, for example, the still very valuable study by Martha Ornstein, *The Role of Scientific Societies in the Seventeenth Century* ([1913] Chicago: University of Press, 1928); also the excellent, very detailed study by Roger Hahn, *The Anatomy of a Scientific Institution: The Paris Academy of Sciences, 1666–1803* (Berkeley: University of California Press, 1971).

In this instance, the Académie commission ran into problems, and after a good deal of procedural wrangling and charges of unfairness from Pouchet the commission disbanded. A second commission was then established in 1864 to deal with the same issue, but once again the contestants could not even come to agreement on procedures or the open-mindedness and fairness of the committee members. Pouchet once again withdrew, claiming that members of the commission had announced in advance their positions on spontaneous generation. If the issues involved were merely those of method and experimental outcome, would the controversy have spilled over from the substantive issue of generation to the procedural ones of the commission's composition and procedures? I contend that the controversy was indeed a controversy, because so little consensus existed that reason and experience would not be competent to resolve the differences. The differences themselves lay outside the science as well as within it.

One contemporary observer, the British paleontologist Richard Owen (himself an opponent of spontaneous generation), spotted the factors in play at once and was not at all surprised by the outcome. He noted in 1868 that "Pasteur, like Cuvier, had the advantage of subserving the prepossessions of the 'party of order' and the needs of theology."[37] The philosopher might argue today, with hindsight, that the debate could have been resolved on the issues of method and results alone, but this abstraction from history would not serve our understanding of why there was a debate or controversy at all and what interests were served within the controversy.

In fact, the willingness of those outside the immediate field of inquiry – in this case those studying problems of generation of microbes in biology – to join the conflict would seem to give further support to the view that the controversy transcended the knowledge claims involved and included instead interests of both a professional and political nature. How, for example, would the philosopher handle Cuvier's success in his debate against Lamarck and Geoffroy St. Hilaire a generation earlier?[38] In this debate over evolution no real claim can be made, even in retrospect, that Cuvier's science was better than that of his opponents. Instead, as one reflects on that debate and recognizes its place within the French system, Cuvier's victory is seen to have come about because of his own position and

[37] Cited in Farley and Geison, "Science, Politics, and Spontaneous Generation," p. 167.

[38] The details of Cuvier's favored position have been studied by Camille Limoges in a major look at the Musée d'Histoire Naturelle. The materials have not yet been published.

his own power within the structures of the state and of science during the First Empire.

The political structures of controversies may be, and in this case I believe were, at least as important as the cognitive and epistemological elements. The debate over spontaneous generation in France is not understandable without recognition of the enormous power wielded by the Paris-based Académie des Sciences, particularly the ancillary features of access and control that it conferred upon French scientists. "Judiciary proceedings" of the sort that it established were really unlike anything that existed in Britain or Germany at the time, and therefore in neither place could the expectation of achieving closure of a debate be as widely held.

The Pasteur–Pouchet debate was carried out in the France of the Second Empire, and Louis Bonaparte was widely recognized as a vigorous opponent of republicanism, atheism, and materialism. In the context of his empire, religious and philosophical issues were most decidedly political issues as well. For example, in order to meet learned groups had to receive permission from the government, and an agent of that government sat in on their sessions in order to record any indiscretions that members might make or any tacit political moves that the organizations might adopt. When the fledgling Société d'Anthropologie was established in 1859, its members were troubled by the state's intervention and found discussions of human development and potential human evolution greatly suspect.[39]

When in 1862 a member of the Société d'Anthropologie, Mme Clemence Royer, translated Darwin's *Origin of Species* into French, she set the stage for what I believe was a critical, if somewhat hidden, element of the spontaneous generation debate.[40] Mme Royer herself was a materialist, an atheist, and a republican, and her preface to Darwin's *Origin* casts Darwin as an updated Lamarckist, implicitly reflecting Lamarck's own interest in spontaneous generation. The implied materialism of Darwinism (celebrated in Britain by Tyndall

[39] A doctoral dissertation studying the formation, development, and intellectual role of the Société d'Anthropologie was recently completed by Joy Harvey: "Races Specified, Evolution Transformed: Race, Evolution, and Social Evolution in the Société d'Anthropologie de Paris, 1859–1902," Harvard University, 1983.

[40] See the reports of Joy Harvey, the paper by John Farley, "The Initial Reactions of French Biologists to *Darwin's Origin of Species*," *Journal of the History of Biology* 7 (1974): 275–300; Robert E. Stebbins, "France," in Thomas F. Glick, ed. *The Comparative Reception of Darwin* (Austin: University of Texas Press, 1974), pp. 117–67; R. E. Stebbins, "French Reactions to Darwin, 1859–82, Ph.D. diss., University of Minnesota, 1965; Yvette Conry, *L'Introduction du Darwinisme en France au XIXe siècle* (Paris: Vrin, 1974).

and in Germany by Büchner), and the linkage between spontaneous generation and Darwinism applauded by Mme. Royer, were, of course, anathema to the French intellectual elite of the day.

Pouchet became a member of the Société d'Anthropologie and gained support from its members for his belief in spontaneous generation. Other support was forthcoming from Littré and the positivists, another of the suspect groups of the Second Empire. The links backward in time were seen leading not only to the philosophical enemy, materialism, but to the Revolution itself and the roots of republican sentiments. Darwin's reception in France among the scientific elite was negative; they, indeed, were engaged in an effort to overcome Darwinian evolution. They recognized in Pasteur's opposition to spontaneous generation a means of eliminating the basis for evolution itself. Thus there were philosophical, political, and religious interests at work in giving structure to the controversy over spontaneous generation. Of the commission established in 1862, all the members were Catholic, and some members had announced their decision on spontaneous generation before examining the submissions of the contestants.[41] Similarly, in 1864 all the members were united in advance in agreement with Pasteur, and Pierre Flourens, the secretary of the French Académie, indicated in advance that he saw in Pasteur a clear means of refuting Darwinism.

Pasteur himself recognized what was at stake, and in his well-known "victory" lecture at the Sorbonne in 1864 he outlined the battle between materialism and religion in which he believed he was engaged: "What a triumph, gentlemen, it would be for materialism if it could affirm that it rests on the established fact of matter organizing itself, taking on life of itself."[42] But having indicated what spontaneous generation would gain if it were victorious, he backed away from this all too obvious linking of his science with established religious views and made instead the explicit claim that these religious and philosophical attitudes did not affect the nature of his science itself. In this context he spelled out what might be seen as the "norm" of objectivism in the sciences: "Neither religion, nor philosophy, nor atheism, nor materialism, nor spiritualism belongs here ... As a scientist it does not much concern me; it is a question of *fact*."[43]

[41] These points were brought out by Pennetier, *Pouchet et Pasteur*, and by Farley and Geison, "Science, Politics, and Spontaneous Generation."

[42] Drawn from the excerpt translated in Farley and Geison, "Science, Politics, and Spontaneous Generation," p. 189. The original may be found in *Oeuvres de Pasteur*, Posteur Vallery Radot, ed. (Paris: Masson, 1922), 2:328–46; see also my recent study, E. Mendelsohn, "The Origin of Life and the Materialism Problem," *Revue metaphysique et de morale*, 1985 (in press).

[43] Farley and Geison, "Science, Politics, and Spontaneous Generation."

But did Pasteur really mean to claim that knowing what the stakes were and what interests were involved would in no way affect his behavior as a scientist? Farley and Geison suggest that Pasteur was not immune to judgment and choice based on interests.[44] They note that in the course of his experiments on spontaneous generation he suppressed as "unsuccessful" all those experiments in which life had mysteriously appeared. Success was measured by the adherence to the explanation that Pasteur had prefigured in his own mind. They note that in his memoir of 1861 that was the basis for his winning the first Académie prize, Pasteur had achieved "successful results" in less than 10 percent of the trials he undertook. And the other 90 percent? "I did not publish these experiments, for the consequences it was necessary to draw from them were too grave for me not to suspect some hidden cause."[45] Were it mere scientific dispute, Pasteur's behavior might well have been other in this instance. Indeed, Farley and Geison show that earlier in his life, while he was still concerned with molecular asymmetry, and late in his life, after the end of the Second Empire, Pasteur entertained a tentative belief in abiogenesis, or spontaneous generation, and had actually engaged in some attempts at the artificial creation of "life."

Pasteur himself had gained explicit Imperial recognition and support beginning as early as 1851 and during the 1860s had achieved a personal acquaintance with the emperor. In 1866 he dedicated a book to the emperor, and in 1867 he used his position of authority at the Ecole Normale to campaign against those students who were critical of Louis Napoleon. This is not to claim that Pasteur was not "doing science" but rather that Pasteur's science tacitly involved interests of a professional and of a broadly political, social, and religious sort.

But there are two anomalies worth mentioning. First, Félix Pouchet himself. Although he was clearly an outsider to the scientific establishment to which Pasteur, on the other hand, early had gained entrance, he was in no way heterodox in his beliefs. He was distinctly orthodox in his religious commitments, and he explicitly notes that it was God's hand that was at work in the spontaneous generation he found to occur. He did not then believe in a natural (not involving God) process of the creation of life from nonliving matter, so-called abiogenesis. Further, he was not a Darwinist, but in the context of the religious-philosophical debates of the time he was explicitly linked with just those ideas that he would have rejected, all this because of his own commitment to and work upon spontaneous generation.

The second anomaly concerns the British physicist John Tyndall,

[44] Ibid., p. 190.

[45] Ibid., p. 191, cited from *Oeuvres*, 2:236.

who had become deeply involved in the controversy over abiogenesis. The debate, which occurred a decade after the Pasteur–Pouchet confrontation in Paris, involved T. H. Huxley and John Tyndall, on the one side, and Charlton Bastian, the proponent of spontaneous generation, on the other. Although the debate itself would be interesting to examine in some detail in that a number of its elements present interesting contrasts to the French situation, the point that I want to note is Tyndall's own vision of his role. What, one might ask, is a physicist doing in the middle of a biological controversy? A recent student of the debate put it very directly.[46] For Tyndall, the spontaneous generation controversy was only "one battle in a much bigger war." And that larger struggle was the one between materialism and idealism, with Tyndall casting himself in the role of "the field marshal of the materialist forces." Tyndall's limited aim was "to clear away this Bastian fog." For spontaneous generation, in his view, was opposed to the principles of materialism. Tyndall, and Huxley with him, had a very "political" conception of science in that they saw the advancement of science paralleled by, and one of the factors causing the advancement of progressive thought in society itself. Thus, in Tyndall's view, such questions as the origin of being and of life were extraneous to science and carried with them the question of "vital principles" linked more to the domain of religion than to rational thought. Strange, then, that the Frenchman Pasteur could see in spontaneous generation the threat of materialism, whereas the Englishman Tyndall could see in spontaneous generation the shadows of vital forces and the links to religion. Each had carried into his scientific judgment an explicit element of his religious and political interests, even though these became manifest in contradictory interpretations of the scientific substance.

Parenthetically I might note that the British controversy, unable to be resolved in an easy and orderly manner, spilled over into France, where still another commission of the Académie was established to examine Bastian's claims. In 1877 the commission was unconvinced, and once again the debate on the French scene was "closed." Although Bastian is recorded to have gone silent after the combined thrust of the French Académie and the vigorous attacks of Huxley and Tyndall, he revived his beliefs in spontaneous generation somewhat later in his life.[47]

That debates can reach a type of formal closure, as was achieved in France in 1867 (and reasserted in 1877), is demonstrated, but the

[46] See Farley, *Spontaneous Generation*, pp. 121–41.
[47] Ibid., pp. 152–4.

structure of that type of closure is more formal than real. We know that Pasteur in 1872 attacked a number of French scientists for their support of a "German" theory backing spontaneous generation, which he found especially repulsive since it came so shortly after the Franco-Prussian war. It is obvious from this that the French closure did not reach widely beyond the borders of France and that adherence to that closure was cast by Pasteur in patriotic garb. Although it is in no way identical to the experiences of Soviet scientists at the time that closure was reached in the genetics debate in 1948, the sense of authority to be wielded by the Académie and its representatives had a familiar tone, if not similar strength, to the authority of the Soviet party. As one Soviet geneticist wrote in 1948:

> So long as two schools of thought in Soviet genetics were recognized by the party and controversies between them were reckoned to constitute a fruitful discussion of theoretical problems, I have been obstinate in defending my own views ... But now, having convinced myself that the principles of Michurianism in Soviet genetics have been approved by the central committee of the CPSU, I no longer find it possible for myself as a party member, to persist in these opinions which have been declared erroneous by the central committee of our party.[48]

The spontaneous generation controversy is instructive in that it clearly involved an important issue in the development of the biological sciences but was by no means limited to that issue alone. Although some have claimed it a debate ostensibly limited to scientific issues and to be carried on within the scientific community, it involved many scientists from outside the specialties necessary for narrow scientific judgment. Further, it had overtly philosophical, religious, and political elements as the interests that prefigured and prestructured the debate and the judgments made within it. There was also a politics to the controversy involving the position and the place of participants in the institutional and professional structures of the time. And finally, here, as in so many other debates, even though an element of closure was achieved, no clear, cognitive resolution was gained among the contestants themselves.

Vivisection

Medical and scientific experimentation using living animals (commonly called vivisection) became the focus of a serious controversy in

[48] In a letter to *Pravda* following the August 1948 meeting of the Lenin Academy of Agronomic Sciences, where Lysenko was officially declared the "winner" after the direct intervention of the Communist Party of the Soviet Union.

Britain during the last quarter of the nineteenth century.[49] Although it did not involve disagreement or dispute over theories or explanations in science, as had the two controversies discussed earlier, it went right to the core of the procedures of science and focused on the nature of the research process and experimental activity. In this it was parallel in many ways to the recent debates that have surrounded recombinant DNA experimentation.[50] Societal sentiments and public norms were explicit and focal elements in this controversy, and the battles themselves took place within public forums and public political structures. Parliament, the courts, voluntary associations, and the press became the site for many confrontations, as did the traditional scientific forums, the medical and scientific societies and their learned publications.

Since political and social structures were specifically involved, it is important to point out that Great Britain should be differentiated from continental Europe in the decades of the 1860s, 1870s, and 1880s. Whereas France, throughout the first part of this period, was influenced by the deeply conservative Second Empire, and Germany, still retreating from the revolutionary movements of 1848, was preparing for the era of Bismarck's leadership, Britain was undergoing a series of democratic reforms (the delayed reaction to the continental upheavals of 1848). By tradition, as well, Britain had developed a plurality of institutions rather than the strongly centralized structures of France or the authoritarian ones of Germany. In addition there were many areas of social regulation in Britain that were largely extragovernmental, and it was within this realm that the controversy over vivisection was most successfully pursued.

In the years just after the development of anesthesia, a group of British physiologists and others working in areas of experimental medicine took stock of the state of their professional discipline and realized that it had fallen behind the achievements that had been accomplished on the continent – in France by François Magendie and

[49] Much of the basic information on the vivisection controversy is drawn from Richard D. French, *Antivivisection and Medical Science in Victorian Society* (Princeton: Princeton University Press, 1975); the interpretive comments, however, are my own; see also James Turner, *Reckoning with the Beast: Animals, Pain, and Humanity in the Victorian Mind* (Baltimore: Johns Hopkins University Press, 1980).

[50] See my recent paper "Frankenstein at Harvard: The Public Politics of Recombinant DNA Research," in E. Mendelsohn, ed., *Transformation and Tradition in the Sciences* (Cambridge: Cambridge University Press, 1984), pp. 317–35. For an account of the debate within science, see Clifford Grobstein, *A Double Image of the Double Helix: The Recombinant DNA Debate* (San Francisco: Freeman, 1979).

Claude Bernard, and in Germany by Karl Ludwig and numerous other experimenters. The British, led by John Burdon-Sanderson and Michael Foster, both of whom had spent time working in continental laboratories, had a very direct and simple aim – to catch up with their counterparts in France and Germany and improve the opportunities for and quality of animal research in Britain.[51] Their interests were directly professional, and their desire to use living animals was fully consonant with the current practices in Europe.

On the other side of the nascent controversy were segments of the public who wanted to extend the efforts at animal protection from the marketplace and recreation field to the laboratory. They were part of the broad movement of humane reformism then widespread within Great Britain and represented the interests of the middle class wanting to see the enforcement of norms and mores that they considered part of the "proper" way of living.[52] Many of them had been active in the Royal Society for the Prevention of Cruelty to Animals (RSPCA) and its ancillary and rival organizations and had already developed experience in using these organizations as part of broader efforts at political reform and societal regulation.

The procedural focus of the debate was over the mode of regulation of experimentation. Should it be internal and left to the scientific community to oversee? Or should it involve formal and perhaps even legislative controls? This controversy within a controversy added an additional political element to the broader debate in that it brought out in some quarters fears of government intervention and in others distrust of experts and their activities. Even within the medical profession there was division; some practitioners were suspicious of laboratory research and expressed doubt as to the usefulness of medical experimentation to actual therapy. Vivisection, in the eyes of some medical people, represented an activist or interventionist approach in medicine as compared to a responsive or therapeutic mode. On the other side, however, the 1860s and 1870s were a period of intense activity in physiology and were the period of the establishment of a number of important new chairs of physiology and of laboratories for physiological experimentation. Physiologists them-

[51] See French, *Antivivisection*, Chap. 3. See also Gerald L. Geison, *Michael Foster and the Cambridge School of Physiology: The Scientific Enterprise in Late Victorian Society* (Princeton: Princeton University Press, 1978), and Brian Harrison, "Animals and the State in Nineteenth Century England," *English Historical Review* 88 (1973): 786–820.

[52] See Turner, *Reckoning with the Beast*; also Brian Harrison, "Religion and Recreation in Nineteenth-Century England," *Past and Present* 38 (1967): 98–125; A. W. Moss, *Valiant Crusade* (London: 1961), a history of the RSPCA.

selves, almost all trained in medicine, found their work at the boundary between medicine and the natural sciences. They were anxious to achieve clearer professional and intellectual identity. Some among the broader group of scientists, probably fearful of public reactions, proposed within the British Association for the Advancement of Science in 1871 a code for monitoring animal experimentation. They were undoubtedly responding to that organized segment of the public, acting through the RSPCA, who called for "the suppression of experiments upon animals made for discoveries in science when conducted with torture."[53]

The RSPCA, for its part, had achieved an interesting position within Britain. Drawing its constituency largely from the upper middle classes and having gained a strong following in Parliament, it had, since its founding in the early decades of the nineteenth century, come to be the unofficial keeper of public morals where animals were concerned. Because of legislation passed early in the century, the RSPCA had the right to prosecute those found to be mistreating animals. Most often these were workers driving horses or mistreating animals in the marketplace or setting animals up for cruel games such as cock fights. The role of the RSPCA was significant: during the decade 1830–9 they brought some 1,357 prosecutions, and by the time of the controversy over vivisection (1870–9) the figure had risen to over 23,000. They had moved from being at the margin of social influence to being very much at its core. They embodied upper middle-class social values, biases, and interests, and although they had grown conservative as an organization as their influence grew, they had maintained a clear position of moral hegemony where animals were concerned.

Richard D. French, in his excellent study *Antivivisection and Medical Science in Victorian Society* (1975), has identified the focal points of the debate as it came to a head in the early 1870s.[54] In 1872 a *Handbook of Physiological Experimentation* was published providing a description of the classic vivisection experiments that had been carried out on the continent. Rather than receiving attention only in specialist publications, it was attacked publicly in the press and referred to as representing a "moral contagion" from French and German sources. But an even stronger outcry was precipitated by a defense of the volume that appeared in a letter to the *Times* by E. Ray Lankester, one

[53] French, *Antivivisection*, p. 46. For interesting comments on the role of the RSPCA, see Harrison, "Religion and Recreation." He provides interesting details on their role in enforcement.

[54] Page and chapter numbers given in parentheses in the following section refer to this source.

of Britain's leading "public" scientists and a student of T. H. Huxley. Lankester cast himself as an apostle of experimentation: "If you allow experiment at all, you must admit the more of it the better, since it is very certain that for many years to come the problems of physiology demanding experimental solution will increase in something like a geometrical ratio instead of decreasing" (p. 51). The image of experimentation on living animals increasing at this geometrical ratio, bringing back memories of the earlier Malthusian debates, brought a public clamor. Even several medical journals backed away from the notion, explicitly in fear of the public criticism that it was causing. The incident that precipitated the major moves in the controversy came, however, at a meeting of the British Medical Association in Norwich in August of 1874. A French experimenter on alcohol had been invited to demonstrate his research in which he induced epileptic fits in dogs by injecting them with absinthe. Just as he was about to proceed with his experiments, several members of his audience, including physicians, raised questions about their propriety. As if to give the experiments some legitimacy, the chairman of the meeting took a vote, and the majority voted that the demonstration should proceed. It was in response to these demonstrations that the RSPCA decided to interfere and prosecute those connected with the meeting. One of the rules for RSPCA prosecutions was that two physicians had to have witnessed the acts of cruelty to animals. Two Irish physicians who had been present at the meeting willingly stepped forward to testify against their fellow professionals. The daily press covered the story and indicated its lack of confidence in the conscience of the physiologists. The *British Medical Journal,* on the other hand, came forward with a hard-line defense and indeed went on to criticize those physicians who were willing to testify on behalf of the RSPCA; the editors of the journal were particularly incensed at the breaking of professional ranks.

At this stage an alliance was formed between those with a general interest in humane animal treatment, organized around the RSPCA, and those who proposed a ban or at least curtailment of vivisection experiments. The political power of the former group was added to the zeal of the latter. In the course of this move, responsibility for evaluating one type of scientific experimentation was transferred from the laboratory to the judiciary, and it became apparent that one form of closure in the vivisection debate could be a ruling by the courts. In the case of the alcohol experiments, the ruling was inconclusive, since the French experimenter had returned home and several technicalities had not been properly fulfilled. But even as the court had given a mixed ruling, so the medical community too have given a mixed response. Their establishment was split over whether or not to give full support to live animal experimentation.

There are two additional elements that were brought to the debate at this stage. A prominent establishment scientist, albeit an outsider to physiology, Richard Hutton, the mathematician and sometime theologian, joined the fray against vivisection. As a member of the senate of the University of London, he had tried to block vivisection experiments at Browne's Institute at the university. And in his outspoken letters and talks he gave indication of a fairly deep division that existed within the British scientific establishment at large.

When Frances Power Cobbe, the Anglo-Irish feminist, author, and philanthropist, took up the antivivisection cause, she attempted, by prodding the RSPCA to more vigorous prosecution, to form a more powerful advocacy group, and she advocated the passage in Parliament of antivivisection legislation. Since the physiologists, in her view, were motivated by self-interest, she wanted to remove decisions concerning vivisection from their influence and control. By contrast, at the very same time the editors of *Lancet*, the prominent medical journal, said that they wanted any procedures of regulation or control to be removed from public control and sheltered from what they identified as "the power of lay ignorance." Frances Cobbe launched a semipublic campaign, including petitions to notables, pressure on the RSPCA, and a general development of "agitational tactics" borrowed from the other prominent crusades of the day – the temperance, antislavery, and women's suffrage movements. In linking antivivisection to these other moral crusades, she led her organization into cooperation as well with antivaccination groups and those who had organized in opposition to the Contagious Diseases Act (see French, Chap. 8).

The element of distrust of the self-interest of the scientists became even more obvious in the links that the antivivisection movement had established with the feminist and suffrage efforts.[55] Women from the middle and upper classes formed approximately 40 percent to 60 percent of the antivivisection committees. At the time, these active women's groups consistently expressed a general distrust of medicine. They had sought social sanctions against physicians; they had suggested that women could do without medical attention from practitioners; they expressed resentment at the exclusion of women from the medical profession; and they claimed that women were mistreated at the hands of male doctors. Antivivisection, then,

[55] For a study of the class basis of antivivisection and other related movements, see Harrison, "Religion and Recreation," pp. 103–19.

became part of a much broader movement with strong implications of opposition to experts and distrust of their judgments. With this broadening of the political context for the debate, what were the implications for resolution or closure of the debate?

But at the very same time that the public organized its activities and found allies within a divided scientific community, a "scientist lobby" was being formed. Its membership, including Charles Darwin and T. H. Huxley, was made up largely of influential "insiders" in the science establishment of the day. Their moves were adroit, and they avoided head-on confrontation with the opponents of vivisection and instead opted for a policy that included elements of appeasement and a strategy of heading off any legislative ban on live animal experimentation. Even though this group of scientists was not itself directly connected with animal experimentation, its members were influenced in their actions by their perception of the interests of the professional community of scientists, interests that were particularly at variance with public or legislative interference in the conduct of scientific research (pp. 70–1). They took the step of drafting an alternate parliamentary act that they hoped would gain greater support than the one being brought forward under the aegis of the RSPCA and the antivivisectionist movement. The act itself was drafted by Charles Darwin's son-in-law and introduced into Parliament by one of the influential scientific members, the chemist Lyon Playfair (pp. 72–3). The scientists' legislation proposed licensing of experiments on living animals, supervision of this experimentation by government representatives, and accountability of the scientists to the licensing authority. By adopting this mode of insider politics and the move to legislative action, and by working with several of the influential members of the RSPCA, they were able to draw some of the fire of the opponents of vivisection and find an apparent compromise.

The controversy, as related by French, moved toward a type of closure when a Royal Commission was appointed to examine the legislative needs in the area and to provide a report to Parliament. The commission in this case (unlike the French body) was quite balanced in its composition and included several figures who had questioned vivisection. They conducted hearings that brought forward almost all points of view and indicated the very real division that existed within the scientific and medical communities, as well as the strong opposition in lay groups. Their final report was cautious, and perhaps because of this was quite widely accepted. It led in 1876 to the Cruelty to Animals Act. In the course of its passage, members of Parliament were lobbied by groups from both sides of the debate.

Moderate politicians sensed that some sort of legislation was necessary in order to shield the government from further lobbying, and particularly from the vigorous criticisms of the antivivisection movement and its allies. It was apparent in the course of the moves toward legislation that nonbiologists did not share a common outlook with biologists in this area. It was also noticeable, however, that the physiological experimenters and medical scientists, who in this case were very anxious to establish their new field (experimental physiology) themselves had very little direct political strength. The strength that did exist came in part from the medical establishment, which had a clear set of interests of its own that were only partially coincidental with that of the experimenters; its members were willing to see some legislative control of animal experimentation. The bill, as it finally passed, represented the consensus among "insiders" in the scientific community who believed that a bill was needed to protect experimental medicine from harassment by the antivivisection movement (and only secondarily to protect animals used in experimentation). Ultimately the insiders found themselves in common cause with moderates in the antivivisection movement who were willing to have experiments continue under supervision and license rather than have them totally banned. But the scientists engaged in what might well be called "prudential acquiescence."

A modified form of closure was achieved with the passage of the legislation. Vivisection was to be regulated and licensed under the supervision of the home secretary, who, in turn, was given broad discretion. But the regulation and licensing was swallowed up by government agencies, and the home secretary, utilizing his discretionary powers, appointed an advisory group of experts to oversee the procedures of licensing and regulation. All members of this group, it turned out, were scientists. In principle, lay power was gained over scientific experimentation. In fact, the scientists gained the "shield" that they needed. Vivisection actually increased dramatically after the legislation was passed, and one of the paradoxes of the new act was that this increase was carefully recorded. From 311 licensed live animal experiments in 1880, French's data show, the total rapidly rose to 95,000 in 1910 and over 950,000 in 1938 (pp. 72–3, and see Fig. 16, p. 394).

In turn, the antivivisection groups themselves did not give up or dissolve, but continued their opposition. But now the "center ground" had been cut away from them, and the moderates concerned with the protection against cruelty to animals had been won over by the legislation that was passed. The fundamental conflict in interests and outlook on the part of the critics of vivisection and the proponents was, of course, not resolved, nor was agreement achieved. The controversy itself had clearly involved the mixed interests of a professional scientific community, of public sentiment and humane considerations, and of direct political action and agitation in the context of a broader movement toward social reform. Although the subject was the conduct of scientific experimentation the debate was ultimately a political debate. In the end it gained a "political" resolution in which the activity itself was removed from the explicit terrain of the experts to the forums of public accountability, the courts, and finally the legislature and administrative authorities.

Conclusion

The cases that I have examined support my introductory claims that controversy in the sciences is best understood in terms of both the intellectual content of the discussion and the context of social interests that guided the participants. These interests ranged from the narrowly professional through the philosophical, metaphysical, and religious to the broader societal. Further, these interests are not accidental intrusions into an otherwise ordered rationality of science but are natural elements of scientific debate.

The nature of closure or resolution, when it is achieved, comes very much within the context of the conflicting interests. Indeed, the forms of closure and resolution are often themselves determined or influenced by the interests actually involved in the conflict. The failure of a full resolution should not cause surprise. That the actual parties to debates and controversies seldom shift their basic views or commitments, even when a formal closure might be achieved, is compatible with the type of commitment that has raised a disagreement into a controversy. But even without a formal means of ending a debate, the patterns of institutionalization in science allow a continuation of new scientific activities. New generations, new problems, and new professional contexts in time supersede an old debate. It also should be pointed out that failure to take part in a resolution need not be linked – indeed, should not be linked – to poor science or incompetent scientists. Recall that Albert Einstein never altered his basic belief in the deterministic character of the physical universe, despite the support for its indeterminacy developed in quantum mechanics. "God doesn't throw dice with the universe," was his comment on the matter.

I think it has become clear enough that the cases examined are not special or unique ones but rather contain elements found broadly in other debates and controversies. The developing research on numerous other incidents of this sort in history show similar results, especially when historians cast their own research broadly enough to allow elements of interest to be noticed. For example, several recent studies of the development of Mendelian genetics and the opposition to it at the turn of the century by the biometricians Weldon and Pearson bring out these points with some clarity.[56] Weldon and Pearson never fully accepted Mendelism in their own lifetimes, even though the new hereditarian science was making an impressive record for itself on the experimental level and was recruiting important adherence within the scientific community. Further, it can be noted that the biometricians went outside biology for support, with those supporting them being more interested in the extracognitive elements of the debate than in the details themselves. The biometricians were involved in an internal professional dispute and were engaged in a struggle to control the Royal Society's commission on evolution and through it the "formal" discussions of what elements and factors would be used to explain the evolutionary process. They were engaged in a vigorous defense of orthodox Darwinism against the Mendelian attempts to supplant it with a hereditarian explanation of evolution. But here the social interests commanding the debate for Pearson and the other biometricians becomes clearer. Darwinism was widely used in ideological causes during the latter decades of the nineteenth century and in political debates at the time in which Pearson was involved.[57] But perhaps most important for the controversy over the statistical bases of evolution and heredity is the fact that biometrics was seen by its developers as providing the rational basis for eugenic intervention. Eugenics, a fairly widespread movement in the latter part of the nineteenth and early part of the twentieth centuries, was a fascinating mixture of biology and social doctrine, and recent studies of it have indicated the extent to which it embodied the social interests of a specific sector of British society.[58]

But for Pearson, orthodox Darwinism, which he had built biometrics to defend, served in addition as the basis for his theory of social change and indeed legitimated his social and political views: "No great social chance ever occurs with a leap; no great social

[56] See Mackenzie and Barnes, "Scientific Judgement"; A. G. Cook, "William Bateson, Mendelism and Biometry," *Journal of the History of Biology* 6 (1973): 1–36.

[57] See Robert M. Young, "Evolutionary Biology and Ideology: Them and Now," *Science Studies* 1 (1971): 177–206.

[58] Lyndsay Farrall, "The Origins and Growth of the English Eugenics Movement, 1865–1925," Ph.D. diss., Indiana University, 1970; also Mackenzie, *Statistics in Britain, 1865–1930*, esp. Chap. 2, "Eugenics in Britain," pp. 15–50.

reconstruction which will permanently benefit any class of the community is ever brought about by a revolution. It is the result of a gradual growth, a progressive change, what we term evolution. This is as much a law of history as of nature."[59] For Pearson, then, the defense of biometrics was nothing short of a defense of his fundamental social philosophy and went far beyond the narrower intellectual efforts to defend new statistical techniques or the esoteric social interests involved in the building of a new discipline.

Similarly, in the debate between August Weismann and Herbert Spencer over the inheritance of acquired characters (a debate in which the intellectual details have been well outlined by Frederick B. Churchill), it is clear that we cannot understand their argument without becoming aware of the social interests that were a part of the positions they advanced.[60] Herbert Spencer's unwillingness to join the growing consensus against the Lamarckian concept of the inheritance of acquired or use characteristics can only be understood when we see that he needed the mechanism of the inheritance of use characteristics for his whole system to work. His biographer, J. D. Y. Peel, makes the point bluntly: "Spencer's Lamarckism is indispensable to his concept of race."[61] Had he abandoned use inheritance, Peel continues, "it would have driven a wedge between biological and social cultural or superorganic evolution, and so nullify the major premise of the entire synthetic philosophy."[62] For Spencer, then, the inheritance of acquired characteristics was a linchpin in his system of social evolution and also provided the idea of an evolutionary process connecting the physical world to the biological and to the social worlds. Use inheritance is exactly what happened in culture, in his view, and what he ended up proposing was a "culturized" or "historicized" biology. Spencer also needed the concept for his view of sociology itself, the new science of society that he was so important in helping build. In his view, the improvement of the race depended on people's own efforts at improvement. These efforts, in turn, had to become organic within the race, something that could happen only through inheritance of acquired characteristics. And this view of self-improvement was the cornerstone for one of his key political

59 Quoted from Pearson, *The Ethic of Freethought* (1888, p. 363), in Mackenzie and Barnes, "Scientific Judgement," p. 204.

60 Frederick B. Churchill, "The Weismann–Spencer Controversy over the Inheritance of Acquired Characters," in E. G. Forbes, ed., *Human Implications of Scientific Advance: Proceedings of the Fifteenth International Congress of the History of Science* (Edinburgh: Edinburgh University Press, 1978), pp. 451–68.

61 J. D. Y. Peel, *Herbert Spencer: The Evolution of a Sociologist* (New York: Basic Books, 1971), p. 143.

62 Ibid., p. 143.

ideas – the opposition to the welfare state. For if self-improvement were to be discouraged, as he believed it would be in a welfare state, then the human race would have no chance to inherit the improving mechanisms that his notion of acquired characteristics would allow it to do. Thus, in the debate in the 1880s between the believers in the inheritance of acquired characteristics and the doubters, there was choice, and for Spencer it was clear; he chose a biology useful to his broader social visions and commitments.

We could multiply examples of this sort, but already we have looked at biological and statistical materials. The distinctions so carefully cherished within the history of science between "internal" and "external" fade into the background, and a much subtler picture of relationships between intellectual elements and social interests emerges. A greater understanding of issues of controversy and closure is gained through a recognition of the profoundly social nature of those activities that we call science.

4

Controversies involving science and technology: a theoretical perspective

RONALD N. GIERE

Why study controversy?

Widespread public controversies involving science and technology have been a prominent feature of American public life since World War II. At first these controversies were concerned mainly with military technology: fallout, the test ban treaty, the antiballistic missile. More recently, controversies have developed over civilian technology (nuclear power, the supersonic transport) and even over the conduct of scientific research itself (race and IQ, molecular biology). These controversies have contributed substantially to public awareness of the increasing role of science and technology in modern societies and have quite naturally become the subject of scholarly research. Indeed, case studies of particular controversies provide an important part of the literature of the new academic field of "Science, Technology and Society" (or "Science and Technology Studies").[1] After a decade of such research, it is important to reconsider the question. Why study controversy?

One motivation for such case studies is fairly clear. It is obvious that science and technology interact with other components of society in many important ways, but the nature of these interactions is

I would like to thank the Hastings Center and the directors of the Closure Project – Dan Callahan, Art Caplan, and Tris Engelhardt – for providing me the opportunity to participate in this project and thereby to clarify my own thinking on issues with which I have been less systematically concerned for many years. It would be impossible for me to document the influences that led to this paper's achieving its final form, so I will simply thank all my fellow participants for the many hours of intellectual stimulation that they provided. A few specific attributions will be recorded in later footnotes. The author would like to acknowledge the support of the National Science Foundation.

[1] For a good overview of this field, see Ina Spiegel-Rosing and Derek de Solla Price, eds., *Science, Technology and Society: A Cross-disciplinary Perspective* (Beverly Hills, Calif.: Sage, 1977).

elusive. However, the points of interaction tend to be obvious in the context of specific controversies. Moreover, the record is mostly public and easily accessible. Major controversies, then, can be regarded as a microcosm in which the interactions between science or technology and the rest of society can conveniently be observed and studied.[2]

Although the literature is dominated by case studies, there have recently been several attempts to move beyond the particular cases toward more general accounts of the nature and dynamics of controversies involving science and technology. This work utilizes the concepts and methods of the social sciences, particularly sociology and political science.[3] It attempts, for example, to discover general patterns in the behavior of various types of participants in controversies, particularly scientists. Hypotheses as to why and how controversies spread – from local to national to international level, for example – are discussed. Nevertheless, even this more theoretical work shares with case studies a solidly descriptive orientation. It thus largely forgoes any attempt at evaluation.[4] Yet one of the main reasons for interest in these controversies is the desire for evaluation, for normative conclusions.

The distinction between descriptive and normative concerns can be developed by looking briefly at the notion of controversy itself. According to standard dictionaries, a controversy is a protracted dispute among individuals or groups over some issue of mutual concern. This definition is useful because it locates controversies within the broad category of social phenomena. It implies that the notion of a controversy cannot be reduced to a merely logical category – the existence of contrary propositions. Nor can it be reduced to a

[2] I think this paragraph represents fairly a widespread attitude, exemplified, for example, by Dorothy Nelkin and her coworkers in the Cornell Program for Science, Technology and Society. A number of these case studies have been summarized in Dorothy Nelkin, ed., *Controversy: Politics of Technological Decisions* (Beverly Hills, Calif.: Sage, 1979).

[3] I am thinking particularly of Robert Rich and Allan Mazur, both members of the Closure Group. For Rich, see Chap. 5, this volume. In addition to his written contributions to the Closure Project (see Chap. 9, this volume), Mazur's work is available in his book entitled *The Dynamics of Technical Controversy* (Washington, D.C.: Communications Press, 1981). This book, unfortunately, was completed only at the end of the project, and I received my copy after I had completed the first draft of this paper.

[4] As far as Mazur is concerned, this sentence overstates the case. Mazur is personally and to some extent professionally concerned with normative questions. See, for example, Chap. 3 ("Separating Disputes over Facts from Disputes over Values") and Chap. 9 ("Controlling Technology") of his *Dynamics of Technical Controversy*.

mere psychological or even cognitive category – the existence of contrary beliefs.[5] For there to be a controversy there must be a substantial amount of social interaction – an active dispute. It is thus entirely appropriate that controversies should be studied by social scientists.

Being a social phenomenon, any controversy has a history. There is a period of time before any dispute over the issue in question takes place, followed by a period during which the controversy begins. Then comes a period of active development; eventually, the controversy ends and there is no longer a dispute. Since there seems to be more interest in endings than in beginnings or middles, let us focus on endings.[6] The investigation of how a particular controversy ended or even of how types of controversies typically end may be regarded as primarily a descriptive enterprise. It is just a fact that some controversies end in particular ways. But one may also investigate the ending of a controversy from a normative or evaluative standpoint. Was the controversy resolved correctly (or fairly, or in accordance with appropriate procedures)? Since the word *resolve* is natural in such contexts, I will use it when some evaluation is implied, and restrict talk of "endings" to purely descriptive contexts.[7]

As long as one is concerned only to develop a descriptive account of controversies, there is no fundamental difference between controversies involving science or technology and many other types of social disputes. The main sociological difference is that in scientific

[5] In papers prepared for the Closure Group, Ernan McMullin defined controversies as disputes over matters of belief (see Chap. 2, this volume). This may be true of controversies over scientific issues, but it is not true of controversies in general or even of controversies "involving" science and technology. Some of the most notable controversies are clearly over matters of action or policy – for example, the siting of a power plant or the setting of standards for exposure to toxic wastes.

[6] I cannot pass over Garland Allen's question of why there is more interest in the end of controversies than in their beginnings or development (see Chap. 6, this volume). Could it be the "interests" served by controversy? Many studies indicate that the existence of controversy generally works against the established powers and in favor of "dissidents." Could the focus on how to end controversies reveal an establishment bias? Would a project that focused on the creation and development of controversies ever have been funded?

[7] It seems that the term *closure* was used in the original description of the "Closure Project" simply to avoid conjuring up memories of "conflict resolution," a subject that flourished briefly a generation ago. But this apparently harmless linguistic dodge ended up causing more confusion than it was worth as members of the project struggled to give meaning to a term that had no established currency in this area. It was never agreed, for example, whether or in what circumstance "closure" should carry normative as well as descriptive force. In the end, I found it preferable simply to abandon the term altogether.

controversies members of a particular social group, scientists, play a prominent role. That some of the issues in question are scientific issues may be of minor importance. But if one is to speak normatively of the resolution of a controversy, the nature of the issue is crucial. There are very few, if any, issue-neutral criteria for evaluating the resolution of a dispute. Whether a given resolution is correct (or fair, or whatever) depends on the kind of issue in question.

All students of controversies involving science and technology agree that these controversies incorporate several different types of issues. Moreover, there is near unanimity on the types of issues involved: scientific issues, ethical or value issues, and political or public policy issues. Indeed, one of the reasons for engaging in case studies is to see how these different types of issues interact in controversies involving science or technology. Normative concerns, however, may best be served by considering separately the type of resolution appropriate to each sort of issue and then examining criteria for the proper resolution of the more complex controversies.[8] Of course there are also controversies over how to properly resolve each of the three types of issues, and I could not attempt to resolve these (meta) controversies here. I can, however, point out the implications of the major (meta) viewpoints for the more complex sorts of controversies in question. And I can at least suggest the most promising way to go.

Lest anyone misunderstand or be put off by the distinction between descriptive and normative concerns, I should point out that these two types of concerns are not incompatible and may even be complementary. A good descriptive account is necessary if one is ever to apply normative categories to real cases in a useful manner. Conversely, keeping in mind the possibility of applying normative categories may suggest richer and more adequate descriptive categories. Both types of interaction will be illustrated as we proceed.

Scientific issues

Throughout the first half of the twentieth century, the dominant view of science in the English-speaking world was a strongly positivistic form of empiricism. This view was not confined to a few philosophers but pervaded the sciences, from physics to sociology. Within the positivist framework, the separation of scientific issues from all others

[8] Here I am following the original strategy of the Closure Project, which began with presentations by Ernan McMullin on scientific controversies (see Chap. 2, this volume) and Tom Beauchamp on ethical controversies (Chap. 1, this volume). Robert Rich's attempt to do the same for public policy issues occurred at the end of the project (see Chap. 5, this volume).

was ensured. Indeed, one of the primary motivations of positivist philosophers was to distinguish science from nonscience, particularly the intellectual irrationalism of romantic metaphysics and the sociopolitical irrationalism of World War I. To put it crudely, in the positivist view science deals with verifiable propositions formulated in terms of "operationally defined" concepts. Everything else is, strictly speaking, "nonsense."

Since about 1960, the positivist view of science has come increasingly under attack. The revolt was led by philosophers of science who appealed to the history of science to argue that the positivist picture simply did not fit science as it has actually been practiced. The antipositivist movement was joined by both intellectual and social historians of science and, most recently, by sociologists of science. To some extent the attack has been misdirected, because positivists were not attempting to provide a description of science as it is actually practiced. They wanted to show that the conclusions of science could be rationally justified. For this purpose, only a schematic reconstruction of some aspects of science seemed to be required. In particular, the social organization of science was generally ignored. On the other hand, if the reconstruction on which the claim of rational justification is based is too dissimilar from science as we know it, the claim to have justified "science" is itself questionable.

Those who now reject positivism either ignore the problem of providing a "rational justification" for scientific conclusions or else explicitly reject the possibility of there being any such thing. The extreme view is that the validity of scientific claims relative to the norms of the scientific community is in principle no different from the validity of any other type of claim relative to a corresponding community (e.g., the metaphysical claims of a religious community). Each community has its own norms for validating claims (e.g., experimentation or textual exegesis), and there is no way of justifying the claims of one community as "more valid" than any others.[9]

Each of these extreme views of science has clear implications for both a normative and a descriptive approach to controversies involving science and technology. If one adopts the extreme antipositivist position, then the claims of scientists carry no normative force outside the scientific community. In the evolutionist–creationist controversy, for example, one cannot say that the evolutionists are right

[9] Here I am thinking particularly of the "Edinburgh School" of sociologists of science, represented, e.g., by Barry Barnes's *Scientific Knowledge and Sociological Theory* (London: Routledge & Kegan Paul, 1974). Among members of the Closure Group, Everett Mendelsohn probably comes closest to the extreme antipositivist viewpoint, though just how close I am not really sure.

and the creationists mistaken. The controversy cannot be resolved on a rational basis, because the two communities have different standards for assessing evidence. The only differences between these two communities that could legitimately be resolved are those contained in proposals for public policy – for example, the teaching of evolution in public schools. Here the relevant norms are those of the political system and questions about their legitimacy fall to judges and, eventually, to legal scholars, political theorists, and political philosophers.

Implications for descriptive inquiries are no less evident. If one thinks that the scientists have reason on their side, then one must seek explanations of why otherwise reasonable and often scientifically trained individuals are so adamant in their defense of creationism. If, on the other hand, scientific claims have no special status outside the scientific community, no such explanations are necessary. Thus, even case studies, which are a paradigm of "purely" descriptive investigation, may contain substantial normative presuppositions.[10]

Although I reject extreme forms of positivism, I would argue for the superiority of scientific procedures in deciding questions about the empirical world. It would be very difficult to state minimally necessary and sufficient conditions for according this status to scientific procedures, but sufficient conditions are relatively easy to state. One set, which seems to accord well with the views of scientists, is the following:

1. Scientific claims are either true or false, in the sense that they either do or do not correctly represent some aspect of the real world.
2. The methods (or procedures) of science permit the truth or falsity of many (though not necessarily all) scientific claims to be reliably (but not infallibly) determined. Scientific claims to knowledge are justified by scientific procedures.
3. The reliability with which claims are made can often itself be reliably determined.

These conditions are jointly strong enough to give scientific claims the desired normative clout. Whether they are too strong is another matter.[11]

[10] In her study of the evolutionist – creationist controversy, Dorothy Nelkin clearly assumes that the biologists are basically correct, an assumption that shows up in the way in which she structures her inquiry. See *Science Textbook Controversies and the Politics of Equal Time* (Cambridge, Mass.: MIT Press, 1977). A shorter discussion of the controversy appears in *Controversy: Politics of Technical Decisions*.

[11] In his initial discussion of ethical and moral controversies (Chap. 1, this volume), Tom Beauchamp introduced the notion of "sound argument closure."

It should be noted that these conditions refer to the claims and procedures of science, not merely to claims made by scientists. The difference is crucial, particularly when considering claims made by scientists operating outside the scientific community, as is the case in typical controversies involving science and technology. The normative force of scientific claims rests not on the personal virtue of scientists but on the procedures that define the scientific enterprise.

There are different kinds of scientific claims, and some of these differences are relevant here.[12]

1. *Factual claims.* In 1965–6 geophysicists discovered regions of alternating magnetism in the sea floor on both sides of the Pacific–Antarctic Ridge. Moreover, these regions formed symmetrical bands parallel to the ridge. These facts were not particularly easy to obtain, but no responsible geophysicist seriously doubts the truth of the claim that these regions do exist in the sea floor of the southern Pacific Ocean.[13] This is a good example of a scientifically significant factual claim. In general, a true factual claim correctly describes some existing state of affairs in the actual world.

2. *Theoretical hypotheses.* The geophysical facts just described were not discovered randomly. Indeed, their existence had been predicted three years earlier by Fred Vine and Drummond Matthews, and the

This occurs whenever a controversy is resolved by exhibiting a sound argument leading to the correct conclusion. It was Beauchamp's contention that few moral controversies are resolved in this manner. However, a controversy developed within the Closure Group over whether sound argument closure occurs in science. Here I am in agreement with Engelhardt, Feinberg, Mazur, McMullin, and others that it does and, indeed, is the standard form of resolution for scientific issues. I have avoided using Beauchamp's categories, however, because they seem to me to obscure more than they clarify and to inhibit communication with nonphilosophers who are not accustomed to operating in the "formal mode." I would be happier saying that scientific issues are resolved by "procedural closure," which was another of Beauchamp's closure categories. It is the characteristics of scientific procedures – namely, that they are fairly reliable indicators of the truth – that provides the basis for the resolution of scientific issues. That the result of successfully applying these procedures can be summarized in the form of a sound argument seems to me relatively unimportant, though true. It is only in special philosophical and critical situations that one's attention is properly directed to the argument form rather than to the procedures themselves. Here it is merely diversionary. Most of this controversy could have been avoided if the term *closure* had never been introduced into the investigation.

12 The following distinctions roughly parallel those introduced by Ernan McMullin in his initial paper prepared for the Closure Group (see Chap. 2, this volume).

13 For the facts of this example I have relied entirely on Henry Frankel's presentation, "The Continental Drift Debate" (see Chap. 7, this volume). During the latter stages of the Closure Project, this case served as a paradigm of a scientific controversy that did not have a strong ethical or political overlay.

verification of this prediction was the strongest evidence to date for the hypothesis of sea floor spreading and the plate-tectonic model of continental drift. Since this discovery, the hypothesis of continental drift has been regarded as firmly established. According to the positivist conception of science, a theory is simply a conjunction of factual claims. But in the case just described, the theoretical hypothesis is more than just a summary of actual facts. It includes possible facts as well. The platetectonic model encompasses not only the actual structure and history of the earth but possible structures and histories as well. It tells us, for example, what would be found if the Pacific–Antarctic Ridge were fifty miles west of where it really is. Putting the point in its most general and provocative form, a theoretical hypothesis describes the causal structure of a kind of system.[14]

3. *Methodological claims*. Scientists make claims not only about the world but about the conduct of scientific inquiry itself. These may be fairly specific, as, for example, in the claim that prospective studies are better than retrospective studies for establishing causality. But the claims may be quite general, as, for example, in the claim that one should seek the explanation of biological phenomena in the chemical composition of biological organisms.

There is a rough correlation between the type of claim being made and the certainty with which its reliability can be determined. Reliability is greatest for factual claims, generally less for theoretical claims, and least for methodological claims. But contrary to most empiricist (including positivistic) theories of science, this correlation is not perfect. Our theoretical conclusions concerning the geological structure of the earth are far better established than our factual claims about the surface of Mars. Our theoretical knowledge of reactions in the core of a nuclear reactor is greater than our factual knowledge of what happened at Three Mile Island.

Contemporary scientists seem reluctant to acknowledge the existence of controversies within science over scientific issues. They would reserve the term *controversy* for related personal or emotional issues such as who deserves credit for a new discovery.[15] James

[14] A more extensive exposition of similar views on the nature of theoretical hypotheses and causal systems was presented to the Closure Group by Frederick Suppe.

[15] This point of view was expressed by Gerald Feinberg. I do not really know to what extent he was expressing the scientists' idealized perception of science or the actual practice as it would appear to an outside observer. However, I am quite sure that it is mistaken to describe every disagreement among scientists over a scientific issue as a "controversy."

Watson's publication of *The Double Helix* provides a good example. Biologists would say that prior to Watson and Crick's final investigations there was a "difference of opinion" about the structure of DNA, but not a controversy. The controversy was over the means by which they acquired some of their information and over the propriety of Watson's publishing so "personal" an account of the story. There was no controversy over whether they had uncovered the actual structure of DNA.

Returning to geology, the possibility of continental drift was seriously debated for fifty years between Alfred Wegener's publication of *The Origins of Continents and Oceans* and the verification of the Vine-Matthews prediction. It is probably stretching a point to refer to this protracted debate as a "controversy." But the point is worth stretching because it provides some insight into the nature of controversies in general. In particular, what should we say about the way in which this controversy was resolved? What one says clearly depends on one's theory of science. In line with the account sketched above, this controversy, which included factual, theoretical, and methodological issues, was finally resolved through the normal procedures of science, which eventually produced decisive evidence in favor of continental drift. No doubt there were important psychological and social forces at work that led some people to take one side or the other during the long course of the debate. And no doubt the availability of funding from various sources had much to do with who discovered what and when. But in the end these factors ceased to be relevant to the overall issue of whether or not continental drift occurs and by what mechanism. In short, the controversy was resolved through the use of rational (or "cognitive") procedures, which, in this case, succeeded in producing a decisive conclusion.[16]

This example also illustrates the importance of distinguishing between the resolution of a controversy through the operation of legitimate procedures and the ending of the controversy as a social phenomenon. In this case, as in most scientific cases, a few scientists failed to realize or refused to admit that the major issues had been resolved. Thus the controversy continued to exist as a social phenomenon, though on a much reduced scale. The "end" came later. It is a tribute to the relative objectivity of scientific procedures that

[16] I think that both Frankel and McMullin could agree with this statement. However, they are both willing to take a much broader view than I as to what counts as a "rational" or "cognitive" factor. Frankel, for example, follows Lauden in emphasizing the "problem-solving ability" of the drift hypothesis. I would argue that problem solving is not an end in itself but may be an important indicator of the truth.

holdouts are generally few in number and recognized as holdouts.

It may be objected that the recent revolution in geology is atypical. It makes an "evidential" or "cognitive" interpretation of scientific procedures look better than it is. As contrary cases, one might cite the dispute between the Mendelians and biometricians at the beginning of this century or the current debates between sociobiologists and their critics. In both of these cases, one can label the dispute a "controversy" without any stretching of the term. And in both cases, social factors are prominent.

The Mendelian–biometrician controversy concerned matters of theory and methodology, issues more difficult to resolve than mere factual disputes. And during the first two decades of the century, when the controversy was the strongest, there simply was not sufficient evidence to resolve the issue. When the resolution came, it was not through evidence but through a new unifying theory, the genetic theory of natural selection, which was later firmly established. But before the scientific issue was resolved, there was a "correct" scientific conclusion, although its advocates tended not to be heard. The correct conclusion was simply that the evidence was not decisive. One could not, with full justification, take either view.[17]

It is a fairly well-established fact that humans dislike uncertainty. And scientists are only human. Thus, even though the evidence is inconclusive, many scientists tend to favor one view over others. In many cases the favored view is that of one's teachers. But even when the socialization process is not strongly biased, scientists use their own best judgment as to which view is correct. The trouble is, of course, that individual judgment is subject to all sorts of influences that have little to do with the scientific issues. And when there is a clear connection between the scientific issue and a question of public policy, it is to be expected that some scientists' judgments as to which theory is correct will be biased by their understanding of which theory best supports the policy that they favor.[18] It is thus not at all surprising that there should be a strong correlation between opinion on the Mendelian–biometrician debate and opinion on matters of social policy such as eugenics.

[17] For an account of this controversy that emphasizes social and political factors, see particularly "Biometrician versus Mendelian," Chap. 6 in Donald A. MacKenzie's *Statistics in Britain, 1865–1930: The Social Construction of Scientific Knowledge* (Edinburgh: University of Edinburgh Press, 1981).

[18] Mazur has shown the correlation between scientists' opinions on scientific issues (e.g., cancer risk from low-level radiation) and their opinions on policy issues (e.g., nuclear power). See Chap. 9, this volume, and *Dynamics of Technical Controversy*, Chap. 2, "Disputes between Experts."

These comments may be applied to the case of sociobiology as well. They also apply to a wide variety of controversies involving science and technology. Students of such controversies have often remarked that one finds scientists on both sides of major controversies. Some have worried that this tends to undermine the authority of science among the general public. If it is true that science contains procedures for resolving scientific issues, why do we find scientists testifying on both sides of major controversies? Sometimes this is due to a confusion of issues, as when biologists or doctors discuss when human life begins. "Human" is not a biological but a moral category, so the issue is really not a scientific issue at all. But in many cases (e.g., in analyzing the potential for major accidents at a nuclear power plant), the issue is scientific, and still the scientists disagree. The reason is simply that the procedures of science have not yet yielded a clear resolution of the scientific issue. But individual scientists have opinions on the issue, and some are willing, even eager, to voice those opinions.[19] However, in the absence of a clear scientific resolution of the issue, individual scientific opinions are subject to all sorts of influences, both scientific and otherwise. Disagreement is all but inevitable.[20]

The preceding discussion has concentrated on scientists and scientific issues, to the exclusion of engineers and engineering issues. Most of what has been said, however, can be applied to engineering issues as well. The typical engineering problem is to design a system that meets given specifications. Whether a proposed design is successful or whether a system that has actually been produced meets the specifications are fundamentally scientific questions. Nor are such questions merely factual. They are more like theoretical issues in that one is concerned with the future behavior of the system in a variety of different circumstances, perhaps more than can actually be tested. In any case, there are well-established engineering procedures for resolving such issues.

[19] This is why it is so important to distinguish between the conclusions justified by scientific procedures and the opinions of scientists. If scientists were forced to refer more to actual experiments and theories, as in some "science court," rather than speaking as authorities giving conclusions, the resolution of disputed scientific issues would be much more likely, although the final conclusion might often be that the issue is really still uncertain. This is Mazur's position as well.

[20] It should be said that politicians and policymakers are at least as adverse to uncertainty as scientists. They do not want to be told that the answer to some important question is not available. It is much more difficult to formulate policy when the result of adopting various options is not known. Thus scientists in the public arena are likely to find themselves strongly encouraged to give a definite answer when none is justified and then find that answer taken as definitive.

Value issues

Everyone agrees that value issues play a major role in controversies involving science and technology. In the Laetrile debates, for example, a major issue is whether patients have the right to any medication they choose, even if it has been reliably established that the medication is ineffective or, worse yet, possibly harmful. Do citizens, on the contrary, have a right to be protected from those who would promote useless drugs for their own profit? Again, should people living near a nuclear power plant be compensated for the extra risk that they suffer against their will? Is it fair that some should suffer greater risk when all benefit equally from the generation of electricity?

The view taken concerning the nature of value issues also has important implications for both descriptive and normative accounts of these types of controversies. For example, one might think that value questions have "right" answers – that there are truths about values to be discovered. Or one might agree with various ethical theorists that there are other ways to justify some ethical views as uniquely rational. In either case one will take a very different view of controversies than if one thinks that it is impossible to find a correct answer to value questions.

At the moment the situation for the student of controversies seems to be this. Whether value issues have correct answers or not, there is no widespread agreement on what they are or on how to discover them. Thus, as a practical matter, one is effectively in the position of proceeding as though no right answers exist. This is a very difficult situation for one who wishes to construct an account of the resolution of purely ethical issues. But the vast majority of controversies under investigation do not pose value questions in a pure form. More typically, the value issues are embedded in a public policy issue. A good strategy would be to focus on the public policy aspects of the controversy and hope that the value aspects will be adequately dealt with in this context. That is the strategy that I will follow here.[21]

Public policy issues

Most of the celebrated controversies involving science and technology may be seen as controversies over issues of public policy. Often a

[21] Among members of the Closure Group, Gil Omenn was most forceful in urging the restriction of value questions to explicit public policy contexts (see Chap. 19, this volume). I believe that he based this recommendation on the practical ground that the narrower the policy question, the more likely it is to be resolved by standard policy procedures. I suspect that this view has a theoretical basis as well.

widespread controversy exhibits itself in the form of many small controversies focused on fairly local public policy issues. The nuclear power controversy, for example, has surfaced in the form of local controversies over the siting and licensing of particular nuclear plants. It has also been exhibited in many debates in all branches of the federal government – the administration, the Congress, and the courts. The future of the breeder reactor program, for example, now rests with the administration and the Congress. Similarly, the Laetrile controversy was fought out in numerous state legislatures, in the Food and Drug Administration, and in the courts.

In the United States it is possible to regard the whole governmental structure, at all levels, as containing a complex set of mechanisms for resolving controversies concerning issues of public policy. Broad controversies over the general direction of governmental policy are settled by presidential and other national elections. More specific controversies, such as those concerning taxes and federal spending, are resolved by votes in the Congress. Other controversies are settled by local governments or the courts, including the Supreme Court. Some controversies are resolved by regulatory agencies.[22]

In thinking about public policy controversies, it is essential to distinguish between specific resolutions of particular policy questions and the ending of the more general controversy. Particular policy issues may be resolved by established governmental procedures without thereby ending the controversy. The Food and Drug Administration resolved the Laetrile controversy by making it illegal to transport the drug across state lines. But that did not end the controversy. Proponents of Laetrile shifted their focus to state legislatures, which had the power to permit the manufacture and sale of Laetrile within their own states. Regulations issued by the Occupational Safety and Health Administration (OSHA) have been challenged in the courts all the way to the Supreme Court. It is both a virtue and a defect of the American system that it permits controversies to be shifted from one part of government to another. The virtue is that it gives those who are dissatisfied a place to go. The defect is that it allows controversies to be prolonged to the advantage of those with the means to pursue them.

Consideration of public policy issues reinforces the positions on scientific and value issues that I have just asserted. Governmental procedures resolve specific questions as to what should be done. This provides a resolution of value conflicts that may be focused on a

[22] Here I am adopting the general viewpoint of the Closure Group's political scientist, Robert Rich (see Chap. 5, this volume). On this view, dispute over public policy issues is the normal state of affairs, and the procedures of government provide the mechanisms for resolving normal disputes.

particular policy question. But governmental procedures do not attempt to resolve scientific questions as to what is true about the world. Thus, state legislatures did not vote on whether Laetrile is effective but only on whether to permit it to be manufactured, sold, and used. The courts did not rule that exposure to benzene at levels of one part per million is not harmful but only that OSHA had not provided specific evidence that exposure at such levels is harmful. The issue before the courts was not scientific but a question of policy, namely, Should permissible exposure levels for benzene be set according to what is feasible or according to what has been proved harmful? That is not a scientific issue; it is primarily a value issue. If it is scientifically uncertain whether levels below ten parts per million are harmful, should producers and consumers pay the cost necessary to protect workers against the possibility, or an unknown probability, of harm? Who should bear the burden of the uncertainty?[23]

[23] Whether it is always possible to separate scientific issues from policy issues is a question that was never really resolved within the Closure Group. I think it is possible, as in the examples cited. So does Mazur. One source of confusion on this point might be that sometimes policy questions are framed so that they sound like scientific questions. Policies regarding low-level exposure to radiation and other substances provide a good example. The dose–response relationship for most low-level exposures is not known, nor is it clear how to go about discovering the relationship. The range of possibilities, however, is known. There might be a threshold dose below which there is no response. Or there might be no threshold and thus no completely safe exposure level. And if there is no threshold, the dose–response curve might have different shapes – e.g., linear or perhaps stronger than linear. Moreover, for each of these possibilities one can make rough calculations of the consequences of that possibility being true, say in expected numbers of cancers or deaths, for any given exposure level. Even supposing that policymakers know what consequences are "acceptable," it may seem as if they still must decide which dose–response model is correct in order to set permissible exposure levels. This puts policymakers in the ridiculous position of having to resolve a scientific issue that cannot currently be resolved by scientific procedures. But that is not their job. Their task, one well within their authority, is to set permissible exposure levels. They may attempt to justify their decision on the permissible level by pointing out that the chosen level has acceptable consequences if the linear model is correct. They may even say that they have "based" their decision on the assumption that the linear model is correct. But the "assumption" of the linear model in this case cannot be one based on scientific evidence but on policy considerations. For example, it might be argued that this assumption protects workers from the possibility that there is no threshold. Opponents of the proposed regulation may reply that this increases costs to producers and consumers. These are both policy considerations – as they should be. The question of which possibilities to protect against, and at what cost to whom, is a policy question – not a scientific question. Thus even where the issues are so closely intertwined, as here, the separation of scientific and policy issues, and their modes of resolution, is

Finally, it has been said that Americans are obsessed with process in government and law. Indeed, some have claimed that we are in danger of stagnating in our own processes. Whether the latter is true or not, the fact that a society with democratic and egalitarian ideals should be especially concerned with governmental processes supports some of the positions suggested earlier. These are (1) that value issues are strongly exhibited in many policy issues, and (2) that there is no right answer to a value issue. In addition it need only be realized that no complex policy decision can even approximate perfect equity; some people will bear a disproportionate share of the costs while others enjoy a disproportionate share of the benefits. Thus, some people are going to be justly displeased with almost any resolution of a policy issue. But it is possible that almost everyone is satisfied that the process that led to the outcome was fair and proper. Maintaining agreement on the process when agreement on the outcome is impossible is very important to the functioning of a democratic society.

Types of controversies involving science and technology

It is doubtful that all of the controversies involving science and technology that have attracted scholarly attention form a homogeneous group. One would expect there to be some notable differences. Moreover, any differences would be especially important if one is concerned with the normative question of how to properly resolve such controversies. Appropriate procedures for resolution must depend on the type of controversy at issue. Yet very little attention seems to have been given to going beyond the obvious fact that these controversies combine scientific, ethical, and public policy questions.

A recent anthology of case study summaries, for example, groups the cases under the following headings: (1) "Efficiency *vs.* Equity," (2) "Benefits *vs.* Risks," (3) "Regulation *vs.* Freedom of Choice," and (4) "Science *vs.* Traditional Values." Yet the editor is quick to admit that all of the cases "involve multiple issues." Thus, at the very least, the categories are not exclusive, and no reasons are given for regarding these categories as important or in any way fundamental.[24] Nor are the more theoretically oriented studies in that volume more helpful.

possible. For an example of a possible victim of this kind of confusion, see the much-cited article "Science and Trans-science," *Minerva* 10 (1972): 209–22, by Alvin M. Weinberg.

[24] Dorothy Nelkin, "Science, Technology, and Political Conflict: Analyzing the Issues," in *Controversy: Politics of Technical Decisions.*

The circumstance of a controversy's being predominantly local rather than national in scope, for example, seems hardly a distinction on which to hang substantial normative claims regarding proper procedures for resolution.

Approaching the problem from a more normative perspective, it is tempting to regard purely scientific controversies and purely public policy controversies as two ends of a continuum involving both types of issues. But if we adopt the position that scientific questions and public policy questions are fundamentally different in kind, then no such continuum exists. The "mixed" controversies must resolve into one of the other two or belong to a distinct third category.

In spite of having varying degrees of familiarity with over a dozen "mixed" controversies, I do not feel at all competent to propose a comprehensive typography of such disputes. I will therefore confine my remarks to just two types of controversies. They are at least logically distinct, and together they encompass a large number of the paradigm cases. I will assume without further argument that scientific issues, value issues, and public policy issues are distinct and can, in principle, be separated.

Controversies over science and technology policy

Some controversies "involving" science and technology center directly on the activities of the scientific and technological community rather than just on issues in which scientific and technical knowledge is relevant. One such controversy is whether research priorities should be determined solely by scientists themselves or whether some direction from outside the disciplines is appropriate. At issue are questions about the level of government funding and how funds should be dispersed. The paradigm of such a debate is the recombinant DNA controversy. Of course many types of issues are raised, directly or indirectly, by research on recombinant DNA. Here the concern is solely with the issue of whether there should be any regulations on how such research is carried out and, if so, what the regulations should be and who should administer them.[25]

The view that the truth of scientific claims is to be decided by scientific procedures does not imply that science is or should be a totally autonomous activity. Even if they are directed primarily at the discovery of truth, the activities of scientists take place within a larger social context. These activities are thus subject to the same kinds of social constraints as the activities of any other group. One of the constraints on any group is not to impose a risk of harm to individuals

[25] See "Limits of Scientific Inquiry," *Daedalus* (Spring, 1978).

or groups unless the imposition of such risk is permitted by law (or general custom). Thus, if the actual conduct of research on recombinant DNA does impose a risk of injury to members of the society (including the researchers themselves), it is quite proper that social procedures should be invoked to protect society from these risks. In any case, the question of whether the risks to the general population are worth taking is a matter of public policy, to be decided by appropriate procedures for making such policy. Scientists have no more right than any other group to claim that they alone should set such policies.

This controversy is complicated by the fact that the people in the best position to assess the nature of the potential risks are those who are engaged in the research and have a strong interest in continuing to do research. If there are considerable uncertainties about the possible risks, then there will be a strong temptation to substitute the informal judgments of individual scientists for an objective assessment of the uncertainties involved. And, as noted above, informal judgment, however well informed, is subject to personal bias. Ideally one would have requested that the scientists provide well-justified scientific conclusions about the nature of the possible dangers, the range of uncertainty regarding the realization of these dangers, and about the nature and uncertainties regarding possible benefits of the research. Then the public could have struggled with the policy question of whether restricting the pace of research in this area is a reasonable price to pay for greater protection from the unknown but possible dangers. Only a very careful study of this particular case could reveal how far from the ideal the actual history was. In fact it seems that the voluntary moratorium lasted long enough for the development of new evidence that pretty well ruled out the most feared sorts of possibilities.

In terms of the present analysis, one of the major lessons to be learned from the recombinant DNA controversy is this: Although scientific procedures are the sole arbiters of what we should regard as true of the world, the search for new scientific truths is only one among many values that must be honored in the society at large. The pursuit of scientific knowledge is not an ultimate or intrinsic value. Thus, when this particular value comes into conflict with other social values, deciding which values take precedence is a matter of social policy, to be determined by established social procedures.

Technological decisions

A number of well-known controversies involving science and technology focus on what may be called "technological decisions." As the term will be used here, a technological decision is a decision to develop or employ a specified technology in a given context for a stated purpose. In principle, a technological decision could be made by an individual, group, corporation, or governmental body. The present discussion will be restricted to decisions by public bodies. A decision by a public utility to build a nuclear power plant, for example, is a technological decision. So was the decision by the federal government to begin the development of commercial nuclear power. The future of the breeder reactor program rests on a technological decision. (Decisions not to develop or employ some technology – e.g., the supersonic transport – are also technological decisions.)

Since many controversies of interest focus on technological decisions, any general understanding of such controversies requires some further characterization of such decisions. How we characterize them will necessarily influence our view of what is at issue in these controversies. The literature on controversies surrounding technological decisions includes many references to, but little in the way of a general characterization of, technical decisions. The implicit view seems to be that a technical decision is either a decision made primarily (or perhaps solely) on the basis of technical information, or a decision made by experts, or both. On this view, a technological decision is understood in terms of some characteristics of a process that has as its termination a decision to do one thing or another. That is, it is a process in which a decision is made by experts using technical data.[26]

There is another way of characterizing decisions that abstracts from the process of decision making and focuses instead on the structure of the decision to be made. This is the approach taken in standard treatises on "decision theory." There is no logical incompatibility between these two models of decision making. Indeed, they may complement one another for both descriptive and normative purposes. In describing a particular decision, for example, one may ask how and when the decision was made and which individuals or groups influenced the final choice. But one may also inquire about various features of the structure of the decision, such as the set of alternative choices considered. One might even go on to inquire why a decision was structured in a particular way. Why, for example, were some possible alternatives not considered?

The differences between the two models in normative contexts are even more striking. Probably the major normative question to be

[26] For an example of a recent study in which the process model of decisions is presupposed, see Dorothy Nelkin, *Technological Decisions and Democracy* (Beverly Hills, Calif.: Sage, 1977).

asked of any particular decision-making process is whether it was fair. Did everyone who has a stake in the result have an appropriate opportunity to influence the final decision, or was the decision-making process biased in favor of one group or one set of interests? In general, a process model allows one to ask whether the decision was "made correctly" but not whether the decision itself was "correct." That is something that, in terms of this model, could only be judged retrospectively, if at all.

The structural model, by contrast, was designed primarily to answer the question of which choice is correct. Here "correct" means something like "rationally justified relative to all the information available when the decision was made." Thus it could turn out that a decision reached by a fair and legitimate procedure is not rationally the best decision. And conversely, a corrupt process might yield a rationally correct decision. One might hope that good decision-making processes tend to yield rational decisions, but unless the decision-making process has been designed in light of criteria for rational decisions, that may be just a hope.

Since most of the existing literature on controversies over technological decisions focuses on the decision-making process, some further discussion of the structural model is appropriate here. This model does have some important implications for our understanding of such controversies. The structure of many decisions is simple in outline, though perhaps very complex in detail.

For there to be any need for a decision, there must be more than one course of action open to the decision makers. If there were no other options, there would be no decision to be made. The alternative to building a nuclear plant, for example, might be to build a coal-fired plant at the same site or not to build on that site at all. The structure of the decision to be made includes all of the options.

Decisions are not made in a vacuum, but in a world in which many other things happen, including some things caused by whatever decision is made. The price of oil may increase no matter what kind of power plant one builds; a nuclear plant may release large amounts of radiation into the environment; a coal plant may cause acid rain. These "possible states of the world" are also part of the structure of the decision to be made.

The set of options, together with the possible states of the world, constitute what decision theorists call the "possible outcomes" of the decision. In general, a possible outcome is the combination of having chosen a particular option in one of the possible states of the world. To have built a nuclear plant in a world where oil prices are rising is thus a possible outcome of a utility company decision.

Values are represented in the structure of the decision to be made in the form of a value ranking (or value measure) of the set of possible outcomes of the decision. It is neither options nor states of the world but outcomes that are the objects of our value judgments.

Finally, our scientific knowledge relates the options to the possible states of the world and thus to the outcomes associated with each option. Sometimes our knowledge is so meager that all we really know about some outcomes is that they are possible. A better situation is to have reliable knowledge of at least the probabilities of the possible states of the world relative to the different options. Best of all, of course, would be to know for each option just which outcome would occur. In this case, the "rational" choice is obviously that option known to lead to the highest value of all obtainable outcomes.

Even this brief sketch of a structural model suggests important implications. First, since scientific knowledge and values play distinct roles in the structure, the model implies a simple relationship between scientific issues and public policy issues in the context of technological decisions. Scientific procedures yield whatever relevant knowledge is to be had. Political procedures provide the relative evaluations of the possible outcomes. Then, if we are lucky, the resulting decision will have a structure for which decision theory yields a rationally best, or least rationally satisfactory, decision. I say "if we are lucky," because it seems to me that the decision theorists' dream of providing criteria of rational choice for any arbitrary decision structure is only a dream. Not all combinations of outcomes, values, and knowledge yield rationally best or even satisfactory choices. But some clearly do.

It follows that any approach to technological decisions that concentrates on the process of decision making to the exclusion of structural considerations must misunderstand the nature of such decisions and the controversies they generate. The decision-making process is all-important only if the issue is strictly one of determining a policy when there is no uniquely justifiable policy choice. Roughly, if there is no possibility of correctness in the choice, fairness in the decision-making process is everything. But sometimes our scientific knowledge, together with politically generated value rankings, does imply a correct choice. Then the problem is to ensure that the decision-making process produces the right decision.

Although the structural model allows a separation of knowledge from values, it also insists that both be present in any decision structure. Thus, although technology may have some "value-free" aspects, there is no such thing as a value-free technological decision. This conclusion also has implications for how we think about controversies concerning technological decisions.

According to the standard analysis, there are "technical decisions" and there are "political decisions." What has happened since the mid-1960s, on this analysis, is that technical decisions have become increasingly politicized.[27] But if one thinks solely in terms of the decision-making process, all this means is that technological decisions have become controversial and that the controversy has been carried out in the political arena. This is not a very enlightening explanation of what has been happening. On a structural view of technological decisions, such decisions have always been political in the sense that they incorporate a set of values that are expressed through political processes. On this view, the most likely source of the increasing controversy is a divergence in relevant values between those who were in fact making the decisions and sizable groups in the population at large. This difference in values then expressed itself in the political arena, which is exactly the right place to carry out such conflicts. The decisions became controversial because the value assumptions implicitly built into the structure of these decisions became controversial. Thus, as long as there was general agreement on the values to be served by new technologies, as there was in the years immediately following World War II, controversies were few. When the value agreement broke down, controversy increased.[28]

That contemporary controversies have their main source in value differences is of course an empirical hypothesis. But it is a hypothesis with considerable theoretical support, and this makes it worth considering, even if it should in the end prove to be empirically mistaken. Nor does it tell the whole story. That is, the structural model of decisions gives no hints as to why value conflicts developed. That is another issue. But it does suggest that the conflict in values focuses primarily on the results of technological decisions and not, as many scientists have feared, on the value of science or scientific knowledge itself.

Ethical theory and technological decisions

The idea that many controversies involving science and technology are rooted in value differences is hardly new. Indeed, it no doubt explains why many moral philosophers have been attracted to the

[27] Here I am thinking particularly of Dorothy Nelkin's analysis, as expressed, for example, in *Controversy: Politics of Technical Decisions*.

[28] For a good case study and theoretical analysis emphasizing the role of values in technological controversies, see Robert E. Snow and David E. Wright, "Analysing Symbolic Dimensions of Technological Disputes: The Michigan Container Controversy," *Science, Technology and Human Values* 29 (1979): 11–23.

study of these controversies. In this climate, a structural, or decision-theoretic, approach to understanding technological decisions requires a special defense because it runs counter to current trends in ethical theory. The conflict cannot be eliminated, but it can be localized, and that is my objective here.

From the standpoint of traditional ethical theories, the fundamental feature of standard decision-theoretic models is that they are *consequentialist* in nature. That is, an action is not chosen because it is morally "right" but because of the value of its consequences relative to the value of the consequences of its alternatives. Decision-theoretic models are thus potentially in conflict with any nonconsequentialist ethical theory that evaluates the correctness of an action independently of its actual or possible consequences. And nonconsequentialist views are currently fashionable among ethical theorists.[29] The conflict is only potential, however, because the application of these opposing points of view may be restricted so as to avoid actual conflicts.

In advocating a decision-theoretic model for the evaluation of technological decisions, one is not committed to applying similar models to every possible decision. Such models seem particularly appropriate for technological decisions but hardly applicable, for example, to a decision by a jury as to whether or not to acquit an accused felon. Objections to a decision-theoretic analysis of technological decisions must be made on much narrower grounds.

The standard example of a consequentialist ethical theory is, of course, utilitarianism. Contemporary ethical theorists, however, have provided many reasons for rejecting utilitarianism, and it might be supposed that some of these reasons apply equally well to decision-theoretic models. I will examine briefly one recent attempt to develop this connection. Its failure again shows that ethical theorists must narrow their focus if they are to object to decision-theoretic models on ethical grounds.

The purpose of any ethical theory is to provide a general justification for regarding some actions as morally correct. According to utilitarianism, the correct action in any situation is that which will lead to "the greatest good for the greatest number." One powerful type of argument against utilitarianism consists in showing that to apply the "greatest good" principle in a specific context, an agent must make prior moral judgments that cannot themselves be justified

[29] Foremost among current nonconsequentialist ethical theories is, of course, the "contractarian" theory developed in John Rawls's *Theory of Justice* (Cambridge, Mass.: Harvard University Press, 1971).

by the utilitarian maxim. So the utilitarian maxim fails to provide the ultimate justification for any action.

Some ways in which the application of the greatest good principle requires prior moral judgments are relevant to decision-theoretic models as well. No real agent, for example, can consider all possible courses of action or evaluate all possible consequences of even one option. In decision-theoretic terms, the number of outcomes that can be considered and evaluated in any real context is relatively small. But any option not considered represents an implicit value judgment – namely, that it would not turn out to be the correct action if it were considered. Nor can it be supposed that the elimination of options is based on a prior assessment that includes those options. Such a supposition leads us back to the set of all possible options that, it is agreed, could not be considered by any real agent.[30]

Such arguments would be fatal to the use of decision-theoretic models if, like the utilitarian maxim, they were supposed to provide an "ultimate" justification for a decision, in the sense that no value presuppositions would be necessary to set up the structure to be considered. But advocates of decision-theoretic models, unlike ethical theorists, may reject any claim to providing justifications for decisions that are ultimate in this sense. They may admit that any decision structure representing a real decision by human agents presupposes many value judgments. Moreover, this admission does not imply moral dogmatism. Any particular value presupposition could be brought explicitly into the structure of the decision at issue. It is only the simultaneous incorporation of all presuppositions that is impossible.

The distinction between the structure of the decision actually considered and the presuppositions of that structure can be useful in the study of controversies surrounding technological decisions. Sometimes controversy focuses on aspects of the decision recognized as relevant by all parties to the dispute. The health risk posed by low-level radiation is a clear example of such an issue. At other times the controversy may focus on what are in fact presuppositions of the decision as structured by one side of the dispute. In such cases the parties in the dispute may appear to be talking past one another. This is because what one side considers a relevant outcome in the decision

[30] This paragraph is basically a summary of an argument in Alasdair MacIntyre, "Utilitarianism and Cost–Benefit Analysis," in Kenneth Sayre, ed., *Values in the Electric Power Industry* (Notre Dame, Ind.: University of Notre Dame Press, 1977). I have discussed this argument in somewhat greater detail in "Technological Decision Making," in Michael Bradie, ed., *Reason and Decision* (Bowling Green, Ohio: Bowling Green State University Press, 1981).

does not even appear as a possibility in the decision as implicitly structured by the other side. The consequences of significantly decentralizing electrical power generation may represent such an issue.

A third sort of objection does reveal a real conflict between some ethical theorists and advocates of decision-theoretic models. There are strong and weak forms of this objection. The strong form insists that some relevant options can never be evaluated morally in terms of their consequences. The weak form only questions whether all relevant outcomes can be evaluated on a common scale, as required by decision-theoretic structures. Some outcomes, it is claimed, are morally incommensurable. Typical examples are outcomes that involve equity or justice, as compared to outcomes involving mere utility. How, for example, can one weigh the inequity in the distribution of health hazards to those living near a proposed power plant against the reduced cost of electricity for all consumers served by that plant? (The strong objection, of course, is that inequity is wrong, no matter what the benefits for others.) How, it is often asked, can one put a dollar value on a human life?[31]

At this point, I think, ethical theory must give way to the facts of life. In a large technological society, it is impossible to avoid making decisions that imply tradeoffs such as those between justice and utility. Ethical theorists have two choices. They can sit on the sidelines deploring the immorality of modern society, or they can attempt to deal constructively with the problem of how best to make such decisions. The problem deserves all the attention it can get.

Conclusion and prospect

The point I have been most concerned to make throughout this chapter is that the considerations of fairly abstract normative questions is highly relevant to the study of controversies involving science and technology. Moreover, this remains true even if one's primary concerns are mainly descriptive. It applies all the more, of course, if one's concerns are explicitly normative.

Beyond urging the importance of normative considerations, I have also suggested some specific normative viewpoints. Foremost among these is the distinction between scientific and policy issues and the autonomy of scientific and policy procedures. This viewpoint has

[31] The "strong" objection is suggested by K. S. Shrader–Frechette's emphasis on the "equal protection" clause of the U.S. Constitution in *Nuclear Power and Public Policy* (Dordrecht: Reidel, 1980).

far-reaching implications for many questions related to contemporary controversies, questions such as the possible value of a "science court."[32]

From a broader perspective, current scholarly interest in controversies surrounding science and technology is not motivated simply by a desire to understand a recent social phenomenon. It is also motivated by the suspicion, if not the belief, that the existence of these controversies is symptomatic of deeper problems. By studying the controversies one may gain some insight into the deeper problems and thereby contribute to their solution. My analysis suggests that the immediate cause of the controversies is a sharp difference in values among various groups in the society. Moreover, the persistence of controversy suggests that the normal political processes for dealing with these value differences are not functioning as well as they should. This leaves us with the questions of why the value differences exist and why the political system is not handling them more effectively.

There are, no doubt, many factors that have contributed to the value differences that now exist in the United States and many other Western nations. One of these is surely the rapid rate of social change since World War II. And one of the causes of social change has been the development of science and technology during this period. Thus, it is natural that "value stresses" produced by social change focus on scientific and technological developments. Whether these speculations are or are not substantiated by further study of our recent past, the more important question is whether our social and political system is dealing with these stresses as well as it should.

There are those who argue that recent controversies surrounding scientific and technological developments pose no special problems for our current political system. Perhaps the system needs improvement, but no special improvements are needed to deal with the sorts of issues raised by science and technology.[33] Others, of course, disagree. They hold that the existence of widespread controversy surrounding scientific and technological developments is itself evidence of special weaknesses in the political system. The problem, on

[32] See Allen Mazur, "Science Courts," *Minerva* 15 (1977): 1–14.

[33] This "conservative" position was often voiced by Robert Rich and Gil Omenn in discussions within the Closure Group. Omenn, in particular, argued that the political system has been dealing for years with highly complex economic arguments. Why are other scientific issues any different? The answer, as the recent vogue for "supply-side economics" demonstrates, is that economics simply is not sufficiently well developed as a science to separate it from social policy, or, indeed, from ideology.

this view, is to pinpoint the weaknesses and to devise ways of correcting them.

I cannot contribute much to this debate here. However, since I incline toward the latter opinion, I will close by mentioning one further point in its favor. This is that the time scale appropriate to developments in science and technology is much longer than the typical time scale of social and political procedures.[34] It has taken twenty years for the drawbacks of commercial nuclear power to become widely recognized. A similar period of generous support for cancer research has not yet produced the desired "cure." It will take a decade of research and development just to determine whether synthetic fuels can be produced at a significant rate. Yet all members of the House of Representatives and one-third of the Senate must stand for election every two years. There is a presidential election every four years, and a change in presidents may bring major changes in the administrative operations of government. The same is true of state and local governments. This makes it very difficult to provide the kind of long-range planning, development, and oversight required for sound science and technology policy and for good technological decision making. Nor is this type of difficulty confined to the public sector. The most widely accepted analysis of the recent poor performance of the U.S. automobile industry is that the need to show profits in the short run (i.e., quarterly) prevented the industry from devoting sufficient resources to long-range planning and development.

One could adduce many other reasons for the same general conclusion – namely, that our social, economic, and political systems are not well suited to dealing with the potentials of modern science and technology. However, whether this is indeed so and what can be done about it is a topic for another project.

[34] Here I am elaborating on a point made in discussion by Gerald Feinberg.

5

Politics, public policy-making, and the process of reaching closure

ROBERT F. RICH

In writing constitutions and framing governments, the authors focus on *the structure of governments* (i.e., which institutions and offices of government will be formed), *the powers assigned to those who will govern, the limits to be placed on rulers,* and *rules or criteria for reaching official decisions* (e.g., that the "majority rules," or that a presidential veto of a congressional action will stand unless two-thirds of the Congress votes to override the veto).

In the United States, particular attention has been paid to the rules for reaching official decisions. These rules have been a topic of debate or discussion since the original debates over the U.S. Constitution. In characterizing the traditions of American liberty, John P. Roche notes:

> First . . . there is . . . the proposition that when all the fighting is over and all the pluralities have had their licks, a public policy which incorporates "justice and the general good" will emerge. Taken for granted here is a willingness on the part of all participants to play by the rules. A faction may disagree about substantive matters – about the content of particular items of policy – but it will not overturn the cardtable and shoot the other players. In short, there will be procedural consensus. (Roche, 1967)

These rules of decision making providing for "procedural consensus" are not designed to dictate specific substantive outcomes; instead, they are meant to formulate a process that will ensure fair, equitable, and "democratic" outcomes.

The process specifies the mechanisms that will be employed by political and organizational actors in their official interactions. A framework – institutional and procedural – is provided for debating and ultimately acting upon issues on the public agenda. In democratic societies, the mechanisms usually consist of a process for "negotiating" outcomes or decisions.[1]

In the American tradition, we have cared a great deal about the way in which decisions are made and the way in which decision-making rules affect different groups and individuals. Although decision-making processes are oriented towards majority rule, the forefathers were very concerned about protecting minority rights. In *The Federalist Papers*, Hamilton notes:

> If a majority be united by a common interest, the rights of the minority will be insecure. There are but two methods of providing against this evil: the one by creating a will in the community independent of the majority – that is, of the society itself; the other, by comprehending in the society so many separate descriptions of citizens as will render an unjust combination of a majority of the whole very improbable, if not impracticable . . .
>
> The second method will be exemplified in the federal republic of the United States. Whilst all authority in it will be derived from and dependent on the society, the society itself will be broken into so many parts, interests, and classes of citizens, that the rights of individuals, or of the minority, will be in little danger from interested combinations of the majority. *In a free government the security for civil rights must be the same as that for religious rights.* It consists in the one case in the multiplicity of interests, and in the other in the multiplicity of sects. *The degree of security in both cases will depend on the number of interests and sects.* (*The Federalist Papers*, most probably written by Alexander Hamilton, quoted in Roche, 1960–7, p. 32)

These excerpts reveal a concern for the process of decision making that was built into the way in which political institutions were structured, and into the federalist system as a whole.

It might be argued that as a polity we have concentrated more on developing procedures for decision making than we have on evaluating the substantive outcomes that are produced when the procedures are implemented and effectively utilized. In this context, it is worth underscoring some key principles of American government:

Principle 1. A legislature should be the supreme organ of government. The legislature, not a strong executive, is the foundation of democratic government.

[1] At the same time, the mechanisms reflect societal values (or at least a consensus on societal values) on the processes that are acceptable for making decisions and bringing about change.

Principle 2. It is critical to keep written records of the operations of the institutions of government. We are to be governed by a rule of law, not a rule of men.

Principle 3. Separation of powers (along several different dimensions) is critical. Individuals should not hold more than one public office, and there is an important separation between federal, state, and local authority.

Principle 4. Indirect election of officials is generally considered to be the best policy for assuring adequate citizen representation. Jefferson argued that a responsible public elects "wise" men to represent it. Wisdom should guide the decisions to be taken, as opposed to a principle of trying in each case to assess what the majority of one's constituency feels and then acting upon that "consensus."

Our founding fathers were concerned that government be plain, simple, and governed by written rules, laws, and procedures (Roche, 1967). Hence, the powers of government should be greatly limited, and one must ensure against the arbitrary exercise of authority by any one institution or organ of government (Roche, 1967).

The fact that the structure of the United States government emphasizes individual rights, written laws, limited government, sovereignty of decentralized units of government, and indirect representation certainly affects the way in which decisions are structured and the outcomes that grow out of the structure of decision making. A high value is placed on procedural consensus and on a political system that allows all interested parties to be heard: a system of pluralism.

Assumptions about conflict

The structure of American government reflects a belief (or perhaps a fear) that conflict will occur and will have to be "resolved" (or at least dealt with). Conflict could occur over several types of political issues:

- Over accumulation of power by one branch of government or one single individual; in this case, the principle of separation of powers would be violated.
- Over substantive constitutional or procedural questions (after all, our Constitution is a rather vague document that leaves a great deal of room for discretion and interpretation); in this case, conflict might be over procedures and rules of decision making; it might be over procedures for negotiating or settling conflicts; or it might be over a particular substantive issue.
- Over protection of the rights of individuals or segments of

society (e.g., issues concerning civil rights, individual rights, due process, nuclear waste disposal, the right to take Laetrile, or worker protection).

It should be noted that conflicts that were foreseen in the Constitution (in some sense predicted) and therefore taken into account in specific decision-making rules and procedures tended to be conflicts primarily over procedures or process. In particular, there was a concern, growing out of the English experience, to avoid the "arbitrary" exercise of authority by one man or organ of government. As already noted, there was also a real fear of majority rule and the potential power of one faction of society to force its rule upon another. The preferred response was to provide for decision-making processes that balanced various interests. In this context, John Adams notes that

> Nothing short of an independent power above the [factitious people] able to check their majorities ever can keep them within bounds. It is the interest and the policy of the people for their own safety *always to erect and maintain such a power* . . . Power must be opposed to power, force to force, strength to strength, interest to interest, as well as reason to reason, eloquence to eloquence, and passion to passion. (Quoted in Zoltan Haraszti, *John Adams and the Prophets of Progress* [Cambridge, Mass., 1952], p. 220)

Institutional responses to conflict

Since conflict could be foreseen, institutional mechanisms were set up (or evolved in the early periods of the Republic) to deal with disputes and certain political issues:

1. *Judicial review.* In *Marbury* v. *Madison* (1803), Chief Justice John Marshall established the critical principle of judicial review.

> The powers of the Legislature are defined and limited; and that those limits may not be mistaken, or forgotten, the Constitution is written. To what purpose are powers limited, and to what purpose is that limitation committed to writing, if these limits may, at any time, be passed by those intended to be restrained? . . . It is a proposition too plain to be contested, that the Constitution controls any legislative act repugnant to it; or that the legislature may alter the Constitution by an ordinary act.

With this famous opinion, Marshall altered the principle of legislative supremacy. The judiciary (primarily the Supreme Court) was to be the ultimate arbitrator (tribunal) for any constitutional question, whether technical or broad. Indeed the Supreme Court has become very involved in the fabric of United States politics through the implementation of this principle of judicial review. Some scholars would argue that the court only takes on the role of umpire between competing interests; others would argue that the court is responsible for initiating and fostering public policy and social change. The only formal restraint on the court's power is the "self-restraint" of individual judges. Justices might make private decisions as to how much policy-making authority they want to take on.

2. *The principle of separation of powers.* The Constitution established three branches of government with different areas of authority and with very limited discretion. Conflict can occur among branches of government over lines of authority and over substantive issues. The clear notion was, however, that conflict would be limited by drawing very clear lines of authority. In addition, our forefathers provided for checks and balances to facilitate smooth operation and limit controversy.

3. *The principle of checks and balances.* Conflict would in some cases be resolved (almost by an invisible hand) by one branch of government providing a check to the power and authority of the other branch of government. This principle establishes mutual mechanisms for accountability built into the very fabric of government.

Two very critical implicit rules of conflict resolution are also inherent in our system of government:

Creation of the market system. Following Adam Smith, it was assumed that the government should interfere as little as possible in the free marketplace and with voluntary transactions. The marketplace is self-regulating and is the best arbitrator of competing economic interests [the invisible hand]. As Milton Friedman suggests,

> The existence of a free market does not of course eliminate the need for government. On the contrary, government is essential both as a forum for determining the "rules of the game" and as an umpire to interpret and enforce the rules decided on. What the market does is to reduce greatly the range of issues that must be decided through political means, and thereby to minimize the extent to which government need participate directly in the game. The characteristic feature of action through political channels is that it tends to require or enforce substantial conformity. The great advantage of the market, on the other hand, is that it permits wide diversity. (Friedman, 1962)

Thus one has established an implicit rule for conflict resolution: Government is to establish some rules of the game and to enforce the rules, but only on a very narrow range of issues. Thus, in the public sector, controversy should be possible only over a narrow range of issues.

An emphasis on individual rights and property rights. Wherever there is a conflict, individual rights are to be emphasized over all others. This is directly correlated with the notion that wherever possible, government should be carried out at the local level in the most decentralized fashion. It has always been thought that the most democratic form of government is the one that represents the smallest number of people. After all, representation can be most genuine when fewer competing interests have to be taken into account.

An overall view of the political process

In specifying institutions and procedures for reaching decisions, one is concerned with the process of allocating scarce resources in a given polity. "A political system, therefore, will be identified as a set of interactions, abstracted from the totality of social behavior, through which values are authoritatively allocated for a society" (Easton, 1965, p. 57). The "allocation process" may be the product of influence of interest or pressure groups, or alternatively it may reflect "elite" or "corporate" interests. The classical pluralistic conception of politics specifies that policy-making represents the resolution of a conflict between several different groups, each of which is competing for the same pool of available resources (Laswell, 1951).[2] Pluralists view each group as having a "legitimate claim" and add that there is no equitable or rational basis on which to make an a priori choice among competing interests. Consequently, decision making represents a process of negotiation and compromise.[3] This characterization of policy-making helps to account for how decisions are reached. Further refinements view politics as a process by which interest groups form coalitions and compete for scarce resources from the strength of the new, loosely formed "team."

Thus, the study of political decision making can be viewed as the inquiry into how a number of actors, each with a "legitimate claim" to resources and each with a power or constituent base of his or her

[2] The role that special interests play in American politics was recognized as early as James Madison's statement, in *The Federalist Papers*, on the importance of divergent groups.

[3] The "elitist" school of theorists has concluded that certain groups of interests consistently dominate the decision-making process (e.g., Mills, Prewitt). However, even among "elite" theorists, a role is still assigned to negotiation and compromise.

own, are able to reach a point where an official governmental action can be taken.

> They [the actors] are seen as engaging, in a variety of bargaining and coercive maneuvers to exert their will, and the "pulling and hauling" which results is seen as the essential characteristic of the decision process. Outcomes occur in this view as consequences of an elaborate bargaining game which spins them out, as often as not, as a by-product of action taken in the course of play. What happens is rarely intended and more rarely preferred by any one of the actors individually. The stakes of the game are high, and the play is commensurately rough. (Steinbruner, 1974, pp. 140–1)

The development of policy, from this perspective, is viewed as dependent upon the ability of individuals or groups to bring about some official actions through one or more of the following instruments of political decision making:[4]

Coercion
Pressure, cajolement
Bargaining, negotiating
Charisma (the ability to charm someone into doing something)
Invocation of a common core of cultural values, norms, or traditions (another form of leadership)

The policy-making process is seen as an "open system" that allows for the input of the multiple interests of individuals and groups. Policy-making is not dominated by the state. On the contrary, it is response to the guidance and pressure of individuals and groups outside of government.[5] As David Easton (1971) points out, there are some basic assumptions at the foundation of studies of the political process:

> The term [*political process*] signifies, first, that policy arises out of a situation consisting of the interaction of various social elements; second, that policy is not a final product but an aspect of an on-going interaction among the various elements of the social situation; and third, that of the vast variety of activity involved in political situations, that of the persons within the governmental and party structure is only a manifest and small

[4] David Truman (1951) is a leader in assuming that the process of negotiation and bargaining between competing groups produces the "best possible and most legitimate" policy outcomes.

[5] Although there are alternative views of how decisions are made (e.g., rational actor models, sociodemographic and economic models, cybernetic models), this view represents the dominant conception of decision making from a political perspective.

part, when compared with the importance of non-governmental social groups. (p. 172)

Working toward the public interest

It is fair to assume that governmental structures and operating "rules of the game" are designed to achieve the public interest. The "public interest" is the vague outcome toward which all public sector actions are oriented. Presumably a great deal of conflict and controversy in democratic polities focuses on alternative means for achieving the public interest. Thus, in finding the means for resolving a conflict, one is inherently forming mechanisms for resolving such controversies. In studying this process, we must assume that there are legitimate mechanisms in place that transfer the private interests of groups and individuals into something with the adjective "public" before it. It is not just any interest, but it is the "public interest." In the United States, there is a variety of mechanisms for transforming private interests into public interests:

1. *Through elections.* A variety of points of view are put forward, and it is assumed that the person who receives the majority of the votes is also given a license to begin to implement his or her conception of the public interest. This conception is also related to the Jeffersonian notion that one elects wise men to office who will be allowed to exercise their best judgment in reaching public decisions. (The problem with this conception is that minority interests are suppressed.)

2. *Through referendums.* A particularly controversial issue is put forward for a direct public vote (e.g., Proposition 13, siting of nuclear plants, disposal of nuclear waste). The public is being asked to "resolve" a controversial issue through legitimate means of direct referendum. (The problem with this mechanism is the same as through elections – minority interests are suppressed.)

3. *Through executive orders.* An executive of the federal, state, or local level who is elected to office with the Jeffersonian ideal in mind, makes a judgment. The President of the United States often makes these types of judgments, especially in matters of foreign policy (with or without the direct consent of Congress). (The basic problem with this mechanism is that the judgment of the President may not really be in line with a societal consensus on basic values.)

4. *Through the legislative process.* Duly elected officials pass legislation that presumably reflects some form of resolution among competing groups. The basic assumption behind the legislative process is that each "interest" in society has an equal chance to be represented

through some group. Groups are the basic vehicle by which interests are represented. Thus, one joins a group to have the interest represented. (The problem with this perspective, as Lockard points out, "despite our talk about Americans being natural joiners, it is in fact difficult to get any group organized to act. For the vast majority of any group to participate in a social action is abnormal. People are not activists by nature; their natural state is social motionlessness rather than social action." Thus, to rely on groups to represent interests and to reflect the viable conflict resolution mechanism probably contains several distortions.)

5. *Through legislative hearings.* Any legislation is assigned to a committee or subcommittee for "formal consideration." In many cases hearings are held that in theory are designed to inform committee members and help provide a basis for reaching a decision. The hearings often provide a forum for all interested parties to present their points of view. This "mechanism" is one that also has the potential for many distortions. Often legislative hearings are "staged" to present carefully selected and edited points of view, as opposed to providing an open forum that gives concerned and interested parties an opportunity to present their perspectives and any evidence they want to offer. In a "staged" process, legislators only confirm the points of view that they already hold.

6. *Through Supreme Court decisions or other courts.* Presumably, through the principle of judicial review, the Supreme Court serves as an umpire among various interests. The highest tribunal of the country resolves broad constitutional questions. History shows us that this resolution is only ephemeral. Some of the best examples of change have to do with the area of intergovernmental relations and integration/desegregation. In these areas the court has appeared to "contradict" itself over time. At one point in time (or in one era), the court decides that racial relations should be governed by the principle of "separate but equal" [*Plessy* v. *Ferguson* (1896)]; half a century later the court decides that this principle is unconstitutional [*Brown* v. the *Board of Education of Topeka* (1954)]. Due to the vague nature of the Constitution, justices are given sufficient procedural latitude to allow radically different substantive interpretations of constitutional questions consistent with what the public will accept at any given point. Indeed, Archibald Cox argues that justices struggle to make decisions that they believe will win the confidence of the public and hence lead to voluntary acceptance. Cox believes that the Court cannot defy the "popular mandate expressed by the political branches" (Cox, 1979, p. 34). Justices will make decisions that will be acceptable to the general public. "Acceptable" may mean challenging beliefs and attitudes, but

it does not mean that fundamental beliefs and attitudes are totally rejected by the court. If they were rejected, the public presumably would call on the other branches of government not to enforce the court's decisions (an appropriate part of the "checks and balances" system).

7. *Through traditions.* Many times the conflict is avoided or "resolved" by acting in a manner that is consistent with ways in which things have been done in the past.

8. *Through ritualistic processes.* Sometimes it is important to evoke symbols (e.g., constitutional conventions) in order to diffuse or even help a controversy.

9. *Through the judgment of experts.* On highly technical and scientific issues, expert opinion is often sought to help influence the resolution of the controversy or even to dictate a "solution." Experts are often used as part of formal advisory committees. The advisory committees are given special status, and it is often assured that their views will play a major role in formulating public policy. The committees appointed by the National Academy of Sciences are an example of the use of expert advisory bodies having a major influence on public policy making.

These mechanisms for translating private interests into public interests are obviously not mutually exclusive alternatives. It is, of course, possible to combine these in trying to mediate conflicts and controversies.

In most of the cases listed above, a procedure is outlined for settling controversies among competing interests. "Settlement" simply reflects the fact that the majority of stakeholders are willing to allow the decision-making process to "move forward." No substantive decision may actually be taken; alternatively, there may have been a conscious decision "not to decide" (not to formulate a policy). The important point is that procedures for negotiating a settlement among competing interests are in place. (Indeed, the fact that some procedures of this kind were necessary was foreseen by our forefathers and discussed in *The Federalist Papers*.)

In this context, it is also worth noting that one can think of settling disputes at two different levels: a micro- and macrolevel. "Microlevel" disputes may be over particular procedures or rules; they may also be over a specific case of a more general principle (e.g., siting of a specific nuclear plant, as opposed to an overall policy on nuclear siting). The mechanisms outlined above are well suited to settling these microlevel types of disputes.

Macrolevel disputes involve much broader questions concerning many competing interests. In order to negotiate a settlement for this type of dispute (e.g., a safety policy for offshore drilling of oil, a policy for protecting the public when transferring nuclear waste), an agreement will have to be reached, and several significant questions may be left open. For example, when the Carter administration formulated the energy policy, it realized that in order to negotiate a settlement within Congress, certain questions would have to be left open (not included in formal legislation). Again, we are dealing with a particular type of settlement or "closure."

Resolving political controversies with a scientific content

Having specified how the structure of American government was, in part, designed as a response to the fears of the framers concerning conflict, it is especially appropriate to examine whether all types of conflict are amenable to the "traditional" solutions. For example, is there something special about scientific controversies, or, rather, political controversies with a scientific content? As Jean-Jacques Salomon points out, there is no reason to believe that policy with a scientific content is necessarily handled any differently from other forms of politics and policy (Salomon, 1974).

In any society it is important to understand how political controversies are resolved. It is also appropriate to examine whether or not the established mechanisms and "ground rules" are successful in working with competing interests, competing claims for scarce resources, and a broad range of conflicts.

Attempting to resolve controversies in a political environment characterized by distrust and demands for antipaternalism is complex and difficult. In recent years, controversies that have involved "scientific issues" or simply have had a "scientific content" have not been resolved in any sense of the word. Public officials who have the formal responsibility for reaching decisions (e.g., legislators, political executives, the courts) have found it difficult to integrate their own individual and ethical social values with "facts" from scientific experts. As matters now stand, public officials have been left to rely largely on intuition and ad hoc procedures. Politicians, especially legislators, have been reluctant to choose between the consequences of the "Faustian" bargains facing them: between gaining a competitive edge in international trade, and limiting the mobility of citizens who do not believe that there is a "true energy crisis"; between producing a potentially valuable source of energy (nuclear), and unknown but potentially serious risks to the environment; between increasing the speed with which American manufacturers produce a highly efficient engine, and easing the pressure on meeting environmental standards; between increasing the speed with which the oil pipeline between California and Texas is completed, and meeting local safety and environmental standards or regulations (Rich and Rydell, 1979).

One might logically ask the question: When "science" enters into the political arena, whose rules of deliberation and negotiation are accepted and followed – the rules of scientists, or the rules of politicians?

It is often observed that there is a gap between the "culture of science" and the "culture of politics" (C. P. Snow, 1959, 1961). Scientists have well-developed procedures for working on so-called scientific controversies. There are rules for settling controversies. These specify what types of evidence are acceptable, how to present challenges to "state-of-the-art findings," and which institutionalized procedures will be emphasized to decide which claims of competing groups will be upheld.

"Closure" or "settlement" in science can be characterized by some of the following principles:

The best evidence to be presented are data derived from controlled experiments.

Findings should be related to existing "state-of-the-art" data.

Challenges are made by replicating experiments and showing how new techniques and new findings replace the existing wisdom in the literature.

The leaders in a given field of knowledge are the best arbitrators of disputes.

Peer review committees and expert advisory bodies are best suited to decide among competing scientific points of view.

It is worth noting that in science one is not dealing with a "negotiated settlement." Instead, by using well-developed canons of research and evidence, a procedure is legitimately developed for settlement of disputes without having to rely on negotiation. Scientists characterize this process as a "rational" one. Because the procedure is considered to be "legitimate" by all competing parties, the results derived generally go unchallenged.

In a polity, as already noted, the rules of the game are quite different:

Each stakeholder should have a voice and should be heard in a public forum.

Each stakeholder should have the opportunity to legitimately affect decision-making procedures.

The best evidence in a system of negotiated settlement is often political influence and the ability to use political pressure.

There are no fixed procedures for presenting challenges to existing paradigms.

Settlement or closure is reached through negotiation or, alternatively, through the exercise of "raw power."

Thus, when science enters into the political arena, it is not necessarily a unique or unusual event from the policy or political process perspective. It is, however, a very different experience for scientists who are accustomed to different rules and procedures for "playing the game." Most of the time, scientists must play by the political rules.

In the DNA controversy, scientists had to appear before community planning bodies and city councils and play by the political rules.

In the Laetrile controversy, the debates occurred in state legislatures and state departments of health.

In the area of generic drugs, consumer groups debated the issue in public bodies.

In the debate over the World Health Organization's marketing code of infant formula, scientific issues and controversies were not considered. Instead, primarily political concerns (e.g., the domination of multinational corporations, the manipulative role of corporations) were focused on.

Although scientific bodies have also taken up these disputes, their decisions have not governed the procedures for deciding upon these questions in the political arena. Public policy has not generally been influenced by these deliberations.

It would, however, be unfair to say that science has not had a major role in helping to define how political problems are addressed. Jerome Weisner, for example, points out that the "state-of-the-art" within a scientific discipline can and does affect policy-making. If the secretary of defense knows that a weapon system is being developed, it will potentially affect arms control negotiations (Rich, 1979).

Consequently, we have observed the phenomenon of passing the buck. Critical issues are not resolved by elected officials; they are, instead, assigned to judicial or quasi-judicial (regulatory) bodies for "resolution." The courts are then put in the position of having to balance strongly held competing sets of values. In such cases (e.g., granting permanent landing rights to the Concorde; nuclear siting; and Laetrile) the federal circuit courts have often acted in contradictory ways. In these cases, the buck has been passed to the Supreme Court. Over the last few years the Supreme Court has tried to select a "narrow issue" or technical area of law on which to base its decision. When the Supreme Court acts in this manner, the political and adjudicatory process has come full circle. None of the official channels were utilized to set a consistent policy. The courts are clearly uncomfortable in being the final arbitrators in such cases, and one can legitimately ask the question: Should the courts be in the business of formulating science policy? On the other hand, one could also simply take the position that one policy is not more "scientific" than any other policy merely because it deals with issues of "science"; there are policies whose subject matter is scientific and technical, and these policies are merely one aspect among others of a country's general policy (Salomon, 1974).

The courts have been coopted to play a major role in defining procedures for settling disputes and in arbitrating substantive controversies (Nelkin, 1979). As we have noted, the courts have tended to define issues very narrowly and have made few attempts to settle "facts."

The Laetrile controversy

The Laetrile controversy is one in which one can observe the phenomenon of passing the buck at all levels of the political decision-making process. Legislation to legalize Laetrile was debated in forty-nine states; twenty-two states have adopted legislation to legalize Laetrile. Yet Laetrile is not available for sale on the open market in any of these states. Indeed, most legislators – even those who passed the bills to legalize it – do not believe that Laetrile is efficacious. That was not, however, the issue in passing the legislation. The legislation was passed on the basis of "freedom of choice" issues, as well as concerns about "big government intervention in our lives." However, the legislators knew that they themselves could not put Laetrile on the market.

Departments of Health (bureaucrats or regulators) were responsible for devising regulations concerning the legalization of Laetrile. In almost every state, the Department of Health adopted the Federal Drug Administration (FDA) "model regulations" or wrote more stringent regulations of their own; in either case, the effect was to delay the sale of Laetrile on the open market for seven to eight years. The legislators who had originally passed the legislation could have challenged these actions, through public hearings or in the press; there were, however, no such challenges (see Rich, 1980, for further details). This is clearly an instance of passing the buck in a case where the legislators were uncomfortable in publicly stating, during the legislative hearings, that the evidence shows that Laetrile is neither safe nor efficacious. Instead, they expressed this same belief by not challenging the actions of officials from the Department of Health.

The Laetrile case is most interesting for what it suggests more generally about accountability and responsibility in the public sector. Two forms of accountability appeared to be operating for these types of issues: (1) short-term responsiveness to the pressures and demands of the public; and (2) paternalistic concern (long-term) for defending the interest the public sector wants to protect. As Friedrich suggests, the experts need to be in a position to decide on issues that require expertise. Although this standard of accountability is exercised by bureaucrats (and some governors) and accepted by legislators, the public often believes that bureaucrats are acting autonomously and "abusing their power" (Friedrich, 1975).

The Laetrile case suggests that our thinking about responsibility and accountability needs to be reexamined. One perspective contends that the public interest is defended through the legislative mandates of its elected representatives (Finer, 1975). Our analysis suggests that the public interest is defended through the actions of elected representatives in the policy-making or implementation phases of the problem-solving process; as long as the elected representatives consider the actions to be "legitimate," then the public interest is being defended. This reformulation assumes that the legislators are aware of what their options are during the implementation phase and what implications follow from action or inaction. If elected officials are willing to accept the professional judgment of experts, then in Finer's terms they are acting responsibly.

The Laetrile case also suggests that traditional assumptions concerning the legitimacy of paternalism need to be reexamined. The Laetrile proponents are most concerned about the interference of big government in the lives of citizens. Yet the very legislators willing to pass the Laetrile Act were also willing to allow public officials to take actions that ensure the continued removal of Laetrile from the market.

Conclusion

The cases of political controversy with a scientific content help us to understand sociopolitical change mechanisms. Institutionally, we have not established a set of mechanisms that we have reached a consensus on as a society. Passing the buck seems to have become the modus operandi for public officials. Politicians (elected officials) do not want to be on record as taking responsibility for pushing one side of very controversial or potentially dangerous issues. Similarly, the courts do not want to take on this role. We have very little political or legal "tradition" to guide us in these matters. Public opinion is divided, referendums offer little recourse. Hence, we are at a standstill. The standstill leads us to grasp for temporary "solutions," those that allow us to move forward while making the least commitment to the polity as a whole.

It is not that we lack the institutional or "traditional" mechanisms to deal with political controversy. Indeed, as the earlier parts of this essay suggest, they are well in place. It is, instead, that we have difficulty in adopting these mechanisms to the types of issues (complex, technical, dangerous) that face us as a society.

In the case of scientific controversies, one will have to give some thought as to how traditional political mechanisms may be molded to affect the peculiarities or uniqueness of science. It is not sufficient to say that there is a "gap" in the cultures or to say that scientists think differently from policymakers. It is natural for policymakers to believe that science is not a special area that deserves unique or special attention. Instead, thought will have to be given to how scientific rules of settlement can be used to influence the ultimate policymaking system.

References

Cox, Archibald. 1979. *The Role of the Supreme Court in American Government.* New York: Oxford University Press.

Easton, David. 1965. *A Framework for Political Analysis.* Englewood Cliffs, N. J.: Prentice-Hall.

1971. *The Political System: An Inquiry into the State of Political Science.* New York: Knopf.

Finer, Herman. 1975. "Administrative Responsibility in Democratic Government." In Alan Altschuler, ed., *The Politics of Federal Bureaucracy,* pp. 425–32. New York: Dodd, Mead.

Friedman, Milton. 1962. *Capitalism and Freedom.* Chicago: University of Chicago Press.

Friedrich, Carl J. 1975. "Public Policy and the Nature of Administrative Responsibility." In Alan Altschuler, ed. *The Politics of Federal Bureaucracy,* pp. 414–25. New York: Dodd, Mead.

Laswell, Harold D. 1951. *Politics: Who Gets What, When, How.* New York: McGraw-Hill.

Nelkin, Dorothy, ed. 1979. *Controversy: Politics of Technical Decisions.* Beverly Hills, Calif.: Sage.

Rich, R. F. 1979. "The Pursuit of Knowledge." *Knowledge* 1: 6–30

1980. "The Political Implications of Laetrile: Who Gets What, When and How." In Gerald Markle and James C. Petersen, eds., *Politics, Science, and Cancer: The Laetrile Phenomenon.* Boulder, Colo.: Westview Press.

Rich, R. F., and Randy Rydell. 1979. "Who Is Making Science Policy?" *Science*, July/August, 1979, pp. 18–22.

Roche, John P. 1967. *Origins of American Political Thought*. New York: Harper & Row (Torchbooks).

Salomon, Jean-Jacques. 1974. *Science and Politics*. Cambridge, Mass.: MIT Press.

Snow, C. P. 1959. *The Two Cultures and The Scientific Revolution*, The Rede Lectures. Cambridge: Cambridge University Press.

1961. *Science and Government*, The Godkin Lectures. Cambridge, Mass.: Harvard University Press.

Steinbruner, John D. 1974. *The Cybernetic Theory of Decision*. Princeton: Princeton University Press.

Truman, David B. 1951. *The Governmental Process*, New York: Knopf.

6

The role of experts in scientific controversy

GARLAND E. ALLEN

The eugenics movement in the United States between 1905 and 1940 provides a useful case study of the role of "experts" in a scientific controversy with obvious social implications. I have chosen to focus my attention on this topic because it falls most closely within my own area of research interests and is also the one in which my familiarity with the primary source material is the greatest. Throughout this chapter, I will make explicit references to various other case studies that were examined in the course of the Closure Project, thereby hopefully providing a basis for some generalizations about the role that experts play in scientific controversies. By employing a specific historical methodology (Marxism) for the analysis of a detailed case history (eugenics) with respect to a single question (the role of experts), I hope to suggest some ways in which future research on topics such as closure might be carried out.[1]

Let me say at the outset that to me the function of this or any of the other case examples discussed in the Closure Project is to help us

I would like to thank a number of people who read or heard earlier versions of this paper and offered valuable suggestions. In addition to specific members of the closure group – Everett Mendelsohn, Ernan McMullin, and Allan Mazur – I would like to thank Arthur Caplan for a number of valuable organizational suggestions, and especially Randy Bird for a rigorous, thoroughgoing, and insightful critique. Part of this work was carried out with a grant from the National Science Foundation, grant no. SES 8026095.

[1] I will use the term expert to refer to individuals who understand and/or have the capability of understanding technical details and overall conceptual aspects of an issue. This technical capability may or may not be the result of formal training, at least with respect to the particular issue at hand. This is, of course, a purely epistemic definition of "expert." There is also a nonepistemic definition that sees "experts" as technically trained individuals who have learned to play certain rules within an academic "discipline." In this sense most experts draw heavily on the authority and/or prestige of science in general, or individual fields in

arrive at valid sociohistorical explanations for the generation and closure of controversy. My view is that we should seek to understand general causes in history. I appreciate the value of descriptive, "tell-it-like-it-is" narratives, but I think the most important function of the historian is to seek more wide-ranging explanations where possible. At certain points in the investigation of any subject the descriptive material must come first, and I therefore emphasize the value of preparing detailed case histories. At the same time, I feel that it is valid and even crucial to seek some more general explanations. I am unhappy with the notion that the only thing that can come out of the Closure Project is a description of many more or less unrelated kinds of closure, the "plurality" concept of historical explanation. I do not wish to juxtapose description and explanation as contradictory or mutually exclusive processes. Obviously, to some extent, descriptive case histories must precede more general theoretical discussions, but both are important. If we can make no generalizations about how public-oriented scientific controversies are closed, then we are as much as saying that the history of closure follows no patterns, no laws, and is only a series of specific, unrelated events. Few, I think, would be satisfied with this conclusion.

General thesis

The role of scientific experts, in the long run, is of minimal importance in the closure of socially oriented scientific (technical) controversies. Such controversies are not scientific to begin with but result from the confluence of various social, political, and economic forces that use the process of debate over technical issues to attain their own social, political, or economic ends. In the words of Ernan McMullin, a debate that is nonepistemic (that is, nonscientific) to begin with cannot really be closed by the resolution of epistemic (that is, substantive, scientific, or technical) arguments, however powerful they may be (see Chap. 2, this volume). This principle was illustrated specifically in the Laetrile and homosexuality cases. However, this thesis applies only to given short-term outbreaks of a controversy. Over the long haul, epistemic factors do prevail, because reality is the ultimate testing ground. But in terms of specific situations and specific time periods, it is the resolution of the social and political, not

Footnote 1 (*cont.*)
particular, and are frequently found as participants in one or another controversy (these might be dubbed the "professional experts."). The eugenics experts I refer to were either eugenicists with some genetic training or geneticists with an interest in eugenics.

the technological and scientific issues, that produces closure, no matter how much the controversy itself appears to be epistemic in nature. Epistemic arguments do influence the controversy to some extent, in different ways and to different degrees in each special case. But in the sorts of controversies that were studied in the Closure Project, epistemic arguments did not appear to be major determiners of closure.

Controversies are often the focal point for social conflicts that are coming to a head for their own historical reasons. The social forces that cause these conflicts may be either large-scale (such as general economic decline or the rise of civil conflict) or they may be small-scale and specific (such as an increase in the incidence of cancer or renewed interest in atomic power when oil reserves are declining). In either case, however, the same basic social forces and processes apply. In such contexts, scientific experts are used to buttress arguments within a controversy, giving legitimacy to one or the other side, helping to win public support for one special interest group or another, or, in some cases, camouflaging the real reasons for the dispute, which are economic or sociopolitical in nature. In looking at examples of such controversies, it is absolutely crucial to look at the development of events in terms of the forces that bring the controversy to the fore. Only in this way is it possible to understand the role of the scientific expert, who sometimes is cognizant of the sociopolitical issues at stake and sometimes is not. Scientific experts quite often fall at one of two extremes: On the one hand, knowledgeable advocates who know the sociopolitical scene, including who is paying the bill for their advocacy, or the future rewards (prestige and power, in particular) that will be reaped for their services; on the other hand, naive pawns who see themselves as disinterested (whether they truly are or not) seekers after truth, and who are more often than not victimized by social and political circumstances that they do not fully understand and thus cannot adequately control.

Let us now turn to an analysis of the role of scientific experts in the rise, perpetuation, and ultimate decline of eugenics in the United States.

The role of scientific experts in the American eugenics movement, 1900–1940

General overview

Eugenics, the application of principles of heredity to the improvement of the human genetic stock, developed into a widespread and important movement in both the United States and Europe during the first half of the twentieth century. The movement found particularly fertile ground in the United States in the period just prior to and following World War I. Eugenics was viewed by many people as the application of rational, scientific procedures to the improvement of human society. If, as many eugenicists claimed, human social problems such as feeblemindedness, unemployment, alcoholism, and insubordination were genetically determined, then the most rational approach to social reform would be to eliminate these problems at their source by selective breeding. In practical terms this meant increasing the birthrate among people thought to be superior (positive eugenics) and decreasing the birthrate among people thought to be "inferior" (negative eugenics). What eugenicists considered superior or inferior, however, was based on subjective criteria. Most generally, eugenicists classed as superior those people descended from Nordic or Aryan stock or those from the wealthy classes. Conversely, inferior people were those from eastern European, Mediterranean, Asian, African, native American, or Jewish stock, along with the chronically poor. Eugenicists were alarmed by what they called "differential fertility" – the apparent fact that "inferior" people were producing their kind at a much faster rate than "superior" people. To American eugenicists, the old Anglo-Saxon, Nordic stock was in danger of being swamped by a massive increase in the number of hereditary degenerates. In eugenic terms this means that "good" genes were increasingly being outnumbered by "bad" genes. Continuing to let this happen, according to some eugenicists, amounted to nothing short of "race suicide."

We can divide the history of the American eugenics movement into two periods: an early, or "old" eugenics movement, which developed from 1900 to 1925, and a later, or "new" eugenics movement, which began emerging between 1925 and 1940. What differentiated the old from the new movement, at least most visibly, was the emphasis that old-style eugenicists placed on heredity. Using the newly discovered Mendelian laws of heredity, old-style eugenicists were often simplistic and dogmatic in their arguments that one or another social or behavioral trait in human beings was genetically determined. Their racial and ethnic biases were often blatant, and they tended to see Anglo-Saxons and Aryans as the pinnacle of social and cultural evolution.

By contrast, new eugenicists placed more emphasis on social reform, saw heredity and environment working more closely together, and began to look beyond proposals for sterilization of individual family members as the solution for social problems. New eugenics embodied the notion of improving the human social condition in a broader context. Real improvement would come only through reforms in the social environment, in conjunction with reforms in the genetic breeding stock. The best genes, they realized, could not reach full potential in an unfavorable environment. The new eugenicists emerged in the period from 1925 to 1940, reacting against the naiveté, overt racism, and simplistic genetics of the old eugenicists. Especially after World War II, their issues came to center more and more on controlling large-scale population growth, especially among the poor, both at home and abroad, as a way of abating the many problems that they saw emerging from "overpopulation."

In the United States, the old eugenics movement may be said to have achieved visibility for the first time in 1910, with the establishment of the Eugenics Record Office (ERO) at Cold Spring Harbor in Long Island, New York. The director of the ERO and its parent organization, the Station for the Experimental Study of Evolution, also at Cold Spring Harbor, was Charles B. Davenport, one of the earliest supporters of Mendel in the United States. In direct control of the ERO's everyday operations was Harry H. Laughlin, a former teacher of agricultural science at Northeast Missouri State College at Kirksville.[2] Between 1910 and 1915, eugenics, under the leadership of Davenport, Laughlin, David Starr Jordan (first president of Stanford University and well-known evolutionist and ichthyologist), among many others, became a widespread popular as well as scientific movement. The movement's popularity is indicated by the increase in articles on the subject appearing in popular magazines throughout this era. For example, as summarized from the *Reader's Guide to Periodical Literature*, the number of articles on eugenics was eleven in 1910, sixteen in 1911, forty-five in 1912, back down to fifteen by 1913, and remained steady with a slight increase to about twenty per year through the mid-1920s. In addition, a number of textbooks and popular books on the subject appeared between 1914 and 1920: Madison Grant's best-selling *Passing of the Great Race* and Lothrop Stoddard's *The Rising Tide of Colour against White Supremacy*. Paul Poponoe's and Rosewell Johnson's *Applied Eugenics* was the most widely used textbook on eugenics in the United States between 1918 and 1933.[3] In 1914 the American Genetics Association boasted that forty-four colleges offered courses in, or at least dealing in part with, eugenics. By 1928 they gloated over the fact that the number had swelled to 376. There were three international congresses of eugenics,

2 Randy Bird and Garland Allen, "Sources in the History of Eugenics: The Harry Hamilton Laughlin Papers in Kirksville, Missouri," *Journal of the History of Biology* 14 (1981): 339–53.

3 William Provine, "Geneticists and the Biology of Race Crossing," *Science* 182 (1973): 790–6.

one in 1912 (London), a second in 1921 (New York), and a third and last in 1932 (also in New York). The American Eugenics Society (AES), founded in 1921, sponsored "Fitter Family" contests throughout the country at state fairs and launched a number of educational and propaganda programs, including a prize for the best "eugenics sermon" contributed each year. Other organizations, such as the Race Betterment Foundation (Battle Creek, Michigan) and the Eugenics Research Association, all contributed to both the collection of genetic and eugenical information and the propagandizing of the eugenics cause.

The most notable achievements to which the eugenicists pointed were their roles in establishing U.S. immigration quotas in 1924 for Mediterranean, eastern European, and Jewish people.[4] Kenneth Ludmerer has studied in some detail the input of eugenicists into this legislation through testimony and written presentations to the House Committee on Immigration and Naturalization.[5] In addition, eugenicists, particularly Harry Laughlin, sought sterilization legislation. Targeted for such sterilization were those claimed to be "habitual criminals," "sex perverts," "chronic alcoholics," "feebleminded," and "hereditary criminals," among many others. Indeed it can be said that the heyday of the old eugenics movement occurred around 1924–5. At that time old-style eugenicists were riding high on a wave of financial and popular support.

Between 1925 and 1940, however, the old eugenics movement began to decline. Geneticists dropped away from the movement silently or, in a few cases, not so silently, offering strong criticism of the way eugenicists ignored or distorted genetic data for their own ends. More important, the statements of many eugenicists with regard to ethnic and minority groups became an increasing embarrassment for the movement in the public arena. For example, Madison Grant, in the *Passing of the Great Race*, stated that "dwarf stature, peculiar mentality and ruthless concentration on self-interest" was characteristic of Jewish people and was "being engrafted upon the stock of the nation."[6] Catholics, he found, were "ignorant and destitute people who have been for ages in contact with the highest civilizations, but have failed to benefit by such

[4] The vehicle for restriction was the Johnson Act of 1924, sometimes known as the Immigration Restriction Act, which set specific quotas by nationality, based on percentages of people of a given nationality residing in the United States at the time of the 1890 census.

[5] Kenneth Ludmerer, *Genetics and American Society* (Baltimore: Johns Hopkins University Press, 1972), esp. chaps. 4 and 5.

[6] Madison Grant, *The Passing of the Great Race* (New York: Scribner, 1916), p. 28.

contact, whether physically, intellectually, or morally." Negroes, on the other hand, were hereditarily determined to be "willing followers who ask only to obey and to further the ideals and wishes of master races."[7] This phase of the movement declined precipitously after 1930. By 1940 and the beginning of World War II, the "old" eugenics movement had suffered virtually complete closure. Symbolically, the Eugenics Record Office was closed on December 31, 1939.

The genetics of the old eugenics movement

To understand the origin, prominence, and eventual closure experienced by the old eugenics movement, we will begin by examining the biological (mostly genetic) arguments that eugenicists employed in the first three decades of the twentieth century. A good example of the "scientific" side of old eugenics can be found in the work of Charles B. Davenport, although the work of a number of other contemporary biologists (E. G. Conklin of Princeton, W. E. Castle of Harvard, and S. J. Holmes of Berkeley) illustrates the point equally well.

Davenport is a useful example, not only because of the extensiveness of his published writings on eugenics but also because he was an eminently respected member of the United States scientific community in the first three decades of the twentieth century. He was not an amateur scientist or a zealous crackpot, like many connected with the eugenics movement throughout its history. Davenport had received his doctorate from Harvard and held a prestigious post at the University of Chicago. He was a member of the National Academy of Sciences, a trustee of the Marine Biological Laboratory at Woods Hole, Mass., and the director of the Carnegie Institution in Washington's two research laboratories (the Station for Experimental Study of Evolution and the Eugenics Record Office) at Cold Spring Harbor. In addition, Davenport had carried out genetic studies on a wide variety of animal species and had published voluminously in the most prestigious journals. His were among the clearer attempts to ground eugenical arguments in the scientific framework of genetics.

It is impossible to understand Davenport's commitment to eugenics without also considering his commitment to the then young Mendelian theory of heredity. Particularly important had been the attempts to apply Mendelian theory to an understanding of human heredity. For example, from analysis of many family pedigrees, geneticists had shown by 1910 that the inheritance of a number of

[7] Ibid., p. 78.

human traits did in fact follow Mendelian patterns. The ABO blood groups, the blue-brown eye color pattern, and two physiological abnormalities – phenylketonuria (a defect in the metabolism of the amino acid phenylalanine) and alkaptonuria (a defect in the metabolism of the amino acid tyrosine) – were all known to be Mendelian traits. Similarly, by 1910, the genetics of sex-linked traits such as red-green color blindness and hemophilia had also been shown to conform to Mendelian expectations.

For the first time in the history of biology, a general theory of heredity was emerging. Recognition of this fact prompted optimistic and enthusiastic predictions about the social value to be derived from new findings in genetics. Not only did it include the prospect of an agricultural revolution (which did, in fact, occur in the 1920s and 1930s in the United States) but also the possibility of eliminating many of the chronic ills of human society that eugenicists thought were genetically determined.

With his eugenic orientation, Davenport was especially interested in the inheritance of not merely clinically definable traits such as the ABO blood groups, but the human personality and behavioral traits. He argued that traits such as laziness, violent temper, worthiness, matter-of-factness, reading habits and feeblemindedness were Mendelian dominants. Conversely, thalassophilia ("sea lust," or the desire to go to sea), alcoholism, shyness, shiftlessness, and mechanical, artistic, and musical abilities were all Mendelian recessives.[8] How did Davenport justify such conclusions? To Davenport, inheritance meant Mendelian inheritance – traits being determined by single genes in a dominant or recessive way. The basic data from which he worked were pedigree charts with data gathered by a whole host of different methodologies, from examining medical records to personal interviews with individuals and family members.[9] From the pedigree charts Davenport attempted to deduce the Mendelian inheritance pattern for a particular trait. His way of determining such features as dominance or recessiveness was highly simplistic. If a trait skipped a

8 Charles B. Davenport, *Heredity in Relation to Eugenics* (New York: Henry Holt, 1911), esp. Chap. 3.

9 It should be apparent that numerous pitfalls lie in the way of any sound analysis of pedigree charts when the data are collected in a nonuniform and highly subjective way. For example, can interviews with family members about other family members be considered reliable? To what extent is this problem compounded by dealing with subjective traits such as "worthiness," "pauperism," "alcoholism," or "matter-of-factness"? Such methodological problems did not deter most eugenicists, including Davenport, from constructing elaborate and often highly detailed pedigree charts of hundreds of families.

generation, it was considered recessive, whereas if it appeared in every generation it was considered dominant. Davenport often pooled offspring data from several different marriages (sometimes within, but often outside the same family) in order to obtain larger numbers and therefore more significant ratios. This is a practice that would have been unacceptable to any laboratory geneticist at the time.

Consider the following example of Davenport's reasoning process, in claiming that thalassophilia is genetically determined:

> Thus we see that thalassophilia acts like a recessive, so that, when the determiner for it (or the absence of a determiner for dislike) is in each germ-cell the resulting male child will have love of the sea. Sometimes a father who shows no liking for the sea . . . may carry a determiner for sea-lust recessive. It is theoretically probable that some mothers are heterozygous for love of the sea, so that when married to a thalassophilic man half of their children will show sea-lust and half will not.[10]

Davenport's method of argument was not by using direct evidence but by reasoning and analogy. He first refers to the existence of "nomadism" as a hereditary trait. Then, he argues that, since thalassophilia can be considered a form of nomadism and since nomadism is hereditary, it is a simple step in logic that sea lust must therefore be hereditary, too. Davenport goes on to draw an analogy between thalassophilia and the inheritance of comb size in fowl: "It is possible . . . that the irresistible appeal of the sea is a trait that is a sort of secondary sex character in males in certain races, just as a rose comb is a male characteristic in some races of poultry."[11] By 1919, the inheritance pattern for rose comb was a well-established Mendelian trait. By making the comparison between human beings and poultry, Davenport assumed that superficial similarity in patterns of inheritance between two quite different species implied similarity in genetic causality. More important, he virtually discounted the effect of environmental factors in molding human behavioral traits.

Davenport's genetic determinism led to some obvious conclusions about the solution to social problems. The source of such problems was not environment but "bad genes." He urged that philanthropists donate their funds to eugenics, not to charity, which would only perpetuate hereditary degeneracy. In a report to the committee on eugenics of the American Breeders' Association in 1909, Davenport

[10] Davenport, *Naval Officers, Their Heredity and Development* (Washington, D.C.: Carnegie Institution of Washington, 1919), p. 31.

[11] Ibid., p. 29.

insisted, "Vastly more effective than the million dollars to 'charity' would be ten million to eugenics. He who, by such a gift, should redeem mankind from vice, imbecility and suffering would be the world's wisest philanthropist."[12] In public and private, Davenport belittled social reform. He apparently was fond of telling the parable of a man who found a bitter gourd and watered and tended it carefully to produce a delicious vegetable. That man was, Davenport claimed, like the trustee of a rehabilitation hospital for the insane. Poverty and lack of social or economic success were de facto the phenotypic expression of genotypic inferiority. In 1912 he advised the National Conference of Charities and Corrections that social reform was futile, since "the only way to secure innate capacity is by breeding it."[13] To Davenport the comparison between breeding humans and breeding strains of domesticated animals or plants was self-evident.

Racial, ethnic, and class bias in eugenical arguments

From its very inception in the works of English biometrician Francis Galton in the 1880s, the old eugenics movement had had a strong racial, ethnic, and social-class bias. For example, Galton regarded the "sentiment ... against the gradual extinction of an inferior race" as being "for the most part quite unreasonable."[14] Other eugenicists expressed similiar views. W. E. Castle of Harvard argued by analogy that crossbreeding between human races might produce the same kind of a misfit hybrid as crossbreeding between a thoroughbred and a draft horse: "It is a particular combination of qualities which makes a race horse useful and a different combination which makes a draft horse useful. Crossing the two will produce neither one type nor the other. The progeny will be useless as race horses and they will not make good draft horses. For such reasons wide racial crosses among men seem on the whole undesirable."[15]

In his book *Heredity in Relation to Eugenics* Davenport passed judg-

[12] Davenport, "Report of the Committee on Eugenics," *American Breeders' Association Report* 6 (1909): 94, as quoted in Mark Haller, *Eugenics: Hereditarian Attitudes in American Thought* (New Brunswick, N. J. : Rutgers University Press, 1962).

[13] Ibid., p. 65.

[14] Francis Galton, *Inquiries into Human Faculty* (London: Macmillan, 1883), p. 100.

[15] William E. Castle, *Genetics and Eugenics* (Cambridge, Mass: Harvard University Press, 1916), p. 233. Davenport argued along the same lines that cross breeding between a white and a black person would produce "disharmony": The offspring might have the "long legs of the Negro and the short arms of the Caucasian and thus have to stoop more to pick up a thing off the ground."

ment on all sorts of non-Aryan peoples. Of the Jews he wrote, "There is no question that, taken as a whole, the hordes of Jews that are now coming to us from Russia and the extreme southeast of Europe, with their intense individualism and ideals of gain at the cost of any interest, represent the opposite extreme from the early English and more recent Scandinavian immigrant with their ideas of community life in the open country, advancement by the sweat of their brow, and the uprearing of their families and the fear of God and the love of country."[16]

Similarly, Harry H. Laughlin, Davenport's right-hand man at Cold Spring Harbor, worked steadily and persistently to keep Jews, southern and eastern Europeans, and the Irish out of the United States. As he wrote in his conclusion to the report of the New York State Special Commission on Immigration and Alien Insane, "High-grade Jews are welcome, and low-grade Jews must be excluded."[17] Further in the same report he wrote, "Racially, the country will be liberal if it confines all future immigration to the white race, then, within the white race, if it sets up differential numerical quotas which will admit immigrants in accordance not with external demand but on the basis of American-desired influence of such racial elements on the future seed-stock of America."[18] Laughlin further distinguished himself by devoting considerable research energy to showing that recent immigrants and "aliens" had been responsible for much of the crime in America during the period between 1890 and 1920. Laughlin also carried out an extensive correspondence with the leaders of the Nazi eugenics organization, the Gesellschaft für Rassenhygiene, men such as Eugen Fischer and Theodor Lenz, and in 1936 was awarded an honorary medical degree from Heidelberg University for his contributions to eugenics, particularly eugenical sterilization.[19]

From the inception of the movement, eugenicists were proponents of the belief in Aryan and Nordic superiority and were virtually paranoid about the rise of other racial and ethnic groups. The goal of the movement, according to eugenicists, was to maintain "racial purity" and to eliminate both individuals and groups with a high propensity for "undesirable" traits.[20]

[16] *Heredity in Relation to Eugenics*, p. 219.

[17] Harry H. Laughlin, *Report of the Special Commission on Immigration and Alien Insane* (Submitted as a Study on Immigration Control to the Chamber of Commerce of the State of New York, 65 Liberty Street, New York, April 16, 1934) p. 17.

[18] Ibid., p. 18.

[19] Bird and Allen, "Laughlin Papers"

[20] Allan Chase, *The Legacy of Malthus* (New York: Knopf, 1977), esp. Pts. 1 and 2.

Between 1921 and 1924, Harry Laughlin testified at great length before the House Committee on Immigration and Naturalization (headed by Representative Albert Johnson of Washington) in the deliberations that led up to the passage of the Immigration Restriction, or Johnson Act. Laughlin presented a vast number of statistics, which were later published both in the *Congressional Record* and as part of his endless output of scientific and popular treatises, purporting to show the innate rates of crime, insanity, feeblemindedness, and degeneracy in various non-Aryan stocks. Laughlin consistently presented himself as a dispassionate and objective scientist with no ax to grind. As he stated in his 1924 congressional report, *Europe as an Emigrant-exporting Continent*, "I made this biological investigation and put the facts on record here for the benefit of the Committee which must draw its own conclusion. I am not here as an advocate for or against any race. Indeed, my position with the Carnegie Institution of Washington would prevent me from standing as an advocate or special pleader."[21] Only one other biologist was brought in to testify before the committee, Herbert Spencer Jennings of Johns Hopkins University. Jennings was brought in at the insistence of several representatives on the committee with immigrant constituencies, but was given only a few minutes to make an oral presentation. The remainder of his critique was presented in writing. Jennings was highly critical of Laughlin's so-called biological studies, but Laughlin's presentation had the greater exposure, and hence greater influence.[22]

Scientific opposition to eugenics

Especially as a result of the publicity given to eugenical claims during the time of the immigration restriction debates of 1923 and 1924, several biologists, principally geneticists, attacked the eugenicists for their wholesale perversion of or, more frequently, ignorance of modern genetic findings. One of the most vocal of these critics was Raymond Pearl of Johns Hopkins University, himself at one time an avid supporter of old-style eugenics.[23] Pearl attacked eugenics openly in an article in H. L. Mencken's influential journal the *American Mercury*. Entitled "The Biology of Superiority," the article took

[21] Laughlin, *Europe as an Emigrant-exporting Continent: Hearings before the House Committee on Immigration and Naturalization, 68th Congress, First Session* (Washington, D. C.: U.S. Government Printing Office, 1924), p. 1318.

[22] Ludmerer, *Genetics*, Chap. 5. Ludmerer's study details the committee's deliberations and the influence of eugenical ideas on the committee's decision.

[23] See Garland Allen, "From Eugenics to Population Control in the Work of Raymond Pearl," *Science for the People* 12 (1980): 22–8. A more complete version of this article is still in press: "Old Wine in New Bottles: Raymond Pearl's Transition from Eugenics to Population Control."

eugenicists to task for their hasty generalizations and propagandistic tendencies.[24] Pearl pointed out that much of modern eugenics is based on the idea of "ancestral heredity" derived from Francis Galton. Several modern discoveries in genetics, he emphasized, had undercut Galton's basic notions: Johannsen's pure line selection experiments with beans, and East and Jones's inbreeding and outbreeding experiments. Johannsen's theory emphasized that it is impossible to determine with any certainty the genotype of an organism from an inspection of its phenotype. East and Jones showed that continual inbreeding of organisms from generation to generation produces a less viable and sickly stock. Inbred lines could be rejuvenated, however, by outcrossing – that is, by breeding them with organisms from another hereditary line.[25] Thus, close inbreeding, the animal counterpart of the eugenicists' "racial purity," was shown to be genetically deleterious. The failure of eugenicists to account for these two new findings, Pearl argued, made most of their biological arguments outdated and erroneous.

Additional criticisms of eugenics came from H. J. Muller, who launched a broadside attack on the old eugenicists at the Third International Eugenics Congress in New York in 1932. In a speech entitled "The Primacy of Economics over Eugenics," Muller castigated eugenicists for making sweeping statements about genetic causes of social phenomena while at the same time ignoring economic influences on social welfare.[26] Other eugenicists, such as T. H. Morgan, quietly withdrew any support that they had once proffered to the eugenics movement but made no public attacks.[27] There is no evidence of much direct confrontation between geneticists and eugenicists, either on a public platform or through an exchange of letters or views in professional or popular journals.

By and large, eugenicists and their critics tended to write for separate audiences. Eugenicists took their popular writings to newspapers and major mass publications, such as *Good Housekeeping,*

[24] Raymond Pearl, "The Biology of Superiority," *American Mercury* 2 (1927): 257–66.

[25] Edward M. East and Donald F. Jones, *Inbreeding and Outbreeding* (Philadelphia: Lippincott, 1919), esp. Chaps. 6 and 7. East was a eugenicist and went to great lengths to show that the conclusions derived for domesticated animals and plants had to be applied cautiously to humans. Nonetheless he contrived an argument to show that race crossing in humans was on the whole undesirable (see p. 244, 255, 262–5).

[26] Hermann J. Muller, "The Dominance of Economics over Eugenics," *Scientific Monthly* 37 (1933): 40–7.

[27] See Garland Allen, *Thomas Hunt Morgan, the Man and His Science* (Princeton: Princeton University Press, 1978), Chap. 6, esp. pp. 227–34.

Saturday Evening Post, Century Magazine, Atlantic Monthly, and the like. In addition, eugenicists had a number of publications of their own, including the *Journal of Heredity* and *Eugenical News*. Anti-eugenicists, when they wrote at all, tended to publish in more technical or eclectic journals (*Scientific Monthly* or *American Mercury*) or special-interest publications such as Catholic weeklies. Debate between experts was never widespread or public. Eugenicists had biologically trained "experts" to whom they always referred (Davenport, Laughlin, Conklin, and Castle, among others). Anti-eugenicists had their own group of experts to whom they listened (Morgan, Muller, Jennings, or L. C. Dunn). The closure faced by old eugenics was thus not primarily caused by scientific criticism, which was never very widespread or open. The objections of bona fide geneticists was not irrelevant, however. As more and more biologists simply refused to become involved with eugenics, the movement lost the momentum that it had originally possessed. The very fact that biologists had originally begun teaching eugenics in college courses must have helped give the movement some impetus. As the number of these courses peaked by the late 1920s and declined thereafter, one important means for inculcating the ideals of eugenics became less available. This to some degree hurt the movement. It was not, however, the main cause for the movement's decline.

The funding of eugenics

Historians of eugenics have often pointed out that the funding for the movement at the turn of the century came from wealthy philanthropists. However, few historians have shown in detail the extent or the nature of this funding. In addition, few have looked to the history of the philanthropic foundations that were organized in the early decades of the century and their role in the growth of eugenics. The source of funding and the conditions under which it was given become important considerations in understanding how the eugenics movement got under way and was perpetuated. Likewise, withdrawal of that funding was a major factor in the movement's closure.

To establish the Station for the Experimental Study of Evolution at Cold Spring Harbor in 1904, Charles B. Davenport had received financial backing from the Carnegie Institution of Washington (CIW). The CIW was established by Andrew Carnegie in 1902 to foster technical and scientific research. Davenport, when he came to organize the Eugenics Record Office (ERO) a few years later, approached Mrs. E. H. Harriman, widow of the railroad and telegraph tycoon E. H. Harriman, for support. She provided the original site, an estate of 74.85 acres with buildings worth $80,680, all next door (conveniently) to the experimental station in Cold Spring Harbor. Along with yearly endowments for operating expenses, the Harriman contribution to the ERO amounted to $246,842 between 1910 and 1917. In addition, between 1910 and 1917, the ERO received a total of $21,650 from John D. Rockefeller's personal account.[28] This money went largely for training of eugenical field-workers and for their expenses while collecting field data (mostly family pedigree information) at a number of state hospitals and institutions.

In 1918 the ERO was transferred to the Carnegie Institution of Washington as part of its Department of Genetics. At this time Mrs. Harriman gave the office an additional $300,000 endowment. The Carnegie Institution picked up the rest of the bill, amounting to approximately $25,000 a year between 1918 and 1939. Thus, from 1918 to 1939 the Carnegie philanthropies budgeted a total of nearly $475,000 to the operation of the ERO. This comprised about 16 percent of the total budget that the Carnegie Institution of Washington contributed to the study of heredity and evolution in general (see Table 6.1).

The ERO was, however, only one of several eugenics organizations and institutions in the United States at the time, and the Carnegie philanthropies only one of several foundations contributing money to eugenics. In addition to providing early support of the field research program, Rockefeller monies went independently into a number of other eugenics-related projects not directly connected with the ERO. For example, in 1913, the same year that the Rockefeller Foundation was chartered in New York, John D. Rockefeller, Jr., provided his own personal funds to set up the Bureau of Social Hygiene (BSH).[29] The purpose of the bureau was "the study, amelioration, and prevention of those social conditions, crimes, and diseases (including venereal diseases especially) which adversely affect the well-being of society."[30] An offshoot of the bureau, called the Criminalistic Institute, was set up in January, 1912. With advice from C. B. Davenport and Henry Goddard (of the Vineland Training Center for the Feeble-Minded in New Jersey), John D. Rockefeller, Jr., in conjunction with the courts of New York City, developed a process whereby women

[28] "Notes on the History of the Eugenics Record Office, Cold Spring Harbor, Long Island, New York," unpublished summary, Cold Spring Harbor, December 31, 1939, p. 4.

[29] Barry A. Mehler, "Sources in the Study of Eugenics, Number 2: The Bureau of Social Hygiene Papers," *Mendel Newsletter* 16 (1978): 6–11.

[30] Inventory of the Bureau of Social Hygiene Papers, Rockefeller Archives Center, Tarrytown, N.Y., p. 1.

Table 6.1. *Expenditures for eugenics by Carnegie Institution of Washington, 1918–39*

Year	Budget for Genetics Dept. (S.E.E. and ERO)	Sum from Genetics Dept. budget expended for ERO		Percentage from Genetics Dept. budget expended for ERO
1918	$ 84,790.00	$ 25,000.00	[Individual	29.5
1919	95,910.00	26,836.00	ERO budget	28.0
1920	109,129.00	30,785.76	(exact)]	28.2
1921	125,974.03			
1922	125,205.00			
1923	121,290.00			
1924	124,055.00			
1925	129,125.00			
1926	125,960.00		[Estimated for	
1927	131,510.00		the 14 years	
1928	133,727.67	283,429.63	at 15% or	15.0 (est.)
1929	139,380.00		approximated	
1930	143,666.67		(estimated)]	
1931	146,384.99			
1932	158,460.00			
1933	143,550.00			
1934	141,242.00			
1935	135,780.00	22,903.15	[Actual figures	16.8
1936	144.135.00	25,256.78	by chief clerk	17.5
1937	138,980.21	20,180.00	of ERO]	14.5
1938	145,745,00	20,943.37		14.4
1939	143,220.00	18,680.00		13.0
Total	$2,887,220.10	$474,014.69		16.4

Source: The "Official Records" in "Notes on the History of the Eugenics Record Office, Cold Spring Harbor, Long Island, New York." Mimeographed report, Eugenics Record Office, compiled by Harry H. Laughlin, 1939, p. 5. Laughlin Papers, Northeast Missouri State University, Kirksville, Mo.

arrested on criminal charges were subjected to intense eugenical investigation (interviews, family histories, and medical examination) before being sent to trial.[31] Particularly important in the thinking of those involved with the Criminalistic Institute was the importance of hereditary factors in determining social behavior. Davenport and one of the ERO-trained field-workers, Katherine B. Davis, became chief advisers to Rockefeller on the operation of the institute. In addition, the Bureau of Social Hygiene kept up with and monitored eugenical research and programs in European countries – particularly England and Germany.[32]

One of the most important organizations for the propagation of American eugenics, if not in direct financial contributions, at least in terms of contacts and organizational role, was the American Eugenics Society (AES). From its founding in 1923, the AES was closely connected to the ERO and to Davenport, Laughlin, Irving Fisher (a Yale University economist), and Henry Fairfield Osborn for leadership. The AES regularly published *Eugenical News*, a periodical filled with popularized reports and accounts of eugenics research, legislation, and zealous propaganda. Other channels by which the AES spread "the eugenical word" included exhibitions at state fairs; films and educational workshops for teachers; liaisons with other organizations such as the National Educational Association (NEA); and Fitter Family contests (with prizes for the "most eugenical" couples); and even a contest for the best eugenical sermon.[33]

[31] The project originally involved only women because it was aimed at determining the causes of, and therefore presumably solutions to, female prostitution in New York. Later the BSH began collecting information on male criminals as well.

[32] Other foundations, less prominent than the Carnegie and Rockefeller foundations, also contributed to the funding of eugenics projects. The Race Betterment Foundation, in Battle Creek, Michigan, had been set up by the Kellogg cereal family. The foundation provided funds for specific eugenics projects, most notably the publication of reports and surveys emanating from the ERO. Other organizations that contributed to the support of eugenics included the Eugenics Research Association (ERA), founded in 1913 with a select group of members accepted by invitation only. The ERA donated approximately $5,000 a year to the encouragement of eugenical research and to the publication of material such as *Eugenical News*. In addition, the Pioneer Fund was founded in the late 1920s by Harry Laughlin, Madison Grant, and Wycliffe Draper, heir to a Massachusetts textile machinery fortune. (The Pioneer Fund is still in existence and contributed $179,000 between 1967 and 1977 to the work of William Shockley and Arthur R. Jensen.) The total assets of the Pioneer Fund in 1972 were listed as $1.4 million dollars, according to the *St. Louis Post-Dispatch* (December 11, 1977, p. 6).

[33] Copies of the winning sermons for 1926 can be found in the American Eugenics Society Papers, American Philosophical Society, Philadelphia.

What is clear is that eugenics was funded from the beginning by extremely wealthy sources. It had the support of two of the largest and most well-to-do foundations of the time, the Carnegie Institution of Washington and the Rockefeller Foundation. In addition, it was funded by a number of wealthy families, such as the Harrimans, Kelloggs (the cereal family from Battle Creek, Michigan), and Osborns. (Henry Fairfield Osborn was a wealthy paleontologist and the president of the American Museum of Natural History; his nephew Frederick Osborn, a railroad executive and financier, became a key leader of the AES after 1932, and later of the Rockefeller-funded Population Council.)

It might be argued that there is nothing surprising about the fact that such causes as eugenics were funded by the wealthy and elite. After all, they are the only people with substantial resources to support any movement. This, of course, is true. However, wealthy people pick and choose among the causes that they support. During the same period in which eugenics was growing (1910–25), Franz Boas, the eminent anthropologist, sought support from the same individuals and foundations for the establishment of a Museum of African History and Culture. He was turned down flat across the board.[34] It is not a singular revelation of Marxist history to point out that ruling classes give money to the social movements that they see supporting their own basic interests. That those interests fall along class lines, which are obviously socioeconomic in nature, is also clear. Now, how did funding of eugenics fit ruling class interests in the period 1910 to 1935? In other words, why were wealthy, ruling-class elites in this period more interested, for example, in supporting eugenics than a museum of African culture? Was it something more than narrow ethnocentrism, or were there economic, social, and political factors that made eugenics an attractive movement to support at the time?

The social composition of the eugenics movement

Two major social groups supported the American eugenics movement during the first four decades of the twentieth century: upper-class ruling elites, and middle-class professionals. We have just discussed the upper-class elites, so we do not need to dwell upon

[34] Edward H. Beardsley, "The American Scientist as Social Activist: Franz Boas, Burt C. Wheeler, and the Fight for Racial Justice, 1900–1915," *Isis* 64 (1973): 50–66.

them here. More interesting for our concern with "experts" is the social background and role of the middle class eugenics experts. Who were they? What did they see their role to be as experts, and how can we use a Marxist analysis to shed light on the role that eugenics experts (particularly the proeugenicists) played?

In a series of suggestive articles, Donald McKenzie has shown that the idea of eugenics was clearly enunciated and avidly expounded by many representatives of the professional middle class in Britain. Such was also the case in the United States. Most of the eugenicists came from Anglo-Saxon, white, Protestant, and moderately to well-educated groups, the so-called professional "expert" class. Men such as William E. Castle, Edwin Grant Conklin, Edward M. East, Herbert Spencer Jennings, David Starr Jordan, Harry H, Laughlin, Samuel J. Holmes, and Vernon L. Kellogg were from the Midwest (predominantly from Ohio, Illinois, and Kansas); others, such as Davenport, Raymond Pearl, Henry Fairfield Osborn, his nephew Frederick Osborn, Madison Grant, and Prescott Hall, were from the East (New England and New York). With the exception of the Osborns, Grant, and Hall, most practicing eugenicists (as opposed to financial supporters) were from lower middle-class to middle-class families whose ancestors had been farmers, teachers, or ministers. A few, such as the Osborns, Grant, and Hall, were from wealthier elite families. Most saw their role as eugenicists in a rational and progressive light, as bringing modern science to bear on larger social issues and policies. They were enlightened apostles of genetics with the technical understanding to contribute (they thought) to such thorny social problems as immigration, crime, alcoholism, poverty, and war. Most of the experts were moderately conversant with Mendelian genetics, at least as it was understood in the period between 1910 and 1915, and a few, such as Castle, Pearl, East, Jennings, Conklin, and Holmes, were active geneticists who contributed to, as well as kept up with, developments in genetics and cytogenetics in the 1920s and 1930s. All were Darwinians but derived their eugenical ideas and inspiration from Francis Galton rather than from the social Darwinists of the preceding generation (indeed, both Davenport and Pearl had actually spent approximately six months each with Galton and his disciple Karl Pearson at University College in London).[35]

Although brief and necessarily sketchy, this description of the background of American eugenicists and the history of the American

[35] Donald Pickens, *Eugenics and the Progressives* (Nashville: Vanderbilt University Press, 1968); also, James Weinstein, *The Corporate Ideal in the Liberal State: 1900–1918* (Boston: Beacon Press, 1971).

eugenics movement will serve as the basis for a Marxist analysis of how and why such movements arise and the sociopolitical function they serve at certain points in history. Again, let me emphasize that my focus will be not on analyzing the entirety of the eugenics movement but on investigating the role of experts (that is, of eugenicists with some genetic training and of geneticists with an interest in eugenics) in the development and ultimate demise of the movement.

Analysis of the role of experts in the American eugenics movement, 1910–1940

We can begin by posing four general questions regarding the old eugenics movement:

1. Why did ruling elites support eugenics to the extent that they did, at the time that they did? Why was eugenics attractive to its wealthy supporters, and what ends did they see it leading to? In other words, what were the self-interests that led many ruling-class elites in the United States to support eugenics in one form or another between 1910 and 1935?
2. Through what processes were ruling-class philanthropists led to support eugenics activities in particular, as opposed to other causes? Did philanthropists go out and find eugenicists, or did eugenicists go calling on ruling-class leaders, or was there some confluence of factors that led each group to the other?
3. Once the movement was going and had received support from above, what role did scientific experts – that is, the individuals who spoke for (or occasionally against) eugenics – play in the perpetuation or demise of the movement?
4. How, and under what circumstances, did the old eugenics movement face closure? What form of closure was it, by what mechanism was it brought about, and what role did experts play in effecting closure?

Question 1. Why did the ruling class support eugenics to the extent that it did in the first decades of the twentieth century? To answer this first question, we must keep in mind that eugenics came to prominence in an era of intensified class struggle. The cause of this intensification lies in a number of socioeconomic changes in the United States associated with rapid industrialization between 1870 and 1910: (1) increased concentration of labor in urban areas; (2) increased immigration as a result of intense labor demand; (3) massive exploitation by "robber barons" of the Gilded Age; and (4) militant union-organizing, especially in mining, railroads, and steel. Increasingly, in response to these developments, ruling-class leaders began to move away from laissez-faire to a more planned and regulated form of capitalism (so-called managed capitalism). It is against this background, involving a significant shift in the evolution of capitalism itself, that the widespread response of ruling-class leaders to the eugenics movement must be viewed.

The Marxist perspective plays a crucial role in emphasizing this connection. Eugenics, in its origin and rise to prominence, was first and foremost a response to the social conditions resulting from industrialization; the form, the particular perspective and the ideology in which eugenical theory was cast, was a direct consequence of the new philosophy of regulated capitalism. Eugenics is, after all, the *regulation* of human reproduction itself. Space does not permit us to analyze both the social conditions and the new economic philosophy thoroughly. However, a brief outline will give some idea about how economic and social changes led to support of something like eugenics by the wealthy elites.

Progressive economic thought, in particular the notions of scientific planning and the regulation of economy and big business, was in part a response to high-level monopolization in the 1880s and 1890s, the accompanying decline in the overall profit rate, and growing worker opposition to vast abuses in the workplace. Progressive thought emphasized the necessity of planning not only in the economic sphere but also in the social sphere. The formation of large business-community planning groups, such as the National Civic Federation and the National Association of Manufacturers, betokened the influence of such thinking on government and corporate mentality, respectively. Although such ideas were anathema to one segment of the ruling class (the older, more conservative sector), it was attractive to another segment, which clearly saw the social problems (especially bolshevism) that might result from persistent economic instability such as that symbolized by the crash of 1929. Thus, eugenics was merely the biological counterpart of scientific management. Nowhere was this more clearly stated than by a leading British eugenicist, Sir Arthur Keith, as reported in a newspaper article from 1930. The heading of the article reads: "Rationalising Mankind/'Big Business' Methods in Evolution/Eugenic Reform." The article then states, in part:

> That the time is now favourable to a new phase in the history of mankind, when human nature might be rationalised as "big business" has been rationalised, was the opinion expressed by Sir Arthur Keith, speaking at yesterday's meeting of the International Federation of Eugenic Organisations at Hinson St., Mary, Dorset.
>
> "If eugenists have their way," said Sir Arthur, "and

ultimately I have no doubt they will obtain the ear of statesmen, then a new phase in the evolution of mankind will be initiated . . . If eugenists have their way human history enters a new, or eugenic phase.

"The public are becoming familiar with rationalisation in the world of big business; instead of being controlled by circumstances, business men take steps to control circumstances to plan."[36]

It is interesting to note that in this regard eugenics was to the progressive era of twentieth-century capitalism what social Darwinism had been to the more laissez-faire era of mid-nineteenth-century capitalism. Social Darwinism was an eminently laissez-faire biological theory of society, in an era still dominated by laissez-faire economics. Eugenics was an eminently regulative and scientifically planned biological theory of society, in an era beginning to be dominated by progressive social planners. I cannot demonstrate conclusively that the shift in economic theories caused a corresponding shift in biological explanation of social problems (this would require another paper in its own right). But the shift on both levels at more or less the same time (the economic shifts seem to have begun earlier; social Darwinism, for example, was still a prominent theory in 1900) is suggestive of something more than chance.

As mentioned earlier, the change in economic thinking was the product of an increasingly intense class struggle in the United States in the period between 1890 and 1930. In 1890 the ruling class, especially the wealthiest industrial sectors, enjoyed considerable position and influence. There were few checks on their financial dealings or on the labor policies (especially concerning wages and working conditions) that they created. But the Haymarket riots, the Homestead strike, the Pullman strike, and the Populist revolt, to mention only a few militant movements between 1880 and 1910, betokened the rising unity and power of the working class. The years 1900 to 1920 saw the increasing organization of labor: the rise of the International Workers of the World (the IWW, or "Wobblies"), trade unionism, and the founding of the Congress of Industrial Organizations (CIO), as well as the militant, revolutionary Seattle general strike of 1919.[37] Furthermore, the Bolshevik revolution of 1917 had

[36] From the Laughlin Papers, Pickler Memorial Library, Northeast Missouri State University, Kirksville, Mo.; the clipping has as a heading, "Romeike Curtice," 35 Shea Lane, London: "Extract from *Birmingham Mail*," and is stamped with Laughlin's date stamp, September 13, 1930.

[37] Ludmerer, *Genetics*, pp. 23, 100. See also, Harvey O'Connor, *Revolution in Seattle* (New York: Monthly Review Press, 1964).

helped to link many immigrants (especially those from eastern Europe) in the mind of the public with communism and the "Red Scare," and with the increasing radicalization of workers in general, and the IWW in particular.[38] It is true that many of the radical union leaders were immigrants who had found the land of milk and honey less utopian than they had expected. They had used their experience in the European trade union movement to channel the growing but sometimes politically and organizationally naive American labor movement into a tougher, more militant, and more unified position.

There is no question that the American ruling class saw a threat to their class interests in the labor movement and other socialist-oriented mass organizations. In response they used one of the classic techniques of those in power – "divide and conquer" – to maintain their position. The eugenics movement was one way – not the only way, by any means, but certainly one way – of implementing this strategy. Nothing works to keep people apart more effectively than to convince one group that others with whom it might ally itself are inferior, degenerate, and unreliable. As a form of what Allan Chase calls "scientific racism," eugenics accomplished this goal admirably. With what appeared to be the "newest" scientific backing, eugenical ideas created a biological, and thus an "innate," hierarchy of social groups that mirrored the socioeconomic hierarchy of capitalist society at the turn of the century. The clearest way to understand why the ruling elites gave such enormous financial backing to eugenics in the early twentieth century is to recognize that they were responding to the rising tide of popular, working-class clamor for more control over the workplace and over the social and political institutions that governed the working class. The fact that many eastern and southern Europeans, together with Jewish workers, were prominent in the militant leadership of the trade unions in the early 1900s may also explain why the Johnson Act targeted those groups for the most severe restriction in immigration quotas.[39]

Question 2: Given this background, how was the eugenics movement fostered by the ruling class, and what role did experts play in generating and sustaining ruling class interest in eugenics? The traditional answer to this question (for example, in the expositions of Haller or Ludmerer), stresses the role of the eugenicists in initiating and pushing the movement – that is, it credits the experts with generating and sustaining the momentum. In this interpretation the ruling class is

[38] Ludmerer, *Genetics*, p. 108. See also Robert Wiebe, *The Search for Order, 1877–1920* (New York: Hill & Wang, 1967), pp. 288ff.; also Graham Adams, *The Age of Industrial Violence* (New York: Columbia University Press, 1966), pp. 1–9.

[39] Ibid., pp. 104–5.

seen as rather passive and, although not uninterested, rather dilettantish in their support of such philanthropies. In contrast, a Marxist analysis emphasizes a dialectic between experts and the ruling class in which each rather clearly (sometimes, of course, more subconsciously than consciously) uses the other. In the context of class struggle, the Marxist analysis sees that ultimately the ruling class had the upper hand in this dialectic, primarily because they controlled the purse strings. The experts were dependent on ruling class support and ultimately became pawns in the hands of the philanthropists, to be used or discarded as necessity demanded. The model for this sometimes subtle and complex interaction between expert and ruling-class philanthropist might go something like this. Initially, the experts brought together a coherent plan and program to both explain and correct the problem of increasing social "degeneracy." Such an initial contact person in the early days of the American eugenics movement was C. B. Davenport. For example, his initial contact with John D. Rockefeller, Jr., in 1911–12 was prompted by Davenport's desire to find funds to support the eugenical field-workers operating out of the newly formed ERO.[40] The program that eugenics experts such as Davenport developed was derived epistemically from their knowledge of Mendelian genetics, biometry, and animal and plant breeding. It was also derived nonepistemically not only from their desire (sincere, for the most part) to bring the science of genetics to bear on human social problems and from their conviction that scientific planning and expert know-how was essential in the formulation of social policy but also from their need to find research funds. That their eugenics program also fit into their own and ruling class's sense of the superiority of people of white, Anglo-Saxon ancestry was a convenient and necessary bonus. The program, however, without support in the form of money, and a platform from which to reach the public, amounted to little. Davenport and other eugenicists went directly to the wealthy philanthropies and offered up their wares. Davenport himself may have been sincerely concerned about social problems such as crime, prostitution, unionization, and bolshevism, but he was also aware that John D. Rockefeller, Jr., Mrs. Harriman, and Andrew Carnegie shared similar concerns.

In attempting to gain support for his views, Davenport emphasized to his benefactors the role that heredity played in governing such social behaviors. As an expert he played on the social problems that concerned those very groups from whom he sought support. And, as

[40] Correspondence between Charles B. Davenport and John D. Rockefeller, Jr., Bureau of Social Hygiene Papers, Rockefeller Archives Center, Tarrytown, N.Y.

the record shows, some of the wealthiest ruling-class leaders were receptive to Davenport's arguments. In addition to supporting views of their own superiority, eugenical arguments supported their growing fear of the uncontrollable masses, the dissatisfied immigrant workers, and, worst of all, of social and economic upheavals like the Bolshevik Revolution. Even prior to 1917, the fears of mass uprisings of workers was prominent. The published, but even more frequently the private, writings of eugenicists and wealthy elites are peppered with derogatory statements about socialism, workers, and peasants, and (after 1917) included also bolshevism.

It is important to point out that both the expert eugenicists and the leaders of ruling-class interests, such as the Rockefeller and Carnegie philanthropies, saw the eugenics program as forward-looking and progressive. It fitted into the progressivist philosophy of rational and scientific planning (as opposed, for example, to the environmentalist-oriented philosophy of social work, which the ruling class and eugenicists alike described as subjectivistic "bleeding heart", and nonscientific). Thus, while today we might describe the overall program espoused by the eugenicists as conservative, even reactionary, in their own day eugenicists regarded themselves as in the forefront of progressive thinking. They saw themselves cutting through irrational sentimentalism and applying the new findings of science to the solution of age-old social problems. For their part some ruling-class leaders, especially those who had been won to the necessity of abandoning laissez-faire for planned capitalism, saw themselves in the same progressive mold.

What I am suggesting is that eugenicists went to the wealthy elites of their day with an ideology that blamed the victims of social and economic oppression for their own problems. Such ideas were attractive not only because they shifted attention away from the possible social and economic causes of poverty, crime, and social unrest, but also because they built on existing, or created new divisions among the dissatisfied masses themselves. As an ideology, eugenics could have a long-lasting effect, because it fostered distrust and suspicion among ethnic, racial, and class groups.

Although the social elites pushed and directed the movement once it was brought to their attention, it was the experts who took the initiative in developing eugenics as we know it. What some of the experts who at first supported the movement (for example, Morgan or Jennings) failed to realize, of course, was the way in which their wares – their eugenics program – was to be used. Some of them, at least, were naive about the larger social forces – that is, the class struggle – that had promoted the program to begin with. This accounts for the fact that some of the more politically naive geneticists (including both Morgan and Jennings) joined the movement initially but later defected when the racial and class biases, especially with regard to immigration restriction debates in the 1920s, became overt and widely publicized. It also explains the other side of the coin: why the experts who remained with the movement (such as Davenport, Laughlin and Grant) were not only more overtly racist from the beginning but remained overtly so throughout the movement's heyday. Racism and class struggle were inherent in the very epistemic questions on which eugenics was built.

Question 3. What role did eugenics experts play in the perpetuation and ultimate decline of the movement? A clear answer to this question comes from examining the fate of Harry H. Laughlin and the ERO at Cold Spring Harbor (of which, recall, Laughlin was superintendent, under Davenport's general aegis, from 1910 until 1940). Sociological studies by Ralph Huitt, David Farnsworth, and David Truman suggest that more often than not congressional committees hold hearings to justify decisions already made, rather than to gain information for the formulation of a new policy or piece of legislation.[41] I would like to use this notion as the basis for a scenario that sketches out how experts like Laughlin become involved in public policy controversies involving some aspect of their science. It is immaterial for the usefulness and applicability of the scenario whether the expert in question is aware of or naive about his or her political role or about the social and political forces that have brought particular controversy to the fore. The scenario is also independent of whether the expert appears before a congressional committee, even though Laughlin was particularly involved in this activity. If we substitute ruling class for congressional committee, the scenario has much broader applicability and, I think, shows the same forces at work. That is, the ruling class had its mind made up prior to the appearance of eugenicists on the scene: working-class people, Jews, eastern and southern Europeans, and blacks were a dissatisfied and dangerous lot, and were, of course, biologically inferior. Eugenicists simply provided the seeming legitimation of that notion, just as experts legitimate congressional decisions that have, in effect, been decided before any testimony is given. In either case the determining feature is which experts are selected and given a platform.

[41] David Farnsworth, *The Senate Committee on Foreign Relations* (Urbana: University of Illinois Press, 1961); Ralph K. Huitt, "The Congressional Committee: A Case Study," *American Political Science Review* 48 (1954): 365; and David B. Truman, *The Governmental Process* (New York: Knopf, 1959). These were all first called to my attention in Frances Hassencahl's valuable "Harry Hamilton Laughlin, Expert Eugenics Agent for the House Committee on Immigration and Naturalization, 1921–1931", Ph.D. diss., Case Western Reserve University, 1970, p. ii.

Laughlin was appointed "expert witness" in 1920 to the House Committee on Immigration and Naturalization by the committee's chairman, Congressman Albert Johnson of Washington State. Johnson had made a name for himself politically in the northwest as a newspaper editor with an aggressive anticommunist and antiunion stance. Through his newspaper, the *Grays Harbor Washingtonian*, in Hoquiam, Washington, Johnson opposed Japanese immigration and directly attacked all attempts at unionization in the lumber industry in the Northwest. He was especially hostile to the IWW and tried to introduce legislation into Congress making membership in any revolutionary organization grounds for the deportation of aliens.[42] Johnson was thus ideally suited (from the ruling-class point of view) for his job as a major political leader against "radical" eastern European and Mediterranean immigrants.

Laughlin was, of course, eager to serve as an expert witness. Johnson appointed him to this position, which included "franking privileges," in 1921. Laughlin appeared before the committee three times. His total testimony lasted three days, and in transcription occupied several hundred pages of House documents. As Frances Hassencahl has pointed out in her detailed study of Laughlin and the congressional hearings, Laughlin was "hired" (Hassencahl's term) by the committee to buttress their own innate but still vaguely formulated antiimmigrant views.[43] The committee's decision with regard to the 1924 immigration proposal was already made by the time Laughlin appeared. Indeed, it had been made by the time Johnson became chairman of the committee. Laughlin's testimony provided what could be taken as scientific authority for already existing racist views. All but the two Jewish members of the seventeen-member committee in the Sixty-eighth Congress (Samuel Dickstein of New York and Adolf Sabath of Chicago) were convinced restrictionists, as evidenced by the fact that in addition to Laughlin, the other experts invited to testify were all restrictionists: Madison Grant, Lothrop Stoddard, Robert De Courcy Ward (of the Immigration Restriction League), and Gino Speranza (who, despite his Italian name, was a noted antiimmigration publicist).[44] In the face of this stacking of witnesses, freshman congressman Emanuel Celler of New York (not a member of the immigration committee) complained that no opposing views had been presented at all. He suggested that the committee call the noted anthropologist at the Smithsonian, Arles Hrdlicka, who was known to oppose the eugenical arguments on race and

[42] Hassencahl, "Laughlin," p. 230.

[43] Ibid.

[44] Ludmerer, *Genetics*, p. 231. pp. 109–11.

nationality. Hrdlicka was not invited, but Herbert Spencer Jennings, anti-eugenics professor of biology at Johns Hopkins, was. Although Jennings had originally been sympathetic to eugenics, he had grown increasingly critical of its bias and lack of scientific credibility. Jenning's testimony, however, was scheduled to speak on the afternoon session of January 4, 1924, along with five other speakers, giving him only a few minutes to criticize Laughlin's mammoth collection of tables, data, pedigrees, and restrictionist conclusions. Jennings's testimony had no impact. The bill was sent to the floor of the House by a committee vote of 15 to 2 (Dickstein and Sabath voted against it). In Congress, the Johnson Act was passed by a large majority in both houses.[45]

What Laughlin did not realize was that despite his seeming success, he was merely a pawn in an elaborate economic chess game. Although he nurtured the distinct illusion that his eugenical arguments were determiners in the passage of the Johnson Act, it is clear that actually he served merely to rationalize in a "scientific" way (a powerful form of rationalization nonetheless) decisions already made on quite different grounds. As evidence of this, it is interesting to note that increasingly, between 1924 and 1931, Laughlin tried to use his congressional connections to gain support for eugenics research, but to less and less avail. He had served his purpose. Finally, in 1931, his appointment as eugenics expert to the House Committee on Immigration and Naturalization was terminated. With it went also money, prestige, and power.

Laughlin, like many experts who actively seek involvement in public policy questions, was aware of a certain interplay of social and economic forces in his appointment and role as eugenics expert to Congress. For example, he explicitly attempted to counter the argument that he represented the industrial and business side of the immigration issue. In his congressional testimony, Laughlin said,

> I must state also that as a member of the Carnegie Institution of Washington, if I were biased in favour of the immediate financial interests that created the institution, I should perhaps advocate open immigration in order to get cheaper labor, which might increase the immediate profits of Carnegie investments. But I am not doing that. I am here simply as a scientific investigator to present the facts to the gentlemen of the committee, with the hope that the facts and their analysis might be of some use.[46]

[45] Ibid., p. 110.
[46] Laughlin, *Europe as an Emigrant-exporting Continent*, p. 1318.

The comment is interesting. First, the fact that he brought this matter up at all suggests that there must have existed some general level of criticism in Congress as well as outside, that special interests were strongly at play in the immigration issue. Second, he directly confronts the biases attributed to him as an employee of the Carnegie Institution and dismisses the criticism by showing that he is taking the opposite of the expected stand. Third, he invokes the image of the expert as a "disinterested scientist" presenting facts *as they are*, the eternal seeker after truth. And finally, Laughlin assumes that by virtue of his scientific authority he is in a position to sway or change opinion – in other words that he is directly influencing major political events.

Question 4. How did the eugenics movement reach closure, and what form of closure was involved? Closure in the old-style eugenics case involved both epistemic and nonepistemic factors. Epistemic factors reached closure earlier than nonepistemic factors, meaning that debate on the genetical arguments behind eugenics was closed among specialists long before the eugenics movement per se faded from sight.

By the mid- to late 1920s a number of geneticists (T. H. Morgan, Raymond Pearl, H. S. Jennings, H. J. Muller) in the United States, as well as some abroad (such as J. B. S. Haldane, Lancelot Hogben, Otto Louis Mohr), had advanced serious arguments against the scientific credibility of eugenics. Although the critics were not as publicly or as doggedly vocal as the eugenicists themselves, they nonetheless made their point on more than one occasion. Most cogently, the critics simply pointed to the outdated character of eugenical arguments and to the racial bias of many of the most prominent eugenicists. The growing estrangement between eugenics and bona fide genetics took its toll in the long run. Yet it was not until the 1930s that ruling-class supporters began to become aware of the fact that eugenics no longer had the aura of backing from the legitimate scientific (particularly the genetic) community that it had once possessed.

Instrumental in turning the tide of ruling-class opinion away from "old-style" eugenics were Raymond Pearl and Frederick Osborn, both long-standing supporters of and contributors to the eugenics movement. Pearl's attacks on eugenics in the *American Mercury* (see n. 26) in 1926 laid out the epistemic issues clearly and succinctly. The main thrust of his argument was the failure of eugenicists to keep up with modern developments in genetics. Osborn, like Pearl, had become disenchanted not only with the extreme and overt racism of the old eugenicists but also with their lack of scientific rigor. Osborn pointed out that old-style eugenicists like Davenport (but, curiously, he did not mention Laughlin in this regard) refused to admit that environment shaped human personality traits in any significant way. The extreme hereditarianism of the "old-timers" was out of touch with reality. Osborn was in a fortunate position to convey his (and Pearl's) views to the ruling-class philanthropists, particularly the Rockefeller foundations (there were several different Rockefeller funds giving money to eugenic projects in the 1930s). In a memo to the "Rockefeller interests" in 1933, Osborn wrote,

> Particularly in this country under the leadership of Davenport at Cold Spring Harbor, the relation between genetics and eugenics was over-stressed, and studies in the mechanism of human heredity were carried forward rapidly and published as evidence in the cause of eugenics without sufficient experimental support in regard to their application to man. At the same time that this development was going on the side of genetics, the American Eugenics Society was formed by Irving Fisher, Popenoe, Johnston, Henry Fairfield Osborn, and others, and started on a sort of Galtonian propaganda without having enlarged their base of factual and experimental studies. Excellent and carefully considered proposals for research were outlined but were not carried out, and gradually the American Eugenics Society came to be run almost entirely by Leon Whitney, the paid Secretary, who was far more in the way of being a promoter and propagandist than a scientist.[47]

Carefully arguing, as did Pearl, against the lack of scientific quality, not the ultimate aim, of eugenics, Osborn was persuasive to the Rockefeller interests. It is important that he chose to focus on the epistemic issues, for it reveals that he continued to believe in eugenic aims while attacking the most neutral aspect of the eugenics argument, its biological or genetic basis. More subtle and polished than either Laughlin or Davenport, Osborn knew that it is always better tactics to win an argument on epistemic rather than nonepistemic grounds, even though the nonepistemic grounds may to him have been just as valid and perhaps ultimately the most important.

But what about the form of closure per se that the eugenics movement experienced? Several forms of closure were visited on the movement, occurring at different times and affecting different aspects of eugenical practice. Within biology and genetics, closure began to

[47] Frederick Osborn, "Memorandum on the Eugenics Situation in the United States," May 24, 1933, written for "the Rockefeller interests" (note penciled in at top of the page in Osborn's hand). Frederick Osborn Papers, American Philosophical Society, Philadelphia.

occur in the mid-1920s, at least publicly, when a number of well-known geneticists openly attacked eugenics as scientifically unsound. Still, as we have seen, that did not immediately end the controversy. The form of closure was not resolution, for the hard-line eugenicists (Davenport, Laughlin, Conklin, East, Grant) did not accept the criticism. In fact, they dismissed it. Yet, as more geneticists (and biologists in general) abandoned the movement, it became increasingly isolated from the scientific community. Although this isolation may not have been apparent to the lay person at the time, the repetitiveness and general sameness of the eugenicists' writings from the mid-1920s onward is an important indication that the field had lost a fruitful contact with legitimate biology.

Ultimately a form of resolution closure was responsible for the decline of the eugenics movement. Unlike the scientific positions described in the other case studies in the Closure Project, with the exception perhaps of the Laetrile case, eugenics was based on distinctly bad – that is, erroneous – science. The fundamental biological (genetic) concepts on which the field was based were shown to be incorrect. The discredited concepts included the idea that single genes produce single traits; that personality and social traits are inherited in patterns of simple dominance or recessiveness; that it would be possible to eliminate deleterious genes from the human population by rigorous selective breeding in just a few generations; and that inbreeding within "superior" stocks produces better offspring than outbreeding to very different stocks. By 1925, all of these eugenical ideas had been demonstrated to be tacitly untrue and this fact ultimately hurt the movement in the place where it was most vulnerable: its scientific credibility. Having been built on the image of applying the newest findings in biology to improving the human social condition, eugenics was hard-pressed to maintain that image in the face of strong criticism from well-respected members of the biological community. The simple fact that eugenics did not (and could not) work wrought a kind of resolution closure on the movement in the long run.

But the eugenics movement suffered another kind of closure, namely, abandonment. It was abandoned on several levels. First, it was abandoned by some of the geneticists themselves. Later it was also abandoned by its ruling-class supporters when the movement had served their purposes. Abandonment by withdrawing funds can be one of the most effective and immediate forms of closure that exists.

A case might be made for considering that eugenics underwent a change of focus; whether or not one wishes to consider that a form of closure is a matter of definition. By change of focus I mean that eugenical ideals did not die out or experience closure; what was closed was the old-style format in which eugenical ideas were stated. As I have mentioned, old-style eugenics became transformed (through the aegis of workers such as Pearl and Osborn) into the more sophisticated idea of population control, a transformation that has been reported elsewhere.[48] Nonetheless, I do want to emphasize that in presenting a case for the highly contextual analysis of closure, we encounter variations in the major categories established elsewhere in this volume. Thus, "abandonment closure" can mean abandonment by the participating experts or, in the case of eugenics, abandonment not only by some participants but by the chief funders as well. Certainly these two forms of abandonment have a number of significant differences, as well as some fundamental similarities. I thus want to sound a word of caution: We must not apply the terminology of closure too rigidly or perfunctorily. As a general taxonomy of types of closure, the classification given by Tom Beauchamp (Chap. 1, this volume) can be very useful; if applied too categorically or simplistically (or, especially, out of sociopolitical context), it can be an impediment rather than an aid to understanding the general problem of closure.

Conclusion

I have tried to show how a Marxist analysis of the development of the American eugenics movement helps us understand two important facets of those scientific controversies that have a social implication: (1) the role of experts in the origin, perpetuation, and demise of the controversy; and (2) the way in which the controversy was closed (that is, what form of closure it experienced). I have argued that both epistemic and nonepistemic factors were involved in the closure of the eugenics movement by 1940. Traditional historians have suggested that eugenics declined largely because (1) eugenicists became propagandists more than scientists, and their ideas, as a consequence, alienated more and more of the public; and (2) the public was disturbed by the similarity between – indeed, in many cases, the sameness of – American eugenics writing and that of the promoters of

[48] Chase, *Legacy*; Linda Gordon, "The Politics of Population: Birth Control and the Eugenics Movement," *Radical America* 8 (1974): 61–97; and "Birth Control and the Eugenists," *Science for the People* 10 (1977): 8–15.

Nazi race hygiene in the 1930s. The fact that many American eugenicists (such as Laughlin, for example) were ardent promoters of Nazi ideas of racial purity was taken as a key factor in turning public opinion away from eugenical ideas.

Without denying the importance of either of these two explanations I have argued that the movement rose and fell *primarily* because it served a distinct *need* of the ruling elites in the United States between 1900 and the mid-1930s, not because it served any *need* of the American people. A dialectic between eugenicists and their ruling-class backers was instrumental in getting the eugenics movement going. But when the movement had accomplished the purpose of the ruling-class, it was dropped. After 1931, Laughlin lost not only his prestigious congressional appointment but ultimately his job. The files of the Carnegie Institution of Washington and the C. B. Davenport Papers at the American Philosophical Society in Philadelphia are peppered with increasingly uncomplimentary remarks about Laughlin's and Davenport's eugenical pronouncements, particularly in the period after 1931. Especially after Vannevar Bush took over the leadership of the Carnegie Institution of Washington from John C. Merriam in 1938, Laughlin was given the distinct impression that his days were numbered. Indeed, he was forcibly retired on December 31, 1939, at the age of 59. The doors of the ERO were closed at the same time, and, symbolically, the nerve center of the American eugenics movement (old-style eugenics) ceased to function. (In this case, institutionally, closure was certainly by decree, or what we might call cloture). At any rate, Laughlin was discarded when his task was finished, and especially when he became an embarrassment (one of his most embarrassing acts was to accept gratefully in 1936 the honorary M. D. degree from Heidelberg University, then under Nazi control; the degree was given for Laughlin's work in improving the human "seed-stock"). Experts are used in this fashion as pawns. Seldom are the experts prepared to be dropped as fully as Laughlin was and sometimes, if they are not embarrassments, they are not dropped so summarily. It is important to see that Laughlin was dropped not by his public but by those who hired him. There are passing references among biologists or nonbiologists in the 1920s and 1930s about the racial bigotry of Laughlin's writings. But there was no concerted effort from eugenicists or the general public to remove Laughlin from power. No doubt there was some outcry and it may have been to that growing public disenchantment with old-style eugenicists that the ruling-class responded. In any event, it is clear that Laughlin rose or fell, throughout his career as an expert in eugenics, on the decisions (and that included, of course, funding) of his wealthy benefactors. This, I suggest, is to a very large degree the position of most, if not all, of the experts who become involved in such social policy-oriented technical controversies.

7

The continental drift debate

HENRY FRANKEL

The sixty-year-old controversy among earth scientists about the reality of continental drift that ended about fifteen years ago stands in stark contrast to the other controversies addressed by participants in the Closure Project. Unlike most of the other controversies, that of continental drift was not a public policy controversy with technical or scientific components. Nor was it a scientific controversy having significant political, economic, or social aspects. It was a controversy among earth scientists about problems within the earth sciences.

Despite the fact that the drift controversy did not involve issues of public policy, it has an interesting dynamics of its own. The major aim of this chapter is to offer an analysis of its dynamics. In the second section of the chapter, I will give a capsule history of the controversy – long enough to provide a defense of my analysis but short enough not to overburden the reader with too much detail. The analysis of the overall controversy is offered in the third section of the chapter. In short, I argue the following:

1. Proponents of competing theories attempt to provide solutions to a common nest of problems that constitute the subject matter of the controversy.
2. Proponents of competing theories attempt to bring up difficulties with their opponents' solutions.
3. Such attacked solutions are defended against the difficulties raised by altering the solutions so as to avoid the difficulties or showing that the difficulties are ill founded. Sometimes the attacked solution is simply replaced by a new solution that avoids the difficulties.
4. Closure of the controversy comes about when one side enjoys a

Research on this project has been supported by the National Endowment for the Humanities and by the National Science Foundation History and Philosophy of Science Program.

recognized advantage in its ability to answer the relevant questions. In light of points 2 and 3, this comes about when one side develops a solution that cannot be destroyed by its opponents.

In the final section I compare the drift controversy with some of the other controversies discussed in this volume. I argue that once a distinction is made between the scientific and public policy aspects of such controversies, the sort of analysis offered in my third section can be employed in analyzing scientific controversies that accompany public policy controversies but cannot be used to explicate the public policy controversy. However, I should add that the most exciting action in these dual controversies, generally, is in the controversy over what policy to pursue.

Historical account of the continental drift debate

Wegener's theory of continental drift

The general problems solved by the drift theory. Although two Americans, Howard Baker and Frank Taylor, developed versions of drift theory, Alfred Wegener was the first to propose an extensive version. Wegener first presented his theory in a 1912 article and expanded the defense of his theory with the 1915 publication of *The Origin of Continents and Oceans.*[1]

The basic aspects of Wegener's theory are easily summarized. He regarded the continents as ships of light sialic material floating upon a heavier basaltic material that formed the sea floor.[2] He hypothesized that the continents underwent horizontal displacement relative to one another by plowing through the denser basaltic ocean floor. During the Carboniferous era all the continents had been united, forming a supercontinent that he called "Pangea." Pangea began to break up during the late Cretaceous or early Tertiary. The Atlantic

[1] The first edition of *Die Entstehung der Kontinente und Ozeane* appeared in 1915 and was followed by four more editions in 1920, 1922, 1929, and 1936. The third and fourth editions were translated into English: *The Origins of Continents and Oceans*, tr. J. Dkerl (London: Methuen, 1924) and *The Origin of Continents and Oceans*, tr. J. Biram (New York: Dover, 1966). Unless otherwise specified, all references to this work are to the 1924 English translation of the 1922 German edition.

[2] Sialic rocks are relatively rich in silicates of aluminum, whereas simatic rocks are richer in silicates of magnesium, iron, and calcium. Sialic rocks are not as dense as simatic ones, and it was generally agreed that the continents were made up of sialic rocks, whereas the ocean basins were taken to be primarily of sima, as was the material underneath continental sial.

continued to open during the Eocene, with the separation of North and South America from Europe and Africa, and Asia moved away from Antarctica and southern Africa by migrating northward and rotating counterclockwise. By the beginning of the Quarternary, Australia finally split off from Antarctica, the Americas migrated westward, Asia continued to drift and rotate, and India, although always attached to Asia, migrated more rapidly than Asia and partially collapsed into it, forming the Himalayas.[3] Moreover, Wegener suggested that polar wandering, that is, a shift of the geographical poles relative to the earth's surface, occurred along with continental drift.

He argued that his theory offered solutions to a number of empirical problems within the earth sciences. Some of the more important ones were as follows:

1. *Why do the contours of the coastlines of eastern South America and western Africa fit together so well, and why are there many similarities between the respective coast lines of North America and Europe?* Here he simply posited that the continents had originally been united and subsequently broke apart.

2. *Why are there numerous geological similarities between Africa and South America, and others between North America and Europe?* He appealed to similarities in the Cape mountains of South Africa and the sierras of Buenos Aires, numerous similarities in the huge gneiss plateaus of Brazil and Africa and the Pleistocene terminal moraines. He also spoke of the continuity of the three major systems of folds between North America and Europe, namely, the Armorican, Caledonian, and Algonkian. In addition, he cited similarities in the geological structure of India, Antarctica, Australia, New Zealand, and New Guinea in further support of his hypothesized Pangea.

3. *Why are there many examples of past and present-day life forms having a geographically disjunctive distribution?* Here Wegener argued that the distribution of the life forms had become disjunctive through the separation of the respective land areas with the breakup of Pangea.

[3] A partial geological time scale is in order: The most recent era is the Cenozoic, followed by the Mesozoic and Paleozoic. The Cenozoic is subdivided into the Quarternary and the Tertiary. The Quarternary began 2 million years ago with the Pleistocene period. The Tertiary extends back approximately 70 million years. Beginning with the most recent period, the Tertiary is made up of the following periods: Pliocene, Miocene, Oligocene, Eocene, and Paleocene. The Mesozoic era is made up of the following three periods: the Cretaceous, which began about 135 million years ago; the Jurassic, beginning about 190 million years ago; and the Triassic, which started about 225 million years ago. The last two periods of the Paleozoic were the Permian and Carboniferous; the latter began about 350 million years ago and the former about 280 million years ago.

4. *Why are mountain ranges usually located along the coastlines of continents, and why are orogenic regions long and narrow in shape?* He hypothesized that the leading edges of drifting continents crumbled as they were compressed by the resisting ocean floor. The Andean range from Alaska to Antarctica was Wegener's most spectacular example. He also claimed that mountain ranges formed when continents collided. The Himalayas formed when India slid into and under Asia.

5. *Why does the earth's crust exhibit two basic elevations, one corresponding to the elevation of the continental tables, the other to the ocean floors?* Wegener argued that "there simply were at one time two undisturbed primal levels" that have remained relatively unchanged once they reached isostatic equilibrium, since, according to his theory, the major diastrophic disturbances were horizontal rather than vertical.

6. *How can we account for the Permo-Carboniferous moraine deposits found in South Africa, Argentina, southern Brazil, India, and in western, central, and eastern Australia?* Wegener's solution was to suppose that the respective continents had been united during the Permo-Carboniferous with the South Pole, located in southern Africa.

Wegener's defense of his theory. Besides showing how his theory solved these empirical problems, Wegener argued that some of the solutions offered by his theory were preferable to those proposed by proponents of the competing fixist theories.[4] I consider two examples – namely, his treatment of fixist solutions to the problems of the disjunctive distribution of life forms and the formation of the Permo-Carboniferous Ice Cap.[5]

[4] The fixists were those who argued that the continents have not changed their position relative to each other through any sort of horizontal displacement. Some argued that former continents had sunk into ocean basins, but they still did not maintain that continents have shifted position through horizontal displacement relative to each other. There were even some fixists who allowed for a wandering of the poles. Again, however, they did not posit relative horizontal displacement of the continents. With polar wandering, either the axis of rotation of the earth was supposed to have shifted, or the whole outer skin of the earth was assumed to have rotated. In contrast to the fixists were the mobilists, or "drifters." They maintained that there had been horizontal displacement of the continents relative to each other. Some of them also posited polar wandering. The modern view, plate tectonics, which developed out of the continental drift theory, does not include the idea of polar wandering. Finally, it should be pointed out that the division of earth scientists into fixists and mobilists, or drifters, is not the invention of historians of science, for various earth scientists referred to fixists and drifters beginning in the early 1920s.

[5] I have examined these two controversies in some detail. See H. Frankel, "The Permo-Carboniferous Ice Cap and Continental Drift," *Proceedings of the Ninth*

The two major fixist solutions to the problem of disjunctive life forms were the land bridge hypothesis and the permanency theory. Those fixists who postulated the existence of former land bridges or submerged continents to account for disjunctive distributions had saddled themselves with a conceptual difficulty: There was no known mechanism for sinking the supposed land bridges into the ocean floor, once they had served as migratory routes, that was consistent with the principle of isostasy. The permanentists, who agreed with Wegener's criticism of the land-bridgers, attempted to explain the spread of disjunctive flora and fauna without resorting to the postulation of former land connections; they appealed to existing land bridges like the Bering bridge and Central America, in conjunction with changes in sea level, as well as allowing for greater mobility of life forms than admitted by land-bridgers. According to Wegener, the permanentists really had no solution whatsoever; they could account for only a few of the many cases of disjunctively distributed life forms. Moreover, Wegener argued that his solution was biologically superior to the land bridge alternative. Drift offered a better account of those cases where past life forms of land mass A and B resembled each other even though A and B are presently at different latitudes and enjoy quite different climates. Here Wegener cited *Glossopteris*, the late Paleozoic flora of the southern hemisphere and peninsular India. *Glossopteris* deposits had been found on all of the southern continents as well as India and had often been associated with glacial deposits. In addition, he argued that only his account could make sense out of the puzzling past and present Australian fauna. Beginning with Wallace's tripartite division of the Australian animal kingdom, Wegener claimed that the nature of each group, as well as the differences among the three groups, could be well explained only by his drift theory. The oldest element, according to Wegener, shows affinities with the fauna of India, Ceylon, Madagascar, and South Africa and originated when Australia was still connected with India. The second faunal element, marsupials and monotremes, shows kindred relationships to South American forms but is quite different from Asiatic ones. Its members originated when Australia was still attached by Antarctica to South America. The third and most recent element, mainly placental mammals, shows similarities to Asiatic forms and came from Asia across islands in recent times. Wegener took the drift theory solution to this aspect of the paleobiogeography

International Congress of Carboniferous Stratigraphy and Geology, vol. 1, pp. 113–20, and "The Paleobiogeographical Debate over the Problem of Disjunctively Distributed Life Forms," *Studies in the History and Philosophy of Science* 12 (1981):221 – 59.

problem "to be of quite considerable importance to the question of displacement." Thus Wegener argued for the superiority of his solution to the problem of life forms having a geographically disjunctive distribution by pointing out that his solution accounted for more aspects of the problem than either fixist alternative.

Wegener also thought his solution to the problem of the origin of the Permo-Carboniferous Ice Cap superior to the fixist alternatives. Indeed, he thought he had the only solution. Of the fixist solutions, the only one that Wegener thought even worth considering was the polar wandering solution proposed by several fixists. They suggested that the ice cap could have formed if the South Pole had been located at its apparent center during the Permo-Carboniferous. Wegener, however, brought up an anomalous difficulty with their solution that his solution did not face.

> If we merely place the South Pole in the conceivably best position (50° S. 45° E.) in the midst of these traces of ice, the remotest traces of land-ice in Brazil, India and Eastern Australia would possess a geographical latitude of not quite 10°, that is, a complete hemisphere of the earth was buried beneath the ice, and had therefore also the polar climate necessary for that purpose. The other hemisphere, however, the Carboniferous and Permian deposits of which are in most areas very well known, does not show sure traces of glaciation, but on the contrary at many places the remains of tropical vegetation. It need scarcely be said that this result is absurd.[6]

Wegener also thought his having the only solution to be a weighty reason for accepting the reality of drift, it being "of great importance" in deciding "the correctness of the displacement theory."[7]

Besides pointing out problems that his drift theory was able to solve, as well as the advantages of at least some of the solutions that his theory had to offer compared to the competing fixist theories, Wegener admitted that his drift theory introduced a new problem, namely, that of providing a mechanism for propelling the continents through the sea floor as they underwent horizontal displacement. Of the known forces, there were none of sufficient strength to push the continents through the basaltic sea floor. Wegener, however, suggested two mechanisms: tidal force and pole-flight force. *Pohlfluct* is a differential gravitational force caused by the elliptical shape and the spinning motion of the earth. The continents would tend to move toward the equator through the action of centrifugal force brought

[6] Wegener, *Origins*, p. 85.
[7] Ibid., p. 98.

about by the spinning and shape of the earth. Thus, the continents supposedly would flee from the poles, since the equatorial radius of the earth is larger than the polar radius. Tidal force, on the other hand, was taken by Wegener to provide the requisite westward drift of the continents. These tidal forces were actually stresses brought about by the gravitational action of the sun and the moon. Wegener claimed that these tidal stresses, which slowed the earth's diurnal eastward motion, would act most strongly on the surface of the earth. As a result, their action would lead to "a slow sliding motion of the whole crust or of the individual continental blocks" in a westerly direction. Wegener, at least by the late 1920s, was not too sanguine about his solution to the mechanism problem, but it was the best he had to offer.

> The Newton of drift theory has not yet appeared. His absence need cause no anxiety; the theory is still young and still often treated with suspicion. In the long run, one cannot blame a theoretician for hesitating to spend time and trouble on explaining a law about whose validity no unanimity prevails. It is probable, at any rate, that the complete solution of the problem of the driving forces will still be a long time coming, for it means the unravelling of a whole tangle of interdependent phenomena, where it is often hard to distinguish what is cause and what is effect.[8]

The fixist response to Wegener's drift theory

The fixists responded to Wegener in three characteristic ways. First, they pointed out the difficulty of finding an adequate mechanism for the drifting of the continents. Not only did they argue that Wegener's solution was wholly inadequate, but they claimed that it was highly unlikely that any answer to the problem would be satisfactory. Moreover, they augmented this difficulty by arguing that even if there were sufficient forces to propel the continents through the sea floor, the continents never would have been able to survive such a journey and remain intact, since sea floor material is stronger than continental material. Second, they ferreted out difficulties with some of the solutions offered by Wegener to empirical problems. Here they typically brought up anomalies that showed up when his solutions were used and argued that much of the support for his solutions was based upon unreliable or incomplete data. Third, they improved their own solutions in response to Wegener's criticisms.

[8] Wegener, *Entstehung* (1929), p. 167.

The mechanism objection. The most serious of all the attacks on Wegener's theory was over developing a mechanism for the drifting of the continents. There were, I believe, three reasons why this criticism carried such weight. First, the mechanism difficulty was directed at the central element of Wegener's theory, namely, that the continents had displaced themselves horizontally by plowing through the sea floor. This idea was at the heart of Wegener's theory; it served as a crucial premise in every one of his empirical solutions except the one concerning the two basic elevations of the earth's crust. Second, only drift theories proposed lengthy horizontal movements of the continents. Third, this mechanism objection was actually more extensive than Wegener had suggested. Wegener viewed the problem as simply a matter of finding forces sufficiently strong to propel the continents, but critics were quick to point out that the problem also concerned, among other things, the question of whether the continents could survive such vast movements without crumbling, regardless of whether there were sufficient forces to propel them.

During the 1920s, numerous drift opponents raised this mechanism objection, but the most important antidrifter to voice the objection was Harold Jeffreys. Jeffreys was perhaps the most influential geophysicist in Great Britain during the first half of the twentieth century, and he was wont to dismiss the drift theory because of the mechanism objection. As far as Jeffreys was concerned, having the continents plow their way through the sea floor would be like attempting to thrust a leaden chisel into steel.

Wegener and some of his followers offered a reply to this objection, but their answer carried little weight. Basically they argued for a distinction between "residual rigidity," or strength, and "rigidity," or plasticity. At the 1926 Symposium of the American Association of Petroleum Geologists (AAPG), W. van der Gracht argued as follows:

> This . . . [objection depends upon] . . . the usual misconception concerning the physical properties of matter which apply to this problem: a confusion of "*rigidity*" with "*residual rigidity*" or "strength." The sima has greater "rigidity" than the sial, but the latter, as a whole, evidently has greater resistance against long enduring stresses . . . Whether a substance is "hard" or "soft," like steel or lead, has not necessarily anything to do with its "strength." In looking at similies, substances should be used whose properties are more within our grasp, because we can handle them under room temperature, and ordinary pressures and time intervals. So I shall refer again to beeswax and pitch. Pitch has great "rigidity" . . . but extremely little "strength,"

> while beeswax has little rigidity, but considerable strength. We can perfectly well press a chisel of soft beeswax into a block of hard pitch, *provided we push in our chisel slowly enough*. That is what happens both in an isostatic adjustment and at the front of a continent floating forward into the sima.[9]

However, this ploy gained no additional support for Wegener. First, as Wegener admitted, there was not any solid data on viscosity. Second, by the time of the 1926 AAPG symposium, lesser mortals than Harold Jeffreys had expanded the "mechanism" objection. At the symposium, Bailey Willis and William Bowie argued that seismological studies indicated a sima with great residual rigidity (strength) and rigidity; that tidal forces would act more strongly on the sima than the sial, since the former was denser than the latter; and that if the continents were to drift, their trailing edges would split apart through the action of strong tensional forces. Third, as far as Jeffreys was concerned, the distinction between "strength" and "rigidity" missed the boat. Jeffreys wholeheartedly agreed with the distinction, but he believed that sima had greater rigidity and strength than sial. Thus it would not matter how long small forces acted on the sial; it could not move through the sima. And if there were sufficient forces, the sial would just crumble.

Drifters had another, much more potent answer to the mechanism difficulty: They had Arthur Holmes.[10] Holmes, perhaps the only British earth scientist willing and able to take on Jeffreys, won six major prizes in earth sciences, including the Vetlesen Prize, was a major force in the development of the Phanerozoic time scale, and wrote one of the most influential textbooks on physical geology in the English language. In 1928, 1931, and in various editions of his textbook throughout the 1940s, 1950s, and 1960s, Holmes provided drifters with a mechanism. He originally believed in the contraction theory of mountain building. Contractionists, including Jeffreys, argued that the earth had been contracting due to an overall loss of heat. They suggested that mountain ranges would develop as the crust folded upon itself, since the mantle contracted more quickly

9 W. Van Watershoot van der Gracht, "Remarks Regarding the Papers Offered by the Other Contributors to the Symposium," in *Theory of Continental Drift: A Symposium on the Origin and Movement of Land Masses both Inter-continental and Intra-continental, as Proposed by Alfred Wegener*, ed. W. Van W. van der Gracht (Tulsa: American Association of Petroleum Geologists), p. 225.

10 For a detailed discussion of the development of Holmes's ideas about continental drift, see H. Frankel, "Arthur Holmes and Continental Drift," and "Biogeography before and after the rise of seafloor spreading," *Studies in the History and Philosophy of Science* 15 (1984): 141–67.

than the crust. By the mid-1920s, Holmes had rejected contractionism. He argued that the discovery of abundant amounts of radioactivity made the hypothesis that the earth has been losing heat untenable. Although Holmes was quite impressed with several of the drift theory solutions, especially the one concerning the formation of the Permo-Carboniferous Ice Cap, he thought Wegener's mechanism for the drifting of the continents wholly inadequate. Holmes saw a way to solve the mechanism difficulty once he came to view radioactivity as an active force. He suggested that escaping heat would bring about convection currents that could make the mantle "fluid." These currents would arise under continents, since sial is more radioactive than sima. As a result, the currents would split the continents apart and create a new ocean basin in the wake of the drifting continents. The continental blocks would be carried along on the backs of the horizontally moving convection currents. Finally, the laterally moving currents would descend into the interior of the earth at the peripheral regions of the continent, where they would meet weaker convection currents from oceanic regions. Here the ocean floors would sink because of certain phase changes brought about by the convergence of the convection currents. Holmes argued that this mechanism appealed to a force of sufficient magnitude to move the continents and would allow them to move through the sea floor because of the effect of the convection currents. Even Harold Jeffreys viewed Holmes's mechanism as an improvement over that of Wegener, but Jeffreys still considered it unreasonable. In 1931 Jeffreys remarked, "I have examined Professor Holmes' theory of subcrustal currents to some extent, and have not found any test that appears decisive for or against. So far as I can see there is nothing inherently impossible in it, but the association of the conditions that would be required to make it work would be rather in the nature of a fluke."[11]

Wegener, on the other hand, thought better of Holmes's proposal and suggested, in his 1929 edition of *The Origin of Continents and Oceans*, that it should be pursued. Unfortunately, there was little way to test Holmes's proposal when it was introduced.

Debates over the problems of disjunctively distributed life forms and the formation of the Permo-Carboniferous Ice Cap. The other two modes of response by the fixists were to bring up difficulties with Wegener's solutions to the empirical problems addressed by his theory and to

[11] H. Jeffreys, in "Problems of the Earth's Crust: A Discussion in Section E (Geography) of the British Assoc. on 28 September 1931 in the Hall of the Society," *Geographical Journal* 78 (1931): 453.

eliminate the force of his attacks upon their own solutions by showing that they were mistaken or by improving or developing new versions. The best way to illustrate these two fixist responses is to trace the debates between Wegener and other drifters with the fixists over who had the best solution to the problems of the formation of the Permo-Carboniferous Ice Cap and the existence of life forms having geographically disjunctive distributions. Not only will this exemplify these two ploys on the part of the fixists, but it also will show how the drifters responded to the difficulties and improvements brought up by the fixists.

The two paleoclimatologists of fixist persuasion who brought up objections to Wegener's solution to the ice cap problem were A. P. Coleman and C. E. P. Brooks.[12] Coleman, a highly respected Canadian climatologist and geologist, raised two difficulties with Wegener's solution during the 1920s and early 1930s, and Brooks, an eminent British climatologist, brought up a third objection in the mid-1920s and continued to repeat it into the late 1940s.[13] These objections, which became the standard attack against Wegener's solution, were as follows:

1. Wegener's solution conflicted with standard accounts of glaciation. Most climatologists agreed that in order to have glaciation there had to be massive amounts of snowfall and that abundant amounts of moisture were required to produce enough snow. However, given Wegener's grouping of the southern continents around the South Pole in one large land mass, they argued that the central region of the postulated massive area would have been as dry as the Gobi Desert in Central Asia, since it would have been far from any existing ocean and thus would have had no source of moisture available for abundant snowfall.

2. The tillite data indicated that the massive ice cap had united about the South Pole at sea level. However, given Wegener's reconstruction, it was reasonable to expect that it would have formed at a

[12] The positions of the major participants mentioned in this aspect of the drift controversy break down as follows: among the drifters were Wegener, Holmes, Alex du Toit, and George C. Simpson; the fixists were A. P. Coleman, C. P. E. Brooks, Charles Schuchert, and Bailey Willis.

[13] For A. P. Coleman, see "The Wegener Hypothesis: Discussion (Geological Section) British *Association for the Advancement of Science,*" *Nature* 111 (1923): 30–1; "Ice Ages and Drift of the Continents," *American Journal of Science* Ser. 5, 7 (1924): 398–404; "Permo-Carboniferous Glaciation and the Wegener Hypothesis," *Nature* 115 (1925): 603; *Ice Ages, Recent and Ancient* (New York: MacMillan, 1926); and "Ice Ages and the Drift of Continents," *Journal of Geology* 41 (1933): 409. As for C. E. P. Brooks, see *Climate through the Ages* (London: Ernest Benn, 1926) and the 1949 revised edition (New York: McGraw-Hill).

higher elevation. Thus the elevation of the glacial deposits was anomalous with Wegener's solution.

3. According to Wegener's solution, North America was located near the equator during the Lower Carboniferous when the Permo-Carboniferous ice sheets spread over Gondwanaland, and, therefore, any evidence of Permo-Carboniferous glaciation in North America would be anomalous for drifters. Three such cases had been discovered, and their existence was often referred to in the literature. The troublesome cases were the apparent conglomerates in Oklahoma and Kansas and the famous Squantum tillite near Boston.

Fixists also developed their own solution to the formation of the ice cap. Indeed, this was Brooks's greatest contribution to the debate, for in 1926 he developed a solution that was mathematically sophisticated enough even to impress Sir Harold Jeffreys. Charles Schuchert also presented a very similar fixist solution in the same year.[14] Both Brooks and Schuchert utilized the "standard" fixist geographical map of the Upper Carboniferous, with its extensive, horn-shaped Gondwanaland and the Tethys and Volga seas. Extensive glaciation in the Southern Hemisphere would come about through

1. The diversion of the whole of the equatorial ocean current into the Northern Hemisphere, which thereby became abnormally warm
2. An extensive elevated continent along the equator but extending much farther into the southern than northern latitudes
3. A southern ocean shut off by land barriers from all warm currents
4. Possibly a general refrigeration that might be due to the presence of abnormally large quantities of volcanic dust

They had an answer for the assumed glaciation of North America through the interaction of the mild conditions brought about by the Volga Sea with the Arctic Ocean, which extended down as far as the southern tip of Greenland. Finally, Schuchert argued that his solution was climatologically superior to Wegener's drift alternative, and he, along with Bailey Willis, who was a vehement fixist, claimed that the solution was preferable to Wegener's, since it did not require the geophysically unpalatable claim that the continents had undergone large-scale displacement.

Wegener and two of the drift theory's most important supporters, Alex du Toit and Arthur Holmes, all responded to these various criticisms, and Holmes addressed himself to the Brooks-Schuchert-Willis alternative. Du Toit was one of South Africa's most respected earth scientists. He was considered a superb field geologist, became

[14] C. Schuchert, "The Paleogeography of Permian Time in Relation to the Geography of Earlier and Later Periods," *Proceedings of the Second Pan-Pacific Science Congress* (1926), pp 1079–91.

president of numerous South African geological societies, and was a foreign member of the Royal Society. Moreover, he was Wegener's most vehement defender in the southern hemisphere until the 1950s. In 1921 du Toit developed a slightly different reconstruction of Gondwana. According to his reconstruction, the present continents were not united, only very close to one another.[15] This reconstruction was recognized by Coleman in 1924 as "a wise variation since it provides open water for evaporation," and it thereby served as a rebuttal to Coleman's first criticism.[16] It also allowed for making sense out of the sea level glacial deposits. It is interesting to note, however, that Coleman never referred to du Toit's reconstruction after 1924. Perhaps he felt that it was not worthwhile to pursue it because of the other objections. Wegener and Holmes were especially active in rebutting the third objection, although Holmes mostly repeated Wegener's rebuttal. Wegener admitted that if "all or any of these conglomerates are truly glacial, they would be in flagrant contradiction" to his theory, and he then attempted to shift the burden of proof to the fixists by arguing that these apparent glacial indicators stood in sharp contrast to the rest of the abundant fossil evidence from the United States of a tropical climate during the Permo-Carboniferous.[17] Holmes brought up four objections to the fixist alternative solution. His first two objections were geological and geophysical.

> [Brooks has] adopted as a working hypothesis "a great plateau in the interior of Gondwanaland, rising gradually to an elevation of 10,000 feet." This assumption of great height certainly eases the meteorological problem, but it has no geological justification. Moreover, it doubles the difficulty of the physical problem, for now we should have to explain, first a great thickening of the sial of Gondwanaland and then its total disappearance from the very extensive oceanic areas that now intervene between the existing southern continents.

His latter two objections were climatological. First, he emphasized the fact that even Brooks was unhappy with the theory of the glaciation of India, and then he argued that the inability of both Wegener's and Brooks's solutions to account for the nonglaciation of Antarctica was "certainly more damaging to [Brooks's] attempt at a solution than it is to the less rigid hypothesis of continental drift."[18]

Drifters also received aid from Sir George C. Simpson, who was the

[15] See, for example, A. du Toit, "Land Connections between the Other Continents and South Africa in the Past," *South African Journal of Science* 18 (1921): 120–140.

[16] Coleman, "Ice Ages," p. 402.

[17] A. Wegener, "My Theory of Continental Drift," in van der Gracht, *Theory*.

[18] A. Holmes, "A Review of the Continental Drift Hypothesis," *Mining Magazine* 40 (1929): 340.

most respected British climatologist to defend Wegener's drift theory. He was a fellow of the Royal Society, president of the Royal Meteorological Society (1941–2), and a recipient of the Symonds Gold Medal in 1930. Simpson offered support for Wegener's theory at the 1923 meeting of the Geological Section of the British Association, where he reportedly claimed that Wegener's theory "was a wonderful one from the meteorological point of view, as it explained the marked changes of climate given by the geological record."[19] But Simpson's greatest contribution came in his 1929 Alexander Pedler Lecture, *Past Climates,* wherein he developed a meteorologically sophisticated solution to the general problem of glaciation that even offered an account of interglacial periods – a factor lacking in Brooks's account. Simpson made it quite clear that his meteorological solution depended upon Wegener's theory of the displacement of the continents.

Simpson's contribution effectively closed the debate until the rise of paleomagnetism. Du Toit's *Our Wandering Continents* appeared in 1937; Wegener elaborated upon his solution in the 1929 edition of *Die Entstehung der Kontinente und Ozeane;* Holmes defended drift in his 1944 *Principles of Geology;* but nothing of real importance was brought to bear on the debate that would give a decided advantage to fixist or mobilist solutions to the problem.[20] Holmes nicely summed up the status of the debate in his review of the 1949 symposium on "The Role of the South Atlantic Basin in Biogeography and Evolution" held by the Society for the Study of Evolution by saying,

> Meanwhile there remains the most serious enigma of all: the Permo-Carboniferous glaciation. Dunbar points out that the Paleozoic glaciations in low latitudes present "a problem still unsolved, unless we accept continental drift." But if we accept continental drift only to explain these and other still older glaciations, it becomes *ad hoc* hypothesis. As such, it may still be justified as a stimulant to research, but it may also stand in the way of progress by distracting attention from the real problem. Can the meteorologist not come to our assistance and tell us whether or not widespread equatorial and low-latitude glaciation is possible while high latitudes for the most part enjoy a genial climate?[21]

19 For G. C. Simpson, see "The Wegener Hypothesis," *Nature* 3 (1923): 30–1, and "Past Climates," *Manchester Mem. Lit. Phil. Soc.* 74 (1929): 1–34. The quoted passage is from the 1923 work.

20 A. du Toit, *Our Wandering Continents* (Edinburgh: Oliver Boyd, 1937); A. Holmes, *The Principles of Physical Geology* (London: Nelson, 1944).

21 A. Holmes, "The South Atlantic: Land Bridges or Continental Drift?", *Nature* 171 (1953): 671.

Unfortunately, the meteorologist could not come to Holmes's aid.

As already pointed out, Wegener argued for the superiority of his view over both fixist alternatives, and he was particularly hard on the permanentist solution in claiming that it really provided no solution whatsoever.[22] Both land-bridgers and permanentists raised difficulties with Wegener's solution and provided improved solutions of their own. Indeed, the ensuing debate exhibited the same moves and countermoves on the part of both fixists and drifters. However, in this case I believe that the permanentists were able to launch a stronger attack against the Wegenerian solution than was mounted by fixists against the drift solution to the question of the formation of the Permo-Carboniferous Ice Cap. Nevertheless, no party gained anything approaching a clear-cut victory, and the debate continued into the 1960s – it was not settled until after the confirmation of sea floor spreading and plate tectonics.

Two typical difficulties raised against Wegener's solution were as follows: Land-bridgers claimed that the circum-Pacific similarities were just as compelling as the Atlantic ones but that they were much more difficult to explain with Wegener's positioning of the continents than if he had left the continents fixed in their present positions. Others brought up a specific difficulty with Wegener's account of *Glossopteris*. They appealed to the apparent *Glossopteris* finds in Russia and argued that Wegener's reconstruction of Pangea made it extremely difficult to account for the presence of Russian *Glossopteris* – more difficult than if he had left the continents in their present position.

But, the most vehement attack against the drift solution came from George Gaylord Simpson, the Harold Jeffreys of the American paleontological community. Simpson was a forceful defender of permanency theory throughout most of his career. He developed a highly sophisticated permanentist solution to the disjunctive distribution problem by greatly expanding upon the work of another prominent American paleontologist, W. D. Matthew, and raised numerous objections to both drift theory and the doctrine of sunken land bridges. I briefly outline the major aspects of Simpson's contribution to the debate.[23]

1. He developed his solution to account for the disjunctive dis-

[22] The major participants of this aspect of the overall drift controversy who are discussed or mentioned in the text are (1) drifters: Alex du Toit and Arthur Holmes; (2) fixists: W. D. Matthew and George Gaylord Simpson.

[23] See, for example, G. G. Simpson, "Mammals and the Nature of Continents," *American Journal of Science* 241 (1943): 1–31 and "Probabilities of Dispersal in Geologic Time," in "The Problem of Land Connections across the South Atlantic with Special Reference to the Mesozoic," *Bulletin of the American Museum of Natural History* 99 (1952): 163–76.

tribution of mammals, although there were occasions when he claimed that it supported the permanency of continents throughout the whole time involved in mammalian history.

2. He delineated three types of migratory routes: corridors, filter bridges, and sweepstakes routes. Corridors are links between regions where no barrier exists. Filter bridges connect regions that have some similar life forms but quite different faunas. Sweepstakes link together regions of quite different fauna that have a few forms in common. He named the latter route "sweepstakes" to emphasize their indeterministic or adventitious character and to stress the point that he thought sweepstakes were the best way to account for situations where similarities were few, preferable to the ad hoc devices utilized by permanentists to account for recalcitrant cases.

3. He developed quantitative techniques for evaluating the degree of similarity between various fauna and applied it, for example, to the Triassic reptile faunas of South America and South Africa, one of the cases cited by drifters. Simpson argued that, given the available data, no connection was indicated; the similarities were less than presently obtaining.

4. Simpson was contemptuous of much of the literature dealing with the problem and showed little mercy in his treatment of Wegener, du Toit, and various land-bridgers. Throughout his attack he concentrated upon showing that his opponents had grossly overestimated the number of legitimate cases of disjunctively distributed organisms by (1) utilizing unreliable and inaccurate data based upon inadequate taxonomical criteria or mistaken identification of fossil specimens; (2) incorrectly summarizing data drawn from studies other than their own; (3) neglecting to take into account new fossil finds and advances in taxonomy; (4) underestimating the dispersal ability of various organisms; and (5) neglecting to consider the possibility of parallel or convergent evolution. Besides discrediting his competitors, this attack had the effect of decreasing quite considerably the number of legitimate cases of disjunctively distributed life forms. Of course, this was just what Simpson, a permanentist, desired. In short, Simpson argued that drifters postulated corridors where filter bridges or even sweepstakes routes offered a better solution, since the similarities were considerably fewer than supposed by his opponents.

5. Simpson also offered a permanentist solution to the question of the Australian fauna, criticized the drift account, and argued that his solution was superior. He claimed that Australia had been in its present position, and that marsupials had come from Asia, not South America, by way of a sweepstakes route that placentals just hap-

pened not to have taken. Simpson admitted that fossil marsupials had not been found in Asia but pointed out that his opponents had grossly overestimated the similarity between South American and Australian marsupials. Moreover, he objected to the South American connection on the basis that it would have led to the migration of South American placentals to Australia.

Simpson's attack on the drift theory solution to the problem of disjunctive distribution did not go unanswered, for Alex du Toit wrote a rejoinder.[24] Du Toit's major points were as follows. He attempted to sidestep Simpson's critique by arguing that the distribution of mammals was at odds with the distribution of other life forms and had come about after most drifting had occurred. As a result, mammalian distribution should offer little support for drift compared to the distribution of other life forms. But in light of Simpson's attack, du Toit admitted that he had formerly placed too much importance on the drift theory's offering a solution to the overall problem of disjunctive distribution. He suggested that because of the unreliability and incompleteness of the data base in paleobiogeography, arguments could be constructed that supported either drift or fixism. In support of his assessment of the biogeographical data base, du Toit strategically cited the then agreed-upon, erroneous identification of *Glossopteris* flora in Russia, thereby removing the apparent anomaly from drift's solution to the distribution of *Glossopteris*.

Simpson's influence with the paleontological community was tremendous. American paleontologists, almost without exception, were strong supporters of permanency. Moreover, Simpson's influence was international. For example, although Australian zoologists were sympathetic toward the drift theory in the 1920s and 1930s, it fell out of favor after Simpson's analysis. There also were European supporters of Simpson's view. Even Arthur Holmes, in his review of the 1949 New York symposium on Trans–South Atlantic Mesozoic land bridges, admitted that "in recent years the weight of evidence has become less oppressive, and this symposium has left me with the general impression that a few land bridges or linkages by island stepping-stones would probably suffice for the biogeographical problems."[25] Any advantage accrued to the drift theory because of its solution to the problem had been neutralized by Simpson's development of a permanentist solution. If anything, the permanentists had gained the advantage.

[24] A. du Toit, "Tertiary Mammals and Continental Drift," *American Journal of Science* 242 (1944): 145–63.
[25] Holmes, "South Atlantic," p. 671.

Major changes in the drift debate: the development of paleomagnetism and marine geology

In 1950 the British Association for the Advancement of Science held a meeting in Birmingham on the question of drift, and what is most apparent from the report is that nothing new was introduced. Supporters for both positions basically summarized the same arguments that had been offered in the years leading up to World War II. Neither group made much headway in improving the explanatory power of its theory or in changing opponents' minds. It is clear that new developments were needed. However, it was not clear from where these new advances would come. It turned out that they arose from two relatively unexpected quarters, namely, paleomagnetism and marine geology.

Although several areas of paleomagnetism eventually affected the debate, the results having an initial impact came from the so-called directional studies.[26] This area of paleomagnetism, begun in the early 1950s at Cambridge University by S. K. Runcorn and at Imperial College by P. M. S. Blackett, soon appeared to offer a completely new means of determining the previous positions of a land mass throughout different geological periods. However, neither Blackett nor Runcorn initially turned to paleomagnetism in order to further the drift debate. Their original motivation was to learn more about the history of the earth's magnetic field. Blackett, Nobel Prize laureate for his work on cosmic rays, had become interested in magnetism after World War II. This led to his further development and pursuit of the hypothesis that any rotating body possesses a magnetic field because of its rotating. He designed a sensitive magnetometer to test his theory. Although the results were negative, he realized that he could use his magnetometer for paleomagnetic investigations to find out about the history of the earth's magnetic field. Runcorn, who had been a lecturer under Blackett when they were both at Manchester University, shared Blackett's interest in paleomagnetism and set up a

[26] The rise of paleomagnetism is discussed from a slightly different point of view in H. Frankel, "Problem-solving, Research Traditions, and the Development of Scientific Fields," P. Asquith and R. Giere, eds., *Proceedings of the 1980 Meeting of the Philosophy of Science Association* (East Lansing, Mich; Philosophy of Science Association, 1980), pp. 29–40. Those involved with the development or criticism of paleomagnetism and its role in the drift debate who are discussed in this section include (1) drifters: P. M. S. Blackett, S. K. Runcorn (became a drifter in 1956, was fixist before then), E. Irving, and R. A. Fisher; (2) fixists: H. Jeffreys, A. Cox, and D. Doell (Cox and Doell became drifters in the mid-1960s). I would also like to thank E. Irving for discussions of the rise of paleomagnetism and its role in the revival of drift theory.

research unit at Cambridge University. He got the program going in the spring and summer of 1951. Blackett also left for Imperial College, at Manchester, where he and A. J. Clegg set up a research team. They began working after the Cambridge group.

The first among the Cambridge group to realize the possibility of using paleomagnetism to test the drift theory was one of Runcorn's students, E. Irving, who was hired by Runcorn in June of 1951. Irving, a Cambridge undergraduate majoring in geology, always had thought the drift theory a good idea and argued for it at his 1951 presidential address to the Sedgwick Club, the Cambridge undergraduate geology club. Irving understood the relevance of paleomagnetism to the drift question rather early and initiated steps for obtaining samples from India in the fall of 1951, for he realized that India was critical, since it supposedly had drifted from the Southern to the Northern Hemisphere. Upon analysis of the Indian samples in late 1952 or early 1953, Irving reached the conclusion that India had been in the Southern Hemisphere, as Wegener had maintained.

Runcorn did not come to accept the drift theory until 1956. In 1955 he claimed that polar wandering without drift could account for the paleomagnetic data, but he changed his mind the following year and argued in favor of drift and polar wandering. Blackett's acceptance of the drift theory was rather different. By the time the group at Imperial College began working, he had come to view the drift theory as a promising hypothesis. Some paleontologists at Imperial College began to collect samples from India in 1954, and they and Blackett endorsed the drift theory shortly thereafter. By this time the race was on to collect samples throughout the world in order to test the drift theory, and by 1959 paleomagnetists had marshaled an impressive array of data in support of drift.

The basic idea behind these directional studies was quite simple: It was that the ancient *geographical* position of a land mass could be determined on the basis of its ancient *magnetic* position. That is, if an area's position were known with respect to the magnetic pole during a past geological period, its position with respect to the geographical pole could be calculated. However, carrying out this rather straightforward idea turned out to be extremely difficult. To make it work one needed to have (1) reliable paleomagnetic data; (2) a statistical means of analyzing or packaging the data; and (3) a way of interpreting the paleomagnetic data in terms of paleogeography. As it turned out, obtaining reliable data proved to require sophisticated techniques; for statistical method, the development of a statistics that could be applied on a sphere was needed; and interpreting the data in terms of paleogeography depended upon a crucial assumption about the behavior of the earth's magnetic field through geological time. In obtaining data, researchers had to determine issues such as the following: the age of the tested sample; whether the sample had undergone local reorientation after it had formed; whether the remanent magnetism of the sample was stable; and whether the sample had acquired additional magnetization after it had formed. Various researchers worked out a whole list of procedures that were to be followed in order to avoid using unreliable samples.

Fortunately the second issue, that of statistically packaging the data, was solved by Sir Ronald A. Fisher. Fisher developed the requisite statistics in very late 1950 or early 1951, soon after he was approached by Runcorn about the matter; it turned out that Fisher thought the continental drift theory a good idea and had faith in the possibilities of the directional studies. The greatest difficulties, however, surrounded the move from paleomagnetism to paleogeography. The key theoretical assumption involved in the extrapolation was that the earth's magnetic field, when averaged over periods of several thousand years (to get rid of secular variation), is that which would be maintained by a geocentric, axially directed dipole – that is, the geomagnetic and geographical poles, on the average, have coincided. In defense of this assumption, these paleomagnetists appealed to the paleomagnetic work of Jan Hospers, another Cambridge paleomagnetist, who in the early 1950s showed that the poles had coincided for the last 23 million years. They also appealed to paleoclimatological data that proved to be consistent with the assumption, and to the Elsasser-Bullard self-exciting dynamo theory for the origin of the geomagnetic field, because it required the coincidence of the rotational and geomagnetic axes. But there were additional difficulties, for there was a tremendous amount of controversy even among paleomagnetists about precisely what could be inferred about past latitudes and longitudes of given samples and the appropriate landmasses, even if the sample's ancient magnetic dip (inclination) and declination were known. It was generally agreed that the former latitude of a landmass could be determined from its previous inclination. But there were genuine arguments among various paleomagnetists about what if anything could be inferred from a sample's declination. The difficulty was that there was no way to decide from the declinational data whether the given area had rotated or undergone displacement. This was particularly important, since the case for continental drift, or relative displacement of the continents with respect to each other, depended in many cases on longitudinal shifts rather than latitudinal ones. For example, the opening of the Atlantic, with the shift of the Americas and of Africa and Europe, was supposed to be a longitudinal shift. Fortunately, there were latitudinal shifts, for example, with Australia and India, but the general indeterminacy of longitude hurt the paleomagnetic case for drift. It even caused a number of disagreements within the paleomagnetic community. Finally, some paleomagnetists argued that a case could be made for rapid shifts of the geomagnetic pole *on the basis of* the paleomagnetic data.

There is no question that the directional studies failed to bring about closure of the debate. Some fixists suggested that the studies were highly questionable. They supported this claim by pointing out how the drift interpretation depended upon the assumption of the coincidence of the magnetic and rotational axes, how obtaining reliable data rested upon a number of highly technical and exacting procedures, how paleo-longitudes were indeterminate, and how some of the paleomagnetic data did not fit the drift interpretation. Jeffreys, for example, dismissed paleomagnetism in the 1959 edition of *The Earth* with the following remark:

> When I last did a magnetic experiment (about 1909) we were warned against careless handling of permanent magnetism and the magnetism was liable to change without much carelessness. In studying the magnetism of rocks, the specimen has to be broken off with a geologic hammer and then carried to the laboratory. It is supposed that in the process, its magnetism does not change to any important extent, and though I often asked how this comes to be the case, I have never received any answer.[27]

Many fixists simply paid no attention to the results. They just seemed too complicated from the outside, especially to earth scientists who had not been trained in geophysics. Other fixists did not know about the results. This was the case, I believe, with many North American earth scientists. Many of them did not become acquainted with the directional studies until 1960, when Alan Cox and Richard Doell, two highly respected paleomagnetists from the U.S. Geological Survey, wrote a lengthy review of the directional studies.[28] And to make matters worse, the Cox and Doell review turned out to be a "fence-sitting" contribution to the paleomagnetic literature. They cautiously opted for continential drift in the Carboniferous and the Permian, but they argued that the paleomagnetic results from the Mesozoic and

[27] Jeffreys, *The Earth: Its Origin, History, and Physical Constitution*, 4th ed. (Cambridge: Cambridge University Press, 1959), pp. 367–8.

[28] Cox and R. Doell, "Review of Paleomagnetism," *Bulletin of the Geological Society of America* 71 (1960): 645–768.

early Tertiary were better accounted for by assuming rapid shifts in the earth's magnetic field rather than appealing to drift. Moreover, the general tone of their article was extreme caution about supporting drift and drawing inferences about paleogeography from paleomagnetism. For example, the section of their abstract in which they cited their conclusions about drift reads as follows:

> Evaluation of the data summarized leads to the following general conclusions:
> (1) The earth's average magnetic field, throughout Oligocene to recent time, has very closely approximated that due to a dipole at the center of the earth oriented parallel to the present axis of rotation.
> (2) Paleomagnetic results for the Mesozoic and early Tertiary might be explained more plausibly by a relatively rapidly changing magnetic field, with or without wandering of the rotational pole, than by large-scale continental drift.
> (3) The Carboniferous and especially the Permian magnetic fields were relatively very "steady" and were vastly different from the present configuration of the field.
> (4) The Precambrian magnetic field was different from the present field configuration and, considering the time spanned, was remarkably consistent for all continents.[29]

If it was not apparent from the abstract, their endorsement of drift is stated in their third point. It was help like this that drifters in America did not need.

Of course, old drifters welcomed the positive results from paleomagnetism. They argued that the studies provided independent evidence for their position and created a battery of data that was difficult for fixists to explain. Some drifters suggested that the results should be taken into consideration when dealing with problems in their own field and proceeded to follow their own advice. This was particularly so with paleoclimatologists and paleobiogeographers.

Perhaps Great Britain was the only place where the results from paleomagnetism changed people's minds or at least encouraged them to reconsider the drift theory. British earth scientists were much more aware of the results than, for example, their counterparts in North America. Some even found out about them in their courses. Moreover, Blackett and Fisher were giants in British science. Sir Edward Bullard, who was impressed with the paleomagnetic results and converted to the drift theory primarily because of them, also was highly respected. Arthur Holmes, as was to be expected, gave a

[29] Ibid., p. 645.

favorable review of the paleomagnetic case for drift in the 1965 edition of his *Principles of Physical Geology.*[30] Thus, despite Jeffreys's resistance, several highly respected British scientists not only endorsed the findings from paleomagnetism but were quite instrumental in promoting them. Regardless of the reasons, the reception of paleomagnetism was much more favorable in Britain than elsewhere, and the development of this field was a major reason why the drift theory enjoyed a more favorable climate in Britain during the early sixties than it did in North America. In 1964 the Royal Society held a symposium on continental drift.[31] Most of the discussions were favorable to drift. The organizers of the symposium were Blackett, Bullard, and Runcorn.

Marine geology underwent tremendous growth in the early fifties. Although advances in this field initially had much less impact on the drift – fixist debate than the results from the directional studies in paleomagnetism, they, and not the paleomagnetic results, led to the closure of the overall debate. By the mid-1960s the combined efforts of several research teams had turned the sea floors from areas of ignorance into places of backyard familiarity. Some researchers uncovered the vast network of mid-ocean ridges; others continued earlier investigations of ocean trenches. Surveys of the northeastern Pacific Basin revealed a network of gigantic, east-west trending fracture zones. These fracture zones were wholly unexpected. But there was more, namely, the discovery of amazing patterns of magnetic anomalies between neighboring fracture zones. The patterns were made up of lengthy, north-south trending magnetic anomalies exhibiting high magnetic readings and separated by similarly shaped and trending magnetic anomalies displaying low magnetic readings. Moreover, several investigators soon realized that these zebra patterns of magnetic anomalies between neighboring fracture zones could be fairly well matched without too much distortion, if it were supposed that the sea floor between the fracture zones had undergone lengthy horizontal displacement along the fracture zones. Investigation of the sedimentary deposits upon the sea floor revealed two more surprises: The thickness of the sedimentary deposits turned out to be much less than had been expected on the basis of extrapolating from present-day deposition rates and, despite numerous dredge and core samples, researchers had yet to find sediments older than Cretaceous.

[30] *Principles of Physical Geology*, rev. ed. (London: Nelson, 1965).

[31] The published account of the Royal Society Meeting on drift is in *Proceedings of the Royal Society* (A) 258 (1965): 1–323.

All of these discoveries led to a number of unsolved problems within marine geology. Perhaps the most important of them, at least the one that received the most attention, was the question of the origin of the vast system of oceanic ridges. Although the data on ridges were often incomplete and indicated differences, ridges generally were associated with the following: a mid-ocean position; a central rift valley; shallow earthquakes; positive gravity anomalies; a central positive magnetic anomaly; seismic velocities in between standard ocean basin crust and mantle; high heat flows; little sediment; and a relatively short life span. Various hypotheses were proposed to account for the origin of ridges. Many of them were fixist in framework. For example, the late Maurice Ewing, original head and undisputed leader of Lamont Observatory at Columbia University; the late Harry Hess, first-rate theorist and member of Princeton's Department of Geology and Geophysics; and H. W. Menard, one of the major figures at Scripps Institution of Oceanography and former head of the U.S. Geological Survey all proposed fixist solutions.[32] However, there were several theorists who offered accounts of the origin and development of oceanic ridges that had as a consequence the drifting of the continents. Two of the leading exponents of the expanding earth hypothesis, B. Heezen and W. S. Carey, the former from Lamont and the latter from the University of Tasmania, both offered their own accounts of the origin of oceanic ridges that had as consequences the expansion of the earth and a possible mechanism for the drifting of the continents.[33] And even Harry Hess and H. W. Menard provided solutions to the problem that involved continental drift.[34]

In retrospect, Harry Hess's drift solution, which was labeled "sea floor spreading," turned out to be the most important of the various hypotheses. Besides explaining the development of oceanic ridges,

[32] See, for example, John Ewing and Maurice Ewing, "Seismic-refraction Measurements in the Atlantic Ocean Basins, in the Mediterranean Sea, on the Mid-Atlantic Ridge, and in the Norwegian Sea," *Bulletin of the Geological Society of America* 70 (1959): 291–318; H. W. Menard, "Sea Floor Relief and Mantle Convection," *Physics and Chemistry of the Earth* 6 (1966): 315–64; and H. H. Hess, "Serpentines, Orogeny, and Epeirogeny," *Geological Society of America*, Special Paper 62 (1955): 391–406.

[33] Heezen eventually gave up the idea of an expanding earth for a form of continental drift in the mid-1960s, but Carey is still arguing in favor of the expanding earth view. Few others subscribe to his view.

[34] See, for example, H. W. Menard, *Marine Geology of the Pacific* (New York: McGraw-Hill, 1964), pp. 150–1, and H. H. Hess, "History of Ocean Basins," in *Petrologic Studies: A Volume to Honor A. F. Buddington* (Boulder, Colo.: Geological Society of America, 1962), pp. 599–620.

his hypothesis offered a solution to the mechanism, or geophysical, difficulty faced by the drift hypothesis and provided answers to the question of the origin and development of oceanic trenches as well as the lack of Pre-Cretaceous sea floor sediment. Moreover, it spawned two important corollaries, namely, the Vine-Matthews-Morley hypothesis and Wilson's transform fault hypothesis. Confirmation of the former led to the acceptance of Hess's hypothesis, and further development of the latter gave rise to plate tectonics. Of all the theoretical innovations that led to the closure of the drift–fixist debate, Hess's was the most important.

Hess first presented his idea of sea floor spreading in a preprint that he finished in December of 1960.[35] He proposed that sea floor material is shaped along ridge axes as it is forced up from the mantle because of the action of rising convection currents; that it spreads out perpendicularly from ridge axes along a horizontal creating new ocean basins; and that it subsequently sinks into the mantle, forming oceanic trenches along the periphery of ocean basins. His hypothesis offered a solution to the origin of oceanic ridges, for it could account for their median position, rift valley, positive heat flow data, positive gravity anomalies, seismic velocity data, shallow earthquakes, lack of sediment, and ephemeral nature. Moreover, it explained the formation of trenches as well as the fact that no sea floor sediments older than Cretaceous had been uncovered. Trenches represented the descending limbs of convection currents. There were no pre-Cretaceous sediments upon the sea floor because existing ocean basins were no older than Cretaceous.

Hess had not been a drifter until he developed his idea of sea floor spreading. However, since he saw that continental drift was a consequence of his hypothesis but was impressed with the ability of his hypothesis to offer solutions to these important problems in marine geology (as well as others), he became a drifter. But there were two other reasons why Hess was not unhappy about having to endorse the drift theory. First, he had been impressed with the developments in paleomagnetism that indicated relative displacement of the continents. Second, he realized that his idea of sea floor spreading offered drifters a way around the mechanism difficulty.

[35] Hess did not publish his thesis until 1962 (Hess, "History of Ocean Basins"). However, he circulated a preprint of his ideas in December of 1960, and there is little difference between the preprint and the published version; see Hess, "Preprint" (1960). For a much more extensive treatment of the development of Hess's ideas, see H. Frankel, "Hess's Development of his Seafloor Spreading Hypothesis," in T. Nickles, ed., *Scientific Discovery: Case Studies* (Dordrecht: Reidel, 1980), pp. 345–66.

"[My hypothesis] is not exactly the same as continental drift. The continents do not plow through oceanic crust impelled by unknown forces, rather they ride passively on mantle material as it comes to the surface at the crest of the ridge and then moves laterally away from it."[36]

Fred Vine and Drummond Matthews published their version of the Vine-Mattews-Morley hypothesis in 1963 when Vine was a graduate student at Cambridge working under Matthews.[37] Lawrence W. Morley, a Canadian paleomagnetist, independently proposed his own version of the hypothesis in the same year, only to see it rejected by two journals.[38] The hypothesis was the obvious, mutual consequence of two rather tentative hypotheses, namely Hess's sea floor spreading theory and the idea that the earth's magnetic field undergoes repeated reversals in polarity. The beauty of the Vine-Matthews-Morley proposal was that they even thought of relating Hess's idea with the notion that the geomagnetic field reverses itself. They reasoned that if these two hypotheses were correct, then "blocks of alternatively normal and reversely magnetized material would drift away from the center of the ridge and parallel to the crest of it"; they supposed that new sea floor material, like other igneous material, becomes magnetized in a direction parallel to the geomagnetic field upon cooling as it passes through its curie point – the temperature at which it begins to solidify. But there was more to the Vine-Matthews-Morley hypothesis: It provided for two aspects of the overall problem of marine magnetic anomalies, namely, the existence of the large magnetic anomaly associated with ridge axes, and the origin and formation of the zebra pattern of magnetic anomalies in the north-eastern Pacific basin. The Vine-Matthews-Morley hypothesis was the first hypothesis offering a common solution to both these aspects of the overall problem of marine magnetic anomalies.[39]

Although it was somewhat surprising that Harry Hess became the major defender of the drift theory from marine geology, it would have been even more surprising if Maurice Ewing had not played the

[36] H. H. Hess, "Preprint," p. 16.

[37] F. Vine and D. Matthews, "Magnetic Anomalies over Ocean Ridges," *Nature* 199 (1963): 947–9.

[38] L. W. Morley, "Letter," in John Lear, "Canada's Unappreciated Role as Scientific Innovator," *Saturday Review*, September 2, 1967, 45–50. Morley's original paper was, at least for the most part, reproduced in Lear's article (letter to author from L. W. Morley, October 15, 1976).

[39] For a much more extensive treatment of the Vine-Matthews-Morley hypothesis and its reception and acceptance, see H. Frankel, "The Development, Reception and Acceptance of the Vine-Matthews-Morley Hypothesis," *Historical Studies in the Physical Sciences* 13 (1982): 1–39.

same role for fixism. Ewing was a staunch defender of fixism throughout his career. He argued against Wegener's theory as early as 1949, when he presented a paper at the New York meeting arranged by the Society for the Study of Evolution on the question of Mesozoic land bridges across the Atlantic, was quick to attack Hess's hypothesis of sea floor spreading throughout the mid-1960s, and did not come to accept it until most active researchers in marine geology had become drifters.[40] Ewing also presented fixists with their own solution to the question of the origin and development of mid-ocean ridges. And, more important, this solution of Ewing's became the central element of the fixist alternative to Hess's theory, for fixists had a worthy competitor to Hess once Ewing's solution was combined with fixist solutions to problems such as the origin of oceanic trenches and the apparent lack of pre-Cretaceous sediments upon the sea floor. Moreover, some of Ewing's coworkers at Lamont were able to devise an auxiliary hypothesis for this solution that provided fixists with their own answer to the overall problem of marine magnetic anomalies.

In 1959, Maurice Ewing, along with his brother John, proposed that oceanic ridges are formed by the upwelling of basalt through tensional cracks in the sea floor. "The [Mid-Atlantic] Ridge has been built by the upwelling of great amounts of basalt magma along a tensional fracture zone. Presumably the extensional forces and the supply of basalt magma come from convection currents deep in the mantle."[41] The hypothesis accounted for the elevation of the ridge, its seismic and heat flow data, the presence of shallow earthquakes and a central rift valley, and the ridge's apparent ephemeral nature. In addition, Ewing's view was easily linked with fixist solutions to the problems of the paucity and youthfulness of sea floor sediments and the origin of oceanic trenches. The standard fixist solution to the sea floor sediment problem was to suppose that there existed a layer of consolidated sediments directly underneath the layer of unconsolidated sediments. Fixists argued that the amount of sediment upon the sea floor either only appeared to be less than expected on the basis of extrapolating throughout geological time from present-day rates or was only slightly less than expected, since there was another layer of highly compacted sediments underneath the top layer of unconsolidated sediment. Such a claim avoided having to deny the principle of uniformity and was consistent with the seismic velocity data,

[40] However, even Ewing accepted sea floor spreading and the Vine-Matthews-Morley hypothesis before most earth scientists who were not in marine geology.

[41] J. Ewing and M. Ewing, "Seismic-refraction measurements," p. 291.

although it was by no means established by the data. As for the problem of the origin and development of oceanic trenches, there were several fixist hypotheses to choose from, ranging from Marshall Kay's hypothesis, which linked oceanic trenches and island arcs with mountain building and the accretion of continents, to Vening Meinesz's convection current hypothesis, which also offered an account of the formation of island arcs.[42]

Closure of the drift–fixist debate

The single most important factor in bringing about the closure of the drift–fixist debate was the vindication of the Vine-Matthews-Morley hypothesis. Vine himself came to accept the hypothesis in late 1965. At about the same time, some of Ewing's staff at Lamont devised a competing hypothesis to Vine-Matthews-Morley that nicely extended Ewing's solution to the origin of oceanic ridges. But by the beginning of 1966, these Lamont workers switched to Vine-Matthews-Morley and sea floor spreading, for they realized that their new data supported Vine-Matthews-Morley but not their own hypothesis. They publicly supported Vine-Matthews-Morley in April of 1966 at the annual meeting of the American Geophysical Union, where they presented their evidence in favor of sea floor spreading. Even Maurice Ewing came to accept sea floor spreading by April of 1967 and cited (along with his brother) the case for Vine-Matthews-Morley as "strong support for the hypothesis of spreading."[43] Moreover, some of the Lamont seismologists were so impressed with the strength of the emerging case in favor of Vine-Matthews-Morley at Lamont during the first few months of 1966 that they began to analyze their data on earthquakes along the Mid-Atlantic Ridge and several trenches in order to test Tuzo Wilson's idea of transform faults, an idea derived from Hess that made sense only if sea floor spreading took place. Before the year was out, they found that their data supported Wilson's idea and could not be explained by invoking standard analyses of earthquake motions that did not involve sea

42 Both Kay and Meinesz were highly respected earth scientists. Kay was a major figure in the Department of Geology at Columbia. For many years he had been an antidrifter, but he switched around 1967. Vening Meinesz, a leading Dutch earth scientist, was an early believer in convection currents. He was best known for his work on oceanic gravity anomalies. Moreover, Meinesz was highly respected by Hess. Indeed, when Hess began his career, Meinesz was already a highly respected figure, and Hess got the chance to work with him doing oceanic gravity anomaly studies.

43 J. Ewing and M. Ewing, "Sediment Distribution on the Mid-ocean Ridges with Respect to Spreading of the Sea Floor," *Science* 156 (1967): 1590.

floor spreading.[44] Their confirmation of Wilson's ideas was the second most important factor in bringing the debate to a close. Thus, the strength of the case for Vine-Matthews-Morley converted most oceanographers who were participants in the debate to the drift theory and was primarily responsible for the testing of the idea of transform faults, whose confirmation was the second most important factor in ending the debate.

Once the strength of the case for Vine-Matthews-Morley and Wilson was became apparent, the resistance among fixists who knew about the results almost completely disappeared. The central issue was no longer whether drift was a reality; it became a question of working out the implications of sea floor spreading and continental drift. By the end of 1968, the implications already had been worked out: Plate tectonic theory was independently developed by Jason Morgan and Dick McKenzie; researchers, primarily at Lamont, began applying plate tectonic theory and the sea floor spreading theory to determine in quantitative terms the movements of various plates; and the seismologists at Lamont who had originally tested Wilson argued that "within the entire field of seismology there appears to be no serious obstacles to the new tectonics."[45] Application of sea floor spreading and plate tectonic theory quickly spread to other fields.

Old supporters of fixism from areas of the earth science where the debate had flourished before Hess had thought of sea floor spreading came to accept the new tectonics. But they endorsed it because of its successes in marine geology and seismology, not because they suddenly realized that it offered a better solution to problems within their own field. Perhaps nobody expressed this more adamantly than George Gaylord Simpson. In 1971, Simpson wrote the following about the discovery of *Lystrosaurus* and the bearing of paleobiogeographical evidence on the acceptance of the drift theory:

> As far as I have been able to learn, the crucial *Lystrosaurus* fauna now known in Antarctica, Africa, and Central Asia is not known in South America. The real evidence for Triassic union of South America and Africa still seems to be nonpaleontological... I now believe that continental drift did occur (and in other

[44] Wilson, a Canadian geophysicist, was a fixist up to around 1959. He switched to drift in the very early 1960s and ended up being a major contributor to the development of plate tectonics. The development of his idea during this period is nicely discussed by R. Laudan, "The Method of Multiple Working Hypotheses and the Development of Plate Tectonic Theory," in Nickles, *Scientific Discovery*, pp. 331–43.

[45] B. Isacks, J. Oliver, and L. Sykes, "Seismology and the New Global Tectonics," *Journal of Geophysical Research* 73 (1968): 5855.

> biogeographic distributions) in the Mesozoic, but direct fossil evidence is still curiously scanty or equivocal. The Cenozoic evidence indicates that drift had little or no zoogeographic effect in that era as regards most of the continents, including all for which the evidence is considerable or reasonably adequate... The whole subject of plate tectonics is very exciting and has revivified geology. I think it is great. But as inevitably happens some students get over-excited... Now *everything* has to relate to plate tectonics. I don't go into detail here, but it is rather amusing that currently there are *four completely different* plate tectonic "explanations" of the early distribution of marsupials, none of them based on a reasonable balance of evidence.[46]

The closure of the debate in areas of the earth sciences in which the debate had formerly taken place occurred because of advances in plate tectonics. In most cases, these old fixists switched to the drift theory once they understood the successes of sea floor spreading and plate tectonics, and then began applying the new theory of continental drift to problems within their own field.

When the Vine-Matthews-Morley hypothesis was proposed in 1963, it faced four serious difficulties. Vine gave a neat summary of them in 1966 and even said of his hypothesis that "at the time this concept was proposed there was very little concrete evidence to support it, and in some ways it posed more problems than it solved."[47] Two of the four difficulties concerned the two major background assumptions underlying the hypothesis, namely, sea floor spreading and the idea of geomagnetic polarity reversals. The other two difficulties pertained to the lack of data in support of the hypothesis. Those who did not believe in Hess were not going to accept Vine-Matthews-Morley in 1963, since it "in some ways posed more problems than it solved." The reality of geomagnetic polarity reversals was also highly suspect. In 1963 they certainly were not accepted. There was no known mechanism for producing them, and there was the alternative hypothesis of self-reversals. Moreover, both major research groups that unearthed the time scale – the Cox-Doell-Dalrymple group from the U.S. Geological Survey, and the McDougall-Tarling group of the Australian National Observatory – found discrepancies in their time scales, had trouble isolating the shorter reversal periods called "events," and always expressed their conclusions with extreme caution. Indeed, Cox and Doell did not

[46] G. G. Simpson, "Drift Theory: Antarctica and Central Asia," *Science* 170 (1970): 678.

[47] F. Vine, "Spreading of the Ocean Floor: New Evidence," *Science* 154 (1966): 1406.

endorse, in print, geomagnetic reversals over self-reversals until June of 1964.[48] The two data difficulties were that (1) many did not believe that the magnetic anomalies of the northeastern Pacific Basin were associated with a ridge, and (2) it was not thought that there were any existing surveys of spreading ridges surrounded by a zebra pattern of magnetic anomalies. The identifiable ridges were associated with only a large magnetic anomaly along their axis.

Vine's first major application of his hypothesis was in 1965, which proved to be a very fruitful year for Vine, Hess, and Wilson and the fortunes of sea floor spreading and continental drift. Wilson spent much of the year visiting Cambridge, and Hess joined him during the second term. During this time Wilson invented his notion of transform faults, and Vine and Wilson applied it, along with the Vine-Matthews-Morley hypothesis, to the Juan de Fuca Ridge – a short extension of the East Pacific Rise off the coast of Vancouver. Wilson came up with the idea of transform faults when thinking about the kinds of movements that would occur if the sea floor spreading theory were correct. He realized that fault motions had to be transformed in direction: The motion of sea floor material was upward at ridge axes; transformed to the horizontal as sea floor spread out from ridge axes along fracture zones that offset ridge segments from each other; and transformed downward at trenches.[49] Among other things, this idea of transform faults allowed Vine and Wilson to treat the San Andreas Fault as a transform fault connecting the Juan de Fuca Ridge with the East Pacific Rise, off Baja California. Without making such a move, they would have been forced to maintain that the Juan de Fuca Ridge either was not part of the East Pacific Rise or that the East Pacific Rise extended under California. Moreover, they would not have been able to explain the relation between the Juan de Fuca Ridge and the San Andreas Fault, which would have caused some embarrassment, since the ridge and the fault run into each other. This application of both the Wilson and the Vine-Matthews-Morley hypotheses to the Juan de Fuca Ridge was extremely important. First, it offered a solution to the two specific data difficulties that the latter hypothesis had encountered, for Vine and Wilson had regarded the Juan de Fuca Ridge as an actively spreading ridge surrounded by a striped pattern of magnetic anomalies located in the northeastern Pacific basin. Second, in analyzing the

48 Cox and Doell, "Reversals of the Earth's Magnetic Field," *Science* 144 (1964): 1537–8.

49 See, for example, J. T. Wilson, "Transform Faults, Oceanic Ridges and Magnetic Anomalies Southwest of Vancouver Island," *Science* 150 (1965): 482–3, and F. Vine and J. T. Wilson, "Magnetic Anomalies over a Young Oceanic Ridge off Vancouver Island," *Science* 150 (1965): 484–9.

Juan de Fuca Ridge, Vine and Wilson realized that if the Vine-Matthews-Morley theory were correct and sea floor spreading occurred at fairly rapid rates, the magnetic anomaly pattern on both sides of the ridge should be symmetrical. The remaining reason for the importance of this application was that Vine and Wilson attempted to determine the rate of sea floor spreading by applying the geomagnetic polarity time scale to the pattern of magnetic anomalies. Unfortunately they ended up with an irregular rate of spreading that was at odds with the conditions that they had laid down for the development of symmetrical patterns. As it turned out, they were working with an incorrect time scale: The time scale they used did not include the Jaramillo "reversal" and therefore they got a highly irregular spreading rate for the Juan de Fuca Ridge.

Vine became completely convinced of the correctness of his hypothesis immediately upon finding out about the missing Jaramillo event, which Brent Dalrymple brought to his attention at the annual meeting of the Geological Society of America in November of 1965. "We were just talking and Brent told me about the Jaramillo event, which, of course, was actually published, I suppose, in mid-1966, and I was fascinated and then the penny dropped, you know, because I realized that the Jaramillo was on the Juan de Fuca survey and you could then interpret the survey using a constant spreading rate" (personal communication). Vine had all he needed, namely, a ridge with clearly defined linear anomalies arranged in a symmetrical pattern, and the correct reversal time scale. Moreover, the assumption concerning the geomagnetic time scale had become fairly secure. As far as he was concerned, his hypothesis and that of sea floor spreading were no longer dubious.

The major alternative hypothesis to Vine-Matthews-Morley was developed by J. Heirtzler, X. Le Pichon, and M. Talwani of Lamont. It underwent three applications: They applied it to the Mid-Atlantic Ridge in the spring of 1965 and to the Reykjanes Ridge and Northeastern Pacific Basin in the fall of 1965, just after the publication of Vine and Wilson's work on the Juan de Fuca Ridge.[50] They discarded the hypothesis in the winter of 1966, when they came to accept Vine-Matthews-Morley and sea floor spreading.

The Lamont solution, one that was easily grafted onto Ewing's hypothesis of the origin of oceanic ridges, was as follows: "The axial

[50] See J. R. Heirtzler and X. Le Pichon, "Magnetic Anomalies over the Mid-Atlantic Ridge," *Journal of Geophysical Research* 70 (1965): 4031; M. Talwani, X. Le Pichon, and J. R. Heirtzler, "East Pacific Rise: The Magnetic Pattern and the Fracture Zones," *Science* 150 (1965): 1109; and J. R. Heirtzler, X. Le Pichon, and J. Gregory Baron, "Magnetic Anomalies over the Reykjanes Ridge," *Deep-Sea Research* 13 (1966): 436.

anomaly pattern is attributed to fractures of the crust under tension and intrusion of volcanic material" that possesses greater magnetism than the surrounding material, and the "differences between the two types is attributed to a difference in the fracture pattern." "The possibility exists that the volcanic material filling this fracture is an alkali basalt, which is known to be more magnetic than the theleitic basalt that probably forms most of the ocean floor."[51] The Lamont group also offered several criticisms of the Vine-Matthews-Morley and sea floor spreading theories, in addition to providing their own alternative.

The Lamont group's switch to Vine-Matthews-Morley occurred during the winter and early spring of 1966, and this turnabout was almost instantaneous. The two items that initiated such a spectacular endorsement of the Vine-Matthews-Morley and sea floor spreading theories were the Eltanin-19 profile and the indpendent construction of the geomagnetic polarity reversal time scale, with the discovery of the Jaramillo reversal. Walter Pitman, one of Heirtzler's students, was responsible for the analysis of Eltanin-19, and Neil Opdyke, who had performed paleomagnetic studies under Runcorn and was the only Lamont advocate of drift, constructed the time scale.[52]

There were two striking and important characteristics of the Eltanin-19 profile that were relevant to the acceptance of the Vine-Matthews-Morley hypothesis. The symmetry of the profile was tremendous. Pitman and others at Lamont even superimposed transparencies of the profiles, reversing one of the transparencies so as to display the symmetry. The second significant aspect of the profile was the marked linearity of the anomalies. It was just as obvious as the symmetry, and, for that matter, it was the striking linearity that gave sufficient definition to the magnetic anomaly pattern to allow for noticing the symmetry.

At around the same time that Pitman was analyzing the Eltanin-19 profile, Opdyke was constructing his time scale from an analysis of sea floor sediment cores. He found that the record of reversals of the earth's magnetic field had been recorded by sea floor sediments. Opdyke's work was particularly important, for his procedure for constructing the reversal time scale was completely independent from the one developed by the researchers who used potassium – argon dating of volcanic basalts. It gave a completely different

[51] Heirtzler and Le Pichon, "Magnetic Anomalies," pp. 4031, 4028.

[52] W. C. Pitman III and J. R. Heirtzler, "Magnetic Anomalies over the Pacific Antarctic Ridge" *Science* 154 (1966): 1164, and N. D. Opdyke, B. Glass, J. D. Hays, et al., "Paleomagnetic Study of Antarctic Deep-Sea Cores," *Science* 154 (1966): 349–57.

confirmation of the idea that the earth's magnetic field undergoes repeated reversals, even down to matching up the number and duration of the various reversals.

The Lamont group became convinced of Vine-Matthews-Morley when they, like Vine, had a spectacular profile and the missing Jaramillo event. Finding the missing event for themselves was particularly important, since they had not known about Cox, Doell, and Dalrymple's independent discovery of it. Vine told them about it when he visited Lamont at the invitation of Neil Opdyke in February of 1966.

> Walt [Pitman] and Neil [Opdyke] were analyzing their data in the same room. When I walked into the room Neil was pouring over a light-table and he was drawing up the diagram – which appeared in his 1966 paper. Although I think the first thing we talked about was that, I distinctly remember that all Walt's profiles, the Eltanin-19 profiles, were stacked up on the opposite wall, and so I looked from one to the other. Neil said, "Look, Fred, fantastic – we just discovered a new event. We call it the such-and-such event." I said, "Oh, yes, I hate to tell you this Neil, but Cox, Doell and Dalrymple have discovered that event and they named it and presented it." He was just astounded. And I said, "Yes, Neil, it is called the Jaramillo. Moreover, here it is on the Eltanin-19 profile." They both looked at Eltanin and looked back at me. They said, "My God!" (Taped interview, August 1979)[53]

This dropping of the penny was the second one for Vine and the Lamont group. They both had independent confirmation of the Jaramillo event, and Vine had another clear-cut profile. These new profiles allowed Pitman, Heirtzler, and Vine to compute profiles for each ridge displaying the sum pattern of magnetic anomalies, which indicated that ridges had constant spreading rates or uneven rates that varied at the same amount, at the same time.

The Vine-Matthews-Morley hypothesis was accepted upon the analysis of several excellent profiles and the discovery of the Jaramillo event. There were four particular factors involved with these two items: (1) the symmetry of magnetic anomalies; (2) the linearity of the anomalies; (3) the determination of an accurate time scale and attendant grounding of the assumption that the earth's magnetic field

[53] Opdyke has confirmed most of what Vine recalls, except that Opdyke doesn't remember Pitman's profiles being up on the wall. Opdyke added that they had named the missing event (the Jaramillo) the "Emperor" and that they had realized before Vine came through that the Vine and Wilson analysis of the Juan de Fuca Ridge was wrong because Vine and Wilson had been working with an incorrect timescale. (N. Opdyke, taped interview, 1982)

undergoes repeated reversals; and (4) the fact that the profiles from the various ridges had similar shapes when analyzed in terms of sea floor spreading and the corrected time scale. Each one of these factors improved the overall problem-solving effectiveness of the Vine-Matthews-Morley hypothesis compared to its major alternative. Moreover, in no case could the relevant data be dismissed as unreliable or the background assumption as dubious or unfounded. Consider the four factors: The symmetry of the magnetic anomaly pattern for the Juan de Fuca Ridge, the Pacific–Antarctic Ridge, and the Reykjanes Ridge was remarkable; it was difficult to deny its presence. Although the symmetry pattern was consistent with the alternative developed by Heirtzler and his coworkers, it was a corollary of the Vine-Matthews-Morley hypothesis. The situation was similar with the linearity factor. The data was clear-cut. It was so good that it allowed for one-to-one matching in a region on the Pacific–Antarctic Ridge that Pitman and Heirtzler suggested would resemble an offset region in the northeastern Pacific. This is precisely what Talwani and his coworkers suggested would be unfavorable to their alternative; they wanted only similarities, as opposed to precise matchups. The grounding of the assumption that the earth's magnetic field undergoes repeated reversals and the development of an improved geomagnetic reversal time scale improved the overall problem-solving effectiveness of the Vine-Matthews-Morley hypothesis, because it no longer depended upon a dubious assumption. In this case there was no direct loss in the problem-solving effectiveness of the Lamont alternative but a decided gain in the problem-solving effectiveness of Vine-Matthews-Morley. Finally, there was no reason, if the Lamont hypothesis were correct, why profiles from different ridges should have similar shapes when analyzed in terms of the geomagnetic reversal time scale and constant rates of spreading. But if one assumed sea floor spreading, it was just what would be expected, given Vine-Matthews-Morley. In 1963 the Vine-Matthews-Morley hypothesis "posed more problems than it solved." However, by the end of 1965 and beginning of 1966 it solved more problems than the competing hypothesis – which was beginning to pose more problems than it solved.

Why the drift debate closed when it did

Weighty problem solutions

The preceding historical analysis of the drift–fixist debate shows that proponents of either persuasion wanted to show that their theory enjoyed a greater problem-solving effectiveness than the competing theory. But does it provide any hint as to how they attempted to augment the problem-solving effectiveness of their theory, other than by solving additional problems? I think so. It suggests that the most important kind of problem solution, the one everyone wanted on his list of solved problems, would have possessed the following characteristics:

P_1. It would not have appeared on the opponents' list of solved problems.

P_2. It would not have been plagued by various difficulties: The data used to support the hypothesis would have been reliable and complete. The solution would not have faced anomalies. It would not have contained essential premises or underlying assumptions that clashed with well-founded scientific theories, accepted methodologies, or were simply controversial.

P_3. It would have enjoyed evidential support from two or more independent sources or from an area that had not been used to support other solved problems possessed by the theory.

Why is it desirable to have problem solutions characterized by P_1 through P_3? P_1 solutions offer proponents of the respective theory the opportunity or argue that the evidence in favor of the solution in question supports only their theory, since they, but not their opponents, can account for the evidence. They can claim that they have a positive test for their theory: A test that counts in favor of their position but against their opponents' views. This does not mean that they can justifiably claim that they have proved their theory. Their opponents may come up with their own solution; it may turn out that the solution faces unsuspected difficulties; or additional investigations may uncover unexpected difficulties. But until and unless such things occur, proponents of a P_1 solution can argue that it provides a weighty reason for accepting their theory. Problem solutions that are saddled with the sorts of difficulties enumerated under P_2 make easy targets for competitors. If the data base supportive of a problem solution is unreliable, opponents can claim that the empirical support for the solution is undermined. If the data base is incomplete, they can argue that additional support is required. Anomalies suggest the need for alterations of the solution, or defenders of the solution have to show that the anomaly is itself derived from a faulty data base. When solutions contain premises or utilize assumptions that clash with well-supported theories, opponents can argue that there is little reason to accept the solution, since it would require rejection of the well-supported theory. Finally, if problem solutions are at odds with a successful methodology, proponents must show why they have not followed the methodological convention. P_3 solutions that are supported by independent lines of evidence allow for consistency tests through comparison of the results from both tests, whereas P_3 solutions, that enjoy evidential support from areas that have not been used to support other solutions possessed by the theory often allow for consistency tests with the other solutions and increase the range of the theory.

Standard research strategies

Assuming that participants in the drift–fixist debate wanted to show that their theory enjoyed a decided advantage over the competition in its problem solving effectiveness and that attempting to develop solutions with P characteristics was standard procedure among drifters and fixists, then proponents of either side should have engaged in the following sorts of research strategies (S = strategy, T = theory).

S_1. Proponents of T_1 develop a P_1 solution. Proponents of T_1 should argue that their having the P_1 solution is a weighty reason for accepting T_1.

This is precisely what Wegener suggested about his having a solution to the origin of the Permo-Carboniferous Ice Cap. He also made similar claims about his solution to the paleobiogeographical problem when discussing his accounts of *Glossopteris* and the nature and origin of Australian fauna. Old drifters followed the same strategy in discussing the pro-drift results from paleomagnetism.

S_2. Proponents of T_1 develop a P_1 solution. Proponents of T_2 should attempt to mitigate the advantage enjoyed by T_1 for having its P_1 solution. There are two straightforward strategies for bringing this about. Proponents of T_2 can develop their own solution to the problem, thereby removing the T_1 solution from its list of P_1 solutions. They can raise difficulties with the P_1 solution of T_1.

Schuchert, Willis, and Brooks pursued the first strategy in responding to Wegener's P_1 solution to problem of the origin of the ice cap, for they presented a fixist account of the glaciation. G. G. Simpson undertook a similar move when he devised his solution to the origin of Australian marsupials. Coleman followed the second strategy when he brought up several difficulties with the drift solution to the origin of the ice cap. He pointed to the anomalous glacial deposits in the United States and argued that if the continents had been arranged as Wegener suggested, no ice would have formed, since there would not have been enough moisture. Fixists also pointed out anomalies in the drift theory solution to the paleobiogeography problem. They cited the presence of *Glossopteris* in Russia and appealed to the numerous trans-Pacific disjunctively distributed life forms. Simpson stressed the incompleteness and unreliability of the data used to support the drift solution. Moreover, Simpson argued that the drift solution did not fit with accepted views about the migratory abilities and habits of life forms. According to Simpson, if Wegener were correct, life forms having a disjunctive distribution should be the rule, not the exception. Fixists raised questions about the reliability and completeness of the data base used by paleomagnetists to support the drift theory. Fixists also argued that the underlying assumption of the coincidence of the rotational and magnetic axes was controversial. They appealed to anomalous paleomagnetic data and even developed the alternative solution of a rapidly changing magnetic field.

S_3. Suppose that proponents of T_2 were successful in pursuing S_2 strategies. Proponents of T_1 should counter with two strategies: Show that the T_2 solution is plagued with serious difficulties. Remove the difficulties hampering their own solution that were brought up by T_2 proponents by altering their solution accordingly, devising a new one, or showing that the difficulties are themselves not well founded.

A number of drifters followed the first alternative when responding to the fixist development of a solution to the problem of the formation of the ice cap. Wegener argued that there was reason to believe that the glacial deposits in the United States had not been correctly analyzed. Holmes argued that the fixist solution depended upon several highly controversial geological assumptions and claimed that Brook's attempt to account for the glaciation of India simply did not work. G. C. Simpson followed the second strategy and developed a more sophisticated account of glaciation. Du Toit also pursued the second alternative and improved Wegener's solution, in order to eliminate Coleman's criticism to the effect that no ice cap would form if Wegener had his way. When Jeffreys laid out the mechanism objection to drift, and lesser fixists repeated his objections in dismissing various solutions offered by the drift theory, drifters responded in accordance with S_3. Wegener and van der Gracht attempted to show that Jeffrey's objection was based on a confusion between rigidity and strength, whereas Holmes, Hess, Heezen, and Carey all developed alternative solutions to the mechanism problem. Holmes suggested sea floor thinning, Hess hypothesized sea floor spreading, and Heezen and Carey both opted for the expanding earth hypothesis.

S_4. Suppose that proponents of T_1 devise a P_3 solution. Proponents of T_2 should argue that the solution is plagued with difficulties and attempt to devise their own alternative.

This is precisely what happened with the drift solution offered by Blackett, Irving, and Runcorn to account for the paleomagnetic directional data. Fixists argued that the data were unreliable and incomplete. They claimed that some of the paleomagnetic data was anomalous and hypothesized rapid field shifts as an alternative solution. Moreover, fixists used the same sort of tactics to counter Wegener's claims that his solutions to the problems of coastline similarities, geological matchups, and the formation of the Permo-Carboniferous Ice Cap gave the same continental reconstructions. They raised difficulties with the drift solutions and developed appropriate alternatives.

Closure of the drift–fixist debate

Before the spectacular success of the Vine-Matthews-Morley hypothesis, the problem-solving effectiveness of the drift and fixist theories was comparable. Certainly neither of them enjoyed a decided advantage. Drifters finally had a respectable answer to the mechanism question. At least, they had a solution that was consistent with gravitational theory. There was not any direct evidence for convection currents, but fixist solutions such as Ewing's solution to the origin of oceanic ridges also appealed to convection currents. Fixists basically had solutions for all of the problems solved by the drift theory. Some drift solutions were better than the fixist alternatives, but the same may be claimed of some fixist solutions. However, all this changed with the acceptance of the Vine-Matthews-Morley hypothesis: It offered sea floor spreading and continental drift a P_1, P_2, P_3 solution to the problem of marine magnetic anomalies, and fixists decided that they could not alter the status of the solution through appeal to the standard research strategies.

A review of the various stages in the acceptance of Vine-Matthews-Morley is in order. The hypothesis suggested the following claim about the presence of a particular pattern of magnetic anomalies surrounding ridges, the occurrence of sea floor spreading, and reversals of the earth's magnetic field:

R_1. If sea floor spreading and geopolarity reversal occur, then there are marine magnetic anomaly patterns of type V.

Moreover, they claimed that they had developed a P_1 solution for sea floor spreading. In other words, they were willing to claim that sea floor spreading would be the only viable solution to the problem, if there were type V marine magnetic anomalies and if the earth's magnetic field could be shown to have undergone repeated reversals. Thus they were willing to maintain that

R_2. If magnetic anomaly patterns of type V exist and there are geomagnetic polarity reversals, it is reasonable to accept sea floor spreading.

In 1965 the existence of geomagnetic polarity reversals had been corroborated through radiometric studies, and Neil Opdyke developed an independent test of the reversals during the first few months of the following year. Thus the second antecedent condition in R_2 had been confirmed. The year 1965 also saw the development of the Lamont alternative. After examining a number of profiles from the Mid-Atlantic Ridge, they argued that the magnetic anomaly patterns surrounding the ridge were not of type V but of type L, and therefore that Vine-Matthews-Morley had been disconfirmed. However, once the Lamont group had analyzed Eltanin – 19, they decided that the pattern of reversals was of type V and not type L. Moreover, they were unable either to bring up serious difficulties with Vine-Matthews-Morley or to develop an alternative fixist solution that would fit in with Ewing's account. As a result, they accepted R_2 and had to admit that Vine-Matthews-Morley offered sea floor spreading a P_1 and P_2 solution to the problem of marine magnetic anomalies.

Once the case had been made for Vine-Matthews-Morley with the excellent profiles and Opdyke's independent confirmation of the geomagnetic polarity time scale, sea floor spreading had become a P_3 solution. First, the hypothesis of sea floor spreading related three different aspects of the earth's history and provided independent tests for each of them:

> The basis of this revolution [in the earth sciences], for it is no less, is that measurements of 3 different features of the Earth all change in exactly the same ratios. The first of these measurements is that of the direction of magnetic polarity in lava flows. . . . This time is the first of the 3 identical ratios. The second group of measurements is that of the widths of successive strips of magnetic anomalies measured over ocean basins. The third group of measurements has been made by Opdyke & colleagues . . . on deep sea cores. . . . The depths at which successive reversals take place is in a constant ratio for all cores. This is the third of the identical ratios. The essence of the new revolution lies in this identity of the ratios of 3 independent groups of measurements. The proved coincidences are already too great to be due to chance, and are all the more remarkable because one measurement is of time in million of years, one is of horizontal distance in hundreds of kilometers and one is of vertical distances in centimeters.[54]

Figure 7.1 is a diagram showing how the three types of measurement coincided to confirm the theory of sea floor spreading. Each

[54] J. T. Wilson, "A Revolution in the Earth Sciences," *Geotimes* 13 (December 1968): 10–11.

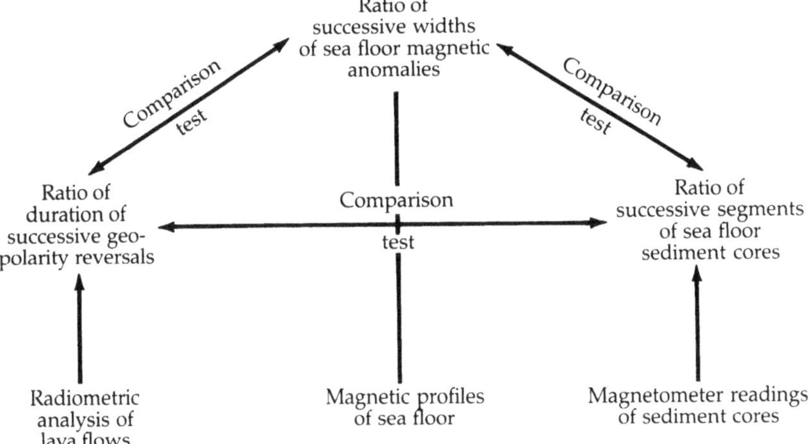

Figure 7.1. Diagram showing how results of three different types of measurement of earth history coincided to confirm the theory of sea floor spreading.

ratio could be determined independently and could be checked by either of the other two ratios, and it was sea floor spreading that made sense out of the fact that the width ratio was the same as the others. Second, sea floor spreading allowed for testing the accuracy and reliability of other drift solutions. For example, comparisons were made between the previous position of the continents as determined by paleomagnetism and sea floor spreading. This cross-check proved quite useful for evaluating the accuracy of the drift solution to the mass of paleomagnetic results going back to the Cretaceous. Similar cross-checks were made on the various drift solutions to the problems of coastline similarities and geological matchups.

It is not surprising that most fixists became drifters after the confirmation of the Vine-Matthews-Morley hypothesis. Implementation of normal research strategies proved futile. Most fixists had to either resort to extraordinary practices or switch to the drift theory. Most of them chose the latter course.

The drift controversy compared with controversies having scientific and public policy aspects

In this final section I should like to contrast the drift controversy with the sorts of controversies that constituted the major aspect of the Closure Project, namely, those having both scientific and public policy aspects. There are two questions that I should like to entertain:

1. Does the analysis of the development and closure of the drift controversy capture the development and closure of public policy controversies having scientific or technological components?
2. Does the analysis of the development and closure of the drift controversy capture the development and closure of scientific controversies that accompany controversies over public policy?

I shall suggest that the answer to the first question is decidedly negative. Public policy controversies follow the sort of pattern outlined in Beauchamp's negotiation closure, whereas the analysis I have offered of the drift controversy is more or less a variety of Beauchamp's sound argument closure (see chap. 1, this volume). As for the second question, I think that it should be answered in the affirmative. There is the same give-and-take among scientists in presenting and criticizing alternative problem solutions as has been outlined in the analysis of the drift controversy: The attempts of scientists to provide more effective solutions than their opponents, to raise difficulties with their opponents' solutions, and to improve their own solutions in response to the difficulties raised are standard operating procedures for participants in such scientific debates. However, it often turns out that participants in scientific debates accompanied by debates over public policy are less willing to accept a competing view, even when it has a decided advantage in its problem-solving effectiveness over their own, than the fixists were to switch to drift theory during the late 1960s and early 1970s. Nor do I think this is surprising. The drift controversy was completely insulated from questions of public policy, and it had no political, social, or moral ramifications hanging on its outcome. As a result, participants in the controversy were less apt to be influenced, qua participants, by their attitudes toward such things as class, race, color, party, or nationality. In other words, nonepistemic factors are likely to play more of a role in scientific controversies that have recognized social, political, and moral ramifications, and the various participants are apt to find themselves quoted in the *New York Times* and on the national news rather than in *Geotimes* and on the weekly science programs of educational television. Use of analyses such as the one developed to account for the drift controversy pinpoints places where participants in the scientific segment of public policy controversies are influenced by epistemically irrelevant factors. Of course, such analyses offer no answers as to the nature of such factors, nor are they designed to handle such questions. These questions are answered through recognizing that the scientific controversy is a component of a controversy over public policy and end with careful historical exegesis. Indeed, the mode of analysis not different in kind from that employed to unearth what happened in any scientific controversy in which nonepistemic factors have played a significant role, although with these particular scientific controversies it is in the accompanying public policy controversy that one begins searching for the source of the nonepistemic factors.

In general, the other case studies in the Closure Project are about analyzing public policy controversies that have scientific or technological components. Typically it turns out that these cases really involve two sorts of controversies: a scientific controversy accompanied by a public policy controversy (discussed in more detail by Engelhardt and Caplan in the introduction to this volume and by Giere, chap. 4). For example, accompanying the scientific controversy about eugenics in the United States during the early decades of this century was a political, economic, and ethical debate over whether the United States ought to have quotas built into its immigration laws, discussed by Allen (chap. 6, this volume). The scientific controversy revolved around whether there was any basis for eugenics within the framework of Mendelian genetics; that is, whether there was a straightforward and determinable relationship between such things as feeblemindedness, alcoholism, perverted sexual behavior, and habitual criminal behavior, on the one hand, and genetic structure, race, religion, and nationality on the other, and whether the methodologies used in data collection and analysis by various participants were sound. The accompanying public policy controversy touched on economic, political, and moral issues. Even if the eugenicists were right, there remained questions such as whether the United States should invoke immigration quotas, and, if so, against whom and whether there should be laws requiring sterilization of the feebleminded, the sexually perverted and the habitually criminal. Or consider the controversy over Laetrile, discussed by Rich (chap. 5, this volume). In essence, there have been two sorts of controversies concerning Laetrile: the scientific one concerning the effectiveness of Laetrile as a treatment for certain forms of cancer, and the public policy one over the conflicting issues of protection of the individual by the government from ineffective treatments versus the right of the individual to pursue his or her chosen mode of treatment. In the political controversy, legislation was passed by various state legislatures, even though many of the legislators believed Laetrile to be ineffective. The legislators dealt with the question of an individual's right to pursue his or her chosen form of treatment, whereas the scientists who undertook the various studies to test the efficacy of Laetrile were concerned with the alleged action of Laetrile on certain forms of cancer.

The differences between scientific controversies such as the one over continental drift (as well as scientific controversies that are accompanied by public policy disputes), and controversies that solely concern public policy, such as the ones concerning the individual's right to use Laetrile for the treatment of cancer or the morality of requiring sterilization of habitual criminals, is a difference in kind. Scientific controversies are over matters of fact and what theory provides the best interpretation of them or offers the best solution to a set of empirical problems. Public policy controversies are over questions of value. Disputants in scientific controversies disagree about how the world goes, whereas participants in public policy controversies often disagree about how the world ought to go, and such disagreements may be central to closure of public policy debates. Participants in the drift controversy, qua participants, were interested in answering the question. "Does drift occur?" not "Ought it to occur?", "Is it economically or politically wise for it to occur?", or "Is drift consistent with certain moral principles?" This basic difference, I think, makes the sort of analysis that I have presented to account for the drift controversy inappropriate for dealing with public policy controversies. The answer to the first question asked in this section is negative. To my mind, some approach such as Beauchamp's negotiation closure is needed. Of course, such an approach could be formulated in terms of a problem-solving effectiveness model, but the sorts of difficulties faced by proposed solutions and criteria for an "ideal" solution would be very different from those outlined in the analysis of the drift controversy.

Conversely, it would make little sense to analyze controversies such as the drift one in terms of negotiation closure.[55] It is absurd to view closure of the drift debate as a matter of negotiation. Imagine drifters and fixists sitting around negotiating just how much each is willing to compromise: "Well, you can have South America, but I put my foot down on Australia." Not only did such conversations not occur, but if they had taken place they would have been as ridiculous as a state legislature's passing a bill decreeing that π is equal to a value of 4 – not that some haven't.

What about the answer to the second question and the differences between scientific controversies that are accompanied by public policy controversies? First, the differences. One difference is that the

[55] This is not to say that negotiation closure cannot occur over scientific matters. It does. But it occurs not over what theory is the better but over policy questions in science, concerning such matters as how to divide funds, whom to invite to a symposium or give computer time to, or even whom to honor with a Nobel Prize.

urgency of closing scientific controversies generally is far greater with scientific controversies that accompany public policy controversies than those that do not. This is to be expected, because in the former case economic, political, or health interests are known to be at stake. Pressure to formulate appropriate policy is directed toward public officials, agencies, and legislatures and is passed on to the relevant scientific community, organization, or appointed task force. Or a sense of urgency may be felt directly by the appropriate or appointed scientific body. The National Institutes of Health, for example, has a formalized procedure for dealing with medical controversies that have significant and immediate implications for health care. The urgency of resolving controversies is not nearly as great when the scientific controversy remains within the appropriate community of scientists and does not have implications for the formulation of public policy. Drifters and fixists, certainly, wanted to know the answer. But there was always tomorrow. Of course, some wanted to rush into print with a "blockbuster" paper, receive foundation grants, and maybe win a prize. However, such pressures are self-imposed and do not have the same force as those imposed by governmental bodies and outside interest groups.

A second difference is that participants in scientific controversies having ramifications in the realm of public policy often find themselves under financial, legal, and moral pressure. The Love Canal controversy is a case in point. Blowing the whistle on one's employer can be distressing. Of course, participants in purely scientific controversies are not immune to such pressures. Occasionally, for example, remarks are made to the effect that it was a good idea for earth scientists in the United States to keep a low profile if they believed in continental drift. But this is quite different from being an employee of a company or governmental agency whose financial interests or official position is contrary to what an employee who is participating in the scientific controversy thinks is correct.

A third difference is that participants in scientific controversies that have decisions about public policy issues riding upon their outcome may be influenced by their views about the public policy issues. Such correlations have been recognized in contemporary cases and show up in the analysis of the American eugenics movement (see Mazur, chap. 9, and Allen, chap. 6, this volume). Indeed, such influences often show up in scientific controversies having ethical, religious, or political consequences.

A final difference is that "media" pressures are directed toward scientists who find themselves involved in a scientific controversy that is part of a general public policy controversy. At the very least, intensive media exposure (which most scientists are not used to), can lead to a clouding of the issues. Such are some of the differences.

Despite these differences, the sort of analysis presented to account for the drift case is appropriate for analyzing scientific disputes having implications for public policy. Competing scientists raise the same sorts of difficulties as suggested in the drift analysis, and they attempt to devise solutions that avoid such difficulties. Admittedly, closure of such scientific controversies often is not as neat and tidy as with the drift case, but this does not tell against the applicability of the suggested analysis. Instead, it illustrates the fact that pressures and influences such as those just described are more apt to be at work than in relatively "pure" cases of controversies concerning scientific questions, such as the drift controversy. It shows that nonepistemic factors have a significant role in controlling the closure of such controversies.

Analyses such as the one presented here of the drift controversy do not display the nature and variety of the sorts of pressures and influences directed against participants in the respective scientific debate, as well as the various relations between the scientific and public policy controversies. Such analyses must be supplemented by historical analysis of these factors and relations. Moreover, any complete analysis of the origin, development, and closure of such controversies must bring out the nature of these nonepistemic factors, for it is the presence and the effect of the public policy debate that differentiate these controversies from the more purely scientific controversies. Indeed, what is usually more interesting about the controversies that involve both scientific and public policy issues is not the scientific content but the way in which the outcome and course of the controversy is affected by the accompanying public policy controversy.

8

How history and politics affect closure in biomedical discussions: the example of the Soviet Union

LOREN R. GRAHAM

The historical past and the present political environment exert heavy influences on discussions of biomedical ethics in all societies. These influences undermine efforts to construct an optimal method of closure that might work equally well in different places, at different times, and on different kinds of issues. The recent discussions about biomedical ethics in the United States (think, for a moment, of abortion and of in vitro fertilization) are incomprehensible outside the context of American social, political, and religious history. Any analysis of how those discussions arose and how they may be closed that does not pay attention to cultural and political factors is certain to go awry.

In this essay I will attempt to show how several controversies over biomedical ethics in the Soviet Union display characteristics resulting from the specific culture and politics of that nation. I will then analyze the ways in which closure on these issues occurs in the Soviet Union, utilizing the typology of different types of closure developed by Tom Beauchamp (chap. 1, this volume). I will conclude by inquiring into the relevance of the Soviet example for the consideration of closure debates in other nations, particularly the United States.

Biomedical ethics in the USSR

In a recent article about discussions of recombinant DNA research in the Soviet Union, I pointed out that the major concerns of leading American and Soviet biologists were different, at least during the formative years of the debate in the early and middle 1970s.[1] The

[1] "Reasons for Studying Soviet Science: The Example of Genetic Engineering," in Linda L. Lubrano and Susan Gross Solomon, ed., *The Social Context of Soviet Science* (Boulder, Colo.: Westview Press, 1980), pp. 205–40.

greatest American concern at that time was the possibility of pathogenic organisms being accidentally produced by recombinant DNA research and then being allowed to escape from the laboratories. The main concern among Soviet scientists was that the debates over recombinant DNA would permit the reimposition of political controls over biology of the type that had ended only after Lysenko's demise a little over a decade earlier. Thus, leading Soviet scientists participating in the debates, such as academicians Baev and Engel'gardt, steadfastly maintained that there was nothing to fear from recombinant DNA, which, according to them, was just one more technology that could be used for good or evil, depending on social motivations. The basic research itself, they insisted, needed no special evaluation or regulation by review boards, especially not by review boards with nonscientists on them, as was often the case in the United States.

The Soviet reaction to the recombinant DNA debate that I have just cited is an illustration of how local (or national) differences can impinge directly upon discussions of biomedicine. In the United States, with its tradition of looking at technology in an economically competitive, legalistic, and regulatory fashion, the main question about recombinant DNA was whether new regulations would be required to handle this special case. The answer given by influential members of the American scientific community was predominantly "yes" from 1974 to 1975: "maybe" from 1975 to about 1978: and, in the early 1980s, "probably not." These answers have been closely tied to hotly debated judgments over just how dangerous the research and resulting technology might be.[2]

In the Soviet Union, on the other hand, with its recent tradition of very close political control of science, researchers saw the reappearance of such control in biological science as a potential disaster, and they did everything in their power to keep the debate quiet so that it would not attract the attention of Communist Party organizations. These Soviet scientists were greatly relieved when their American colleagues finally came around to agreeing with them that fears of recombinant DNA work had been exaggerated. Soviet scientists succeeded in preventing a great debate over recombinant DNA from opening in the USSR until the Americans had come close to closure on their own debate.

In the United States there were also scientists who feared government regulation of science that might issue from the recombinant

[2] For a recent survey of progress in recombinant DNA research and a defense of the view that it can be performed safely, see John Abelson and Eleanore Butz, ed., *Recombinant DNA*, special issue of *Science*, vol. 209, no. 4463 (1980).

DNA debate, and in that sense they shared the anxieties of Soviet scientists. But the differences between the attitudes of scientists in the two countries should not be missed: If some American scientists saw regulation of recombinant DNA as a case of the proverbial camel putting its nose into the tent, the Soviet scientists saw the debates as a case of a camel poking its nose into a tent in which, until a few years ago, it had been a permanent resident.

Recombinant DNA is only one of many aspects of biomedicine now being discussed in the Soviet Union. I would like to analyze the general features of those discussions under three general headings: (1) philosophical issues, (2) social issues, and (3) political issues.

Philosophical issues

In the West, most discussions of biomedical ethics have been based on the assumption that ethics and biology belong to entirely different realms and are different in kind. Ethics deals with values, whereas biology deals with facts. To confuse these two realms is to make a "category mistake," to fail to notice, as A. J. Ayer put it, that ethical statements have no cognitive significance.[3]

Since most Westerners accept the view that ethics and biology are different in kind, ethical advisory committees in Western countries are usually set up in such a way that two distinctly different types of thought are represented. The scientists who are members of these boards are supposed to supply the necessary scientific knowledge, and the moral philosophers, often people with religious backgrounds or even theological training, can supply the necessary knowledge of ethics. In the West, ethics is usually viewed as a set of principles or a methodology useful for guiding science, but not a part of science. To use a metaphor, the science and technology enterprise is seen in the West as a great ship belonging to the material world, whereas ethics is a nonmaterial set of beliefs that is intended to guide the ship's tiller.

In the Soviet Union, on the other hand, ethics and values are considered to be no less a subject of scientific study than is biology. As a Soviet author wrote in the *Herald of the Academy of Medical Sciences of the USSR* in 1977, Marxists believe that, in principle, values can be submitted to "strict scientific research."[4] This assertion of what in the West is called the "naturalistic fallacy" places Soviet Marxists in a

[3] The influence of this particular view on American teaching about ethics is discussed in Douglas Sloan, "The Teaching of Ethics in the American Undergraduate Curriculum, 1876–1976," in the volume he edited entitled *Education and Values* (New York: Teachers College Press, 1980).

[4] A. Ia. Ivaniushkin, "K voprosu o sushchnosti meditsinskoi deontologii," *Vestnik akademii meditsinskikh nauk SSSR* 4 (1977): 48–55.

different position from most of their Western colleagues when they approach problems of biomedical ethics.

The Western tradition of considering ethics and science as totally separate entities is probably not intellectually tenable, but this separation obviously has practical utility. When a new development in science raises ethical questions, the common Western approach is to seek a compromise solution between what the new science permits and what defenders of the old values are willing to accept. This kind of temporary solution has been hammered out in the United States on a whole range of issues, from birth control and abortion to genetic engineering. The absolutist position taken in these debates by some representatives of fundamentalist religions is seen in the United States as the position of one of many interest groups, whereas others, such as women's rights organizations, often have clearly contrasting positions and are therefore a different set of interest groups. The location of the typically American "middle-line compromise" that usually emerges is influenced and even defined by the extremes at both ends. All legislators in the United States know that they dare not offend too many of their constituents, even if many of these constituents hold positions that are intellectually primitive. A person in the United States who maintains that abortion "violates God's will" or one who says that the creation of new life forms by recombinant DNA "is contrary to the laws of Mother Nature" cannot be kicked out of the debate simply because of the nature of his or her argument. The person's influence depends, in large part, on how many active supporters he or she can enlist in the cause.

In the Soviet Union such arguments as the ones just mentioned concerning abortion and recombinant DNA are not considered legitimate, since they are based on officially unacceptable philosophical positions, those of religion, mysticism, and dualism. No doubt these viewpoints have supporters in the Soviet Union, perhaps more than we would guess, but their views cannot be published in the Soviet press, they do not figure in the debates, and they cannot be cited as supporting arguments for solving biomedical dilemmas. Rather than seeking a compromise in biomedical ethics between viewpoints that differ radically in their basic assumptions, Soviet specialists in these areas are expected to find the "correct" solution, using the world view of scientific Marxism-Leninism as their basic guide. The difficulty in finding such solutions when faced with the inherently perplexing problems of biomedical ethics may be one of the reasons that biomedical controversies in the USSR tend to lag behind those in the West.

Since Marxism-Leninism is a materialist doctrine, observers in the

West might mistakenly guess that Soviet writers on biomedical ethics would adopt a scientistic stance in which any procedure in, say, genetic engineering, that seems desirable from a biological point of view would also be deemed socially acceptable. The Soviet position, however, as repeated in numberless articles and books, is that such a stance is a reductionist, vulgarly materialist view that "biologicizes" man, ignoring his social essence. Soviet dialectical materialism contains the concept of "levels of being," with different principles governing, respectively, the physical, biological, and social levels.[5] Within this framework, human societies are considered emergent phenomena that cannot be exhaustively explained in terms of biology, even though they, too, are natural phenomena. As a Soviet writer recently commented, "Although man possesses a social essence, he does not cease being a natural creature."[6]

In the Soviet Union dualistic and scientistic positions are frowned upon alike as being philosophical errors on the opposite ends of the philosophical spectrum. These errors are excluded from discussions of biomedical ethics, and the ultimate morality is seen to be "communist morality," which emphasizes the existing principles of the Soviet state. Thus the potentially radical principles of philosophical materialism are converted into an essentially conservative system that protects Soviet society as it is today. Whatever conclusions Soviet biomedical ethicists reach, they are expected to support Marxism-Leninism, the Soviet state, and the institutions of Soviet society, including the family.

Social issues

During the last decade or so, one of the most common topics of discussion in the Soviet Union has been the "scientific-technical revolution," often called the STR. Through the work of Erik Hoffmann, Paul Cocks, Robert Miller, and others, Western specialists on the Soviet Union have become acquainted with the broad range of changes that Soviet writers believe are occurring in Soviet society as a

[5] The principle of nonreductiveness is an old one in the Marxist philosophy of science and played a prominent role in the writings of L. S. Vygotsky, S. L. Rubinshtein, A. I. Oparin, and, most recently, I. T. Frolov and N. P. Dubinin. See L. S. Vygotsky, *Thought and Language* (Cambridge, Mass.: MIT Press, 1962), p. 51; A. I. Oparin, *The Origin and Initial Development of Life*, NASA TTF-488 (Washington, D. C.: 1968), p. 4; I. T. Frolov, *Genetika i dialektika* (Moscow: 1968), p. 253; N. P. Dubinin, "Filosofskie i sotsiologicheskie aspekty genetiki cheloveka," *Voprosy filosofii* 2 (1977): 46–57.

[6] E. A. Snegur, "Dialektika sotsial'nogo i biologicheskogo v probleme cheloveka," *Filosofskie voprosy meditsiny i biologii* 7 (1975): 27.

result of the advent of this revolution, the applications of the latest developments in science.

In descriptions of the results of the STR, the general tone of Soviet writers has been one of technological optimism. They often admit that the introduction of new technologies has some deleterious side effects, such as pollution, but the official and prevalent view is that socialist societies, as opposed to capitalist ones, have such special advantages in the handling of new technologies and such a deep commitment to human betterment that the STR in the Soviet Union will be an undiluted blessing, indeed, the accomplishment of full communism.

Against this background, Soviet specialists in biomedicine have played a rather special role, since they have portrayed the possible harmful effects of the STR in grimmer terms than other writers. Several of them have, in fact, become natural critics of the official optimists and have even used such alarmist terms as "catastrophe" when speaking of the possible effects of the STR.

A. F. Bilibin, a specialist in medical deontology, wrote in 1979:

> The scientific-technical revolution gives us a great deal, but it also takes away something . . . We must see the threats which are hidden in the mechanization of life and, in general, in our "progressive" technicism. Regrettably, we are inclined to consider only one thing as progress, and that is having lots of technology available . . . Just as in previous times people escaped into mysticism and religion, we often now slip into impetuous and excessive rationalism . . . We must see that the STR, having seized people in its progressive movement, also demands of them great subtlety and an understanding of man's soul.[7]

What is it that Bilibin is worried about in this passage? First of all, as a physician he fears that clinical medicine will be increasingly viewed as a science, contrary to his opinion that medicine is an art in which intuition and human understanding are as important as technical knowledge. And he also worries that the new technology being introduced into hospitals will undermine the authority of physicians, giving the decision-making power to the technicians who read the graphs and computer printouts. His anxieties here are quite similar to those of many American doctors and show that the STR does, indeed, create problems that cut across cultural boundaries. Bilibin warned, "In the final analysis, the work of a clinical physician has a personal

[7] A. F. Bilibin, "Gorizonty deontologii," *Vestnik akademii meditsinskikh nauk SSSR* 5 (1979): 36–7, my translation.

character and is therefore sovereign. We must recognize that the clinical physician has the right of sovereignty in his work."[8] One is reminded of the cries of anguish among American physicians some years ago when it was suggested that decisions about extraordinary care and the use of life-support machines might best be made by hospital lay committees instead of merely by the physician.

Other Soviet physicians have also worried about the impact of the STR on the medical profession and have called for new courses in ethics in medical schools in order to handle the challenges. A. Iu. Ivaniushkin remarked that new developments such as genetic engineering were causing a "multitude of moral collisions" that must be investigated from a Marxist standpoint, since he considered the existing Western literature on the subject inappropriate for the Soviet Union.[9] He was unclear, however, about just how the Soviet approach would differ.

Other fears about the STR were voiced in April of 1980 by A. S. Shiriaeva and A. M. Petrov in the journal of the Academy of Medical Sciences. They pointed out that the STR was creating a new environment, an environment in which radiation from nuclear power stations and mutagens from the chemical industry are having deleterious effects on man. They continued, "if the uncontrolled pollution of the biosphere with mutagens is permitted, then we can speak not only of a threatening 'ecological catastrophe' but even of a possible threat in the form of a genetic catastrophe for man."[10] In the Soviet Union, where for years scientific and technological progress has been equated with social progress, such statements are seen as much more radical than in the United States, where they have been commonplace for years. Only in the last several years has a trickle of public criticism of nuclear power begun to flow in the Soviet Union,[11] and the specialists in biomedicine, with their knowledge of radiation effects on human genetics, have played a signal role in this growing criticism. Shiriaeva and Petrov were careful to point out that the Soviet government is now moving to meet these threats to the

[8] Ibid., p. 37.

[9] A. Iu. Ivaniushkin, "K voprosu o sushchnosti meditsinskoi deontologii," *Vestnik akademii meditsinskikh nauk SSSR* 4 (1977): 49.

[10] A. S. Shiriaeva and A. M. Petrov, "Nekotorye sotsial'nye i mediko-biologicheskie aspekty ekologii i genetiki cheloveka," *Vestnik akademii meditsinskikh nauk SSSR* 4 (1980): 55.

[11] See N. Dollezhal' and Iu. Koriakin, "Iadernaia elektroenergetika: Dostizheniia i problemy," *Kommunist* 14 (1979): 19–28. See also the warnings sounded by N. P. Dubinin in "Genetika na poroge dal'neishikh otkritii," *Kommunist* 18 (1975): 54, and the discussion of them in V. F. Barabanov, *Nauchno-tekhnicheskaia revoliutsiia i sud'by prirody* (Leningrad: 1979), p. 24.

environment, and has created, under the State Committee of Science and Technology, a special "Council on Complex Problems on the Preservation of the Environment."[12] But it is clear that people like Shiriaeva and Petrov do not think enough is being done.

The STR is based on the application of science to all areas of life, and it is therefore inevitable that science will be applied in the Soviet Union to human beings themselves, to the study of their biological and genetic natures. But human genetics is one of the most controversial of all fields in the Soviet Union because of its ideological implications. After a promising start in the 1920s, human genetics fell into eclipse in the Soviet Union for decades, and even after the end of Lysenkoism in 1965 human genetics began to revive much more slowly than molecular biology and animal and plant genetics. In the 1970s, human genetics finally returned to the Soviet Union,[13] although its exponents are careful to avoid taking any stance that appears to imply a biological determination of human behavior. However, in a society where authorities still look suspiciously upon arguments in favor of genetic diversity among humans, the early research results of Soviet human geneticists are controversial. They have begun efforts to monitor genetically the population of the USSR, a research project that may eventually put the human geneticists into conflict with the industrializers and economic planners, who are not overjoyed to hear about the radiation effects of nuclear power or the carcinogens produced by the petrochemical industry. And these controversies are only the beginning of what the study of human genetics may reveal. Soviet researchers are already drawing up maps showing the prevalence of genetic diseases among different nationalities in the Soviet population. They have reported, for example, that thalassemia occurs more frequently in the central Asian republics than it does elsewhere in the Soviet Union. Other genetic diseases have been identified as being especially concentrated in Azerbaijan and Armenia.[14] These discoveries call for special public health measures, and they also raise difficult problems about the relationship of genetics and nationalities, problems that were covered up in the early thirties when human population genetics became a subject too hot to handle. We do not yet know how successful the Soviet geneticists will be in keeping these inherently controversial issues from becoming ideologically charged.

12 Shiriaeva and Petrov, *Nekotorye* . . ., p. 56.

13 For a good overview of the rebirth of human genetics in the USSR, see N. P. Bochkov, "Itogi i prespektivy nauchnykh issledovanii v oblasti meditsinskoi genetiki," *Vestnik akademii meditsinskikh nauk SSSR* 11 (1977): 89–94.

14 Ibid., p. 94

Political issues

The fundamental issue that underlies all of the discussions noted so far is one of political influence and control. Who should debate, and who should make, the decisions about what is permissible in biomedical research and practice? The same sort of question is being asked in the United States, but the answers being reached by Americans are troubling to many Soviet scientists with memories of the tortured history of Soviet biology. These Soviet scientists note with growing anxiety the gradual increases in the West of the influence of nonscientists in making decisions about the wisdom of certain types of biological research. In the United States, Maxine Singer of the National Cancer Institute observed, "Scientists today recognize their responsibility to the public that supports scientific work in the expectation that the results will have a significant positive impact on society ... The scientific community has accepted the counsel of ethicists, philosophers and representatives of the public who long troubled to point out this responsibility."[15] In the United States the inclusion of moral philosophers and lay people on ethical advisory boards and institutional review boards is now widely accepted.[16]

It is to be expected that when a review board in any society is making ethical decisions about scientific research it will be heavily influenced by the predominant values of that society. For the United States, religious and public leaders are the politically logical choices for lay membership; in the Soviet Union the analogous members are Marxist philosophers and party activists. And here the central political issue emerges fully: If the Soviet Union creates ethical review boards that include Marxist philosophers, to advise on the permissibility or limits of biological research and technology, the old question of Marxist ideology and Soviet science takes on a new dimension.[17]

15 Maxine Singer, "The Involvement of Scientists," National Academy of Sciences, *Research with Recombinant DNA: An Academy Forum, March 7–9, 1977* (Washington, D.C.: National Academy of Sciences, 1977), p. 28.

16 The Ethical Advisory Board of the U.S. Department of Health, Education and Welfare that made recommendations for approval or disapproval of *in vitro* fertilization proposals coming to the National Institutes of Health had a membership from diverse professions stipulated by its charter. Of its fourteen members, no more than seven could be scientists, and no more than four could be biomedical scientists. The others were "social or behavioral scientists" and representatives of "other disciplines" and the "general public." The board included a Catholic priest and a philanthropical leader. See "Charter: Ethical Advisory Board," approved December 27, 1976, by David Matthews, secretary of the Department of Health, Education and Welfare.

17 See N. V. Turbin, "Geneticheskaia inzheneriia: real'nost', perspektivy i

To the best of my knowledge, no such review boards have yet been established in the Soviet Union. One reason that the problem of ethical review is less pressing in the Soviet Union than in the United States is that the "projects and grants" approach to funding research is much less developed in the Soviet Union.[18] In the United States, ethical review boards have arisen in the context of the award of federal grants to scientists in research institutions, largely universities. Within the framework of federal support for research by agencies such as the National Science Foundation and the National Institutes of Health, and the assumption of accountability to Congress and the tax-paying public upon which this system is based, it was inevitable that as biomedical research became more controversial the responsible federal agencies would wish to protect themselves from criticism by providing for ethical review by committees on which the public was represented.

In the Soviet Union this powerful motive for ethical analysis of research is not present. Most scientific research there is supported through block funding to whole institutions (rather than to principal investigators, as in the United States). The system of contracts and grants is much more weakly developed in the Soviet Union than in the United States, and even when such awards are given the procedures are quite different. Each scientific institute in the Soviet Academy of Sciences, for example, submits its budget for the coming year to the academy's administrative hierarchy leading to the presidium of the academy; the usual outcome is for each institute to receive an increment every year. Great vacillations in funding are very rare, the concept of public accountability is weak, and a formal system of peer review of research applications of the American type is not deemed necessary. Furthermore, public controversies about science are frowned upon by Soviet authorities. As a result of these characteristics of the Soviet scientific establishment, the Soviet Union still has not witnessed the major controveries over biomedical ethics that have become quite normal in the United States.

opaseniia," *Voprosy filosofii* 1 (1975): pp. 47–56; N. V. Turbin, "Genetika i obshchestvo," *Voprosy filosofii* 2 (1974): 105–15; V. D. Timakov and N. P. Bochkov, "Sotsial'nye problemy genetiki cheloveka," *Voprosy filosofii* 6 (1973): 59–69; I. N. Smirnov, "Metodologiia i mirovozzrenie: nekotorye filosofskie problemy biologicheskogo poznaniia," *Voprosy filosofii* 7 (1978): 57–68; B. A. Grushin, "Dialekticheskii materializm i sovremennaia nauka," *Voprosy filosofii* 7 (1978): 147 – 54.

18 See the discussion in Thane Gustafson, "Why Doesn't Soviet Science Do Better Than It Does?", in Linda Lubrano and Susan Soloman, ed., *The Social Context of Soviet Science* (Boulder, Colo.: Westview Press, 1980), pp. 34 and passim.

Conclusions

If one reflects on the history of the Soviet Union, the fact that discussions of biomedical ethics there tend to lag behind such discussions in the United States seems ironic. After all, in the first two decades of the Soviet Union's existence, Western science was often criticized by Soviet writers for being organized along "closed, caste-like lines," with public responsibility thwarted by a tradition of professional arrogance, industrial secrecy, and individual competitiveness. When the Academy of Sciences and other Soviet scientific institutions were reformed in the late 1920s and early 1930s, one of the major campaigns was to "open up the fortress-like walls" of these institutions to public scrutiny.[19] Committees of lay people made recommendations on needed areas of research and even on the election of new members to the Academy of Sciences. All research was to be included in the economic plan available for public discussion, and scientific research was to be "tied to life."

In contrast to those days, we have now reached a situation in which research on biomedical science is much more open to public scrutiny and political debate in the United States than in the Soviet Union. No person, for example, can today receive a grant from the National Institutes of Health in the United States that involves research on human subjects, laboratory animals, or recombinant DNA unless the researcher has signed statements giving assurance that the research supported by the grant will comply with ethical guidelines drawn up by committees of mixed membership of both scientists and lay people.[20] I am unaware of any procedures in the Soviet Union that intrude public interests into the heart of scientific research to this degree.

Upon being questioned about these differences, Soviet scientific authorities have usually answered that because of the nature of socialist societies as opposed to competitive capitalist ones, there is simply much less need for such regulation in the Soviet Union.[21] (And it is true that private research in industry is less regulated in the United States than federally funded research.) At the present moment, however, this argument seems less and less persuasive. Examples abound showing that biomedicine presents all societies,

[19] See Loren R. Graham, *The Soviet Academy of Sciences and the Communist Party, 1927–1932* (Princeton: Princeton University Press, 1967), pp. 30, 112.

[20] See U.S. Department of Health, Education and Welfare, Public Health Service, Grant Application Form PHS 398 (OMB No. 68-RO249, Rev. 10/79), pp. 3–4.

[21] See, for example, A. A. Baev, "Sotsial'nye aspekty geneticheskoi inzhenerii," *in Filosofskaia bor'ba idei v sovremennom estestvoznanii* (Moscow: 1977), p. 146.

whatever their economic systems, with ethical dilemmas. In both the United States and the USSR, for example, abortions are legal (at the moment). In both countries it is tempting to perform experiments on fetal tissue from aborted fetuses, since much valuable knowledge about the reactivity and characteristics of the human body could be gained in this fashion that could not be obtained in any other way. Should scientific research institutes be established across the street from abortion clinics? Is this question less troublesome in the Soviet Union than in the United States? In the United States the dilemma has resulted in a complex system of guidelines on fetal research, drawn up by a committee of people from a variety of professions, including philosophy and the social sciences. I have not been able to find similar guidelines on fetal research in the Soviet Union, but some sort of de facto guidelines must exist. What does not exist is the sort of public discussion of the issue that has occurred in the United States.

Closure of biomedical disputes in the USSR

Let us look briefly at the means of closing biomedical disputes in the Soviet Union in terms of the typology of four different types of closure given by Tom Beauchamp (Chap. 1, this volume). He calls these four types "sound argument closure," "procedural closure," "natural death closure," and "negotiation closure." The first type of closure comes if a "correct" position is found; the second type occurs if procedures (such as legal ones) arbitrate disputes and allow business to proceed, even if the settlement rests on irrelevant or false grounds; the third type of closure happens when interest in an issue simply diminishes; and the fourth type of closure comes through negotiated compromise resulting from acceptable deviations from ideal positions by several parties. According to Beauchamp, in the United States sound argument closure almost never occurs; procedural closure rarely occurs; and natural death closure is infrequent but not rare. He finds negotiation closure the most interesting possibility; he evidently believes that it happens in the United States rather often, and he hopes that ways can be found for it to occur even more frequently.

In the Soviet Union, by contrast, the type of closure of biomedical debates that Beauchamp finds most rare – sound argument closure – is the most frequent procedure and the officially sanctioned one (although some observers might question whether the argument was, in fact, sound). Conversely, the type of closure that Beauchamp finds most common in the United States – negotiation closure – is very rare in the Soviet Union.

Notice how difficult, in terms of Soviet viewpoints, it would be to accept in ethical disputes the concept of negotiation closure as described by Beauchamp. He says that closure of this type occurs when "an end to controversy is more important than a continuation of fruitless argument in defense of one's 'correct' views." He further says that negotiation closure becomes necessary when several positions collide that are "based on arguments correctly grounded in moral principles themselves defensible by appeal to well-reasoned moral theories." But if one regards moral theory as a subject of scientific study, as philosophers in the Soviet Union do, then contradictions between well-reasoned moral theories should not occur, unless the contradiction is only apparent and can be traced to an area of ignorance in scientific knowledge. I take it that Tom Beauchamp did not mean apparent contradictions of this sort but true contradictions, such as that in abortion controversies between, on the one hand, people who assert that abortion is a sin against religion and, on the other hand, people who say that a woman should always be able to control her own body. This dispute cannot be resolved by an advance in scientific knowledge. In the Soviet Union, both of these positions would be rejected as incorrect – that is, as unsound arguments. The first is based on religion, which is considered inadmissible in public policy questions; the second is based on the principle that the rights of an individual always transcend those of society, also inadmissible in a socialist society.

The fact that the Soviet Union supports sound argument closure does not foreclose changes in policy on biomedical matters. For example, abortion was legal in the Soviet Union from 1920 to 1936, illegal from 1936 to 1955, and has been legal again since 1955. The justification for these differing positions takes the sound arugment form. The argument runs that the interests of society must be in balance with those of the individual, although the higher the level of development of society and the greater its wealth, the more choice should be given to the individual. The prohibition of abortion (except for medical reasons) from 1936 to 1955 was justified by Soviet writers on the basis of the need of the Soviet Union for a larger population during the periods of intense industrialization, war, and the recovery from the enormous losses of war. Both before and after that time the less desperate conditions permitted the greater freedom of legal abortions.

The use of social criteria in biomedical debates, rather than a priori ethical ones, is evident in the very definition of abortion in Soviet medical terminology. In the Soviet Union a pregnancy that is interrupted between the 16th through the 28th week is considered to be a premature birth if the fetus survives until the mother's discharge *from the medical facility*; otherwise, it is considered an abortion.[22]

According to the view given in textbooks in the Soviet Union, morality is based on the historical development of social relations. The article on "Morality" in the *Great Soviet Encyclopedia* states:

> The content and character of human activity in society are ultimately determined by the objective sociohistorical conditions of existence and by the laws of societal development . . . Morality emerges from undifferentiated normative regulations and becomes a particular type of societal relationship as early as tribal society. It goes through a prolonged period of development in preclass and class society, during which its demands, principles, ideals, and values acquire, to a significant degree, a class character and meaning. However, there remain universal moral norms determined by conditions common to all epochs in which people have banded together in societies. Morality becomes most highly developed in socialist and communist societies, where it first becomes common for the society as a whole and then becomes the universal moral system of all mankind . . . Inasmuch as communist morality rejects the idea that either society or individual life are eternal means for the attainment of each other's aims and believes that society and the individual form an indissoluble unity, it also rejects the concept that is characteristic of bourgeois morality–namely, the sacrifice of one moral principle to another (for example, the sacrifice of honesty to advantage, of the interests of one group to the aims of another group, and of conscience to politics). Thus, communist morality is the highest form of humanism.

The rejection in the above paragraph of the sacrificing of "the interests of one group to the aims of another group" makes extremely difficult in the Soviet Union a frankly recognized negotiation closure of the type described by Beauchamp as recognizing that "one's preferred position . . . is . . . impossible." Soviet ethicists believe that the preferred position must be one toward which Soviet society constantly progresses, and the closer that any society comes to being a truly communist one, the more "the preferred" and "the possible" are in union.

In committing themselves to a sound argument solution to prob-

[22] *Great Soviet Encyclopedia* s.v. "Abortion."

lems of biomedical ethics, Soviet ethicists have taken on an intimidating assignment, one that gives them great difficulties when they face specific problems. When a Soviet medical specialist was asked in a newspaper interview if he thought that the cloning of a human being would ever, under any circumstances, be morally permissible, he replied with an almost visible sigh of relief that human cloning was not yet technically feasible, and he called this infeasibility a "fortunate prohibition of nature."[23] The fact that he used the word "fortunate" (*schastlivyi*) showed that he was as disturbed by the prospect of human cloning as many people in other countries, but the sound argument prohibiting it has not yet been developed. Given enough time, and the further development of discussion of cloning in all countries, I am confident that the sound argument on the subject will appear in the Soviet Union, but I do not yet know what it will be.

Conclusions

Discussions of biomedical ethics in the Soviet Union, the preceding analysis suggests, are heavily conditioned by specific historical developments of that country and by unique political and cultural factors. Some readers may conclude that these special characteristics of the discussions in the Soviet Union prove that they are irrelevant to Western discussions of biomedical ethics. The Soviet Union is, they may say, simply too exotic, too extraordinary to be helpful to people trying to understand how biomedical debates arise and how they are closed.

I would caution my readers not to draw that conclusion. In my opinion, a similar analysis to that which I have made of the Soviet Union could be made of the United States – that is, we could examine the closure of biomedical debates in the United States in terms of historical development, political peculiarities, and cultural traditions, just as I have done with the Soviet Union. I have already mentioned, in passing, some of these American peculiarities:

1. The strongly legalistic character of disputes in American society, based on the assumption of adversarial relationships
2. The continuing strength of religious traditions, and the belief that it is legitimate for religious viewpoints to influence policy-making in the area of biomedical ethics
3. The significance of the division in the United States between private interests and public interests

[23] A. Dyban, "Schastlivyi zapret prirody," *Literaturnaia gazeta* (August 23, 1978), p. 13.

4. The strength of the commercial marketplace
5. The importance of lobbies, interest groups, and professional organizations of various kinds
6. The concept of public accountability and the impact of this concept on ideas about review boards and regulations
7. The method of funding of research by contracts and grants to individual chief investigators rather than block funding to whole institutions

Any discussion of closure of scientific disputes in the United States will have to take notice of these characteristics. Furthermore, the importance of these characteristics is particularly great, if, as Beauchamp says, the most interesting and common form of closure in the United States is negotiation closure; the essence of negotiation is the making of agreements and compromises that recognize the legitimacy of the peculiar characteristics (even historical and irrational ones) of the opposing parties.

Historical and cultural factors play major roles in the closure of biomedical disputes, and it is foolish to pretend that the situation could ever be otherwise. Those who hope to find a general method for closure of such disputes that disregards social history are pursuing an unrealistic goal. On the other hand, to take the extreme historical position that asserts that such disputes are entirely social products that cannot be influenced by rational arguments and empirical findings is also mistaken. Biomedical disputes do contain scientific and factual elements that are not social per se, even though the origin and presentation of these elements are socially contingent. Biomedical disputes arise out of the mutual interplay of science and society, and they must be resolved by a method that contains and takes account of this interplay. Tom Beauchamp's negotiation closure is a method that is a product of American society and is particularly adapted to it; it can incorporate both scientific arguments and political ones, and it is the most satisfactory method of closure in our society. But its efficacy in societies with different recent histories is doubtful.

9

Scientific disputes over policy

ALLAN MAZUR

Each year, numerous committees of the National Academy of Sciences (NAS) provide policy-relevant scientific information in an apparently competent and reasonably disinterested manner, with only occasional charges of gross error or bias (Boffey, 1975; Schiefelbein, 1979). Thus for certain policy decisions which should be based in part on accurate technical information, adequate technical advice is obtainable if the pertinent questions fall within the state-of-the-art of our science. Increasingly, however, we see disputes between experts over the accuracy of scientific information. Usually these are complicated arguments, beyond the training of policymakers, so the only people capable of fully understanding them are other experts.

Experts involved in these disputes often have vested interests; for example, they may represent particular corporations, government agencies, or citizen action groups. It is by now well recognized that these experts' positions on what appear to be purely scientific points are correlated with their positions on related policy issues. Thus, scientists who oppose nuclear power calculate higher risks of cancer from a given level of radiation than do scientists who favor nuclear power (Mazur, 1981a).

These scientific disputes produce a clear dilemma for the policymaker. How does one make use of the expert's special knowledge without submitting to the expert's biases?

There are solutions to this problem that are currently being employed. A policymaker with personal biases may selectively depend on experts with similar biases. Or, one may simply ignore the scientific dispute, making policy on other bases, as when the courts adjudicate environmental suits on the grounds of whether or not a government agency has followed proper procedure, as specified in its enabling statute (Bazelon, 1979).

Adjudicating scientific disputes

Another solution, which some regard as more idealized than real, is to determine which of the disputing experts is correct in his or her scientific contentions and then let this judgment provide the technical basis for subsequent policy decision. The usual method of adjudication is to let one or more reputable scientists without vested interests or strong biases in the policy area examine the opposing arguments and then come up with their own pronouncements as to which is scientifically correct. This may occur in a number of formats, including National Academy of Sciences (NAS) committees, inquiry boards, such as those used by the Food and Drug Administration (FDA), peer review of journal articles, or the proposed "science court" (Mazur, 1977). All have the feature that one group of experts, presumed to be relatively objective, judges the claims of other experts who are assumed to be committed to one outcome rather than another.

In judging the efficacy of scientific adjudication, it is important to have a realistic idea of the degree of resolution that is feasible in a scientific dispute. We often visualize a critical experimental test that will settle the issue, but I have rarely seen a policy-relevant scientific dispute that could be resolved in this way, especially within the time span allowed by the pressures of the policy decision. However, after looking in detail at some of these disputes, I believe that both the usual and novel forms of adjudication can greatly clarify the factual arguments, sometimes even settling them in a reasonably definitive manner. I will briefly suggest feasible methods of clarification, providing details elsewhere (Mazur, 1981a).

First, factual claims can be separated from the policy claims in which they are typically enmeshed during public debate. For example, the question of how many cancers are caused by a given population dose of radiation can be separated from polemical statements about whether we should or should not build nuclear power plants. Once separated and stated precisely, claims that are ostensibly scientifically defensible statements often appear to be virtually irrefutable. For example, the claim "Substance X may cause cancer" can be made about nearly any substance without fear of contradiction. A refutation would require that one demonstrate the nonexistence of a carcinogenic effect, which is very difficult. Such claims could be rephrased in empirically meaningful terms.

Once clearly stated, defensible scientific claims are listed, the factual differences between the two sides may not seem as great as had been supposed, and in some cases the small differences that remain are resolvable by slight qualifications in wording or by improved specification of scope conditions or levels of uncertainty. When differences remain, they are now well specified, and proper consideration can be given to future research that might resolve them, though perhaps too late for the particular policy decision at hand.

Occasionally, scientific claims are shown to be groundless, having been based on error or on clear misuse of statistics. For example, Ernest Sternglass's claims of massive negative health effects from very small levels of radiation, which sometimes appear in the mass media, are considered erroneous by anti- as well as pro-nuclear scientists (although these repudiations are not much covered in the mass media).

In many disputes, there will be clear points of disagreement on important scientific facts that are not resolvable because they require knowledge well beyond the limits of our current science. A good example is the question of the precise relationship between radiation exposure and cancer risk at very low levels of radiation, which has been debated for years. A proper adjudication would recognize that this argument cannot be settled with the knowledge at hand, nor is it likely to be settled in the near future. The inability of an adjudication procedure to state "the truth" in this instance can hardly be considered a failure of the procedure. On the contrary, a frank statement that the issue is beyond our knowledge is much preferable to the disparate claims of various experts that each knows the correct answer.

The various forms of adjudication, in use and proposed, all have shortcomings, but any would probably operate reasonably well, keeping in mind the intrinsic uncertainty of scientific knowledge.

Since straightforward ways of clarifying, or even settling, scientific disputes (within the limits of the scientific knowledge of the day) exist, then why do such disputes persist when they are relevant to public policy? There are several reasons.

First, there are many controversies in which reasonable adjudication is never attempted, or when what is attempted takes a long time. In a dispute that I studied about possible harmful effects from the electrical fields around power lines, I found that the principal scientific adversaries had never discussed their differences and were not even clear on what these differences were.

Even when some reasonable form of adjudication has occurred, there are people who do not accept the outcome. Some of these are nonexperts – sometimes policymakers – who are not aware of the outcome, or, if aware, do not understand it and do not trust the process sufficiently to accept the outcome on faith. Also, we may be sure that any scientist whose position (and, by inference, reputation) is discredited by an adjudication will defend him- or herself to the end, arguing that some suitably qualified version of his or her original claim is still viable, even if not for the reasons originally stated. Furthermore, in cases where the dispute rests on a genuinely ambiguous point, perhaps a scientific claim that cannot be settled for lack of crucial data, then each disputant (and his or her allies) may continue to assert that his or her particular claim will turn out to be correct. So adjudication, even when decently done, may have little impact.

Finally, we must not forget that we are dealing first with a political controversy that just happens to have scientific elements. To give this important (but often overlooked) point some emphasis, I have drawn an abstraction that fits most scientific controversies over policy (Figure 9.1). Typically there are two sides, one representing industry, professional organizations, and other establishment fixtures, whereas the other side views itself as a challenger to the establishment. The important action of the controversy is played out within one or more of the main branches of government, whether at the federal or local level or both. The mass media report these events, influence opinion, and may even shape the course of the controversy (Mazur, 1981b).

The essential feature of this abstraction is that it applies equally well to many political controversies that have no scientific or technological component. From this viewpoint, disputes over scientific fact are auxiliary features of what is, at base, a routine special-interest dispute over policy. Thus, even if the factual dispute were settled, the policy dispute would be likely to persist. I recently asked an anti-fluoridationist if she would continue to oppose the water treatment even if she were convinced that it was harmless. She replied that she would still oppose it because she did not want anyone forcing her to have fluoride in her drinking water. Scientific facts are but one of the factors that determine a policy decision, and not necessarily the most important one.

From these considerations, it seems to me that there is a great deal of ambiguity in our questions about closure of scientific disputes related to policy. What do we mean by "closure"? If the scientific arguments are adjudicated to the satisfaction of a broad range of scientists but not to the satisfaction of the original disputants, has there been closure? If the policy decision is made and settled, even though the scientific arguments have not been resolved, is that closure? If neither the scientific nor the policy questions are settled, but after a while the disputants become bored and drop the issue, or die off, is that closure?

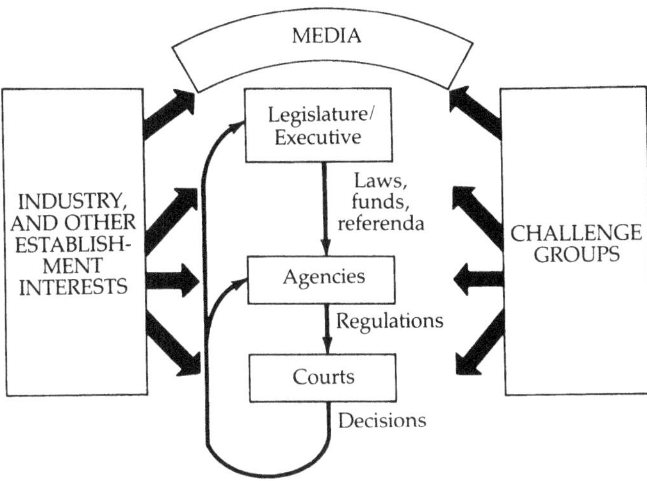

Figure 9.1. Sterile abstraction of many controversies.

I prefer to take a policy perspective and say that closure has been reached once a policy decision has been made (or dropped), independent of whether the scientific arguments have been resolved or not. Closure need not be permanent, for subsequent decisions may alter or even reverse the first policy decision. This view brings the political aspects of the controversy to the fore, leaving the scientific dispute between experts as one of several peripheral matters. It emphasizes the role of the expert as a political actor as well as a scientist, and it shifts attention from the expert's data and theories to his or her values and interests. I would like to take these topics up in the next section of this essay, looking briefly at some descriptive material on the expert as a political person.

The expert as a political person

Our image of scientists and other technical experts has changed over the last several years and become more humanized. In retrospect it is difficult to see how the picture of a totally rational scientist, whose decisions were tied solely to data and calculations, could ever have been taken seriously. Perhaps in earlier times scientists were too distant from our experience to be seen as real people. Now we are used to seeing them right in front of us, in the political arena, on different sides, each claiming that he or she is right and the others wrong.

I want to add some details to our newer picture of the expert as a political person. My focus will be the academic expert, which obviously excludes many other experts. However, descriptive material is more readily available for academicians than for experts outside of that community. It is also useful to focus on academics because they are often seen as freer agents than those who work for corporations or government. Although that is probably true in a sense, it ought to be made clear that academics have their own particular biases.

The social and political values of the modal academic are somewhat more leftist or liberal than the mode of the American population, though not radical, as the mass media of the 1960s used to suggest (Anderson and Murray, 1971). Academicians themselves recognize wide variation among their coworkers, which they often attribute to the individualism of intellectuals but which may be better explained by structural variables such as academic discipline and prestige. Ladd and Lipset (1972) have documented striking variation in the social values of professors from different disciplines in their large (N = 60,000) survey of the academic community taken in 1969. Engineers are less liberal than those in the natural sciences, and social scientists are the most liberal of all. Furthermore, within disciplines, professors at the most prestigious universities are more liberal than those at less well-known schools. The differences are illustrated in Table 9.1.

It is possible that value differences predispose particular groups of experts to take particular positions on technical issues such as nuclear power, the antiballistic missile (ABM), or the supersonic transport (SST) where active opposition is known to be associated with political liberalism. Natural scientists, therefore, might be expected to be more sympathetic than engineers to the antinuclear position, as indeed they are (Mazur, in press).

Apparently, some combination of ideology, academic discipline, prestige, and the social networks one is affiliated with all contribute to the choice of sides in technical controversies. The point may be illustrated with an excerpt from a larger study of political activism among academic scientists (Mazur, 1981a). George Misner and I collected a number of scientists' petitions on ten American political issues of the 1960s and 1970s and extracted the names of scientists who were persistently active, their names appearing on two or more of the petitions. One issue was nuclear power, and many of the multiple signers' names appeared on pro- or anti-nuclear petitions that were circulated in 1976. Three simple "sociological" variables discriminated the pro signers from the anti signers. One was academic discipline; biomedical scientists were almost always on the anti

Table 9.1. *Percentage of professors rated as "liberal" on the Ladd-Lipset scale of political and social values by discipline and school quality*

		School quality			
Discipline	Total	Elite 1	2	3	Lowest 4
Social science	64%	–	–	–	–
Physics	45	66%	47%	28%	33%
Medicine	38	42	30	26	
Mathematics	36	67	45	29	18
Biology	35	54	36	26	24
Chemistry	35	52	42	26	28
Geology	32	46	32	25	17
Engineering	24	39	20	17	16

Note: Dashes represent cases with too few numbers for meaningful percentages.

petition. Among the nonmedical signers, prestige was a good predictor, with members of NAS more likely than nonmembers to appear on the pro petition. Finally, nuclear power position was related to stands on previous political issues, with those who had supported the Johnson presidential candidacy or the deployment of the antiballistic missile (ABM) being relatively more likely to support nuclear power. These results are summarized in Figure 9.2.

The fact that these experts disagree on the wisdom of building nuclear power plants need not in itself cause us concern. If, despite their differing policy preferences, they would agree on basic technical information – the degree of cancer risk from a given level of radiation, the probability of meltdown after a given loss-of-coolant accident – then we could accept their technical advice as divorced from their policy preferences. But too often the facts and values are confounded.

Obtaining closure

As I have said, my own position is that closure is achieved when the policy decision is made, whether or not scientific arguments have been resolved. From this perspective, scientific disputes over policy are to be regarded basically as political disputes involving specialized information that is not well understood by the policymakers. As such, these disputes are fundamentally no different from policy disputes over complexities of law or accounting procedure, which are

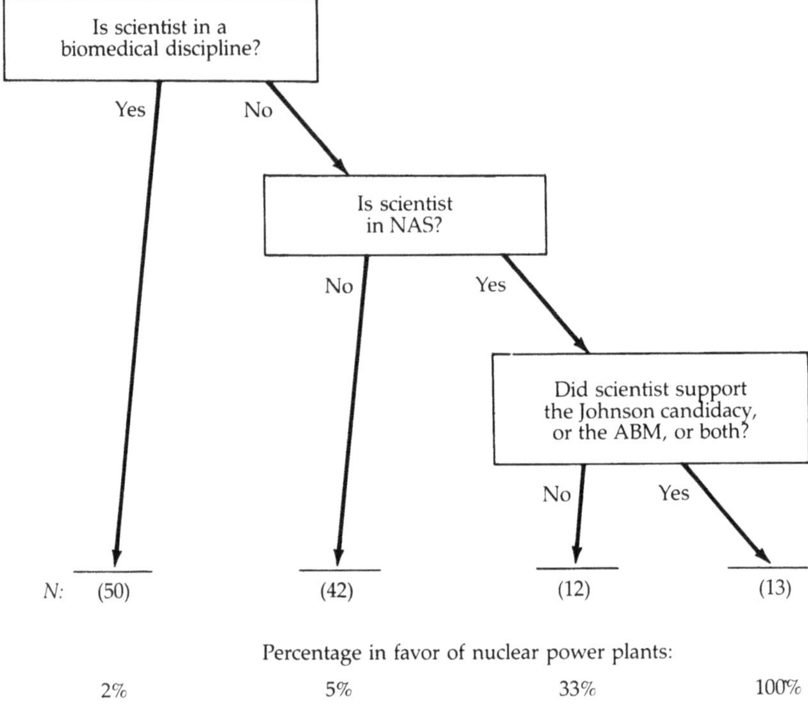

Figure 9.2. Three sociological characteristics were used to differentiate pro-nuclear scientists from antinuclear scientists. Only scientists who signed a nuclear power petition and also appeared on an earlier issue list are included here.

also inaccessible to most policymakers. Although recognizing that policy decisions are often made without regard for the best understanding of these issues, I would like to proceed on the more idealistic belief that policymakers should have access to the best specialized knowledge available.

It is sometimes claimed that the average citizen (and the average policymaker) is capable of understanding most technical issues that are relevant to policy decisions (e.g., Casper, 1976). I do not believe that anyone who has tried to teach statistics to sociology doctoral students would hold that view for long. The typical voter in a referendum on fluoridation or nuclear power seems to have at best a vague understanding of the underlying scientific disputes that have been discussed during the campaign. The confusion of legislators when they hear technical experts make apparently inconsistent claims about factual matters is apparent from congressional hearings. The conclusion seems inevitable that if we are to use specialized knowledge, we must depend on experts.

At the same time, we recognize that experts' positions on factual matters are related to their positions on value issues, sometimes blatantly so. It must also be said that there are many experts who are reasonably unbiased in dealing with scientific material when the related policy matters are not dear to them. Thus it is surely feasible to get a reasonable adjudication of scientific disputes between involved experts by turning to uninvolved experts. Nonetheless, even the uninvolved experts have their own peculiar values and political preferences, which, as with all of us, reflect their particular positions in the society. Thus we want to maximize our access to their expertise while minimizing our dependency on their values.

There seems to be fairly broad agreement, at least within the Hastings group, that factual questions can be distinguished from policy questions (Engelhardt, Giere, McMullin). In Table 9.2, I have sorted many of the controversies that were discussed in the Closure Project according to whether they are primarily factual disputes, policy disputes, or mixed. It is the factual disputes that hold the promise of settlement by "sound argument" (see Beauchamp, chap. 1, this volume) through the adjudication of experts. Policy disputes must be handled differently.

The approach of closure that I prefer is to separate scientific questions of fact from policy questions of value. The rationale here is to let the scientists do what they are especially well equipped to do, which is science, while keeping them away from policy decisions, where they have neither special competence nor a public mandate. Once the factual matters are settled, at least provisionally, then the usual procedures of government can be used to make policy decisions, whether by Congress, the federal agencies, the courts, or by citizens voting in a referendum.

My preference runs counter to the current call for extraordinary public participation in policy decisions regarding science and technology. At one time it was commonly thought that the wisest solutions to technical problems would come from interdisciplinary congresses of scientists, engineers, philosophers, humanists, and clergymen, all pooling their diverse perspectives into a synergistic whole that would somehow be greater than the sum of its part. In time this view was recognized as elitist (laborers and housewives were never included), and there were new suggestions that the participation of ordinary men and women was more important than the involvement of philosophers or clerics. When President Carter appointed a special

Table 9.2. *Fact/policy classification of controversies*

	Political (Narrow Sense) Controversies	
Disputes over facts	Disputes over facts and policies	Disputes over policies
Continental drift	Nuclear power	Human subjects
Detection of gravity waves	Fluoridation	Homosexuality
	Laetrile	
	Love Canal	
	Benzene	
	765-kV power lines	
	Supersonic transport	
	Antiballistic missile	
	Recombinant DNA	

commission to investigate the accident at Three Mile Island, he included a "housewife and mother of six" who lived near the site. When the City Council of Cambridge, Massachusetts, challenged the construction of a laboratory at Harvard for recombinant DNA research, it appointed a much publicized "citizen court" of eight lay persons from diverse occupations and educational levels (Krimsky, 1979), which presumably was a better representative of the city than the City Council. Sweden, the Netherlands, and Austria have made elaborate attempts to involve the average citizen in decision making about nuclear power (Nelkin and Pollak, 1977).

I certainly do not oppose public participation as such, but I do not see why there should be more of it in science and technology policy than in other kinds of policy, or, conversely, less in nonscience policy than in science policy. We never make a point of bringing housewives and blue-collar laborers into formal decisions about the prime interest rate or on whether or not to attack Iran, so why do it when evaluating nuclear power plants and recombinant DNA laboratories?

Moral experts

I have argued that factual disputes should be separated from the policy disputes with which they are often enmeshed and that scientific experts are particularly well qualified to adjudicate the factual disputes. If this is so, are there also special experts – call them "moral experts" – who are particularly well qualified to deal with policy (or ethical) disputes? Presumably these might be moral philosophers, clergymen, or sociologists who study values.

I will deal with this question in a limited way, focusing on a real situation that came up when I was invited by the Nuclear Regulatory Commission (NRC) to participate in a workshop on safety goals. The kinds of questions that were posed in the workshop went like this. If you were an agency delegated with the responsibility of licensing nuclear power plants, as the NRC is, then how would you decide how safe a plant must be in order for it to obtain a license? You are not concerned with setting broad national policy on whether or not there should be nuclear power or on how much there should be. Rather, you are asked to take the NRC's perspective. Some nuclear plants are to be built, and NRC has to approve them. How safe should they be to obtain approval? It is assumed that zero risk is impossible, so the acceptable risk level must be something above zero.

I will consider only one of the many ways that one could state a safety goal. Assume that the risk to an average person of getting cancer in any one year is x. If this person lives next to a nuclear power plant, his or her risk of getting cancer in a given year is larger, say $x + y$. What is the largest y that is acceptable?

It is likely that people living near the plant will have different ideas about this than those living far away, who are at lower risk. How then does the agency arrive at a safety goal, a value of y? Such safety goals are rarely stated so explicitly, but if one wanted to do it, how should it be set?

The kinds of solutions that have been proposed fall into four broad categories (see Table 9.3). First, there is politics as usual. Let y be set through the routine interplay of interest groups acting through normal political institutions, perhaps Congress or the NRC itself (see Rich, chap. 5, this volume). I have argued in favor of this approach, for all its imperfections, because it works fairly well and because the process is already in place and, to a remarkable degree, publicly accepted. However, let us consider some of the other proposals, for these more or less call upon the special skills of moral experts.

The second category of proposals contains all attempts to calculate, in explicit quantitative terms, the net profit to be obtained from alternate policies, or, in this case, alternate safety goals. How costly are cancers to society, and how many are we willing to accept for the benefits of nuclear-generated electricity? Or, put into a comparative mode, if we have alternate methods of generating a given amount of needed electricity, say by coal or nuclear power, what are the risks and other costs of each method, and what risk level would make nuclear power the most desirable (i.e., cheapest) of these options? The value of y could be set so as to maximize net profit, according to these calculations.

Table 9.3. *Proposals for determining "acceptable risk"*

Proposal	Major drawbacks
Politics as usual	Favors special interests; erratic; short-sighted
Explicit attempts to calculate net profit	Quantitative factors are uncertain, often extremely so; ignores equity
Measure risk (+ other costs) vs. benefits	Incommensurate quantities (e.g., dollars vs lives)
Compare alternatives	Usually alternatives are not fully comparable
Revealed preferences	Based on dubious assumptions
Discern public sentiment	Sentiments are not well articulated
Opinion polls	Responses are highly sensitive to questionnaire wording
"Reasonable person" approach	????

It is exceedingly difficult, in practice, to make such calculations in an objective manner. The various quantitative factors, for example, the risk of serious accidents, are often uncertain in the extreme. Also, some forms of calculation require that we apply a common metric to what are essentially incommensurate quantities – for example, setting a dollar value on a life. These are tasks that require substantial expertise as well as substantial moral judgment, and perhaps call for moral experts. On the other hand, a cynic might note that such calculations virtually never contradict the preferred policy of the institution that sponsors the calculation, so the real job for the moral expert is to make apparent the various hidden assumptions in these calculations and to strip bare the biases and value judgments.

The third category of proposal is the method of "revealed preferences" (Starr, 1969). The basic assumption here is that society has already come to terms with various sources of risk, avoiding those that are intolerable and moderating or adjusting to those that are worth the associated benefits. It is assumed that society has reached some sort of equilibrium condition wherein the risks that we accept are, ipso facto, acceptable; otherwise we would avoid them. Therefore, new sources of risk, such as nuclear power plants, ought to be acceptable if y is set so that they offer no greater risks than society already accepts from sources that produce comparable benefits.

I have never found this method defensible. It seems absurd to claim that certain risks are acceptable just because we cannot escape them. The whole notion of an equilibrium requires that individuals have good knowledge of the risks they face and of alternative risks, and that they have real freedom to choose between alternatives. These are dubious assumptions. It is remarkable that such claims enjoy the attention that they currently receive, particularly when they are so obviously a propagandistic attempt to justify nuclear power. Nonetheless, here again is a role for the moral expert, either in discerning the level of acceptable risk for comparable technologies or in debunking other moral experts who discern such levels of acceptable risk.

The fourth category of proposals for setting safety goals goes directly to public sentiment. Since it is the public, variously defined, that must accept the risk, why not ask the public what value of y is acceptable? Opinion polls have been suggested for this purpose.

Opinion polls are well suited for measuring public attitudes about issues that most people have thought about and upon which they have a clear and strong opinion, a good example being their choice of a presidential candidate on the eve of the election. However, when a pollster asks for a presidential choice in June, when the election is still months away, the response is not very meaningful, particularly when the respondent has never heard of many of the names in the running. Similarly, to ask a respondent for an acceptable value of y is to confront him or her with an issue that he or she has probably never considered before, and this is unlikely to produce meaningful results. It is also clear that the particular form in which the question is asked will affect the response; thus I could easily produce questions that sound fair but that would produce, from similar samples of respondents, very different values for an acceptable y. The point is well illustrated by an example from Tversky and Kahneman (1981), who asked two groups of college students to make risk judgments, as illustrated in Table 9.4. Confronted with an impending national outbreak of a disease, the first group preferred policy A to policy B. A second group of students was presented with a choice between policies C and D, which, in mathematical terms, were precisely equivalent to policies A and B, although they were stated in different language. The language difference produced a complete inversion in the results. For such reasons, I contend that the setting of safety goals through opinion polls is unworkable.

Still, there is a grain of appeal in the notion of discerning public sentiment, if it could be done. Here, perhaps, is a role for moral experts, for perhaps they can tell us what the public sentiment is, even though individual members of the public cannot articulate it very well. Thus I, as a sociologist empathetic to American culture, claim that a value for y of 10^{-3} is too high; it would violate public sentiment, or at least it should. However a value of 10^{-7} is plenty low. The proper value of y is somewhere between these extremes, and if the moral experts can narrow this range, then perhaps they can produce a meaningful value. In doing this, I have made a claim that is very like the judicial claim of what a "reasonable person" – presumably like the judge – would do.

The first objection that must be made to this method is that it is exceedingly elitist, delegating to some experts the job of articulating public sentiment. More relevant, though, is a second objection: The experts may not be able to do it. However if this objection is met, if they can do it, then might this not meet the first objection as well? For if the public sentiment is truly discerned, what difference who discerns it?

How, then, do we decide whether or not moral experts are capable of discerning, in a meaningful way, the public sentiment? The catch here is that we have no way of knowing public sentiment independently of what the experts say it is. But we can make some progress by setting simpler tests for the experts. At the least, we would expect

Table 9.4. *Questionnaire bias*

Imagine that the United States is preparing for the outbreak of an unusual Asian disease, which is expected to kill 600 people. Two alternative programs to combat the disease have been proposed. Assume that the exact scientific estimate of the consequences of the programs are as follows:

Choices	Percentage in favor
Choices given to student group no. 1 (N = 152)	
If Program A is adopted, 200 people will be saved.	72
If Program B is adopted, there is a one-third probability that 600 people will be saved and a two-third probability that no people will be saved.	28
	100
Choices given to student group no. 2 (N = 155)	
If Program C is adopted, 400 people will die.	22
If Program D is adopted, there is a one-third probability that nobody will die and a two-third probability that 600 people will die.	78
Total	100

Source: Tversky and Kahneman (1981).

two groups of moral experts, both set to discerning the same public sentiment, to come up with similar answers. One can visualize additional tests to appraise the adequacy of this method, but I will leave this task to the moral experts who might be interested in pursuing it.

The proper role of controversy

Technical controversies tend to be regarded as aberrations, as undesirable disruptions that ought to be ignored, disposed of, or avoided altogether. However, there is another viewpoint that emphasizes their positive function.

This viewpoint begins with the assumption that the growth of both scientific knowledge and technological invention is good and ought to be encouraged. It follows that there must be proper roles in the society for people who will advance science and promote new technologies. At the same time, however, great care must be taken in the distribution of scarce research resources and in the implementation and deployment of innovations. We must minimize waste, health risks, environmental insults, and other costs. These are nearly unassailable assumptions, accepted consensually.

It would be nice if the scientist and the engineer were cognizant of and deeply concerned about the potential risks, secondary consequences, and other costs of their creations. However, it does not seem realistic to me to expect inventors and promoters of innovation to behave in this way. They may take care to avoid obvious problems, but the promoter's role is not compatible with cautious restraint unless there is an obvious reason for it.

If we are to have social roles that promote potentially hazardous innovations, we should have other roles, filled by different people, to counterbalance the promoters, to search for the hazards and other costs that the promoters do not see, or ignore, or actively avoid.

Regulatory agencies fulfill some of this function, but neither the routine work setting of a bureaucracy nor the job motivations of career bureaucrats are conducive to an enthusiastic and imaginative attack on the problem. Furthermore, innovations, by their nature, often fall outside the competence and mandate of preexisting agencies. The congressional Office of Technology Assessment was established to deal with some of these concerns, but it, too, is a bureaucracy and may not be an appropriate setting for the task. NAS committees have been useful in evaluating some hazards that have already been identified, but if we look at salient cases, such as those of nuclear power or DDT, the academy has not been particularly effective in locating problems.

There have by now been numerous instances when the informal process of social controversy has been more effective in identifying and explicating the risks and benefits of a technology than have been any of the formal means that are supposed to do this. Critics attack with great vigor, stretching their imaginations for all manner of issues with which to score points against their targets. Proponents counter-attack, producing new analyses and funding new experiments in order to refute the critics. Each side probes and exposes weaknesses in the other side's arguments. As the controversy proceeds, there is a filtering of issues, so that some with little substance become ignored, and others become prominent. It is true that some unimportant issues gain an inordinate amount of attention for a while, but they usually drop from concern unless they can sustain interest on their intrinsic merits. In the meantime, intelligent people on both sides of the controversy search enthusiastically for new problems, diligently preparing charges and rebuttals, testing the strength of their arguments in open debate. In this manner the controversy functions as a funnel, bringing diverse problems together, and it has the potential to act as a sieve that separates important concerns from those without real merit. This is the proper function of technical controversies and of the experts who participate in them, and to the extent that they are successful in this function, they should be encouraged as an important adjunct to more formal methods of technology assessment.

Unfortunately, technical controversies are usually chaotic, and therefore we fail to achieve the potential benefits available to us. In a controversy, each side is heavily motivated to make the strongest case it can while at the same time finding flaws in the other side's position. When these arguments are juxtaposed, flaws and polemical tricks can be disposed of, and we see the beauty of first-class reasoning in support of one's own ends. However, in most controversies the adversaries never confront one another. When one expert seems to contradict another and we cannot locate the reasons for their disagreement, we become confused. If we could bring order to the chaos so that we might better understand the bases for their disagreements, then the value of such arguments would be enhanced.

References

Anderson, C., and J. Murray. 1971. *The Professors*. Cambridge, Mass.: Schenkman.

Bazelon, D. 1979. "Risk and Responsibility." *Science* 205:277–80.

Boffey, P. 1975. *The Brain Bank of America*. New York: McGraw-Hill.

Casper, B. 1976. "Technology Policy and Democracy." *Science* 1974:29–35.

Krimsky, S. 1979. "Regulating Recombinant DNA Research." In *Controversy:*

Politics of Technical Decisions, ed. D. Nelkin. Beverly Hills, Calif.: Sage.

Ladd, E., and S. Lipset. 1972. "Politics of Academic Natural Scientists and Engineers." *Science* 176:1091–1111.

Mazur, A. 1977. "Science Courts." *Minerva* 15:1–14.

Mazur, A. In press. "TMI and the Scientific Community." Proceedings of the New York Academy of Sciences. 1980.

Mazur, A. 1981a. *The Dynamics of Technical Controversy*. Washington, D.C.: Communications Press.

Mazur, A. 1981b. "Media Coverage and Public Opinion on Scientific Controversies." *Journal of Communication* 31:106–15.

Nelkin, D., and M. Pollak. 1977. "The Politics of Participation and the Nuclear Debate in Sweden, the Netherlands, and Austria." *Public Policy* 25:333–57.

Schiefelbein, S. 1979. "The Invisible Threat." *Saturday Review* (September 15): 16–20.

Starr, C. 1969. "Social Benefit versus Technological Risk." *Science* 165:1232.

Tversky, A., and D. Kahneman. 1981. "The Framing of Decisions and the Psychology of Choice." *Science* 211:453–8.

10

Controversies and the authority of science

DOROTHY NELKIN

Scientists have participated in a remarkable number of controversies during the past few decades.[1] Some of these controversies revolve around technology. Legislation to control the environmental and health risks of technology calls for decisions based on "the best scientific evidence." Government efforts to regulate advances in technology thus engage scientific experts in lively debate over the interpretation of data and the dimensions of risk. Other controversies relate to science itself. The disputes over recombinant DNA research, animal experimentation, in vitro fertilization, or fetal research have inevitably engaged scientists in political activism to protect their own interests.

Still other disputes, including many of those discussed in this volume, have little to do with science or technology at all. Abortion legislation, the decision to ban Laetrile, and the classification of homosexuality by the American Psychiatric Association as a mental disorder are fundamentally ethical, moral, or political issues. Yet, here too science is consistently invoked as the disputes unfold, reflecting the increasing tendency to rely on scientific standards as a basis for legal and policy decisions. Indeed, science is a political resource, called upon as a source of rationality and a basis of consensus in a wide range of policy areas.

Although access to expertise provides a means of legitimating decisions, it can also provide the ability to question them. Thus, controversial public policy questions invariably seem to evolve into scientific disputes, and difficult ethical dilemmas translate into debates over the adequacy and interpretation of evidence or the appropriateness of scientific methodologies. More research, better

[1] For a review of twelve controversies engaging scientists, see Dorothy Nelkin, ed., *Controversy: Politics of Technical Decisions* (Beverly Hills, Calif.: Sage, 1979).

evidence, better expert advice, it is assumed, will help resolve conflict. With greater technical certainty, political or value differences will fade away. To examine this assumption, this essay describes the dynamics of the controversies engaging scientists and analyzes the role of expertise in the resolution of conflict.

Controversies

The proliferation of controversies involving science and technology has several possible explanations. Some contend that it is a manifestation of the "crisis of authority" that began in the 1960s, a reflection of the declining confidence in all modern institutions, and especially the existing forms of political authority.[2] Others suggest that disputes reflect the values and expectations of an educated middle class. Ronald Inglehart has analyzed what he calls a "silent revolution" in the values of this educated class, a shift from an emphasis on material consumption and security towards greater concern with the quality of life and an expectation of greater participation in policy decisions.[3] It is widely assumed that controversies occur because of inadequate information; that people are critical because they are poorly informed.

I would argue that the recent disputes are a means of negotiating social relationships and of sustaining certain values, norms, and political boundaries at a time of important scientific and technological change. Some controversies (e.g., the nuclear debate) represent basic ideological struggles over the political complexion of a "good" society. But even less cosmic disputes express fundamental value concerns. Some develop over a fear of risk, others reflect uneasiness about the social uses of knowledge, the fear that research findings may be used for harmful ends. Other disputes occur because people consider scientific research morally dubious, a threat to their cherished beliefs. Questions of equity or justice arise over the allocation of resources or the distribution of economic and social costs. Finally, scientists are engaged in disputes in which the central issue is freedom of choice. Their contribution to closure of conflict depends on the underlying sources of dispute.

Fear of risk

Because we are constantly deluged with warnings about food additives, polychlorinated biphenyls (PCBs), freon, toxic substances, and

2 Michael Crozier, Samuel Huntington, and Jos. Watanuki, *The Crisis of Democracy* (New York: New York University Press, 1975).

3 Ronald Inglehart, *The Silent Revolution* (Princeton: Princeton University Press, 1977).

radiation – the list is long, and growing – the fear of risk is inevitable. Rather little is known about the sociopsychological factors that determine the public response to technological risks, but clearly, fears are exacerbated when these risks are uncertain, invisible, or unfamiliar.[4] Involuntary risks are especially problematic, as are those that pose delayed threats that are anticipated some time in the distant future.

These are the characteristics of risk that provoked the opposition to both nuclear power and recombinant DNA research.[5] In both, it was the possibility of unlikely but devastating catastrophe that caused concern. In both cases, public fear was exacerbated by the "invisible" and only vaguely understood nature of the risk. How does one know if a lethal gene has been produced or if a nuclear waste storage facility is adequately protected against radiation leakage? In both cases, uncertainly about risk left open large areas for conflicting interpretation. Disagreement among scientists increased public confusion and doubt by revealing a limited ability to predict the impact of science and technology. Finally, in both cases, value-laden questions intruded: What is an acceptable level of risk? Who should assume responsibility for evaluating science projects when they may have an impact on public health? Increased evidence about the nature of risk fails to allay public concern when such value-laden questions prevail.

The fear of misuse

In 1975, an acrimonious dispute took place over a research proposal by Harvard University scientists to study male children with an XYY chromosome pattern. The proposal grew out of a hypothesis about the relationship between this genetic aberration and criminal behavior. This hypothesis was based on observations of a high incidence of the extra Y chromosome among men in prisons. The investigators proceeded with a chromosome-screening program, hoping that better understanding might facilitate the development of remedial and therapeutic intervention. Critics claimed, however, that

[4] W. Lowrance, *Of Acceptable Risk* (Los Altos, Calif.: William Kaufman, 1976); W. D. Rowe, *An Anatomy of Risk* (New York: Wiley, 1977); Baruch Fischoff, Sarah Lichtenstein,Paul Slovic, et al., *Acceptable Risk* (Cambridge: Cambridge University Press, 1981).

[5] This debate has been the subject of a number of books; see, for example, J. Goodfield, *Playing God* (New York: Random House, 1977); S. Stitch and D. Jackson, eds., *The Recombinant DNA Debate* (Ann Arbor: University of Michigan Press); also, *Daedalus*, vol. 107, no. 2 (1978), special issue on *Limits to Scientific Inquiry*. For documents, see U.S. Department of Health, Education and Welfare, National Institutes of Health, *Recombinant DNA Research* (Washington, D.C.: U.S. Government Printing Office, 1976).

XYY research was more likely to perpetuate damaging assumptions about the genetic origins of antisocial behavior and would lead to pernicious mechanisms of social control.[6] Similarly, pervading the debate over the origins of homosexuality was a fear that "scientific" definitions of homosexual behavior would imply that etiology is a relevant criterion for social policy and that this could lead to abuse or, at the very least, to a justification and perpetuation of existing social and economic inequities.

Similar arguments confront almost any area of research that associates genetically mediated characteristics with human behavior. Work on the relationship between race and IQ and, more generally, research in sociobiology raise the specter of stigmatization and social control. Research on genetic engineering evokes eugenic images: A poster at a recombinant DNA debate read, "We will create a perfect race."

Fear of the social misapplication of biomedical research is heightened by the use of medical technologies to subdue aggressive inmates of penal or mental institutions and to treat hyperactive children. Anxiety about such misapplication leads directly to questions about the advisability of seeking certain kinds of knowledge at all, for, it is argued, knowledge cannot easily be disengaged from its uses. Biology, asserts a groups of critical scientists, is a "social weapon."[7]

Moral and ethical concerns

For many critics, it is the moral implications of science and technology that shape their dissent: the fear that science may change the normal state of nature, alter the essential genetic structure of man, or threaten cherished beliefs. At a time when the accomplishments of science have fostered in some a faith in rational explanations of nature, others seek to reinvest educational systems with traditional faith. And even as biomedical research brings about dramatic improvements in medical care, critics question areas of science that challenge traditional values.

The challenge to human gene research is a case in point. The most outspoken critics of this research are concerned with the creation and control of new life forms, a possibility that violates deeply held beliefs about free will and self-determination. They have used anxiety-provoking images of Frankenstein and Faust: "It is no longer science

[6] G. Annas, "XYY and the Law," *Hastings Center Report* 2 (1972); R. Roblin, "The Boston XYY Case," *Hastings Center Report* 4 (1975): 5–8.

[7] Ann Arbor Science for the People Collective, *Biology as a Social Weapon* (Minneapolis: Burgess, 1977).

fiction but its realization," "Scientists hold our genetic future in their hands." The patenting of new microorganisms has crystallized persistent fears that scientific inquiry flaunts cherished ethical principles.

Similar fears continue to arouse fundamentalists who still take issue with the teaching of evolutionary biology in public schools. They feel that it intrudes on their religious beliefs, and they demand that schools provide "equal time" for creation theory. The fundamentalists are a sufficiently powerful force that many educators, biologists, and publishers view their demands as a serious threat to science education.[8]

Concern about the moral implications of science also nurtured the dispute over fetal research and the debate over in vitro fertilization. Such research is highly controversial; closely linked as it is to the issue of abortion, it is perceived by some as a moral affront, an offense to basic beliefs. Although critics feel that human dignity and the integrity of the individual call for stopping such research, scientists argue that the medical knowledge to be obtained is sufficient justification for it.[9] The arguments against genetic manipulation, fetal research, and in vitro fertilization follow a long tradition of criticism of science as disruptive of human values, of nature, of natural law, a tradition that accounts for the emotional character of many recent disputes.

Questions of equity

Given the high financial costs of research, how should priorities be determined? Should funds be allocated for research to develop sophisticated and costly lifesaving technologies or for more basic health care procedures? If a research technique promises social gain but poses individual risk, which should prevail? Conversely, if research will benefit only a select few, should the public pay the cost?

Disputes over questions of equity also occur when citizens in a community become aware that they are to bear the costs of a technology intended to benefit a much broader constituency. This concern is most explicit in controversies over the siting of large-scale technologies. Airports and power plants serve large regions, but proximate neighbors bear the environmental and social risks. Normally, projects are planned and sites selected on the basis of

[8] Dorothy Nelkin, *Science Textbook Controversies and the Politics of Equal Time* (Cambridge, Mass.: MIT Press, 1977).

[9] P. Ramsey, *The Ethics of Fetal Research* (New Haven: Yale University Press, 1975); T. Powledge, "Fetal Experimentation," *Hastings Center Report* 11 (1975): 5–8.

economic efficiency and technical criteria. But community protests raise basic questions of distributive justice: Can any reduction in some citizens' welfare be justified by greater advantages to others? Can the magnitude or intensity of costs borne by neighbors of a major project be reasonably incorporated into benefit–cost calculations? Scientists enter such controversies to assess the seriousness of risk, but the basic issue in such disputes remains one of equity.

Freedom of choice and individual rights

Many controversies engaging scientists concern the appropriate role of government and its impact on freedom of choice. Decisions about the implementation or the regulation of a technology may intrude on individual rights. A process such as fluoridation or a technology such as the automobile air bag mandates universal compliance; everyone must experience the effects of the decision. In other cases, the potential harm of a product encourages government controls. If the sale of Laetrile, saccharin, or cyclamates is prohibited, individuals who want these products are denied the right to buy them. Government imposes regulations on the assumption that individual choices may have social costs, but such constraints may be seen as a means of protecting professional privilege, as unnecessary government paternalism, or as a violation of individual rights. These are the perceptions that have maintained the Laetrile dispute. Indeed, to those who argue their right to use Laetrile as a cancer cure, the very fact of consensus within the scientific establishment about the problems of this cancer therapy has in itself been a source of suspicion.

Science and the resolution of disputes

A fundamental irony pervades the controversies just described. Although questioning the role of expertise and demanding greater participation by those affected by policy decisions, the protagonists in these debates themselves depend heavily on scientific expertise. Whatever social or political values motivate these disputes, they often focus on technical questions that call for scientific expertise. A siting controversy may develop out of concern with the quality of life, but the debate tends to revolve around technical questions: the physical requirements for the facility, the accuracy of the predictions establishing its need, or the precise extent of environmental risk. Concerns about the freedom to select a cancer therapy are reduced to technical arguments about the efficacy of treatment. Moral opponents of abortion or of fetal research engage in scientific debate about the precise point at which life begins. Enviromentalists hire their own experts to question the technical feasibility studies of questionable projects. Laetrile supporters have their own medical professionals to legitimize their claims. Even fundamentalists seeking to have the biblical account of creation taught in the public schools engage scientists to "prove" that "creation theory" is a scientific alternative to "evolution theory."

This is tactically effective, for in all disputes broad areas of uncertainly are open to conflicting scientific interpretation. Decisions are made in a context of limited knowledge about potential social or environmental impact, and there is seldom conclusive evidence to reach definitive resolution. Thus power hinges on the ability to manipulate knowledge or to challenge the evidence that is presented to support particular policies. Technical expertise becomes a resource exploited by all parties to justify their views, to create legitimacy, and to control the terms of debate. In the process, scientific facts, used selectively, converge with political values. Both project proponents and critics use the work of "their" experts to reflect their judgments about priorities or about acceptable levels of risk. Expertise becomes one more weapon in an arsenal of political tools.

The resulting conflicts among scientists are often interpreted as portents of the declining power of the scientific enterprise, as reflections of a crisis of science and an erosion of confidence in science as a basis of rational consensus. How, in fact, has the policy role of expertise changed as a result of conflict? Is the credibility of science at stake? And how has the changing role of scientists influenced the resolution of disputes?

The changing role of scientists

Scientists have clearly played an increasingly important role in the making of public policy. Federal government expenditures for technical consultants is nearly $2 billion a year. There has been a general evolution from ad hoc advisory activity to the institutionalization of an "intellectual technocracy," as federal employment of scientists grew by 49 percent from 1960 to 1970. Congress alone expanded its staff from 2,513 in 1947 to 13,272 in 1976 and included many specialized, science-trained consultants. These changes reflect the complexity of policy decisions but also the tendency to think of scientific expertise as a source of authority, a means to achieve consensus in public affairs.

Scientific standards have universal appeal as an authoritative basis of rational decision making. Enhancing the role of the "knowledge elite" is its monopoly over critical knowledge, the specialization inherent in bureaucratic organization, and the cultural emphasis on rationality, efficiency, and technological progress. Moreover, as Yaron Ezrahi argues, the norms of science have a certain affinity to the liberal democratic image of politics, giving science special force as a model for political decisions based on homogeneous and rational standards. Science has appeared "less blemished than other spheres of activity by transcendental, factional, personal or other publicly inaccessible references."[10]

This image accounts for a long history of proposals that would enhance the role of science in public affairs. In 1932, for example, the physicist, Robert Millikan, proposed establishing "a scientific jury system" to solve the policy problems of a society in the midst of depression. This jury system would discover the true "social facts" from which policy could be determined. Policies were to be established and administered by scientific experts who by virtue of their knowledge were qualified to judge how to improve the human condition. Democracy, in the view of Millikan and many of his contemporaries (Charles Merriam and Michael Pupin) is founded on an objective, knowledge-seeking, scientific approach to the problems of government.[11]

This image of the objectivity and neutrality of science has even survived cases of deception. Well publicized as anomalies, they are blamed more on individual aberration than on the profession and have thus served to reinforce the norms of science as an institution. However, far more damaging to the neutral image of science has been the increased participation of scientists in public disputes. Controversies involve scientists in new policy positions as advisers to citizen groups, interveners in public hearings, and participants in review boards and special commissions. In controversial technological areas, scientists are called upon to represent both pros and cons, and public interest science has itself become institutionalized as a new field devoted to providing expertise and technical representation to very different interests in the course of public disputes. In this context, scientists enter political life less as a source of consensus than as ideologues, adversaries, or as quasi representatives of conflicting interest groups. The notion of neutral science is blemished by factional debates, and its authority tarnished by its use as a means of social manipulation and political control. Metaphors of "the

10 Yaron Ezrahi, "Science and the Problem of Authority in Democracy," in T. Gieryn, ed., *Science and Social Structure, Transactions of the New York Academy of Sciences*, II, 39 (1980): 43–60.

11 R. Millikan, "Science and Social Justice," in R. Tobey, "The New Sciences and Democratic Society," Ph.D. diss., Cornell Unversity, 1969.

mandarins," "the priesthood," or "the new Brahmans" are used to describe the various functions of the scientist in public policy, and for some critics science is less a model for liberal democratic values than a political threat, a means for established institutions to reinforce their influence and to limit effective public choice.[12] The factions that emerge when scientists enter policy disputes have indeed begun to create doubts about the usefulness and credibility of science as a compelling criterion for legitimate decision-making authority.[13] Indeed, a ubiquitous question woven throughout the variety of disputes over science and technology is, Who should control crucial policy choices? Reflected in this question is the fear that complex technological advances with far-reaching consequences have changed the nature of the decision-making process, that scientific approaches to public policy have encroached on the traditional political models, and that deference to scientific knowledge threatens democratic principles, reducing public control of crucial policy choices.

Yet, do these disputes necessarily imply a "crisis" in the relationship of science to politics? Does this portend a declining influence of science in public affairs? On the contrary; the disputes, I would argue, have expanded the political role of scientists, though they have changed the rules of the game. Scientists have tended to assume an adversarial rather than consensual role in public affairs, but their roles have multiplied and their influence increased. Disputes have broadened the scope of scientific research and created new positions in an increasing number of policy areas, providing material benefits to individuals through new consulting and advisory positions. Conflicts have inspired several proposals reminiscent of Millikan's scientific jury. A proposed science court would bring the best judgments of the scientific community to bear on the conflicting technical aspects of controversial issues, thereby providing a "defensible, credible basis for policy decisions."[14] Similarly, proposals have emerged for technical assessment teams, science panels, and even "technological magistratures" to judge the various sides of risk disputes and for a new profession of "Certified Public Scientist" to ensure objectivity as scientists engage increasingly in public affairs.[15] Faced with increas-

[12] See, for example, Spencer Klaw, *The New Brahmans* (New York: Morrow, 1969).

[13] For a discussion of scientific approaches to decisionmaking, see Vince Taylor, "Subjectivity and Science," *Technology Review* 81 (1979): 48–56.

[14] Task Force of Presidential Advisory Group on Anticipated Advances in Science and Technology, "The Science Court Experiment," *Science* 193 (1976): 653.

[15] G. Bugliarello, "A Technological Magistrature," *Bulletin of the Atomic Scientists* 1 (1978): 34–7, and J. C. Glick, "Reflections and Speculations on the Regulation of Molecular Genetic Research," *Annals of the New York Academy of Sciences* 265 (1976): 173.

ing litigation over technological questions, legal scholars have proposed a variety of mechanisms to enhance the technical competence of the courts. These include training lawyers and judges in science and forming a variety of special science advisory groups.[16] All these proposals are directed toward improving the quality of evidence, on the assumption that this will contribute to the closure of disputes.

Problems of closure

As scientists increasingly participate in public controversies, what is the effect on their resolution? Does the tendency to rely on scientific standards contribute to decisions that are acceptable to contending groups? In those cases where the central concerns are simply questions of risk, clarifying the dimensions of risk can facilitate resolution. But in most cases, I would argue, scientific evidence has only limited bearing on the resolution of controversy.

First, many of these disputes have little to do with science at all; rather, questions of aesthetics, of human dignity, of religious beliefs are involved. These are what Lawrence Tribe has called "fragile values": nonquantifiable, intangible, resistant to categorization.[17] The urgency of prompt action in controversial areas may lead to undue reliance on expert forums. Policymakers need answers, but applying the logical deductive methods of science to develop precise definitions can be destructive, diverting attention away from the controversial value questions that are basic to technical disputes.

Second, efforts to develop a factual grounding for the resolution of disputes confront a basic dilemma. In controversial areas, questions of value are difficult to distinguish from facts. Value premises may color scientific findings: The boundaries of the problem to be studied, the alternatives to be weighed, and the issues regarded as appropriate–all may determine which data are selected as important, which facts emerge. And facts themselves are seen through concepts that are derived from theory, preconception and, often, extrascientific considerations. The belief that science is inherently removed from political considerations and that scientists are therefore political celibates is a long-standing one. But in the present social context of science, it is an anachronistic belief. Collecting data for policy purposes requires not only measurement and observation but inter-

16 For a review of proposals to enhance the role of science in the courts, see Sheila Jasanoff and Dorothy Nelkin, "Science, Technology and the Limits of Judicial Competence," *Science* 214 (1981): 1211–15.

17 Laurence H. Tribe, "Ways Not to Think about Plastic Trees," in L. Tribe, C. S. Schelling, and John Voss, eds., *When Values Conflict* (Cambridge, Mass.: Ballinger, 1975), pp. 61–92.

pretation and prediction, activities that scientists themselves recognize as vulnerable to bias.

Third, the use of science to resolve disputes brings scientists into the political culture, where criteria of fairness and equal representation, alien to the scientific ethos, prevail. Ironically, the greater the utility of science in political affairs, the less it can maintain the image of objectivity that has been the very source of its perceived political value. The nuclear debate was a case in point; scientific arguments in this controversial field were strongly identified with political positions vis-a-vis the desirability of nuclear power. In this context, the political value of the technical arguments was neutralized.

Finally, an important issue in most recent controversies is the democratization of political control. The use of scientific rationality to mask basic social choices is a pervasive theme. Thus, scientific experts play a very ambiguous role. For they serve as a source of authority in disputes that are partly based on mistrust of precisely the kind of esoteric expertise that they themselves have come to represent.

The proliferation of controversies presents basic policy choices; such choices require setting priorities and evaluating the public will. This is a political, not a scientific, role. In the absence of social consensus, casting these larger problems of policy into the narrow problem of scientific evidence is unlikely to resolve disputes. The mere accumulation and segregation of technical data or the search for consensus among scientists addresses only part of the problem. The actual means of resolution will vary depending upon the nature of the value issues underlying technical dissent, the degree of polarization, and the urgency of the decision. For closure of disputes requires that outcomes be chosen in ways that are politically legitimate as well as technically sound.[18] This suggests the importance of social experimentation with alternative forms of conflict resolution based on careful analysis of the structure and nature of specific controversies, as well as on the value issues involved.

Conflict persists. Even as individual disputes are closed, the same tensions recur in other contexts. New controversies have erupted over the use of animals in research, weather modification programs, the swine flu vaccine, the use of pesticides, genetic research and its applications, and the study of sociobiology. For perhaps the most important effect of recent disputes is the growing public awareness of the political and value choices that are embedded in technical decisions.

[18] For a discussion of such experiments, see D. Nelkin and M. Pollak, "Public Participation in Technical Decisions," *Technology Review* 81 (1979): 55–64, and Timothy J. Sullivan, *Resolving Disputes through Negotiations* (New York: Plenum Press, 1984).

11

Post-Skinner and post-Freud: philosophical causes of scientific disagreements

ALASDAIR MACINTYRE

Psychology as an academic discipline in the last twenty years seems to have exhibited – I write as an external observer – two very different tendencies. The first has been one of general rapprochement among both experimentalists and clinicians. Theoretical rivalries have been muted, psychologists originally educated in very different milieus into very different standpoints have drawn gratefully upon each others' work, and even when controversy has been extended, a will to at least minimize disagreement has been evident. It is an interesting question how far this represents the outcome of rational progress in a unified science and how far it is a phenomenon to which social psychologists themselves should direct their attention. Perhaps it can mostly be explained by Heider's balance principle, by Festinger's theory of cognitive dissonance, by attribution theory, or by a mixture of all three. But happily I am not going to be concerned here with this tendency but with its counterpart, the persistence of an implacable will not to be assimilated, not to join the psychological melting pot, on the part of two groups of theoretically partisan psychologists: those who follow B. F. Skinner's prescriptions, and those who constitute the central tradition of psychoanalysis. My main concern will be with Skinner's work, and I shall be using psychoanalysis to illuminate it, rather than vice versa.

What I shall be arguing is that Skinner's work, when contrasted with psychoanalysis, reveals – in a way that is quite contrary to Skinner's own intentions – the ineliminability of philosophy from psychological science. But I shall also be arguing that this is not due to

I must express my indebtedness to Dr. Ilona Lappo, whose detailed knowledge of Skinner's work has greatly assisted me. If errors remain, they are the results of my failure to respond adequately to her enthusiastic exposition of Skinner's views.

any deficiency in Skinner's thoroughgoing and hardheaded application of his own principles. Instead, I shall suggest, first, that no systematic psychology can evade philosophical commitment of a particular kind; second, that these philosophical commitments result in the presence of scientifically ineliminable disagreements within psychology – disagreements that can be evaded only by an atheoretical eclecticism; and third, that it is one of the many merits of Skinner's work that it provides a peculiarly suitable test case for this thesis. It will become clear very quickly that I am not in any sense a Skinnerian. But for this reason it is important to stress at the outset my respect for both the magnitude of Skinner's genuine experimental achievement and the importance of his theoretical standpoint. My own point of view is very different from the hostility exhibited both by some psychoanalysts and by Noam Chomsky. If Skinner is wrong, it is in a much more interesting way than most of his critics recognize.

Chomsky's criticism

Let me therefore begin by discussing one of the key misunderstandings in Chomsky's notorious review of *Verbal Behavior*. Chomsky writes throughout that review as if Skinner is offering an alternative and rival type of explanation of certain phenomena, the characterization of which is or could be neutral as between his type of theory and Skinner's. Yet what Skinner insists on throughout is that the items to be explained are not what Chomsky or cognitive psychologists or stimulus–response theorists took them to be.

What Skinner aspires to explain is the occurrence (or nonoccurrence) of particular types of behavioral episodes. The segments of behavior that a Skinnerian theory, if sound, explains are individuated and identified quite differently from the way in which the units that traditional philosophers of mind or contemporary cognitive psychologists or contemporary generative grammarians deal with are individuated and identified. Chomsky, for example, is concerned with the question, What must be the case for a speaker to be able to utter sentences of a certain syntactically definable type? Syntactical regularities are what characterize the data at the most elementary level (elementary, that is, in terms of the structure of Chomsky's theory). But for Skinner, the emergence of syntactical or, as we shall see, phonetic regularities has itself to be explained in terms of what he takes to be the genuinely fundamental data, units of observable behavior.

It is important, then, not to begin in the wrong place. The right place to begin, perhaps, is with a well-known Skinnerian example. One of Skinner's colleagues is sometimes described as having helped

Skinner to solve the problem of how to teach a pigeon to bowl.[1] The pigeon, a marble, and a set of bowling pins were in an enclosed area. At first the pigeon's behavior had no systematic relationship to either marble or pins. But every time that the pigeon looked toward the marble, it was rewarded with a grain of corn; after the frequency of such looks had increased, it was rewarded after, and only after, its beak had touched the marble; and so by successive approximations the pigeon's behavior was shaped until it produced the behavior of striking the marble toward the pins.

Two points need to be made about this well-known Skinnerian example. The first is the obvious one that the only agent of change producing the pigeon's new behavior was the scheduled reinforcement of some behaviors spontaneously emitted by the pigeon prior to the experimental intervention. What shapes behavior is the effect upon the organism of past behavior.

Organisms do, of course, behave as they do partly because of their biological inheritance and structure. But everything in their behavior that is not to be explained biologically is to be explained entirely in terms of the effect upon the organism of the consequences produced by the environment as a result of the organism's past behavior. The point of a Skinnerian experiment such as that with the pigeon is to make this shaping process visible.

A second crucial point is that it is in a way misleading to describe the pigeon as having been "taught" "to bowl." The pigeon certainly "exhibited the behavior of a skillful bowler" (Carpenter, p. 21). But "to bowl at" is a member of the family of intentional verbs. And the implications of or use of such verbs in ordinary language are deeply at variance with Skinner's theory, a fact that Skinner himself emphasizes at some length.[2] Consequently, it is important to notice that the point of the experiment lies in the way in which every term of every causal relationship in the experiment is, or at least easily can be, specified in nonintentional terms. What the relationship is between the intentional characterization of the episode and the nonintentional Skinnerian characterization is left unclear. It would not, of course, be misleading to say that the pigeon's behavior was shaped so that it exhibited the behavior that we would call "skillful bowling," but one implication of this is that hitherto we have mischaracterized such behavior, whether it is produced in and by pigeons or in and by people. To have remarked this enables us to make three central points about Skinner's position.

The first is that Skinner has embarked upon a project that it was

[1] For example, Finley Carpenter, *The Skinner Primer* (New York: 1974), p. 21.
[2] B. F. Skinner, *About Behaviorism* (New York: 1976), pp. 27, 128–9.

extremely important that someone should embark upon. If Skinner had not existed, it would have been necessary to invent him. For the project is precisely that of discovering how much of human behavior can be explained in entirely nonintentional terms and yet without any resort to neurophysiology or biochemistry. Skinner aspires to show how behavior is produced, without saying anything about the underlying mechanisms. Moreover, the initial successes of the project are quite impressive enough to make it important to understand how far Skinner's research program can be carried through. Since generally the limits to any research program cannot be predicted a priori, the only way to discover the limits is to carry through the research program as far as proves to be possible.

Second, Skinner is concerned with how behavior is produced and shaped – that is to say, his central contentions are causal ones. Skinner places unusual emphasis on experimental controls. Boakes and Halliday have stressed that

> the most direct implication of Skinner's approach to scientific enquiry is that rather than test theories about the average rat, one should study the behavior of individual rats . . . by controlling the situation so that the only changes are those initiated by the experimenter. The internal environment is kept as constant as possible by maintaining the animal on a fixed feeding schedule and by holding the experimental session at a fixed time of day. Stray sounds outside the experimental chamber are masked by a steady source of noise. Changes in stimuli, including the presentation of reinforcement, are controlled precisely, remotely and, if possible, automatically.[3]

Skinner's inferences about the causes of behavior are thus not endangered by any of the difficulties that arise from nonexperimental statistical study.

Skinner's own concern has always been with the way in which lack of attention to the individual case may blind us to causal relationships for which there is strong evidence in individual cases, evidence that is concealed by the flattening-out effects of averages and means. But in avoiding this danger, he has also avoided another: Statistical expertise is helpless if the data from which it draws its premises are not adequately collected. Skinner's nonstatistical, experimental evidence is the best kind of evidence that there is on which to rest causal claims. However, there is always a price to be paid for the strength of

[3] R. A. Boakes and M. S. Halliday, "The Skinner Analysis of Behavior," in *Explanations in the Behavioral Sciences*, ed. Robert Burger and Frank Cioffi (Cambridge, Mass., 1970), pp. 361–2.

the warrant that causal claims can derive from rigorously designed experiments; it lies in the further warrant that is required to establish the reliability of inferences from what occurred in the controlled experimental situation to the world outside. More particularly, there is always the question of what would warrant us in inferring from the premise that the occurrence of X was, given the experimental environment, a sufficient condition for the occurrence of Y in the experimental situation to the conclusion that the occurrence of Y in the natural or social world is the effect of the occurrence of X.

Skinner's general contention is precisely of this form. For the kind of causation that is at work in the shaping of pigeon or rat behavior in the experimental situation is precisely the same as that which shapes human behavior in the social world. To say this is not, of course, to suggest that Skinner takes human behavior to have the same causal explanations as does that of pigeons and rats; he has taken care to deny this more than once. But the same kind of causation is involved. Environmental contingencies are the only nonbiological agencies that control behavior. What, if anything, entitles Skinner to extrapolate from even his most impressive particular findings to so very general a conclusion?

It is important that Skinner and his colleagues and pupils have applied his experimental findings in a wide range of nonlaboratory situations, very often with a great deal of apparent success: programmed instruction and a variety of interventions in schools, mental hospitals, prisons, industry, and elsewhere. Consider just one such case, a program experiment directed by Harold Cohen at the National Training School, in Washington, D.C., with about forty male juvenile offenders:

> A new social environment was constructed in which no boy was required to do anything. He could sleep on a pad in a dormitory, eat nutritious if not very palatable food, and sit on a bench all day. But he could greatly improve his lot by earning points exchangeable for more delicious food at mealtimes, admission to games rooms, the rental of a private room or television set, or even a short vacation away from the school. He could earn points by doing simple chores, but much more easily by learning things. Correct responses to programmed instructional materials and correct answers in examinations after studying other kinds of materials meant points.[4]

[4] Described thus by B. F. Skinner in a broadcast talk, "What's Wrong with the Social Sciences?", reprinted in *A Second Listener Anthology*, ed. Karl Miller (London: 1973), pp. 153–4.

Skinner called the results "dramatic." The immediate results were a quite new devotion to and success in learning, with a marked decrease in hostile behavior. The recidivism rates then dropped sharply (from 85% to 25%), but thereafter climbed back so that "after three or four years there was little evidence of any further effect. The boys had been exposed to this exceptional environment for only a few months or at most a year, which was apparently not enough to offset deficiencies in the environments to which they returned."[5]

I have cited this example, taking care to use Skinner's own words as far as possible, in order to make it clear that the questions I am about to raise do not spring from my superimposing a non-Skinnerian characterization on a Skinnerian example. For it takes only a moment's reflection to understand that the success rate of this program experiment is, in fact, almost entirely irrelevant to the vindication of Skinner's general position. For it is not only the effectiveness of the program that is in question; it is also the interpretation of the causal relationships involved. Does the program have its effect, as Skinner claims, only because of the alterations in the environment? Or does it have its effect because of the practical reasoning and the consequent conclusions, deliberations, and decisions of the subjects? Skinner pours scorn on this suggestion, but the design of the program experiment as he describes it does not at all rule out precisely the type of explanation on which he pours scorn. To understand what is at stake here, it is necessary to elaborate a little further upon what Skinner is committed to denying.

Unlike some other behaviorists, Skinner does not deny the occurrence of internal, distinctively mental events and states; what he does deny is that they are causally effective in the production of behavior. And he more specifically denies that behavior can be caused by an agent's deliberate reasoning qua reasoning. In his words,

> We often speak of the consequences of behavior as *reasons*. We cite them in explaining our own behavior: "The reason I went to the bank was to get some money." The term seems more suitable than cause, especially if we have not fully understood the process of selection, because anything which follows behavior does not seem to be in the right place to be the cause of it. Nevertheless, a reason which lies in the future is no more effective than any other future event. It does not become effective because a person "keeps it in mind" or "thinks of it" or "knows the probability that it will occur," for experiences of this

[5] Ibid.

> sort merely reflect on effort to find a prior representation of a future consequence.[6]

That is, reasoning qua reasoning is causally ineffective. The only causes of behavior are the environmental effects of past behavior. Skinner does, of course, allow that not all behavior is overt; there is covert behavior, behavior of the organism not observable by those able only to observe the external surface of the organism. When we talk, that is part of our behavior; and when we talk inaudibly to ourselves that is behavior too, but covert behavior. If we for the moment think of deliberative reasoning as a kind of inaudible talking to ourselves, then we can understand why it would be an error to assert that Skinner does not admit the occurrence of such reasoning as part of our behavior. But what he is committed to denying is its causal effectiveness precisely qua reasoning in the production of other behavior.

Consider Skinner's discussion of what he calls "self-management" in *About Behaviorism:* "Self-management is often represented as the direct manipulation of feelings and states of mind. A person is to change his mind, use his will power, stop feeling anxious and love his enemies. What he actually does is change the world in which he lives. In both intellectual and ethical self-management he analyzes contingencies and may extract and apply rules" (p. 177). But it is crucial to understand that what is characterized by saying "he analyzes contingencies and may extract and apply rules" must itself be a segment of behavior of which is true that "the environment does the selecting: certain conditions in the environment predispose certain actions by the person. And what one repeats is a function of the environmental feedback given on previous occurrences of the same or a similar nature."[7] Thus a segment of behavior characterizable in terms that suggest deliberation may indeed be effective as part of a causal chain whose end product is a change in an individual's behavior, but it is effective qua behavior that in conjunction with other environmental contingencies causes certain effects, not qua rational deliberation. It is not, that is, how good the reasons are which the agent considers, but how effective the structuring of the behavior is, which is, so to speak, the bearer of the reasoning. Or so Skinner seems to be committed to maintaining. Now let us return to the question of the interpretation of the changed behavior of Cohen's young delinquents.

One way to interpret their changed behavior would be as an

[6] *About Behaviorism*, pp. 128–9.

[7] Carpenter, *Skinner Primer*, p. 79.

expression of decisions resulting from reasoning in the form of Aristotelian practical syllogisms. Given that certain types of activity will be rewarded, some or most of the young men argue to themselves that if they do such and such, they will get what they regard as a good. They therefore act *so as to* achieve their good. This interpretation is incompatible with any Skinnerian account, since Skinner rules out all teleological explanation. What evidence would show that a Skinnerian account is true of the delinquents and an Aristotelian account false? It is difficult to answer this question, since, in general, to change the contingencies of the environment, from a Skinnerian point of view, will also be to provide better or worse reasons for action, from an Aristotelian point of view.

One crucial difference between an Aristotelian account and a Skinnerian account has already been implied but ought to be made explicit. It is not just that Skinner denies the causal efficacy of reasoning qua reasoning, whereas Aristotle affirms it. But on an Aristotelian view the international characterization of his or her action by the agent provides the description under which the action is to be explained, if it is indeed the outcome of the agent's reasoning, whereas on a Skinnerian view such characterization is irrelevant. Behavior is to be explained under descriptions that exclude all reference to intentions or reasoning conceived of as causally efficacious qua intentions or qua reasoning. Hence the Skinnerian account of behavior will be framed entirely in nonintentional terms. What is at stake, therefore, in deciding between an Aristotelian and a Skinnerian interpretation of a particular episode is not just how the episode in question is to be explained but also how it is to be characterized. If Skinner is right, then all our intentional characterizations of behavior are out of place and most of our everyday characterizations are as out of place as Aristotle's. So also – and Skinner makes this quite explicit – are the characterizations of even the greatest novelists. Henry James and Dostoyevski are positive sources of error.[8]

This suggests why Skinner is unlikely to be vulnerable to conceptual criticism of the kind that Chomsky tried to mount. A number of writers other than Chomsky have argued that it is at least implausible to suppose that certain characteristically human activities – the writing of poetry, for example – could be explained in Skinnerian terms. What such writers have omitted to notice is twofold: that the behavior in question would have to be recharacterized and redescribed before it would even be a candidate for Skinnerian explana-

[8] *About Behaviorism*, p. 16, and *Particulars of My Life* (New York: 1976), p. 291.

tion, and that Skinner himself has given close attention to this question.

In "Reflections on Meaning and Structure,"[9] Skinner has a remarkable discussion of why Shakespeare may have written in sonnet form, and in "On Having a Poem"[10] he considered the question of poetic creation more generally. What is crucial, however, is the suggestive gap between the particular points that he makes and the general structure of his psychological theory. For whereas in the latter paper he treats "having a poem" as something for which the poet deserves no credit, in a way that is strictly in accordance with his theoretical standpoint, in the latter paper he discusses Shakespeare's possible reasons for choosing one formal structure rather than another in a way that does not seem to commit Skinner to the analysis of "reasons" that arises from his theory.

What is clear is that there are no obvious immediate conceptual barriers to using the account of "reasons" that I cited from *About Behaviorism* to fill out the account of Shakespeare's "reasons" in "Reflections on Meaning and Structure." We can at least glimpse what a fuller Skinnerian account would be like. It may well be that the operant analysis of poetic behavior, if extended, will encounter grave conceptual difficulties, but it has not done so as yet.

To understand this is to recognize that although the extension of Skinnerian interpretation beyond the realm of rigorously controlled experiment involves an essentially debatable hypothesis, nonetheless, once someone has adopted the Skinnerian standpoint, there are no decisive reasons to be adduced for abandoning it. Skinner's basic thesis is irrefutable; there are and there could not be any decisively falsifying observations, given the facts so far. When I say this I mean, of course, something very different from what is meant by those philosophers of science who have argued that no hypothesis of science can ever be decisively refuted, that there are no crucial experiments or observations. Such philosophers of science have generally wished to draw our attention to the fact that when a theory is confronted by some counterexample in the way of an observation or experiment whose occurrence is inconsistent with the predictions yielded by the theory, we always have a choice between, on the one hand abandoning or modifying the theory, and on the other hand developing some auxiliary hypothesis that will enable us to under-

[9] In *I. A. Richards: Essays in His Honour*, edited by R. Brower, H. Vendler, and J. Hollander (Oxford: 1973).

[10] *Saturday Review*, July 15, 1972.

stand how the particular experiment or observation is not after all a counterexample. But it follows that for such philosophers of science we can rank order types of counter-example in terms of the lengths that we would have to go in devising such auxiliary hypotheses, and that is to say that we can rank order types of counterexample from the relatively more decisive to the relatively less decisive end of the spectrum.

When, however, I assert that the Skinnerian need not fear the production of any decisive counterexample, I do not mean that it will be easy for him or her to devise plausible auxiliary hypotheses when candidates are presented. I mean instead that, given the successes of Skinner's experimental production of behavior in the experimental world, the Skinnerian, when challenged by examples of behavior that Skinner's critics claim cannot be explained in Skinnerian terms, merely has to reply, as I suggested in the matter of poetic creativity, that this is a matter of "not yet." A priori the production of a Skinnerian explanation for the kind of behavior that we call "writing poetry" turns out to be no more implausible before the event than was the teaching of a pigeon to produce the kind of behavior that we call "bowling" before the event. We simply do not know in advance what the details of the explanation will be, and – more important – how the segments of the relevant behavior will have to be characterized.

I imagine that the persuasive effect of Skinner's views upon those who become converted to them is of the same kind as the persuasive effect of Darwin's views when they made their first impact. Part of the method of the *Origin of Species* is to present us with a series of difficult cases and then show how each in turn can be explained by an appeal to the principles of natural selection. Part of Skinner's method is to present us with a series of similarly difficult cases and then show how each in turn can be explained by an appeal to the principles of operant behaviorism. At a certain point in both sequences, initial doubts are apt to be replaced by a growing conviction that if the cases so far explained can yield to these principles of explanation, then so will other apparently implausible candidates for such explanation.

Thus Skinner's procedures and methods fit very ill either a positivistic or a Popperian account of natural science. The contrast with a Popperian account is the more obvious: Nothing in the procedures of Skinner and his followers could be described (even in an after-the-event rational reconstruction) as the construction of large-scale hypotheses tested against those examples most likely to falsify them. (This, I take it, is part of what Skinner means when he claims that he abjures "theory".) Skinner proceeds more like a patient inductivist from case to case. But his procedures do not fit a positivistic account easily either, for they involve interesting circularities.

It was of the essence of a positivistic account of the relationship of theories to the data that rendered them testable that the data should be characterizable independently of the theory to be tested. But for a Skinnerian, if observation of the data does not yield the kind of functional relationship that Skinnerian theory predicts, this is a sure sign that the data have not yet been adequately characterized. Or, to put the matter another way, Skinner's canons for experimental procedure are well designed to ensure that only Skinnerian findings can emerge. That is, Skinnerian experimental methods can show which variable-ratio schedule of reinforcement is most effective for producing particular responses over a particular period of time, or how much more effective intermittent reinforcement is than continuous reinforcement in a particular type of case, or even that we cannot for certain types of case produce convincing answers to this type of question,[11] but what they could not show is that Skinner's principles are false. Skinner's theory can be confirmed by his methods, and there are plenty of confirming instances; it cannot be falsified.

I hasten to add that I am not arguing that these circularities of themselves render Skinner's procedures defective. Nonetheless, their presence is an important clue to the status of Skinner's theory, a clue whose significance can only be assessed by turning to consider a parallel type of circularity that arises in a rival body of psychological theory, that which derives from Freud.

Freud

The canonical place that Skinner assigns to the report of a certain type of experiment – Boakes and Halliday ("Skinner Analysis") give an excellent account – Freud assigns to a certain type of case history. As with Skinner, the type of evidence admitted is such that the basic theoretical positions could not be falsified, although they can be and often are confirmed. Popper remarked on this long ago[12] but in so cavalier a way as largely to vitiate its effect on psycnoanalytic theorists. For the use of case histories is such that specific psychoanalytic hypotheses can indeed be tested against the findings of case histories, falsified, to some degree, and reformulated; the work of Melanie Klein and that of W. R. D. Fairbairn are classical examples of

[11] Cf. Boakes and Halliday, "Skinner Analysis," pp. 356–7.

[12] For example, in "Philosophy of Science: A Personal Report," in *British Philosophy in the Mid-Century*, ed. C. A. Mace [London: 1957], pp. 155–91, esp. pp. 2–9.

such testing, falsification, and reformulation. What could not be falsified is the basic theory. And, as in Skinner's case, the basic circularity of the large-scale theoretical argument derives from two closely related features of the theory: The theory specifies what kind of evidence is to be admitted, and the characterization of the data depends upon the theory.

Moreover, as with Skinner's theory, it is the number and kind of the confirming instances that seem to produce conviction in the adherents of the theory. What Freud discovered, as Skinner was later to do, were previously unidentified phenomena: hysterical paralysis, obsessional compulsion, secondary narcissism. And the striking successes of clinical observation and prediction parallel the striking successes of Skinnerian experiments. Unsurprisingly, it is in the extension of Freudian theory from clinical contexts to those of everyday life that Freudian interpretations become highly debatable, in just the way that Skinnerian interpretations do when they are extrapolated from the realm of experiment to that of nonexperimental social life. But just as with the committed Skinnerian, the committed Freudian need not expect to be confronted by any decisive falsification of his or her thesis, and for similar reasons.

These parallels between the theoretical and conceptual structures of Skinner's standpoint and Freud's coexist, of course, with a deep incompatibility between the two bodies of theory. For where Skinner wished to displace and replace our intentional and teleological characterizations of our activities by a systematically nonintentional and nonteleological characterization of behavior, Freud aspired to extend the realm of the intentional and the teleological beyond that to which it had hitherto been restricted. He tells a young woman who suffers from a compulsive, obsessional ritual that prevents her from going to sleep that she is doing what she does with the intention of not sleeping, so that she will not wake up in the night. He treats dreams as expressions of wishes, slips of the tongue as the product of conflicting intentions, and a variety of bodily movements and states not within conscious control as serving fundamental human purposes. Just as Skinner quite correctly understands that if his account is correct, then our inherited notions of freedom, responsibility, and achievement have no application, so Freud, with equal consistency, understands that if *his* account is correct, then we have to be responsible for unconscious thoughts and desires, as well as for their conscious counterparts.

This dramatically brings out the fact that Skinnerian and Freudian bodies of theory are such that, first, the truth of one entails the falsity of the other, and second, there is no neutral tribunal of observation and experiment by appeal to which the issues between them could be judged. The issue of conceptual incommensurability is raised, not for two rival bodies of theory that are historically successive, as with Aristotelian mechanics and Newtonian mechanics, but for two contemporaneous bodies of theory. How ought such an issue to be confronted?

Psychoanalytic theory versus Skinnerian behaviorism

J. L. Austin once suggested[13] that "in the history of human inquiry, philosophy has the place of the initial central sun, seminal and tumultuous: from time to time it throws off some portion of itself to take station as a science, a planet, cool and well-regulated, progressing steadily towards a distant final state." What Austin's metaphor suggests is that any particular academic discipline, with its own specific subject matter and modes of inquiry, can only enjoy the status of being a science if it has already lost that of being part of philosophy and can only retain the status of being part of philosophy if it has not yet achieved that of being a science. The two statuses mutually exclude each other. C. I. Lewis seems to have held a not dissimilar view, and something very like it is clearly presupposed by many other philosophers and by many scientists, too. Nonetheless, every attempt at defining a systematic contrast between philosophy and science so far has had at best a questionable success. And this fact alone encourages me to suggest that what underlies the confrontation of psychoanalytic theory with Skinnerian behaviorism is a set of unsolved philosophical problems.

A useful place at which to begin is one of the rare points of agreement between Freud and Skinner. Both set out to provide a psychology that would, for the moment at least, debar itself from invoking physiological or chemical explanations of behavior. Neither of them doubted that such explanations will one day be available to us. Both of them believed that progress in psychology required for the moment – a very long moment, too – putting that possibility on one side. (It is clear that both are right, and the evidence lies in the extent to which those psychologists who have invoked neurophysiology have so often had to invent the neurophysiology that would fit their psychology.) But it does follow that nemesis awaits both bodies of theory, a nemesis that will arrive precisely when the neurophysiological and biochemical counterparts to their explanations are

[13] "Ifs and Cans," in *Philosophical Papers*, ed. J. O. Urmson and G. J. Warnock (Oxford: 1961), p. 18.

produced. Freud's problem is the more simply stated. Freudian explanations will have to be correlated with their neurophysiological counterparts, but this will be possible only if by then philosophers have solved the problem of how to translate intentional statements into their physical counterparts. Hilary Putnam has written of those physicists who believe that they know how to set about solving those problems that they "leave their increasingly dubious clients with mere promissory notes for future translations, or else they say that the existence 'in principle' of possibly infinite (and not effectively specifiable) translations is good enough"[14] But that latter possibility will not be good enough for those who wish to specify how psychoanalytic explanations and their neurophysiological counterparts are to be related.

Moreover, as Putnam points out in the same article, the identification of physical properties with intentional properties has not yet been shown to be possible either, and the same holds true for the identification of neurophysiological properties with intentional properties – of the property of certain identifiable neurons of mine firing, for example, with the property of my unconsciously wishing not to go to sleep. Until this problem is solved, certain kinds of dualism, all of which derive ultimately from Descartes – one contemporary version is Popper's parallelism, another is Shaffer's straightforward revised Cartesianism – remain plausible. To solve this problem would, if the solution were of one kind, favor a Spinozistic type of ontology; if the solution were of another, an Aristotelian. Traditional metaphysical frameworks specify types of solution to certain contemporary analytic problems, and just that type of analytic problem has to be solved if certain theoretical impasses are to be overcome in a science such as psychology.

It might at first sight be thought that matters are less complex for Skinner than they are for Freud. Since Skinner has already expelled all intentional terminology from the domain of a science of behavior, there is no problem "in principle" about the reduction of his behavioral generalizations to neurophysiological generalities, any more than there is about the reduction of chemical generalizations to physical generalizations. But of course what Skinner has done is simply to transfer all our uses of intentional terminology to the category of what is to be explained. If Chomsky was mistaken in supposing that Skinner's project in *Verbal Behavior* was to be disposed of as quickly as Chomsky's conceptual sleight-of-hand suggested, the

[14] "Reflections on Goodman's Ways of Worldmaking," *Journal of Philosophy* 76 [1979]: 606–7.

linguistic specialist was right in supposing that at this point Skinner faced a problem. The depth of the problem can be brought out by considering why Skinner cannot easily adopt Quine's solution to this problem. This solution is summarized by Putnam as follows: "Quine holds that the failure of mentalistic psychology to reduce uniquely to physics shows that the sentences of mentalistic psychology and of all the discourses that imply mentalistic locutions have no truth value. These discourses are indispensable for daily life, but not truly cognitively significant."[15] Why cannot Skinner solve his problems by gratefully accepting this philosophical offering?

Consider certain types of sentences that no scientist can avoid asserting on occasion: "*P* is true, if and only if *X*'s report that such-and-such occurred is true"; "*X*'s observations were unreliable because he mistakenly believed that his experimental design excluded such-and-such a factor from affecting the behavior of the experimental animal"; "*X* is reluctant to accept the truth of *Y*'s report in view of her own interpretation of *Z*'s observations." All these sentences are ineliminably intentional. But the statements that they are used to make are often offered as providing good reasons for holding certain scientific statements to be true or false. And if my believing or asserting that *P* gives me a good reason for my believing or asserting that *Q* is true, then I must be assigning a truth value to *P* as well as to *Q*. In other words, I am contending that the inferences of actual scientists, including those of Skinnerian behaviorists, would be unintelligible unless certain types of statement made by uttering intentional sentences were understood as having truth values. But in addition to this, the utterance of such statements must be held to be causally effective in any ongoing scientific enterprise such as that of Skinnerian behaviorism. Hence if Skinner is to give a comprehensive account of the functional relationships holding between environmental contingencies and segments of behavior, he must offer us some more systematic account than he has yet given of how those segments of behavior that both he and we characterize as intentional are to be undertood in terms of his theory. Such an account would have to recognize – as Skinner's discussions both of self-management in *About Behaviorism* and of poetic creativity do indeed recognize – that such language is not only used to characterize such segments of behavior. Uses of intentional language are partially constitutive of such segments of behavior. Hence we need to map the systematic relationship of the intentional characterization to those nonintention-

[15] Ibid, p. 613; for Quine's own statement of his view, see his *Word and Object* (Cambridge, Mass.: 1960), Chap. 6.

al characterizations to which Skinnerian theory ought, if it is consistent, to restrict itself. In other words, the very feature of his project that enables him to evade Chomsky's criticisms – that he is not committed to explaining verbal behavior as characterized either by the grammarian or the cognitive psychologist, but only as recharacterized in a quite new way – opens his work up to the challenge: Show us how utterances of intentional sentences and other intentionally constituted segments of behavior can be so recharacterized in a way that captures their character but yet is faithful to Skinner's theory. To meet his challenge, he will have to perform a task precisely analogous to that which the physicalist has so far failed to perform. In other words, Skinner has to solve the same philosophical problem as the psychoanalysts, but at a different stage in his project.

Conclusion

What conclusions we draw from the argument so far will depend in part upon our hunches as to the prospect for a resolution of the philosophical problems in question. But at the very least we can discard the thesis that psychology, by emancipating itself from philosophy in the nineteenth century and becoming a genuinely empirical science, put itself in possession of a method for resolving its internal disputes and arriving at an agreed-upon body of doctrine. It seems rather to be the case that psychology can only evade the burdens of its philosophical inheritance to the degree that it is unsystematic and eclectic, and that insofar as it aspires to be a systematic science it remains inexorably philosophical. And indeed certain philosophical problems turn out to have the importance that they have precisely because of the role that their solution – or lack of solution – has to play in the program of constructing a systematic psychology.

The argument has reached a point at which two larger questions have to be at least entertained. The first is, Ought we now to begin to suspect that the project of a systematic psychology is impossible? Perhaps what the history of the relevant philosophical enquiries ought to teach us is that we have no way of fitting together what we want to say about human beings in the intentional mode and what we want to say about human beings neurophysiologically and biochemically, not because something like Cartesian dualism is true but because there is a deep and ineliminable incoherence in the structure of our most fundamental psychological beliefs. The inability to reach agreement within psychology would for that reason be ineliminable, and the metaphysical victor to emerge from the discussion would be Pascal.

The second, larger question is of a very different kind and points in the opposite direction to the first. For underlying the first question is the thought that perhaps psychology, in its inability to attain an agreed-upon, unified, systematic framework, is very different from many other sciences. But what if instead the condition of psychology is an important clue to the underlying condition of those other sciences? Perhaps whenever conceptual incommensurability seems to appear in a science, what this points to is a family of unsolved philosophical problems, problems whose role in the history of those sciences goes unrecognized because of the way in which we characteristically segregate the history of science from the history of philosophy, and vice versa. But these questions, although unavoidable if the argument of this essay is sound, require a type of treatment that is beyond its scope.

PART II. CONTEMPORARY CASE STUDIES

12

Resolution of the Laetrile controversy: past attempts and future prospects

GERALD E. MARKLE AND JAMES C. PETERSEN

One of the most remarkable features of the Laetrile controversy is its persistence. Although public attention and support of Laetrile treatment clearly reached a peak in the 1970s, Laetrile has a long and colorful history in the United States.[1] Attempts to resolve the controversy surrounding it have been numerous and strikingly unsuccessful. In 1953 the Cancer Commission of the California Medical Association issued a document that concluded that "no satisfactory evidence has been produced to indicate any significant cytotoxic effect of Laetrile on the cancer cell."[2] Ten years later the Cancer Advisory Council to the director of California's Department of Public Health issued a report that found that Laetrile was of "no value in the diagnosis, treatment, alleviation or cure of cancer."[3] Upon the council's recommendation, in 1963 the California Department of Health banned the treatment of cancer with Laetrile.

Agencies of the federal government have also attempted to bring closure to the Laetrile controversy. As early as 1960, the Food and Drug Administration (FDA) conducted seizures of Laetrile and banned its shipment in interstate commerce. The FDA's right to regulate Laetrile was, however, challenged in a series of court cases in the late 1970s. As part of that controversy, the FDA commissioner, in a lengthy 1977 decision, ruled that Laetrile is not generally recognized by qualified experts as a safe and effective cancer drug and that it is

[1] James Harvey Young, "Laetrile in Historical Perspective," in *Politics Science and Cancer: The Laetrile Phenomenon*, ed. Gerald E. Markle and James C. Petersen (Boulder, Colo.: Westview Press, 1980), pp. 11–60.

[2] Cancer Commission of the California Medical Association, "The Treatment of Cancer with 'Laetriles,'" *California Medicine* 75 (1953): 326.

[3] Cancer Advisory Council, *Treatment of Cancer with Beta-cyanogenetic Glucosides ("Laetriles")* (Sacramento: California Department of Public Health, 1963).

not exempt from FDA regulation.[4] Despite this decision, a judge issued a permanent injunction forbidding the FDA from restricting the importation and use of Laetrile. In July 1978, the U.S. Court of Appeals, 10th Circuit, upheld the injunction limiting use of Laetrile to injections given by licensed medical practitioners to terminal cancer patients. The FDA appealed this action, and in June 1979 the U.S. Supreme Court reversed the exception for the terminally ill that had been established by the court of appeals.

The National Cancer Institute (NCI) has also been actively involved in attempts to resolve the controversy. In the 1960s and 1970s, NCI tested Laetrile alone and in combination with beta-glucosidase against a variety of rodent tumors. It concluded that Laetrile was inactive. By 1976 the head of the Public Inquiries Section of NCI informed us that after extensive study NCI felt there was "no sound basis for recommending clinical trials for Laetrile" (R. J. Avery, letter to the authors, March 4, 1976). However, two years later NCI found itself asking the FDA for permission to conduct such clinical trials.

A variety of private organizations with an interest in the cancer field have also made attempts at resolution. Both the American Cancer Society and the American Medical Association tried to label the use of Laetrile in cancer treatment as quackery. Robert Eyerly, chairman of the Committee on Unproven Methods of Cancer Treatment of the American Cancer Society, charged that "the use of Laetrile rather than known, effective cancer treatments is the cruelest of all frauds."[5] In response to our inquiry the American Medical Association provided a statement by their executive vice-president, who asserted, "Many American cancer patients, driven by fear, have become victims of the Laetrile hoax."

When Memorial Sloan-Kettering Cancer Center first got involved in the Laetrile controversy, its president asserted that "this institution can answer the Laetrile question fairly quickly."[6] In fact, however, when Sloan-Kettering released the results of a series of experiments at a press conference, their negative findings came under attack when charges of inaccuracies and suppressed findings were made.[7]

4 Donald Kennedy, "Laetrile: Commissioner's Decision on Status," *Federal Register* 42 (1977): 39768–806.

5 Robert C. Eyerly, "Laetrile: Focus on Facts," *CA. Cancer Journal for Clinicians (New York)* 26 (1976): 5.
Letter received from Susan K. Roberts quoting James H. Sammons, Executive Vice-President of the American Medical Association, March 8, 1976.

6 Nicholas Wade, "Laetrile at Sloan-Kettering: A Question of Ambiguity," *Science* 198 (1977): 1231–4.

7 Richard D. Smith, "Laetrile at Sloan-Kettering," in *Politics, Science and Cancer:*

Despite numerous previous attempts to put an end to the Laetrile controversy, today we find Laetrile's merits being debated by the public, the courts, state legislatures, and the scientific community. Laetrile has become a household word – the subject of extensive media coverage and even an element in popular humor. In our earlier work we stressed the multifaceted character of the Laetrile controversy and that the controversy has a strong sociopolitical basis, as well as a scientific component.[8]

Such a concern with both internal and external factors is essential, in our judgment, in understanding scientific disputes and their resolution. We agree with Blume that those who study such disputes must examine both cognitive and sociopolitical aspects of the conflict.

> Controversies in science seem to offer a research focus permitting concurrent exploration of cognitive and broad social structural factors . . . Epistemological conflicts always, inevitably, have a political aspect. In trying to understand epistemological conflicts, therefore, exclusive focus either (as is common) upon the cognitive dimension, or upon the political dimension, is seen as inherently wrong.[9]

This approach seems especially appropriate for examining the Laetrile controversy, where a sizeable social movement composed of committed individuals and pro-Laetrile organizations has developed. Mendelsohn has observed that the "pseudosciences" or the folk or popular sciences frequently have such a social movement nature. Furthermore, those activities at the fringe of orthodox science and medicine often have political overtones as well.[10]

Our perspective on the Laetrile controversy has also been shaped by Bloor's call for the "strong programme" in the sociology of knowledge. Bloor maintains that the sociology of scientific knowledge should be causal, impartial, and symmetrical. Bloor's appeal for an impartial and symmetrical analysis of scientific controversy is particularly innovative. Too frequently, scholars of science have only

The Laetrile Phenomenon, ed. Gerald E. Markle and James C. Peterson (Boulder, Colo.: Westview Press, 1980), pp. 61–72.

8 James C. Petersen and Gerald E. Markle, "The Laetrile Controversy," in *Controversy: Politics of Technical Decisions*, ed. Dorothy Nelkin (Beverly Hills, Calif.: Sage, 1979), pp. 159–79; James C. Petersen and Gerald E. Markle, "Politics and Science in the Laetrile Controversy," *Social Studies of Science* 9 (1979): 139–66.

9 Stuart Blume, *Perspectives in the Sociology of Science* (Chichester: Wiley, 1977), p. 13.

10 Everett Mendelsohn, "The Social Construction of Scientific Knowledge," in *The Social Production of Scientific Knowledge*, ed. Everett Mendelsohn, Peter Weingart, and Richard Whitley (Dordrecht: Reidel, 1977), p. 21.

sought to explain "error" or deviation while assuming that logic, rationality, and "truth" were their own explanation.[11]

The Laetrile controversy is not an isolated phenomenon but rather the most successful and publicized of a large category of marginal scientific disputes.[12] We believe that the Laetrile phenomenon cannot be understood outside of its social context; moreover, we see the behavior of both sides in the controversy as arising from the same social matrix. Thus, before we can examine attempts to resolve the Laetrile controversy, we first explicate a general framework of social control and influence that is both contextual and symmetrical.

Power and discontent

In *Power and Discontent*, William A. Gamson distinguishes between authorities (those who can make binding decisions in a particular social system) and potential partisans (those who are affected by the outcome of a particular decision in some significant way).[13] Power is then examined in a symmetrical fashion. Gamson sees power as both authorities acting on potential partisans (social control) and potential partisans acting on authorities (influence). Each of these types of power may come in several forms.

Influence may operate through constraint or inducement, the former defined as "the exercise of influence by threat of deprivation," the latter as "the promise of indulgence."[14] The key difference here is whether a disadvantage (e.g., a public protest demonstration) or an advantage (e.g., financial support) is added to the situation of the authorities. Influence may also take the form of persuasion, where some change is produced in the orientation of the authorities without the addition of anything to their situation. Thus communication skills or personal attraction may cause authorities to prefer the outcomes that are preferred by the influencer.

Just as partisans may attempt to influence authorities, so may authorities attempt to contain such influence. Gamson has detailed three types of social control used by authorities. They can

(1) regulate the access of potential partisans to resources and

[11] David Bloor, *Knowledge and Social Imagery* (London: Routledge & Kegan Paul, 1976).

[12] Other elements of the cancer underground include Gerson's diet, Koch's antitoxins, Hoxsey's therapy, the grape diet, the Cresson method, and the Chase dietary method.

[13] William A. Gamson, *Power and Discontent* (Homewood, III.: Dorsey Press, 1968).

[14] Harold D. Lasswell and Abraham Kaplan, *Power and Society* (New Haven: Yale University Press, 1950), p. 97.

> their ability to bring these resources to bear on decision makers, (2) they can affect the situation of potential partisans by making rewards or punishments contingent on attempts at influence, or (3) they can change the desire of potential partisans to influence by altering their attitudes toward political objects.[15]

Gamson further delineates several different strategies, such as insulation and sanctions, by which authorities can contain the demands of authorities. Insulation includes a variety of techniques by which social systems control who enters or leaves them. Sanctions are, of course, various rewards and punishments that authorities have available to them, ranging from promotion to imprisonment. A number of social control techniques, including education, socialization, rehabilitation, indoctrination, and propaganda, are included in persuasion. Here the authorities try to control the desire of potential partisans to influence.

Authorities always attempt to contain, rather than yield to, partisan demands. Several strategies, such as cooptation and participation, allow them to yield some ground without losing control of the conflict. In cooptation, for example, authorities absorb new elements into the policy making structure as a way of dealing with threatening potential partisans. Since authorities typically prefer to limit the access of potential partisans, cooptation is generally used when authorities feel or expect great pressure from partisans. Finally, participation may be a social control device as well as an element of influence. If partisans are permitted to participate in the decision process, they may increase their acceptance of decisions even when the decisions run contrary to what was preferred.

In Gamson's formulation, a variety of factors affect both the likelihood that power will be utilized and the forms it will take. The degree of trust in authorities is an indication of the perceived need for influence. Groups with high trust (confident groups) believe that the systems will produce preferred outcomes even if left untended. Neutral groups are unsure about the outcomes, and alienated groups (those with low trust) are certain that preferred outcomes will not occur if the system is left untended. In Gamson's view, alienation from authorities means that they are regarded as incompetent and stupid in achieving conflicting goals and biased against the group in handling conflicts of interest. They are antiagents of the group, the agents of groups with conflicting goals. Thus, alienated groups believe that influence is necessary, and when this belief is combined with a perception that influence is possible (high efficacy), we have

[15] Gamson, *Power and Discontent*, pp. 116–17.

the optimal combination for mobilizing citizens. Other factors that increase the likelihood of influence attempts are recognition of being affected by an issue, interest group representation, amount of available resources, and the liquidity of the resources.

One of the most useful features of Gamson's writing is the highly structured formal theory that he presents. *Power and Discontent* contains a large number of propositions that relate level of trust to the means of influence or type of social control likely to be used. Although we will not consider those propositions now, we will return to them later. First, however, we will use Gamson's conceptual framework as we look at the dispute over Laetrile. The continuing controversy over this substance, despite numerous attempts to bring closure to the dispute, has obvious implications for many other drugs and medical treatments. Attempts by the FDA to ban or regulate saccharin, dimethyl sulfoxide (DMSO), and Gerovital have already come under attack, and similar protests are likely to increase in the future.

Trust

It is clear that potential partisans in the Laetrile controversy have low trust in the authorities. On specific issues (e.g., efficacy, clinical testing) and on more general ideological issues, advocates are alienated from the political and medical establishment. At the grass roots level, Laetrile proponents do not believe that physicians effectively diagnose or treat diseases; instead, they are more likely to evaluate chiropractors favorably.[16]

The leadership of the movement is highly distrustful of the establishment. In an interview with us (August 26, 1975), Ernst Krebs, Jr., leading theoretician and father of the movement, attacked various opponents of Laetrile. In discussing the FDA, he claimed that "they consistently, consistently and consistently lie in the finest tradition of our beloved president, vice-president and attorney general." In his view the AMA is simply "a wonderful, effective trade union." When asked about an NCI interim report on Laetrile, Krebs asserted that he believed that such reports had no scientific basis. "If somebody were to come to me, and I am very busy, I'd say, 'Look, as

[16] Gerald E. Markle, James C. Petersen, and Morton O. Wagenfeld, "Notes from the Cancer Underground: Participation in the Laetrile Movement," *Social Science and Medicine* 12 (1978): 31–7; Morton O. Wagenfeld, Yvonne M. Vissing, Gerald E. Markle, et al., "Notes from the Cancer Underground: Health Attitudes and Practices of Participants in the Laetrile Movement," *Social Science and Medicine* (1979): 483–5.

a scientist I want you to give me a report on reincarnation. He comes back next week: 'The boss is on my ass – here, take this quickly', and I dictate a report. It's a lot of crap from the armchair. This is the way this stuff is done."

Each and every NCI-sponsored Laetrile study has been treated contemptuously and countered with theoretical and methodological, and sometimes political criticism. The Laetrile studies at Sloan-Kettering are a case in point.[17] There, in 1973, a researcher found that Laetrile tended to inhibit metastasis of spontaneous breast tumors in mice. The results of these experiments were never published, according to a Sloan-Kettering vice-president, because "if we had published those early positive data, it would have caused all kinds of havoc."[18] But by 1975 these data were too controversial to hide. In a pamphlet entitled "Anatomy of a Coverup," the Committee for Freedom of Choice in Cancer Therapy published the findings, along with the following unsigned letter on Sloan-Kettering stationery: "Here are some of the results of Sloan-Kettering's continuing experiments with Laetrile. Due to political pressures these results are being suppressed. Please do your best to bring these important findings to the attention of people. Krebs's theory is very promising, and Laetrile should be tested clinically to see if it really holds water."[19]

Problems at Sloan-Kettering continued. On June 6, 1977, Sloan-Kettering held a press conference at which they distributed copies of two scientific reports scheduled for publication in the *Journal of Surgical Oncology*. Five months later, Second Opinion, a left-wing organization, published a forty-five page pamphlet on Laetrile that included a critique of the two Sloan-Kettering preprints. The results of an independent investigation of Sloan-Kettering's Laetrile research were published in December 1977 in *The Sciences*, the journal of the New York Academy of Sciences. In response to these publications, Sloan-Kettering scientists did modify the two papers prior to publication.[20]

The low level of trust of Laetrile advocates for authorities was also evident in the controversial NCI retrospective study. This research, calling for a post hoc collection and analysis of cases thought to have shown objective benefit from Laetrile, was designed in the latter part

[17] Smith, "Laetrile at Sloan-Kettering."

[18] Wade, "Laetrile at Sloan-Kettering," p. 1231.

[19] Committee for Freedom of Choice in Cancer Therapy, *Anatomy of a Coverup* (Los Altos, Calif.: Committee for Freedom of Choice in Cancer Therapy, 1975), pt. 3, p. 1.

[20] Smith, "Laetrile at Sloan-Kettering." For additional details see Ralph W. Moss, *The Cancer Syndrome* (New York: Grove Press, 1980).

of 1977. Since supplying records for the study might have created legal risks for cooperating physicians or therapists, the FDA agreed to grant limited legal immunity to participants. But the FDA refused to grant Laetrile advocates immunity from ongoing prosecutions for actions unrelated to the retrospective study.[21] As the study took shape, in fact, the FDA initiated a major anti-Laetrile campaign. According to the minutes of the Laetrile Working Group (a planning committee composed of representatives of the FDA, the Center for Disease Control (CDC), and the Armed Forces Institute of Pathology (AFIP), and chaired by NCI), "The NCI and the CDC expressed grave concern over the timing of such a campaign, feeling that it could severely prejudice the conduct of the study."[22] Two weeks later, the NCI, CDC, and AFIP again expressed "severe reservations" over what they regarded as the forthcoming FDA "blitz campaign" against Laetrile.[23]

Cause and effect are not clear here; however, all but one of the major pro-Laetrile organizations eventually boycotted the retrospective study.[24] Since the study was designed to collect as many case histories as possible, such withdrawal proved crippling. Although 455,000 letters were distributed in an effort to reach the estimated 70,000 cancer patients who have used Laetrile, only 93 cases were submitted to NCI for evaluation.[25] Needless to say, the ambiguous findings of the retrospective study, especially with the admitted methodological problems of the study, did little to resolve the controversy.

Influence

"Potential influence," writes Gamson, "is synonymous with the control of resources."[26] In the Laetrile dispute, probably the most effective challenge to medical expertise and authority in American history, advocates have gained control of several different kinds of resources. First and foremost, Laetrile advocates have engaged in claims making. By adopting the tools of science and doing what scientists do, Laetrile proponents have been able to use the consider-

[21] Revised Minutes of Laetrile Working Group, September 23, 1977 (available from NCI).

[22] Minutes of Laetrile Working Group, October 14, 1977.

[23] Minutes of Laetrile Working Group, October 28, 1977.

[24] The Committee for Freedom of Choice in Cancer Therapy supported the study.

[25] Neil M. Ellison, David P. Byar, and Guy R. Newell, "Results of the National Cancer Institute's Retrospective Laetrile Analysis," *New England Journal of Medicine* 299 (1978): 549–52.

[26] Gamson, *Power and Discontent*, p. 94.

able resources of science to legitimize their cause. In fact the early Laetrile debates took place largely within the traditional scientific community. In 1946 Krebs and Gurchot wrote two letters to *Science* outlining experiments (though showing no data) in support of the trophoblastic theory of cancer.[27] Furthermore, Krebs's 1950 paper, seen today as the theoretical exemplar of the movement, was published in *Medical Record*;[28] and Morrone's 1962 clinical experiments on Laetrile's efficacy for treatment of human cancer was published in a reputable journal, *Experimental Medicine and Surgery*.[29] Interestingly, followers as well as leaders of the movement are generally familiar with and are able to cite this early literature as forming the scientific basis of their cause.[30] Even as late as 1970 the McNaughton Foundation, following formal procedures, sought official approval from the FDA to test Laetrile. The important point here is that this activity, which was in the scientific mode, preceded the social movement and turmoil that developed around Laetrile in the late 1970s.

In recent years most pro-Laetrile claims making has been on the fringes or outside of the scientific community. However, advocates have not conceded the experimental domain to the scientific establishment. In 1977 a biologist at Loyola University asserted that Laetrile, when used as part of a megavitamin regimen, effectively controlled mammary tumors in mice.[31] Despite the fact that the paper was read in a nonscholarly setting, was only two pages in length, and that the research had lacked certain experimental controls, the study received national media coverage. Just as the medical establishment had used mouse tumor studies to denounce the movement, advocates now had their own research to use in arguing the Laetrile case.

A more general thrust of Laetrile claims making, particularly in the late 1970s, was a merger of the movement with holistic medicine.[32] Many health food stores around the country and advocates of a wide

[27] Charles Gurchot and Ernst Krebs, Jr., "Growth of Trophoblasts in the Anterior Chamber of the Eye of the Rabbit," *Science* 103 (1946): 25, and Ernst Krebs, Jr., and Charles Gurchot, "Trophoblast Elements in Cancer," *Science* 104 (1946): 302.

[28] Ernst Krebs, Jr., Ernst Krebs, Sr., and Howard W. Beard, "The Unitarian or Trophoblastic Thesis of Cancer," *Medical Record* 163 (1950): 158–73.

[29] John A. Morrone, "Chemotherapy of Inoperable Cancer: Preliminary Report of 10 Cases Treated with Laetrile," *Experimental Medicine and Surgery* 20 (1962): 299–308.

[30] Yvonne M. Vissing, "An Exploratory Analysis of Participation in the Laetrile Movement," M.A. thesis, Western Michigan University, 1978.

[31] Harold W. Manner, "The Remission of Tumors with Laetrile," paper presented at the meetings of the National Health Federation, Chicago, September 1977.

[32] Gerald E. Markle and James C. Petersen, "Social Context of the Laetrile Phenomenon," in *Politics, Science and Cancer: The Laetrile Phenomenon*, ed. Gerald

variety of health causes consider Laetrile as part of the health repertoire. The most important holistic claim of the movement is that Laetrile is a vitamin, known as B_{17}, which is necessary for the maintenance of health and prevention of cancer.[33] Absence of Laetrile, according to one advocate, can "produce headaches, anorexia, bizarre muscular pains, skin changes, anemia, sense of impending doom ... high blood pressure, sickle cell anemia and finally, tumefaction."[34] Thus, cancer is not a tumor disease; it is a metabolic disease in which the tumor is merely an obvious symptom. Just as it takes vitamin B_{12} as well as iron to cure pernicious anemia, and proper diet as well as insulin to control diabetes, Laetrile supporters maintain that vitamin B_{17} – described by one supporter as the "crown jewel in a total diadem of treatment"[35] – and diet will prevent or control cancer. In fact, in public speeches Dr. John Richardson, a leading advocate of Laetrile, now calls cancer "fulminating avitaminosis."

Thus allied with holism, Laetrile advocates have adopted a populist ideology of medicine. Rather than treating cancer by the methods of experts – which Laetrile advocates regard as equivalent to slashing, burning, and poisoning – proponents urge a deprofessionalization of medical care. In the final analysis, individuals are encouraged to be their own physicians. This populist position logically leads to a position that Laetrile advocates refer to as "freedom of choice" in cancer therapy. They declare that cancer patients have a right to choose their form of treatment without interference from the medical community or the government. As Glenn Rutherford, litigant in the class-action suit that went to the Supreme Court, has cried, "You people in authority consider all the rest of us a bunch of dummies ... You set yourself up as God and Jesus Christ all rolled up into one. And we don't have any rights ... Patrick Henry said: 'Give me liberty, or give me death.' Glenn Rutherford says let me choose the way I want to die. It is not your prerogative to tell me how. Only God can do that."[36]

The freedom of choice issue also served as a bridge between the

E. Markle and James C. Petersen (Boulder, Colo.: Westview Press, 1980), pp. 151–74.

33 The claim that Laetrile is a vitamin is rejected by most medical experts and authorities.

34 U.S. Food and Drug Administration, *Laetrile Administrative Rule Making Hearing: Oral Argument Docket No. 77N–0048, Food and Drug Administration (2 and 3 May 1977)*, p. 465.

35 Ibid., p. 353.

36 Ibid., pp. 308, 315–16.

Laetrile movement and the political right wing. In a series of newspaper columns published over a two-year period, James J. Kilpatrick argued that the real point of the Laetrile debate was freedom: "The point is freedom. We lose it by chunks, by bits, by grains. Daily we yield more authoritarian control to the experts."[37] Thus a powerful coalition formed, not only between the cancer underground and holistic medicine but with the American conservative movement as well. This coalition, with its extensive political, economic, organizational, and media resources, won a series of stunning political victories against the medical establishment. The organizational structure through which these resources were channeled, refined, and even created, reflected this coalition.

The Committee for Freedom of Choice in Cancer Therapy, probably the most effective of the pro-Laetrile organizations, was founded in 1972 to aid in the defense of Dr. John Richardson, who was being tried for using Laetrile in cancer therapy. Richardson was an active member of the John Birch Society, as were many of the original officers of the committee. According to a *New York Times* article, the editor of *Choice*, the official publication of the committee, stated that "there are a lot of us Birchers in the Laetrile movement because the John Birch Society has the guts to fight for what it believes in."[38] Pro-Laetrile pamphlets and books are frequently sold at John Birch Society bookstores, and *American Opinion*, published by John Birch Society founder Robert Welch, has attacked the suppression of Laetrile.[39] The committee, which at its peak had about five hundred local chapters with eight thousand members, describes itself as "the nation's major leading advocate of the decriminalization of Laetrile." It seems likely that the promotional material, active political lobbying, and effective use of the courts that have characterized the Laetrile movement may reflect skills gained by the radical right in earlier campaigns against fluoridation and sex education.

From the other side of the coalition, the holistic movement, came organizations that promoted Laetrile as one of many health remedies. Particularly active in this effort was the International Association of Cancer Victims and Friends (IACVF), which by the late 1970s had some fifty chapters around the country and approximately eight thousand members. Schisms within this organization led to the

37 James J. Kilpatrick, "Another Uproar over the Freedom to Choose," *Nation's Business* (May 10, 1976), p. 10.

38 Everett Holles, "Birch Society Members Tied to Smuggling of Illegal Drug," *New York Times* (June 1, 1976), p. 18.

39 Alan Stang, "Laetrile: Freedom of Choice in Cancer Therapy," *American Opinion* (January 1974), pp. 49–66.

formation of another major pro-Laetrile group in 1973, the Cancer Control Society. Other groups that have broken off from IACVF include the Foundation for Alternative Cancer Therapies (1975) and the Cancer Federation (1978), both organizations that promote holistic approaches to cancer therapy. In the late 1970s the National Health Federation, with twenty-five thousand members in one hundred eight local chapters, actively began to promote Laetrile through such means as its "Fund to Stop Government Ban on Laetrile" and its newspaper, *Public Scrutiny*.

Returning to Gamson's influence model, it seems clear that Laetrile advocates rarely attempted to influence authorities through persuasion or inducements. Among the few examples of persuasion attempts were the publication of materials describing Laetrile therapy for distribution to physicians and the scheduling of workshops to describe Laetrile and metabolic therapy to them. Generally the Laetrile movement attempted to add new constraints to the power of the medical establishment. Through state legislatures, courts, and Congress, a broad coalition of Laetrile advocates, health groups, and conservatives attempted to limit the power of the FDA.

Management of discontent

Our discussion of trust and influence has focused on how partisans acted toward authorities. Following Bloor's dictum of symmetry, and specifically explicating Gamson's model, we now examine the obverse: how authorities attempt to control partisans. During any period of discontent the central problem, from the standpoint of the authorities, is the containment of influence. If changes are to come about, authorities want to direct them; if decisions are to be made, authorities want to make them. In the Laetrile dispute, authorities attempted to control the course of events by applying sanctions; specifically, they attempted to prevent, through vigorous prosecution, the importation, sale, and use of Laetrile.

The FDA has regulated the interstate sale and use of Laetrile since the 1950s. From May through July of 1977, the FDA seized 50 tons of apricot pits and various supplies for manufacturing Laetrile, with a value that FDA estimates at three hundred thousand dollars. One plant in Wisconsin raided by the FDA was producing 1,500 kilograms of Laetrile each month. The FDA has maintained its pressure on the movement. Between August 30, 1978, and February 28, 1979, according to FDA enforcement reports, six separate seizure actions were filed against partisans.

The FDA has also imposed sanctions on partisans in a more subtle way. Although Laetrile is legal (or more accurately, deregulated) in twenty-six states, it still cannot be legally purchased in the United States. This is partly because of a tangle of bureaucratic hurdles at the state level[40] but partly owing as well to a broad interpretation of FDA power. The agency threatens to seize Laetrile if any component of the product – labels, cotton, glass bottles, and so on – is shipped over state lines. This threat is enough to scare legitimate pharmaceuticals, especially given the time and cost of legal adjudication.

Laetrile advocates have attempted to use the courts to exercise partisan power. The Chad Green case, on the other hand, shows that authorities are not willing to sanction the private use of Laetrile. In 1978, the Massachusetts Supreme Court ordered the parents of Chad, a three-year-old, to cease Laetrile treatments and begin chemotherapy for leukemia. Rather than comply with the order, the parents fled to Mexico. The child died in November of the following year. The case generated a tremendous amount of publicity and ethical debate, as well as legal action.

Not only have authorities invoked sanctions; they have also attempted to control advocates by limiting their access to legitimate power. This strategy, which Gamson calls "insulation," is most evident in the debate over credentials. The medical establishment routinely attacked the professional qualifications of the leadership of the Laetrile movement. Sometimes advocates with stronger qualifications are also ridiculed. At the 1978 meetings of the American Association for the Advancement of Science, Dr. Dean Burk was characterized as someone who had "spent time" at the NCI. In fact, at retirement, he was head of the cytochemistry section of NCI.

Even so, the FDA claims that few of the researchers and clinicians active in the movement have any "special training in oncology or in the evaluation of drug safety or effectiveness" and that they publish their results in books and pamphlets rather than in scientific journals with peer review.[41] The tactics of the medical establishment are clear: Only specialists operating through professionally approved channels should have the ear of the scientific community. Lack of qualification or evasion of procedure severely damages the credibility of the antagonist.

Gamson would call these credential disputes insulation through selective entry. Here authorities play the role of gatekeeper; only the

[40] Robert F. Rich, "Political Implications of Laetrile: Who Gets What, When and How," in *Politics, Science and Cancer: The Laetrile Phenomenon*, ed. Gerald E. Markle and James C. Petersen (Boulder, Colo.: Westview Press, 1980), pp. 73–98.

[41] Kennedy, "Laetrile," p. 39785.

just are allowed to serve. In insulation through selective exit, authorities disengage themselves from those who support the enemy. In 1977, for example, Second Opinion issued a document that was highly critical of the NCI's Laetrile studies.[42] Ralph Moss, then assistant director of public affairs, for Sloan-Kettering, admitted his role in the publication and was immediately fired.[43]

Only when social control fails to quell the opposition will authorities consider altering their decisions. The legal, popular, and legislative successes of the Laetrile movement, in spite of strong and vigorous sanctions, has forced authorities to accommodate partisans. Such has been the case at NCI. In December 1978, after years of opposition, the institute petitioned the FDA for permission to conduct a clinical trial of Laetrile.

Unable to control Laetrile as an underground substance, unable to diffuse the movement through laboratory studies, authorities attempted to control Laetrile by yielding ground. With the request for clinical testing, Laetrile came under a new and elaborate set of guidelines. For the first time, the substance was treated like any other through standard operating procedure. In so doing officials admitted that Laetrile may have some, albeit low, antineoplastic activity. Authorities maintained that political pressure caused them to test a relatively unpromising compound. Guy Newell, deputy director of NCI, told us:

> It was thought that we would handle Laetrile like we would any other compound in our decision network flow chart. You see it is not a matter of all or none. We have a battery of compounds to go through animal testing. And it really is a matter of prioritizing. Laetrile is on the list somewhere, but we have other compounds that have shown up so much better, and we have only limited human clinical resources, so we pick higher-priority drugs. We never would have gotten down to Laetrile. So Laetrile was really taken out of priority, . . . and I think not for scientific reasons. I think because of other reasons: social, political, human. (Interview, 1979)

To which Neil Ellison, then in charge of the NCI retrospective study of Laetrile, added:

> If you took 100 and you gave everything a priority listing of what you were going to do, Laetrile may be 95 . . . It has some mutagenic capabilities . . . There is some evidence for Laetrile, and it is very small in the preclinical. But . . . it would have a

[42] Second Opinion, *Laetrile at Sloan-Kettering*.
[43] Moss, *Cancer Syndrome*, p. 152.

very low priority, and many other drugs would be tested before it. (Ibid.)

Note the change in tone toward Laetrile. Once it was, for some FDA officials, a "cruel fraud" and a "hoax"; then it became just another rather unpromising, minimally active substance. This strategy, which sought the participation and perhaps even the cooptation of advocates, was a striking shift in the behavior of authorities.

The clinical trial resulting from these actions tested the effect of Laetrile along with a "metabolic therapy" program of enzymes, vitamins, and special diet on 178 cancer patients at the Mayo Clinic, University of Arizona, University of California at Los Angeles, and Sloan-Kettering. At the conclusion of the trial, Moertel reported that "no substantive benefit was observed in terms of cure, improvement, or stabilization of cancer, improvement of symptoms related to cancer, or extension of life span."[44] Furthermore, he reported evidence of cyanide toxicity resulting from the Laetrile therapy.

Patterns of influence and social control

Having examined the role of trust, influence, and social control in the Laetrile controversy, we are ready to briefly consider how some of the propositions presented in *Power and Discontent* apply to the Laetrile controversy. These propositions include efforts to explain the likelihood of influence attempts and the forms of power that will be chosen by both partisans and authorities. We wish to emphasize that data from a single case study, such as our analysis of the Laetrile debate, are clearly inappropriate for testing these hypotheses. Nevertheless, it may be instructive to consider the extent to which actions in the Laetrile controversy are consistent with Gamson's propositions.

The propositions on influence suggest that influence is more likely to be attempted when a group feels highly affected by an issue, when the group controls a large amount of resources, and when the resources are liquid and uncommitted. These rather noncontroversial formulations may well be applicable to the situation of Laetrile advocates, but good data are simply not available. An additional proposition presented by Gamson, however, does seem relevant. It is proposed that "the greater the interest group representation of a solidarity group, the higher the probability of its attempting influence."[45] As we have previously argued, the establishment of

[44] Charles G. Moertel, et al., "A Clinical Trial of Amygdalin (Laetrile) in the Treatment of Human Cancer," *New England Journal of Medicine* (Jan. 28, 1982), p. 201.

[45] Gamson, *Power and Discontent*, p. 146.

several pro-Laetrile organizations in the early 1970s was one of the factors that caused the Laetrile controversy to emerge from obscurity to become a large-scale social controversy during the last half of the 1970s.[46] One might also argue that with the formation of the Committee for Freedom of Choice in Cancer Therapy, new resources were added to the position of partisans, since many John Birch Society members of the committee brought experience and expertise in political lobbying to the Laetrile movement.

Gamson also provides a series of propositions that relate level of trust to the means of influence that groups will choose. He asserts that confident groups (those with high trust) will use persuasion; neutral groups will use inducements; and alienated groups (those with low trust) will tend to use constraints as a means of influence. Although the use of constraints may produce resentment or some form of backlash, alienated groups frequently feel they have little to lose. Authorities are not expected to produce favorable decisions in the absence of influence. It seems clear, as we have already noted, that most of the influence attempts by Laetrile proponents are, indeed, best classified as constraints.

Finally, Gamson also proposes a number of propositions that relate level of trust to the form of social control that authorities will choose. In this view, authorities will tend to use persuasion on confident groups; sanctions and, especially, inducements on neutral groups; and insulation on alienated groups. There are indeed instances of the use of insulation by authorities in the Laetrile controversy. The firing of Ralph Moss by the Sloan-Kettering Institute for his participation in the critical Second Opinion report, or various attempts to remove the medical licenses of physicians who use Laetrile, are good examples. Laetrile proponents also often charged that authorities practice such insulation by keeping pro-Laetrile findings out of research reports or journals. In order to exercise insulation, authorities must have control over a setting. However, the Laetrile controversy has been waged in multiple settings and frequently in arenas where authorities were unable to limit access.

Thus far we have attempted to analyze the Laetrile controversy by utilizing Gamson's model. But we can also use our case study to comment on Gamson's model. We suggest that the addition of another independent variable – claims making – would improve the model. Claims making seems to be a criticial variable in the case of Laetrile and may well be important in many other movements as well. For example, Marxist groups, as proponents of scientific socialism, make claims about the dialectical nature of history. The extent to

[46] Markle and Petersen, "Social Context."

which such activity influences the success or failure of a social movement deserves greater attention.

Resolution

We conclude by trying to identify the conceptual parameters of success or failure, not only for Laetrile but for marginal scientific controversies in general. "Success," writes Gamson, "is an elusive idea."

> What of the group whose leaders are honored or rewarded while their supposed beneficiaries linger in the same cheerless state as before? Is such a group more or less successful than another challenger whose leaders are vilified and imprisoned even as their program is eagerly implemented by their oppressor? Is a group a failure if it collapses with no legacy save inspiration to a generation that will soon take up the same cause with more tangible results? And what do we conclude about a group that accomplishes exactly what it set out to achieve and then finds its victory empty of real meaning for its presumed beneficiaries?[47]

Failure or success, then, results from a set of outcomes that may be independent of one another. Gamson outlines two major clusters of resolution variables. The first of these, "acceptance," refers to the fate of the challenging group as a formal entity. The crucial question is, Do the authorities now accept the partisans as valid representatives of a legitimate set of interests? Acceptance is operationalized by four rank-ordered indicators: consultation, negotiation, formal recognition, and inclusion.

In the Laetrile controversy, authorities consulted with advocates. Partisans were routinely invited to testify before state legislative hearings, as they were at Senator Edward Kennedy's Senate hearings in 1977. Similarly, the NCI consulted Laetrile leaders in the planning of both the retrospective and clinical trials. In Gamson's scheme, negotiation implies continuing cooperation between the two sides. With court battles and criminal prosecutions characterizing the controversy, this option, let alone formal recognition and inclusion, seemed beyond the grasp of Laetrile advocates. Ernst Krebs, Jr., is not yet dean of the Harvard Medical School.

Gamson's second success variable, new advantages, refers to the group's beneficiaries: Did the challenge result in some benefit for a target population? Again, in the Laetrile dispute we do not have the

[47] William A. Gamson, *The Strategy of Social Protest* (Homewood, Ill.: Dorsey Press, 1975), p. 28.

Table 12.1

Outcome of resolved challenges	Acceptance	
	Full	None
New advantages		
Many	Full response	Preemption
None	Cooptation	Collapse

Source: William, A. Gamson, *The Strategy of Social Protest* (Homewood, Ill.: Dorsey Press, 1975).

luxury of historical perspective. The whole thrust of the freedom of choice issue was to increase the patient's power at the expense of the medical establishment. State laws that now deregulate Laetrile are also a significant new advantage. However, these political victories have been largely symbolic. The advocates' goal – that Laetrile be available on demand – has not been realized.

Viewed as variables in such controversies, both acceptance and new advantage are continuous: Different disputes could be scored across the full range of their values. For schematic purposes, however, Gamson treats each variable as a dichotomy, thus yielding the data shown in Table 12.1.[48]

How do we categorize the Laetrile dispute in this scheme? Authorities tried to "collapse" the movement, whereas advocates wanted "full response." Neither side, however, achieved its stated goal. By the late 1970s authorities had yielded advantage to adversaries they had officially labeled as quacks. Partisans, even with such advantage, did not gain much acceptance. By the mid-1980s Laetrile had lost its prominence in the mass media but persisted as a marginal form of cancer treatment.

The maze of countervailing forces suggests that the conflict has resulted in preemption. Partisans gained new advantages such as deregulation and patient rights, but they failed to achieve acceptance. Clearly the proponents of Laetrile had enough resources and the ability to form alliances with powerful groups (e.g., the health food movement, holism, chiropractic, and the political right wing) to ensure new advantages. At the same time, however, the credentials of Laetrile advocates and the alliances that they formed with chiropractors, naturopaths, and other practitioners of marginal medicine continue to prevent acceptance. Thus, many of the very sources of influence for Laetrile advocates are also barriers to acceptance.

[48] Gamson, *Strategy*, p. 29.

13

Federal regulation of Laetrile

ROBERT S. K. YOUNG

Medical science and health care do not exist independently, apart from society. As new applications of knowledge are proposed, some of which will not be totally benign and will be pervasive in their impact, controversies will arise over the distribution of the risks, benefits, and costs involved for different groups that can only be resolved politically. To survive the increasing demands placed on diminishing resources, we must learn to efficiently make the "best" possible political decisions on scientific and technological issues. This seems especially important in health matters, where not only money but life and limb may literally be at stake.

In this paper, I will discuss three issues related to the federal regulation of Laetrile: (1) the federal regulation of drugs; (2) the nature of the scientific and political controversies surrounding Laetrile; (3) the political success of Laetrile. My discussion is not based on formal research and analysis on the Laetrile controversy. I am untrained and unqualified as a medical historian or as a social or political scientist. My presentation is a subjective synthesis of many hours of listening to arguments for and against the unrestricted distribution of Laetrile in two-dozen state legislative hearings, in federal and state courts, and in media "encounters." I have argued against the distribution of Laetrile.

The underlying assumption of the current regulatory scheme is that defective drug products, or improperly – that is, carelessly or ignorantly – administered drug products may needlessly harm patients. Premarket tests can detect defective products, and instructions for safe use can be given. The primary purpose of the federal regulation of drugs, then, is to protect health consumers from needless indirect

The views expressed in this chapter do not necessarily represent the views of the U.S. Food and Drug Administration.

injury by ineffective drugs, or from direct injury by unnecessarily toxic drugs or the ignorant use of drugs. Injury includes not only physical or psychological harm but also the financial harm suffered when consumers do not receive an already paid-for product or benefit. The U. S. Food and Drug Administration (FDA) enforces a standard of performance for drug products that Congress has legislated as the Food, Drug and Cosmetic Act and that courts have upheld on challenge. The law requires drugs to be proved in scientific (adequate and well-controlled) clinical studies to be both safe (not unnecessarily toxic and appropriately labeled) and effective.

Injury caused by neglectful use of a drug is not regulated at the federal level but at the state level. The current federal law regulates two-party transaction: one party, the distributor, promotes and sells a drug to a second party, the consumer. The promotion may be indirect in that the prescribing physician rather than the patient is the target of an advertising campaign. Although the law is broadly written, the agency has never sought to – nor can it, for practical reasons – regulate an individual's possession or use of a drug. Although the agency does not attempt to regulate the private actions of individuals or use of a drug within the context of medical practice, an individual's range of actions may be severely restricted by the agency: for example, by regulatory curtailment of distribution schemes for a drug.

A product that would not be pharmacologically classified as a drug may acquire that legal status by virtue of a medical claim made for it. A promotional claim can be legally made only after evidence to substantiate it has been collected and validated. In biological systems, it is only with difficulty that reliable scientific data that accurately estimates the probability with which a particular sequence of events will repeatedly occur under a given set of circumstances – that is, reproducible results – is generated. In the current regulatory scheme, the approval standard for drugs requires a positive scientific showing that a drug "causes" a particular therapeutic effect. The burden of proof is on the party making the claim. The government is only responsible for a fair and careful evaluation of submitted and other publicly available data. Although this approach runs somewhat counter to our usual sense of justice, it is consistent with scientific methodology and traditions in pharmacology. The grandfather of pharmacology, Paracelsus, observed in the sixteenth century, "All things are poisons, for there is nothing without poisonous qualities." It is the promoter's obligation, then, to demonstrate that the therapeutic benefits of a drug outweigh its hazards.

As the potential for serious harm through product failure increases and the detection of defective products becomes more difficult, more stringent governmental regulation of the product, using higher standards and greater control, is imposed. Let me cite two examples, one from within the health field and one from outside the field. Because life and limb are often at stake in surgical procedures, which tend to be technically difficult to begin with, all states require surgeons to be licensed. Licensure is generally granted upon submission of evidence of specified medical education, including practical training, confirmed by qualifying examinations and evidence of upright moral character. These regulatory requirements set a performance standard to guarantee consumers that they will probably not be harmed by gross ignorance, a lack of skill – that is, failure to perform – or willful malice on the part of the licensed practitioner. There is little guarantee that they will not be harmed by carelessness on the part of the practitioner.

In a similar manner, commercial aircraft are certified as "airworthy" before they are allowed to carry passengers. An airplane that cannot stay aloft in a controlled manner – that is, perform under the usual, and sometimes extreme, conditions of flight – can cause serious harm to passengers. Certification is granted upon the successful completion of numerous tests. Even after a class of aircraft has been certified, each aircraft within the class is nevertheless thoroughly inspected and tested. Frequently, in the event of a serious failure to perform – for example, crash – all aircraft of this class are grounded until it has been determined that the cause of the failure is not generic. Obviously, pilots are not allowed to operate, and passengers are not permitted to ride on a grounded aircraft. The aircraft is considered unsafe because it is undependable. Whether or not such a product can harm directly – by sucking someone into an engine, for example – is a somewhat separate, though related and often serious, consideration.

The advantage offered to consumers by these regulatory schemes is the assurance that they will not be unnecessarily harmed by product failure or by ignorant misuse of a product and that they can reasonably use a product without having to acquire the necessary skill, education, and facilities to test its reliability and safety themselves. Additionally, when use of a product is immediately necessary, time is not needlessly wasted verifying the usefulness of the product and outlining its hazards. The cost of evaluation is paid for by all potential users, the public.

Regulations that serve only to protect the special financial interests of particular groups, such as the owners of airlines and trucking firms, have appropriately been called into question and reexamined.

In the area of health, only unnecessary and unreasonable harm is controlled. Surgeons are not required to perform totally nonmutilating and painless operations. Just how much regulation is enough is a political question. The body politic decides for itself how much unnecessary harm it is willing to tolerate. It also decides procedurally how a regulatory scheme will operate. The present system of drug regulation, though not perfect, appears to be an improvement for the consumer in comparison with the days of the traveling "Indian" medicine shows and their present-day counterparts in countries with weak drug regulatory systems. In times of serious potential large-scale disasters, such as the Three Mile Island reactor failure and the Bhopal disaster, it always appears that whatever regulatory standard had been imposed was not sufficient.

The necessary elements of scientific clinical studies have been described in the Code of Federal Regulations[1] and discussed in the Commissioner's Report on Laetrile.[2] There are no data that "prove" that Laetrile is ineffective. But also, there are no scientific data that support a conclusion that Laetrile is effective as a cure or treatment for cancer. After carefully examining "studies" advanced as support for such a conclusion, the FDA found that each study was critically flawed.[3] The probability that a future study will demonstrate that Laetrile is effective is vanishingly small. Of the tens of thousands of cancer victims who have taken the drug, not one, let alone a number greater than those who recover spontaneously, can convincingly show that Laetrile has itself altered the natural course of his or her disease. Scientifically, there is little controversy over the safety and effectiveness of Laetrile. There are no clinical cancer research scientists who maintain that the present data demonstrate that Laetrile is effective as a cure or treatment of cancer. After being presented with much data and hearing a variety of arguments and conflicting claims by both sides, no (lay) governmental body, such as a court of legislature, has concluded that Laetrile is effective. In fact, these groups take care to state that they do not know whether it is or is not effective. Even the (lay) promoters of Laetrile admit that there is no scientific evidence of the effectiveness of Laetrile. Although they have sold millions of doses of the drug, some of them currently support a clinical trial of the drug to test whether it works. There are a few persons in the general public who do misunderstand the data and

[1] "Refusal to Approve the Application," 210 Code of *Federal Register* 314.111 (1979): 107–9.

[2] "Laetrile," *Federal Register* 42 (1977): 39776–7.

[3] Ibid., 39777–81.

think that the data conclusively support the effectiveness of Laetrile. Most people willingly admit, however, that they just do not know.

A large number of cancer patients have used Laetrile despite the fact that they are not convinced of its effectiveness. They use it in the "hope" that it will be of benefit to them. Most of the patients are, in fact, hopelessly ill – that is, there is no reasonable probability that conventional anticancer treatments will cure their disease. This situation does not mean, however, that they are beyond the help of conventional therapies that can relieve suffering arising from the complications of their disease.

Besides the fact that these patients understand that orthodox medicine has no definitive anticancer therapy to offer them, what motivates them to abandon conventional medicine and use Laetrile? Their principal motivation is their desire to live. Confronted with one of man's most common and dreaded fears, death, which is often equated with cancer, they are powerfully motivated to do all they can to avoid or postpone it. The promoters of Laetrile have skillfully exploited the inherent desperateness of the situation. They unabashedly offer exactly what the patient-consumer wants – a guaranteed nontoxic cure for cancer. Promotionally, conventional medicine cannot match their product, since there is no guaranteed nontoxic cure for cancer. The ironic point is that Laetrile never cures and is toxic.

A number of patients who cannot be characterized as "terminally" ill use Laetrile in lieu of conventional therapy. In addition to a fear of death, there is a pervasive and understandable fear of the side effects of conventional treatments. Surgery is always painful and usually disfiguring, and anticancer drugs are less than ideal, having side effects that are distinctly unpleasant and often disabling. The promoters of Laetrile have skillfully exploited these fears by always describing conventional therapy in pejorative terms – cut, burn, and poison.

The influence of friends and family on patients to use Laetrile should not be overlooked. Anxious over the possibility that they are not providing adequate support and care for a loved one, they leave no stone unturned in the search and trial of any and all possible treatments, including unproved ones. Such a search has the additional benefit of distracting everyone from the anxiety engendered by being forced to confront impending death.

It is extremely difficult for anyone to confront his or her own death without anxiety. Everyone seeks support, and few go it entirely alone. Often friends, relatives, and medical care systems fail to provide the support necessary to enable a patient to come to terms with his or her own impending death. Laetrile users are a distinct cult. They are automatically admitted to a distinct social group, a group that offers diversion and much needed psychological support and reassurance to patients who unfortunately are often stigmatized socially.

Laetrile has succeeded politically much better than previous unproven cancer treatment fads, although some of these, too, enjoyed considerable political success. What is the basis for the success of this movement? Besides the fact that the Laetrile lobby has many of the ingredients for political success – high motivation, commitment, sound financing, and organization – it materially benefits from being a focal point at which a large and diverse group of people who are alienated from the medical establishment converge and with which an even larger group of people sympathize. Laetrile has allowed these groups to place their medical, social, and political concerns on the public agenda. These people are disappointed and dissatisfied, among other things, with the gap between their expected state of health and health care and their actual state of health and health care. The political success of Laetrile, which is a symptom of disorder in the health care system, can only be understood by examining the underlying issues.

1. The good life in America has been overpromoted. Illness attended by real and prolonged suffering and sometimes death, which in many other cultures is recognized and accepted as an inevitable part of life, is not part of the "advertised" way of life in the United States. Some who suffer illness, along with their families, do so poorly, and at times are quite angry because they think they are being "cheated" of benefits vigorously promoted as an integral part of life in America. They fail to realize that life is not, as advertised, a continuous flow of "Miller time."

2. The benefits of medical science have been oversold. We live in a drug culture whose benefits do not extend to cancer. Advertisements repeatedly tell us the good news that so many of life's ills can be painlessly and effortlessly cured by pills that are safe, available, and cheap. The expectation is that all diseases can be taken care of this simply. The public has been misled into thinking that cancer will be cured in our lifetime. Almost daily there is a report of a "new," very promising cancer "cure." Inexplicably, however, the hoped-for benefits, even after years of development, do not accrue to the three hundred fifty thousand who die each year of cancer. Their family and friends somewhat fearfully and anxiously watch them suffer, waste away, and die. From time to time, we collectively watch the ravages of the disease and treatment on public figures, such as Senator Hubert Humphrey and John Wayne, and have our anxieties reaf-

firmed. One cannot but wonder whatever happened to all the reported "cures" and to the medical establishment that takes credit for their development.

3. The medical care system appears to promise care from cradle to grave, but stops somewhat short of the grave. Many "terminally ill" patients feel and probably are "abandoned" at the point when their physicians realize that their condition is hopeless: No definitive cure or extended survival is possible. For the patient and his or her family, this is an enormously upsetting situation, since at the time when they are most dependent on the profession for care, they are cut loose. "Sent home to die," as his right thumb pointed over his shoulder, is the way Judge Bohanan (Oklahoma) put it.

4. Particularly in an obviously hopeless situation, any course of action suggested is expected to be truly beneficial to the patient. Yet it often seems that activities are engaged in for their own sake – that is, in order to be "doing something." Patients and their families shop for doctors and therapies. Physicians administer anticancer drugs, many of which have well-defined, serious toxicities but little or no real possibility of producing objective benefit in terms of extension of life or mitigation of complications of the disease. After experiencing or observing the adverse effects of these drugs with no recognizable benefit, patients and their families rightly wonder about and resent the possibility that they have been used to soothe the physician's psychological need to do "something."

5. The contemporary thrust in interpersonal relationships is to treat each person as a person and as an equal. This has been slow to obtain in doctor–patient relationships. Patients perceive that physicians patronize them, even in situations when the physician really knows no more (or appears to know no more) about what is going on than they do. Patients rightly resent being treated in this way, although they are probably too anxious to express this resentment directly or overtly. Physicians often do not behave as counselors but medical dictators.

6. Although people's expectations of what the government and experts can do are sometimes unrealistic, there are instances when these groups really do underachieve. Some political leaders and medical spokesmen have misled the public as to what they will and can do. For instance, the FDA does not always enforce regulations equitably. In the area of anticancer drugs, the FDA has approved drugs for commercial distribution, and physicians have used these drugs, for types of cancer for which these drugs have no demonstrable benefit, and yet these are among the most toxic drugs on the market. These facts are obvious to ordinary people, and they wonder

quite appropriately what, if anything, the FDA seal of approval means.

The quality of health care that people receive is rightly subject to political control, for the people who use it pay for it. The body politic usually delegates its power to define standards to expert groups, but there is no a priori reason why it cannot set and abide by its own standards. When the public perceives that the government or its experts (in this case, the medical establishment) have failed to enforce standards of quality, it is reasonable for it to withdraw its mandate and revise it or cancel it entirely. The issues that I have described do not specifically relate to Laetrile. They constitute a set of conditions that have allowed a drug like Laetrile to flourish politically, much to the detriment of a reasonable regulatory standard. For the hopelessly ill cancer patient, Laetrile will cure or control as much of the disease as will a good number of approved drugs – that is, none. For those who suffer the real cost of conventional drugs in terms of large bills and toxic side effects, Laetrile is relatively cheap and nontoxic. Those patients abandoned as "terminal" cases can reassert their value as human beings by participating in their own care and taking charge again. Laetrile is a controversy in which those tired of governmental intrusions into their lives with no perceived benefit can strike back and reduce at least part of the government's power over them.

To think that more and more scientific facts will solve the Laetrile problem is naive. There is persuasive substance to the dissatisfactions that the Laetrile promoters use to fuel the movement. Laetrile's political success appears to be well founded on a deep and real sense of alienation of a large part of the public from the medical establishment. Responsibility for this situation is generally in the hands of the establishment. They tend to shape expectations, and they are the ones who have the goods to deliver.

The medical establishment has a revolt on its hands. Laetrile as a fad appears to be passing away. But just as there were other politically potent but unproved "cancer" drugs before it, so there will be others to follow it as long as the conditions that allow these drugs to flourish persist. As a regulator, what concerns me most is the probability that eventually one of Laetrile's successors will succeed in overthrowing the scientific effectiveness (performance) standard for drugs. On the other hand, if the standard is not strictly enforced anyway and the system fails in this way, it should be and will be discarded. The system must demonstrate concretely that it works for the benefit of patients. A propaganda campaign is not an effective substitute over the long haul. The public cannot be deceived perpetually. The take-home lesson is that it is the medical community that needs to be educated, to face reality, to refrain from overselling what it can do, and to do what it can and should be doing. Laetrile and Laetrile-like controversies can be closed by attacking the roots that give them life.

14

Quasi libertarianism and the Laetrile controversy

BARUCH A. BRODY

The Laetrile controversy is an excellent example of a controversy that raises fundamental questions concerning the relation between the individual and the state. Despite the overwhelming evidence that Laetrile is of absolutely no use in the treatment of cancer, many individuals continue to believe otherwise and are vigorously pressing their claim to be allowed the use of Laetrile. From their perspective, governmental bans on the use of Laetrile represent one more instance of a powerful state imposing its views on citizens who hold minority views. The freedom to use Laetrile becomes a civil rights issue. From the perspective of most of the scientific community and from those in the Federal Drug Administration (FDA) governmental bans on the use of Laetrile represent one more way in which the government legitimately protects vulnerable citizens against fraudulent claims used by peddlers of quack cures. To them the ban on Laetrile is part and parcel of the legitimate use of the police power, and the civil rights claim is just a smoke screen to protect the use of fraud.

It seems appropriate to step back slightly from the concrete details of the Laetrile controversy and to examine the more fundamental issues that it raises about the relation between the individual and the state. In what follows, I briefly outline and explain the advantages of a certain approach to political and social issues, an approach that I call quasi libertarianism, because it is an attempt to retain the many strengths of the libertarian position while avoiding some of its obvious weaknesses. I then develop the implications of this position for the whole range of issues surrounding the social control of Laetrile. Finally, I shall point out the ways in which adopting these implications would enable society to fulfill its real obligations to its citizens in a way that avoids some of the practices that have given rise to the civil liberties claims. In short, I shall argue that sound theory leads in practice to our being able to have, and yet eat, the proverbial cake.

Quasi libertarianism

The popular image of the libertarian is that of someone who is opposed to extensive government activity; someone who believes that the only legitimate functions of the state are the watchdog functions of protecting us against force, theft, and fraud. This image is correct as far as it goes, but it leaves out what is truly fundamental in the libertarian picture of man and society. So let me begin by elaborating upon that picture.

To the libertarian, the fundamental fact about man is that he is a rational agent who often chooses to act in certain ways, from among the alternatives open to him, because of his beliefs that these actions are either intrinsically best or best in leading to what he intrinsically desires. Not all of these choices are deliberate, but many are. To the libertarian, man's freedom to act in this way without being restrained or coerced by his fellow human beings is of fundamental importance, and one of the fundamental human rights is the right not to be constrained or coerced from so acting by others. It is this that provides a foundation for the libertarian's opposition to most of the activities of the welfare state. The welfare state, in pursuing its goals of maximizing general welfare and of redistributing wealth, constantly passes laws and regulations that are coercive in that they impose substantial penalties upon our acting in certain ways.

Naturally, the libertarian recognizes that we can sometimes choose to act in ways that violate the rights of others. We can choose to act in ways that threaten their lives or bodily integrity, in ways that coerce them, or in ways that take from them their legitimate property (I will have more to say about property later in this essay). The libertarian ascribes no positive value to our freedom to act in these ways, and we have no right that prevents our being constrained or coerced from acting in these ways. Even if the state, as a minimal state, passes laws and regulations that impose substantial penalties upon our acting in these ways, penalties that are coercive, the libertarian accepts this as legitimate. We have a right not to be coercively prevented from doing things only if we have an original right to do those things.

In theory, these basic libertarian views are independent of any views about the institution of private property and of any views about an individual's right to property. In fact, however, most writers in this tradition, from Locke on, have held that beside our right to life and bodily integrity and our right of freedom from coercion, we also have a right to the "fruit of our labor," to the value produced by our labor. The contemporary libertarian puts that in the form of a right to the value of any unowned commodity upon which we have worked (initial property rights) or to the value that others freely agree either to give to us in exchange for our giving them something else – for example, our labor power for a certain period of time, or to give to us gratuitously (transferred property rights). Because libertarians hold these additional views about property and the right to property, they have a second objection to the welfare state. The welfare state, in order to finance its programs of maximizing the general welfare and of redistributing wealth, takes the property to which we have a right, and this is another reason for its illegitimacy (libertarians, in fact, have to do a lot of work to explain how even the minimal state is to finance its activities).

Naturally, a lot more has to be said by way of explaining and justifying all of these claims, and I cannot do this in the context of this essay. What I do want to say is that (1) it seems to me that libertarianism, understood in this way, is an important doctrine embodying certain correct perceptions about the relation between man and the state, but that (2) it requires modifications in certain ways. Let me elaborate upon these modifications, for it is they which lead to the quasi libertarianism that I wish to advocate and to use to develop a solution to the Laetrile issue.

The first modification that I would introduce into the libertarian account has to do with the question of the distribution of wealth. From the strict libertarian point of view, the only wealth to which an individual is entitled is that wealth that the individual has acquired by his or her labor on unowned commodities or that wealth that has been freely transferred to him or her by others already entitled to it. There is no special pattern of distribution about which individuals can claim that they have a right to get what they would have under this pattern. I would suggest, however, that there are libertarian grounds for modifying this strict conclusion, that there are grounds acceptable to the basic spirit of libertarianism for a modest state-run redistributive program. I cannot present this argument here in full, but I will at least sketch it.

The crucial idea behind the libertarian theory of property, going back to Locke, is that initial property rights are ultimately grounded in entitlements to the value produced by labor, and that all property rights arise from these initial entitlements. But some wealth that exists is simply the initial value of natural resources, and neither Locke nor anyone else in the libertarian tradition has ever really explained why anyone should have an entitlement to that wealth. My argument suggests that the simultaneous existence of both property rights and rights to redistributive welfare arises out of the difference between labor-created wealth and the wealth that is the value of natural resources.

To see how this works, let us imagine an initial position of a social contract. All those forming the contract recognize the existence of equal libertarian rights not to be prevented from or coercively threatened for using the natural resources of the earth. They also recognize that allowing exclusive property rights, which must be over natural resources as well as labor-created values, is economically efficient. What I claim is that such people would agree (1) to allow for the formation of exclusive property rights over natural resources as well as added values; (2) to compensate those who would lose the rights to use the natural resources assigned to property values; and (3) to provide that compensation in the form of socially recognized welfare rights and socially recognized rights to a minimal level of support. This argument, which is in keeping with at least the spirit of libertarianism, is the foundation for one of the modifications that I would make in libertarianism.

The other modification is grounded in very different considerations. The libertarian argument against state coercion and against state redistribution of wealth is essentially that these actions violate the rights of individuals – the right to act freely and the right to property. There is a fundamental question that any theory of rights must confront. The question is this: What is the relationship between not violating human rights and maximizing human well-being? One extreme position is the assertion that respecting human rights is of value only if the act of respect is likely to lead to a maximization of human well-being. This (putting aside the implications of the adjective *human*) is the position of classical utilitarianism from Mill until the present. At the other extreme is the position that respect for human rights is of such importance that no maximization of human well-being can justify violating a human right (without, at least, the permission of the person who has that right). Something close to this position has been held by many authors in the libertarian tradition, most recently by Robert Nozick. For a long time, of course, the utilitarian view on this question reigned supreme. In recent years, as part of the process of giving up utilitarianism, many have gone to the other extreme conclusion.

Without arguing the case in detail here, I would like to suggest that this other extreme is as unacceptable as the first. Much as we must sometimes refrain from violating the rights of individuals, even if that violation would lead to an increase in general well-being, so we must also sometimes pursue that increase in general well-being, even if it means violating the rights of some individuals. Several factors have to be weighed in considering each case separately. Not all rights are as important as others, and the importance of the rights violated must be considered. Not all violations involve as many people for as long a time as do others, and the number of individuals whose rights are violated and the duration of that violation must be considered. Not all gains in general well-being are equally great, and the extent of the gain produced by the violation must be considered.

What are the implications for a theory of the individual and his or her relation to the state of these points about rights and general welfare? Many of the activities of the current welfare state, as opposed to the libertarian state, are justified on the grounds that they are the provision of public goods, goods that cannot be supplied (or at least not efficiently supplied) on the private market. Libertarians object to these activities because they either result in state coercion or in state redistribution of wealth, and both of these violate the rights of the individuals involved. If my previous point was correct, then this straightforward libertarian critique of the state provision of public goods is inadequate. Even if we grant the libertarian points about rights, we still have to recognize that these state activities might be justified if the infringement of rights is minimized and the gain in general well-being is large enough. This, then, is the second modification I propose in libertarianism.

In short, then, quasi libertarianism involves the following views: (1) Any action that either constrains or coerces an individual from acting in a way in which he or she would have acted otherwise and in which he or she had a right to act is prima facie wrong; (2) individuals have property rights acquired initially through labor or derivatively by free gift or exchange, but individuals also have a right to a certain minimal level of support. The state must respect both of these rights if its policies are to be legitimate, and actions that violate these rights are prima facie wrong; (3) one justification for violating rights (even though such a violation is prima facie wrong) is that the violation produces a sufficiently large gain in general well-being and the rights violated are few in number or of lesser importance.

I cannot now do any more by way of explaining, defending, and justifying this position. I shall now employ the basic concepts of this mode of analysis to look at the question of the legitimate role of government regulation in the control of the production and distribution of Laetrile.

Governmental regulation of Laetrile

I should now like to present a series of arguments that justifies an extensive social role for the government in controlling the production and sale of Laetrile. All of these arguments operate within a framework acceptable to the quasi libertarian, and at least one uses a wholly libertarian approach. The first argument – the one that is acceptable to the libertarian – is an appeal to the legitimate role of the government in dealing with fraudulent sales. The second and third arguments, which are arguments acceptable to the quasi libertarian although not to the libertarian, appeal to questions of public goods and of redistribution of in-kind services. We shall see that each of these arguments leads to some state actions, but not necessarily to the ones now employed to control the use of Laetrile.

Libertarians, it must be remembered, are not anarchists. They allow the use of the power of the state to protect citizens against threats to their rights and, in particular, against force and fraud. It is this last role that is relevant in this context. Couldn't one argue that bans on the use of Laetrile are legitimate precisely because they represent state protection against fraudulent sales? This represents the first and libertarian argument for banning Laetrile.

Let us look a little more carefully at the question of fraudulent sales in the libertarian scheme. A freely agreed-upon exchange or sale is, for the libertarian, something that deserves protection, in part because it is a paradigm example of individuals acting in accordance with their free decisions and in part because it is thought that it is likely to lead to a better distribution to all parties of the goods involved. What, however, happens if the exchange or sale is agreed to by one party only because of the false representation of the other party? To begin with, this undercuts the claim that it is likely to lead to a better distribution to all parties of the goods involved. After all, even if individuals are the best judges of how valuable various goods are to them, the judgment of at least one individual in this case is marred by the fraudulent information. Moreover, and perhaps even more crucially, the presence of fraud undercuts the claim that this exchange is an example of individuals acting in accordance with their free decisions. When I am coerced to act in a certain way, I decide to act in that way, but that decision is not my free decision. Similarly, when I act on the basis of the fraudulent information, I decide to act in that way, but this decision is also not my free decision. A whole theory of human action is needed, of course, to explain what is an action in accordance with an individual's free decision and what is an action that follows a decision but is not really in accordance with the agent's free decision, but for our purposes now we can let this matter rest on this level of intuition.

Traditionally, the question of fraud is treated as a matter of private law. An individual who had agreed to a transaction because of a fraudulent misrepresentation had open to him or her both actions in contract meant to nullify the transaction and tort actions for damages incurred. There is nothing in libertarian theory, however, that dictates this private law solution. Given that the fraudulent transaction deserves no particular protection, and given that the person perpetrating the fraud has no right to do so, there is no reason, from a libertarian perspective, for not extending legal protection against fraud to laws banning such activities and imposing penalties upon them. Just such a law seems involved in the Food and Drug Act. Although the initial Pure Food and Drug Act of 1906 protected only against fraudulent misrepresentations on the label of what was in the container of a drug, the Sherley Amendment of 1912 prohibited false claims about therapeutic effects, and this is what is at the heart of the proposed argument for banning sales of Laetrile.

The situation is, however, more complicated. Suppose that those wishing to purchase Laetrile were to admit that they recognize that all scientific evidence available so far strongly suggests that Laetrile has no therapeutic effect. Suppose, in short, that the crucial element of classical fraud of the type banned in the Sherley Amendment (that the individual is deceived) is not present. The current Drug Amendments (passed in 1962) would still ban the sale of Laetrile. They require that those producing the drug must demonstrate both its safety and its efficacy before it is marketed. Leaving aside for now the pros and cons of the much-litigated question as to whether or not Laetrile is indeed covered by that act of 1962, and assuming, for the moment, that it is, can we justify from a libertarian point of view that further action?

To put the question more carefully, let us look at the procedures dictated by the Indiana statute legalizing Laetrile. According to that statute, a patient requesting Laetrile for therapy must sign a form saying that he or she knows that the FDA has banned the manufacture and distribution of Laetrile, that he or she knows that Laetrile is not recommended for use by the relevant medical societies, and that he or she is aware that his or her physician has available alternative forms of treatment, and that despite all of this the patient requests Laetrile. No doubt even more could be added to this form (e.g., that the patient had been informed of additional possible risks of using Laetrile), but let us suppose that such modifications have been made. Does the adoption of such a statute undercut the fraud argument and suggest that any further ban on Laetrile requires additional justification?

One might insist that it does not. After all, why would anyone sign such a form and still request Laetrile unless its efficacy had been fraudulently misrepresented by the sellers? So, isn't the federal ban required to protect against fraud? I think not, and for two reasons: (1) Whatever might have been the case when Laetrile was first promoted, it has become part of the wide underground of unorthodox medical beliefs, and its use today may be owing to a wide variety of philosophical and personal views held by the individuals in question. (2) If all of the information mentioned in the Indiana form is presented to the person in question, and any further information we may want to add is also presented, I find it hard to think of the individual's decision as based upon fraud, as opposed to his or her unorthodox weighing of various claims, no matter what may have been said to the patient by the advocates of Laetrile use. The individual promoter may have lied, but the choice of the individual patient (leaving aside special circumstances, to be discussed later) becomes, in light of the counterbalancing information, a protected decision.

In short, then, the argument from fraud takes us some part of the way to social control of Laetrile use, but not as far as the state currently goes. To begin with, the individual certainly retains the right to seek through the state his or her classical remedies of nullification and of damages, when he or she can show the requisite fraud. In addition, the state can legitimately ban any fraudulent misrepresentation of the value of Laetrile use. Finally, as part of that process of protecting against fraud, the state can ensure that the individual choosing to use Laetrile had represented to him or her the relevant scientifically established facts about Laetrile, so as to counterbalance any mistaken ideas under which the person is operating.

There is another major role for the state in the social control of the use of Laetrile: that is, the gathering of the relevant evidence about the problems associated with it and the effective presentation of that evidence. This role is, of course, different from the state's role in protecting citizens against fraud, since the state could play this informational role without engaging in any protection against fraud. Still, it is worth noting that the process of fraud protection that we have envisaged does in certain parts presuppose that the state has gathered for effective presentation the information concerning Laetrile's effectiveness.

Libertarians, who are defenders of the minimal state, object to this role of the state. Quasi libertarians would have no objection to such a role, however, since this informational role is precisely the sort of thing that quasi libertarians allow, as state actions that greatly increase general well-being while only slightly infringing the rights of citizens. Let me explain each of these points.

The gathering of information about the safety and effectiveness of drugs, and the effective presentation of this information, is a costly process, and the state meets these costs out of tax revenues. Since taxation always involves taking the property to which people have a right, there is a problem for both libertarians and quasi-libertarians about this informational role. Still, we do not have here a very serious infringement of rights, since personal liberties are unaffected (unlike in other cases of state actions), and the taxes used to support these activities are not very substantial. Providing that the state's performance of this function really produces a great increase in general well-being, this would be one of those cases in which quasi libertarians would accept state actions that infringe individual property rights. Now there is little doubt that the central gathering and dissemination of information about the safety and effectiveness of drugs, as opposed to this being done privately, greatly increase general well-being. Most people could not gather this information privately, and even those who could would require great expenditure to do so, and this information is of great value. Perhaps it could be done privately although centrally, and those who wanted it could purchase the information (think of a *Consumer's Report* dealing with drugs). After all, information is not, in the economist's sense, a pure public good, since people can be excluded from using it. Nevertheless, given that uses of information are not rival (someone's using it does not in any way prevent others from using it), it would be a waste to exclude any one, and since no one is excluded, this is a service that must be provided publicly if it is to be provided centrally.

We can conclude then that once we move from libertarianism to quasi libertarianism, the state may legitimately play a central role in controlling the use of Laetrile. It should use its funds to gather and effectively disseminate information about Laetrile's effectiveness; it should provide a forum for private law suits on the grounds of fraudulent misrepresentation; it should ban with criminal penalties such misrepresentation; and it can require as a precondition of Laetrile use that the potential user have the information about effectiveness available. Is this enough? Or should the state go further and flatly ban the use of Laetrile? It would seem that such a flat ban could not be justified from our viewpoint. But there are some further issues to look at.

Suppose that the following argument is put forward in defense of a flat ban on the distribution and use of Laetrile until its effectiveness is established. Normally, when a citizen wants to act in a certain way because the citizen judges it in his or her own interest, then if no one else will be harmed the state should allow the citizen to act, even if the state feels that the citizen will be harmed rather than benefited. The reason why the state should do so is because it should assume that the citizen is more likely to be correct, because people are the best judges of what will be best for them. However, there are cases where this assumption about the wisdom of the individual is unlikely to be true. There are cases in which the individual is operating under conditions (such as stress, fear, or anxiety) in which his or her normal rational capacities are likely to be weakened. In such cases, the state may have to step in to protect the individual from harming him- or herself. Now, individuals facing the diagnosis of cancer are operating under conditions of stress, fear, and anxiety. The diagnosis of cancer is still widely perceived as equivalent to a death sentence, and often as equivalent to a sentence to a painful and lingering death. Such individuals are precisely those who are in need of state protection. But they need more than state protection against fraud. They need to be protected against their own unwise decisions, decisions that grow out of their fear and anxiety. Among other things, they need to be protected against their own free decision to use Laetrile.

It is interesting to note that this argument should be sufficient to undercut the popularity of certain decisions that have permitted the use of Laetrile at least by those who are terminally ill. It is precisely such people, if anyone, who may reasonably be said to need such protection against their own decisions. Even if they are to die anyway, foolish decisions about Laetrile can lead to great harm, and such people are prime candidates for the need for paternalistic protection.

The Constitutional issues surrounding this question of paternalistic intervention have been litigated. A strongly divided Supreme Court in California upheld the ban on the use of Laetrile, and other courts have followed that decision. I am concerned, however, with the moral and philosophical issues, and not with the Constitutional issues. In particular, I am concerned to find out whether or not such an argument can fit into the quasi-libertarian framework.

As it stands, this argument is unacceptable to the quasi libertarian. It presupposes that individual freedom of choice is justified only because it is likely to lead to wise choices, and that therefore freedom of choice is not justified in circumstances that are conducive to unwise decisions. Libertarians and quasi libertarians reject this presupposition. They believe in the value of free choices and decisions, whether wise or unwise, and if the decisions are truly free (not, for example, coerced or induced by fraud), the state must let them stand even if they are foolish. To accept the argument that I have just outlined would be to give up what is fundamental to the whole libertarian and quasi libertarian tradition.

Nevertheless, there is another way of approaching this topic that at least raises some questions for the quasi libertarian. Think, by way of analogy, about the quasi libertarian's social contract concerning property. In it, people are envisaged as giving up their right to the free use of the natural resources of the earth and receiving as compensation a social insurance policy protecting them against the ravages of poverty. The rationality of agreeing to such a hypothetical contract justifies the actual loss of the rights to use natural resources in return for a social insurance policy. Couldn't a similar argument be used in this context of paternalism? Couldn't one argue that rational contractors would give up their freedom of action in contexts in which they are likely to harm themselves, in return for state protection against the ravages of their own foolishness? Wouldn't the rationality of agreeing to such a hypothetical contract justify the actual loss of certain rights of freedom in return for actual state protection?

Although attractive, this argument must, I think, be rejected. The contractors about property know that failure to come to some agreement about the ownership of natural resources, whether private or social, would lead to a disaster. There would be little economic development and grave poverty. Rationality dictates accepting the hypothetical contract. On the other hand, the contractors about freedom know that agreeing to such a contract opens the door wide to a massive invasion of human freedom. The list of contexts for paternalism would only grow and grow. So rationality, I submit, dictates rejecting the second hypothetical contract.

Let me only add that this last point suggests the need for an important research program. Are there any legitimate contexts for this quasi-libertarian paternalistic argument? The answer to that question is unclear, but the context of Laetrile use hardly seems promising.

Conclusion

I have argued that many social roles for the control of Laetrile use exist, even within those frameworks that emphasize human freedom as a fundamental social value. I have also argued that a flat ban on Laetrile use is not one of those social roles. Let me end by suggesting that it is the flat ban that has seemed excessive to many and that has given rise to the great social controversy. Do we really need this flat ban? I submit that even those who are neither libertarians nor quasi libertarians should ponder that question carefully.

15

Judicial deflection of scientific questions: pushing the Laetrile controversy toward medical closure

ROBERT L. SCHWARTZ

It is generally presumed that whatever merits or demerits the judicial process may otherwise possess, that process is superior to scientific debate as a process of obtaining closure[1] in that it provides a precise, definite, and final conclusion to any controversy, within a defined time, and according to a generally understood and formally delineated procedure. It is for this reason that plaintiffs seek resolution of disputes through litigation rather than through academic debates, constitutional conventions, or the vagaries of some other less well-defined processes. The notion that a lawsuit will inevitably lead to the formal resolution of any scientific issues that litigants place before the court is, however, fanciful. It supposes a very narrow view of judicial conduct and the judicial decision-making process: It assumes that the court can, and will, weigh scientific arguments that are cast in an adversary model and then simply declare a winner.

As others have suggested, such a model would not provide an adequate way to resolve scientific disputes, even if the judiciary were willing to undertake such efforts.[2] Fortunately, the courts have

The author is indebted to Nancy Kirkwood, now law clerk to the New Mexico Court of Appeals, for her research into the history of the chemical and biomedical literature concerning Laetrile and amygdalin therapy for cancer, for her enlightening analysis of that history, and for her careful comparison of the medical and political development of the Laetrile debate.

[1] The term *closure* may be used to refer to several different kinds of resolution of social policy and scientific questions. Four senses of closure are described by Tom L. Beauchamp in chapter 1 of this volume. Unless otherwise specified, the term is used here to describe the practical termination of debate about a particular issue of public policy.

[2] See, for example, Milton Wessel, *Science and Conscience* (New York: Columbia University Press, 1980); Robert S. K. Young, "The Federal Regulation of Laetrile," Chap. 13, this volume.

generally recognized their inability to impose solutions to scientific questions. Instead the American judiciary has generally perceived its role as one of maintaining the integrity of the technical decision-making process by requiring those institutions better able to resolve matters of science to do the careful and adequate scientific work necessary for their resolution.

In participating in the Laetrile debate, the courts have been accused of attempting to step beyond the natural limitations of their competence and to resolve a scientific dispute. But in all of the opinions rendered in all of the lawsuits involving challenges to government regulation of Laetrile, no court has ever reached a scientific conclusion; no court has ever engaged in scientific research; and no court has ever attempted a formal and scientific evaluation of the medical value of Laetrile. The courts have never attempted to impose closure on the controversy, which has always been recognized as a scientific dispute.

When *Rutherford* v. *United States*,[3] the most significant of the Laetrile cases, first came into court (1975), the National Cancer Institute (NCI), the U.S. Food and Drug Administration (FDA), the American Medical Association, and other well-regarded scientific institutions with an interest in the issue considered the medical questions to be well settled. The consequence of judicial intervention was not to close the controversy; instead it was to open the controversy and to require the medical community, and especially the FDA, to do the scientific work necessary for a proper resolution of the issue. When the Laetrile litigation began, there was virtually no respectable scientific work on the toxicity or effectiveness of Laetrile. Now, in part as a consequence of the *Rutherford* case and other litigation, the NCI has completed a retrospective study of claimed Laetrile cures,[4] funded a pharmacological and toxicological study of amygdalin, and has entered a phase II study of Laetrile to determine the existence of any antitumor activity or possible favorable symptomatic effects or effects on survival that are a consequence of Laetrile use.[5]

[3] 399 F. Supp. 1028 (W.D. Ok. 1975) aff'd 542 F.2d 1137 (10th Cir. 1976), class certified 429 F. Supp. 506 (W.D. Ok. 1977), aff'd on other grounds 582 F.2d 1234 (10th Cir. 1978), rev'd 442 U.S. 554 (1980), reconsidered 616 F.2d 445 (10th Cir. 1980), cert. denied 101 S.Ct. 336 (1980).

[4] Neil M. Ellison, David P. Byar, and Guy R. Newell, "Special Report on Laetrile: The NCI Laetrile Review," *New England Journal of Medicine* 299 (1978): 549–52.

[5] Charles G. Moertel, Matthew M. Ames, John S. Kovach, et al., "A Pharmacologic and Toxicological Study of Amygdalin," *Journal of the American Medical Association* 245 (1981): 591–4.

Although there has been some doubt about the ethical propriety of the NCI study of Laetrile,[6] no one doubts that this scientific endeavor could lead to closure on this issue. The medical and scientific sophistication of the NCI-funded researchers is well respected by traditional opponents of Laetrile.[7] The researchers, apparent neutrality, which is demonstrated by their conclusion that Laetrile is not toxic in doses recommended by its proponents,[8] will provide this study with credibility among supporters of Laetrile. The purpose of this essay is to evaluate the role of the judiciary in bringing about what will be a laboratory resolution of this issue.

The role of the judiciary in resolving scientific disputes

Courts are not well equipped to make traditional scientific inquiries. They do not contain laboratories, and judges are rarely trained scientists. Although procedures are available that permit the court to employ scientific experts,[9] these devices are cumbersome and not often employed. Judges recognize this inadequacy, but they also confront litigation that requires the resolution of questions of science and technology. The nature of the judiciary's response to these cases depends on the nature of the scientific or technical question presented.

There are three broad classes of litigation that involve scientific questions. When those scientific questions arise merely because of the litigation itself, the courts resolve them. Where the scientific issues are feigned or inconsequential and the lawsuits really implicate underlying policy issues, the courts identify the scientific issues as secondary ones and resolve the policy questions. Finally, the courts have recently been faced with litigation that depends on the resolution of both policy issues and debated scientific questions. In such cases, exemplified by the Laetrile controversy, the courts have been anxious to deflect the issues back into the laboratories of the antagonists and to act to facilitate a scientific closure, rather than a legal one.

6 Mortimer B. Lipsett and John C. Fletcher, "Ethics of Laetrile Clinical Trials," *New England Journal of Medicine* 297 (1977): 1183–4.

7 The NCI tests are now being conducted by the Mayo Clinic, at the Memorial-Sloan-Kettering Cancer Center, at UCLA, and at the University of Arizona.

8 Moertel et al., "Amygdalin."

9 E.g., rule 53 of the Federal Rules of Civil Procedure provides for the appointment of masters, who may be technical experts. Masters may be appointed to prepare scientific reports that the court will accept as evidence. The neutrality of the master generally makes such evidence highly persuasive.

Technical questions arising from litigation

Courts are frequently called upon to resolve questions involving scientific evidence that arise solely because of the prosecution of the lawsuit. For example, questions of the negligence of a physician (or other technical or scientific professional), the cause of death of a homicide victim, and the speed of a car careening through an intersection into an accident all have scientific components. However, in each case the technical question arises only because it is necessary to determine criminal culpability or civil liability. These scientific questions are entirely legal constructs, and they have meaning only because the court must resolve them to assess individual obligations. When the legal proceedings are terminated, the scientific questions become moot and meaningless, just as they would have been *ab initio* if the lawsuit had never been commenced. Because these questions are so integrally related to the resolution of legal policy, it is perfectly appropriate that they be resolved in court. Although some physicians have found this arrangement, especially as it is applied to professional negligence, to be unsatisfactory, there has been little attempt to limit the court's role in resolving these essentially legal quesions.[10]

Policy questions involving fully resolved scientific issues

Some lawsuits that appear to require the resolution of scientific disputes really involve policy questions that deal with established technical or scientific principles. The debate over the propriety of teaching evolution in public schools is an example of such an issue. The scientific basis of that "theory" is fully established, and among the scientific community it has been clearly, definitely, and certainly resolved. Even those who oppose the teaching of evolution in the public schools or wish alternative theories to be presented admit that evolution is the only explanation justifiable in purely scientific terms.[11]

[10] Several recent legislative reforms of the malpractice process have provided for expert panel review of medical malpractice claims before those claims are filed in court. These statutes generally do not require a finding of negligence by the panel before the action can be commenced, though, and they represent the medical profession's response to what they perceive to be general inadequacies in the legal resolution of disputes involving professional negligence, not just the inadequacy of the courts to understand medical testimony. See "An Analysis of State Legislative Responses to the Medical Malpractice Crisis," *Duke Law Journal* (1975): 1417–68, and Kenneth Abraham, "Medical Malpractice Reform: A Preliminary Analysis," *Maryland Law Review* 36 (1977): 489–532.

[11] Even many of those caught up in the current religious revival in this country who oppose the teaching of evolution in the public schools realize that alternative creationist theories have no more basis in traditional science than

Whatever may have been permitted in the *Scopes* trial, today no judge would seriously consider evidence of the scientific merit of that theory in an action challenging an "anti-evolution" law. In 1968, the United States Supreme Court considered a constitutional attack on a statute barring public school instruction in evolution[12] and resolved the legal issue entirely on First Amendment free speech and establishment of religion grounds. Justice Fortas, speaking for the Court, concluded, "The overriding fact is that Arkansas's law selects from the body of knowledge a particular segment which it prescribes for the sole reason that it is deemed to conflict with a particular religious doctrine; that is, with a particular interpretation of the book of Genesis by a particular religious group."[13]

For the Court, the question simply was not one of science; it was one of First Amendment policy. When a feigned scientific dispute is properly cast in the terms of a constitutional application of uncontested scientific knowledge, the role of the judiciary is clear: The court has both the expertise and obligation to make such determinations.

Mixed questions involving unresolved policy issues and unresolved scientific questions

The most difficult questions for the courts have been those questions that include the application of disputed policy to truly disputed scientific issues. The effective denial of access to Laetrile, the legal regulation of nonmedical practitioners of acupuncture, and the administrative determination of the amount of benzene that can appropriately appear in a "safe" workplace are examples of these difficult disputes. Each of these disputes could be resolved entirely on social policy grounds, and in ways that would render the scientific resolution of the problem legally meaningless. For example, the right of privacy could be extended to protect a patient's access to Laetrile or acupuncture treatment, even if those treatments were found to be ineffective or definitely harmful.[14] Similarly, the Occupational Safety

does the theory of evolution. See Robert Lindsay, "Creationists Gather to Try Toppling Darwin's Pedestal," *New York Times* ("Week in Review"), March 1, 1981, p. 203. In fact, in a version of the famous *Scopes* "monkey" trial, a California Superior Court judge refused to allow into testimony any evidence of the scientific validity of evolution or biblical creation. See "Court Study of Evolution Focuses on Language of Guidelines," *New York Times* (March 6, 1981), p. A10.

12 *Epperson* v. *Arkansas*, 393 U.S. 97 (1968).

13 Ibid., p. 103 (footnote omitted).

14 The two leading privacy cases are those that established the qualified constitutional right of a woman to choose to have an abortion. *Roe* v. *Wade*, 410 U.S. 113 (1973), and *Doe* v. *Bolton*, 410 U.S. 179 (1973), have figured prominently

and Health Act of 1970[15] could be construed to prohibit the exposure of any employee to any carcinogen, including benzene, at any level. Aside from the questionable social policy that would be the consequence of the adoption of such principles, such a resolution of these problems might discourage important scientific inquiry in these areas.

Just as a legal resolution of these questions would render the scientific evaluation moot, scientific closure would render the legal policy determination needless. If Laetrile or acupuncture can be proved to be both effective and nontoxic, there will be little objection to their application. Similarly, if careful and well-accepted scientific studies indicate that either one is harmful and ineffective, there will be no demand for their legal availability. It is a well-established principle of American law that a court will not decide a constitutional question unless it is necessary to do so. Thus, when a lawsuit can be resolved on statutory, regulatory, or procedural grounds, the court will refrain from resolving the constitutional issue. In practice, the courts appear to be accepting the potential scientific resolution of policy disputes as another alternative to constitutional determinations. Where a policy closure can follow from either a legal determination or a laboratory conclusion, the courts will attempt to deflect the issue out of the judicial arena and back into the laboratory.

One example of such deflection is the United States Supreme Court's decision in the *benzene* case.[16] In that case the Supreme Court was faced with an administrative determination by the Occupational Safety and Health Administration (OSHA) that the "safe" level of benzene in the atmosphere of the workplace was to be lowered from 10 parts per million to 1 part per million.[17] Rather than determine the scientific propriety of one standard or another, the Court merely reviewed the administrative decision to determine whether or not it constituted arbitrary action. Finding that there was no scientific evidence to suggest that one standard was more appropriate than the

in the arguments made in virtually all recent cases involving access to Laetrile or acupuncture. See, e.g., *People* v. *Privitera*, 74 Cal. App. 3d 936, 141 Cal. Rptr. 764 (1977), rev'd 23 Cal. 3d 697, 153 Cal. Rptr. 431, 591 p. 2d 919, cert. denied 444 U.S. 949 (1979) (Laetrile): *Andrews* v. *Ballard*, 498 F. Supp. 1038 (S. D. Tex., Hous. Dv. 1980) (acupuncture); and, of course, the *Rutherford* opinions cited throughout this chapter.

15 29 U.S.C. Sec. 651 et seq. (1976).

16 *Industrial Union Department, AFL-CIO* v. *American Petroleum Institute*, 100 S. Ct. 2844 (1980).

17 29 CFR Sec. 1910. 1028, 43 Fed. Reg. 5918 (Feb. 10, 1978), as amended 43 Fed. Reg. 27962 (June 27, 1978), cited in *Industrial Union Department, AFL-CIO* v. *American Petroleum Institute*, 100 S. Ct. 2844, 2850.

other, four of the five justices in the majority struck down the new administrative standard.[18] In doing so, the Supreme Court carefully avoided the policy issues, including the application of cost–benefit analysis to occupational health determinations, that were the subject of the arguments made by both sides. Instead the Court's decision has the effect of remanding the question of whether one standard is scientifically superior to another back into the laboratories. The opinion makes clear that the new benzene regulation would be legally sound if it were supported by competent scientific evidence. Of course, if it were thus supported, the policy issue would be resolved outside of the court. By deciding the *benzene* case as it did, the Supreme Court required that the final resolution of that dispute come from the scientific community, not the judiciary.

The Laetrile controversy provides the best example of judicial deflection of a mixed policy and science question out of the judiciary into the laboratories. Before 1975, when the first significant Laetrile litigation – *Rutherford* v. *United States* – found its way into the federal court system, there was no scientifically respectable medical study of either the toxicity or the effectiveness of Laetrile. Now, many years and seven *Rutherford* opinions later, the judiciary, leaving no nationally binding constitutional precedent behind it, has removed itself as a factor in the resolution of the Laetrile controversy, whereas the NCI is in the midst of serious inquiry into the medical merits of the drug.

The legal and medical status of Laetrile in 1975

When Glen Rutherford intervened as a plaintiff in *Stowe* v. *United States* in 1975, he was becoming party to a case that would be the most important judicial intrusion into the Laetrile controversy. He was a sympathetic plaintiff supported by an able attorney. The case was scheduled before a thoughtful and well-respected district court judge who became the first federal judge to analyze seriously the legal issues raised by those seeking access to Laetrile.[19] Although others have raised similar issues,[20] they rarely have been accorded more

18 The fifth justice in the 5 to 4 majority was Justice Rehnquist, who found that the statutory scheme constituted a constitutionally impermissible delegation of congressional authority to an administrative agency. 100 S. Ct., at 2879–87.

19 Ellison et al., "Special Report," and Moertel et al., "Amygdalin."

20 See, e.g., *United States* v. *Spectro Foods*, 544 F. 2d 1175 (3rd Cir. 1976), *Sadler* v. *United States*, 425 F. Supp. 244 (D. Minn. 1979), and *United States* v. *Articles of Food and Drug*, 444 F. Supp. 266 (E. D. Wis.), *Aff'd sub nom. United States* v. *Mosinee Research Corporation* (7th Cir. 1978) (all upholding the FDA's effective ban

than perfunctory hearings at the trial level, and few of the cancer-ridden plaintiffs have remained alive long enough to docket appeals. Rutherford's longevity and his use of the class-action device made the *Rutherford* case the primary legal focus of those concerned with the Laetrile dispute, and it remains the only piece of litigation involving Laetrile to be heard by the United States Supreme Court.

In order to understand the nature of Rutherford's legal claim it is important to evaluate the administrative regulation that led to the effective banning of Laetrile. In addition, because the operation of the legal regulation itself depends on the status of the scientific research, it is important to summarize the scientific data that was available on the toxicity and effectiveness of Laetrile in 1975.

Legal regulation of Laetrile

The FDA treats Laetrile as a "new drug" subject to regulation under the Food, Drug, and Cosmetic Act.[21] A "new drug" is "any drug . . . not generally recognized, among experts qualified by scientific training and experience to evaluate the safety and effectiveness of drugs, as safe and effective for use under the conditions prescribed, recommended or suggested in the labeling."[22] A "drug," under the act, includes anything offered for the treatment, cure, mitigation, diagnosis, or prevention of disease.[23] It is illegal to import or move in interstate commerce any new drug not formally approved by the FDA.[24] The United States may seek an injunction against those who violate the statute and may also seek to criminally prosecute those offenders.[25] In addition, any drug shipped in interstate commerce or imported in violation of the statute is subject to seizure.[26]

For a new drug to be approved under the Act, its sponsor must file an application that includes "full reports of investigations which have been made to show whether or not such drug is safe for use and whether such drug is effective in use."[27] The application will not be approved if the investigations do not include adequate tests "by all

Footnote 20 (*cont.*)
on Laetrile). See also *Rizzo* v. *United States*, 432 F. Supp. 356 (E.D.N.Y. 1977) and *Millet, Pit, and Seed Co.*, v. *United States*, 436 F. Supp. 84 (E.D. Tenn., N.D., 1977), both blocking FDA attempts to limit access to amygdalin.

21 21 U.S.C. Sec. 321 (p) (1) (1976). 22 Ibid.

23 21 U.S.C. Sec. 321 (g) (1) (B) (1976).

24 21 U.S.C. Sec. 355 (a) (1976).

25 21 U.S.C. Secs. 332, 333 (1976).

26 21 U.S.C. Sec 334 (1976).

27 21 U.S.C. Sec. 355 (b) (1) (1976). Several other less onerous burdens are also imposed by the statute. See 21 U.S.C. Sec. 355 (b) (2–6) (1976).

methods reasonably applicable to show whether or not such drug is safe for use under the conditions prescribed, recommended, or suggested," if the tests do not show that the drug is safe, if "there is a lack of substantial evidence that the drug will have the effect it reports or is represented to have under the conditions of use prescribed, recommended, or suggested," or if the labeling is false or misleading in any particular.[28] Essentially, a drug's sponsor must prove that it is both safe and effective for the purposes for which it is sold before it can be approved. The burden of proof is put on the sponsor, who must prove his case by "substantial evidence." Substantial evidence is defined by the statute as "evidence consisting of adequate and well-controlled investigations, including clinical investigations, by experts qualified by scientific training and experience to evaluate the effectiveness of the drug involved, on the basis of which it could fairly and responsibly be concluded by such experts that the drug will have the effect it purports or is represented to have under the conditions of use."[29]

Obviously a drug must be available for use in clinical trials before it can be approved as a new drug. Thus the statute provides for an investigational new drug license (IND) to be issued to those "experts qualified by scientific training and experience to investigate the safety and effectiveness of drugs," but only upon a showing that the drug displays promise on the basis of studies conducted without human subjects.[30] In the absence of an approved new drug application or an approved IND, a drug may be imported or moved in interstate commerce only if it is exempted from the Act.

The only exemption relevant to the case of Laetrile is provided in the 1962 grandfather clause in the Act. That exemption was added by the drug amendments of 1962, which added the effectiveness requirements to the Food, Drug and Cosmetic Act of 1938. These amendments were a consequence of the desire to tighten government control of unproved drugs that followed the Thalidomide scandal in Europe. Ironically, Thalidomide was kept out of distribution in the United States even before 1962 because it could not meet the safety requirements of the 1938 statute. The 1962 grandfather clause exempts from the new drug requirements any drug that on October 10,

[28] 21 U.S.C. 255 (d) (1976). The other requirements must also be satisfied.

[29] Ibid. See also *Weinberger* v. *Hynson, Westcott, and Dunning, Inc.*, 412 U.S. 609 (1973). The new drug approval process also requires administrative evaluation of the ingredients of manufacture, the methods and facilities used in manufacture, the complete labeling, and samples of the drug. 21 U.S.C. Sec. 355 (b) (1976).

[30] 21 U.S.C. Sec. 355 (i) (1) (1976); there are additional requirements. See 21 U.S.C. Sec. 355 (i) (2) and (3) (1976).

1962 (the day before the enactment of the 1962 amendments), was commercially available in the United States, generally regarded as safe "among experts qualified by scientific training and experience to evaluate the safety of drugs," and not the subject of a new drug application pending at that time.[31]

The proponents of Laetrile use have argued that (1) Laetrile is not a drug; (2) Laetrile is not a "new drug," since it is generally recognized as safe and effective; and (3) Laetrile meets the requirement of the 1962 grandfather clause and is thus exempt from regulation under the Act. The last of these three arguments is the only legal argument seriously considered by the courts.

In 1975 the proponents of the use of Laetrile were frustrated by what they perceived to be the unavailability of the new drug approval process. The process is a highly complex one, essentially unavailable to all but the large drug companies. In 1970 a laboratory had sought an IND to prove the safety and effectiveness of Laetrile. The FDA originally approved the IND, but in a very unusual procedure withdrew it within a week of its issuance because of the agency's belief that preliminary testing did not yield any justification for further testing in human subjects.[32] Thus in 1975 the FDA considered the importation or interstate shipment of Laetrile for treatment or research to be a violation of the Food, Drug and Cosmetic Act of 1938.

Although the Food, Drug and Cosmetic Act would not apply directly to the intrastate production and distribution of Laetrile, it serves that purpose indirectly. First, many states have promulgated statutes that make it unlawful to produce, distribute, prescribe, or use any drug not approved by the Food and Drug Administration.[33] Some states, such as California, outlaw the distribution of any "unproven" cancer cure.[34] In any case, the FDA insists that interstate shipment "of anything from apricot pits to bottle stoppers" would bring the final product within the agency's authority.[35] Whether or

[31] 21 U.S.C. Sec. 321 (p) (1976). See also S. Rep. No. 1744 (87 Cong., August 21, 1962), Sec. e, "Transitional Provisions."

[32] This strange series of events is discussed in Michael L. Gulbert, *Vitamin B_{17}: Forbidden Weapon against Cancer* (New Rochelle, N. Y.: Arlington House, 1974), which is itself discussed in the first District Court *Rutherford* opinion, 399 F. Supp. 1208, at 1213.

[33] See, e.g., 35 Purdon's Penn. Stat. Ann. Sec. 78–110 (1977).

[34] West's Ann. Health and Safe. Code Sec. 1707.1 (1979) makes it a misdemeanor to sell, deliver, provide, or prescribe any drug to be used in cancer treatment unless the drug is approved by the FDA *or* a California state board. A very similar statute is found in N.J. Stat. Ann. Sec. 24:64–1 (1975).

[35] Jonathan Spivak, "Laetrile's Message to the FDA," *Wall Street Journal*, Western ed., July 21, 1977.

not the FDA could legally extend its jurisdiction in this way, its threat has been sufficient to dissuade reputable drug companies and physicians from making the drug available for any purpose.[36]

State of scientific research on Laetrile in 1975

Very little formal research on the toxicity or the efficacy of Laetrile appeared before 1976. Laetrile, the popular name for amygdalin, is present in large quantities in apricot pits, rose hips, lima beans, bitter almonds, cherry pits, and many other fruits and vegetables. The first formal chemical evaluation of Laetrile, as amygdalin, appeared in 1965.[37] In 1971, Dean Burke, then with the NCI, published the results of this study with A. R. L. MacNaughton (of the MacNaughton Foundation) and the German researcher M. von Ardenne on the effects of amygdalin on Ehrlich ascites carcinoma cells. They tested cyanide and benzaldehyde as well. Although neither was found to be cancericidal, the combination of the two was, as was amygdalin.[38] One other article on the chemical composition of amygdalin was published in 1971.[39]

A few studies in animals demonstrated little or no efficacy for amygdalin alone or in combination with other substances.[40] Whether

[36] Ibid.

[37] Leo Levi, W. N. French. I. J. Bickis, et al., "Laetrile: A Study of Its Physiochemical and Biochemical Properties," *Canadian Medical Association Journal* 92 (1965): 1057–61. The authors concluded that "the Canadian and the American product are different pharmaceutical foundations, displaying different physicochemical and biochemical properties, and citing then current legal materials" (p. 1057).

[38] Dean Burke, A. R. C. MacNaughton, and M. von Ardenne. "Hyperthermy of Cancer Cells with Amygdalin-glucosidase, and Synergistic Action of Derived Cyanide and Benzaldehyde," *Panminerva Medica* 13 (1971): 520–2. See also Dean Burke, "Effects of Amygdalin, Prinasin, Mardelonitrile, and HCN-benzaldehyde on the Pasteur Effect, Metabolic Death, and Trypan Blue Staining of Ehrlich Ascites Carcinoma Cells," in *Progress in Antimicrobial and Anticancer Chemotherapy: Procreedings of the Seventh Internatio..al Congress of Chemotherapy, 1971*, 1:133.

[39] R. A. Llenado and G. A. Rechnitz, "Improved Enzyme Electrode for Amygdalin," *Analytical Chemistry* 43 (1971): 1457–67.

[40] See, e.g., William Campbell, "Laetrile and Schistosomiasis," *Science* 184 (1974): 588; George Hill, II, Thomas Shine, Helene Hill, et al., "Failure of Amygdalin to Arrest B-16 Melanoma and BW5147 AKR Leukemia," *Cancer Research* 36 (1976): 2102–7; Isidore Wodinsky and Joseph K. Swiniarski, "Antitumor Activity of Amygdalin MF (NSC-15780) as a Single Agent and with Beta-glucosidose (NSC-128056) on a Spectrum of Transplantable Rodent Tumors," *Cancer Chemotherapy Report* 59 (1975): 939–50; W. R. Laster and F. M. Schabel, Jr., "Experimental Studies of the Antitumor Activity of Amygdalin MF (NSC-15780)

because of the negative results in animal studies or because of an anti-Laetrile bias, there have been few clinical studies of Laetrile's efficacy outside of Mexico and the Philippines.[41] Most of the American reports are anecdotal case reports, although one early American study held out some promise for the use of Laetrile in humans.[42]

There were, similarly, almost no studies of the toxicity of Laetrile in 1975. Although one of the early researchers who found Laetrile to be effective also found it to be toxic when taken orally,[43] all of the evidence of the toxicity is anecdotal. The "evidence" of the toxicity of Laetrile included cases of cyanide poisoning from the ingestion of choke cherry seeds[44] and from a diet of improperly processed cassava.[45] Animal studies on the toxicity of Laetrile conducted before 1976 show a toxic reaction only at extremely high dosages.[46] There

Footnote 40 (*cont.*)
Alone and in Combination with Beta-glucosidose (NSC-128056)," *Cancer Chemotherapy Report* 59 (1975): 951–65.

41 The most frequently cited foreign sources are the reports of clinicians. Ernesto Contreras runs a clinic in Mexico, and Manuel Navarro has used Laetrile clinically in the Phillipines. There has been at least one Italian trial, and a number of studies have been reported in Germany, most of which are not clinical. The foreign studies that are often cited include M. Tasca, "Clinical Observations on the Therapeutic Effects of a Cyanogenetic Glucuronoside in Cases of Human Malignant Neoplasms," *Gazzetta Medica Italiana* 118 (1959): 513–59; A. Nahrstadt, "The Isomerization of Amygdalin and Its Homologues," *Archiv der Pharmazie* 308 (1975): 903–10; Manuel D. Navarro, "Five Years Experience with Laetrile Therapy in Advanced Cancer," *Phillippine Journal of Cancer* (1957): 289–307, "Laetrile in Malignancy," *Santo Tomas Journal of Medicine* 10 (1955): 113–18, and "Laetrile Therapy in Cancer," *Philippine Journal of Cancer* 4 (1964): 204–9; P. G. Reitnauer, "Mandelonitrile-gylocosides in Cancer Research and Cancer Therapy: The Amygdalin Problem," *Arzneimittel-Forschung* 22 (1972): 1347–61 (including extensive bibliography); H. M. Summa, "Amygdalin, a Physiologically Active Therapeutic Agency in Cancer," *Krebsgeschehen* 4 (1972): 110–18, abstracted in *Excerpta Medica*: Sec. 16, *Cancer* 24 (1973): 2098; M. von Ardenne and P. G. Reitnauer, "Tumor Hyperacidulation Through Intravenous Glucose Infusion Enhanced by Amygdalin and Beta-glucosidose Application," *Archiv für Geschwulstforschung* 45 (1975): 135–45.

42 John A. Morrone, "Chemotherapy of Inoperable Cancer: Preliminary Report of Ten Cases Treated with Laetrile," *Excerpta Medica*: Sec. 9, *Surgery* 20 (1962): 299–308. Morrone found that the use of Laetrile "provided dramatic relief of pain, discontinuance of narcotics, control of fever, improved appetite, and reduction of adenopathy," as well as possible regression of the malignant lesions.

43 Ibid.

44 "Laetrile: Report under Fire from within," *Science News* 113 (1978): 4.

45 A. Olufemi Williams and Benjamin O. Osuntokun, "Peripheral Neuropathy in Tropical (Nutritional) Ataxia in Nigeria," *Archives of Neurology* 21 (1969): 475–92.

46 Wodinsky and Swiniarski, "Antitumor Activity," and Laster and Schabel, "Experimental Studies."

was simply no evidence that Laetrile was toxic at anything approaching the dosages recommended by those prescribing the drug.

In summary, when the *Rutherford* case commenced in 1975, there was very little reliable data of any kind that would demonstrate that Laetrile was, or was not, either toxic or effective.

The Rutherford complaint

In 1971 Glen Rutherford was diagnosed as having an invasive adenocarcinoma and informed that surgery, including the removal of his rectum, would be necessary. He was upset by the prospects of this treatment, refused the surgery, and went to Mexico, where he was treated with Laetrile. He returned to the United States after several weeks of Laetrile treatment and continued to take Laetrile while at home.

The Mexican clinic continued to deliver Laetrile to Rutherford until 1975, when the courier was arrested and Rutherford's supply of Laetrile was seized. Rutherford then moved to intervene in an action that had been commenced by other terminally ill cancer patients who were seeking authority to import Laetrile for their personal use. By the time the case came up for a hearing in the district court, the two original plaintiffs had died.

In August, Judge Luther Bohanon, of the federal District Court of the Western District of Oklahoma, rendered an opinion that demonstrated his exasperation with the FDA's treatment of Laetrile. He was apparently disturbed by the cavalier defense of Rutherford's earnest arguments provided by the agency and by the inability of the FDA to produce any evidence of toxicity for Laetrile that could justify its own very strong position. As the court pointed out, "Inaction by the FDA constitutes the crux of plaintiff's procedural dilemma."[47] Transparently sharing in Rutherford's frustration with an overbearing FDA, the judge concluded that "the plaintiff Rutherford and those similarly situated are wholly without means or resources to comply with the provisions of [the Food, Drug and Cosmetic Act]" and that "the evidence supported the findings that Laetrile is not a toxic or harmful substance if used in proper dosage, but is on the other hand an alternative treatment of cancer which can be used in lieu of surgery or radiation cobalt."[48] Judge Bohanon did not hide the source of his real concern. The FDA simply had no reason – in science or in logic –

[47] 399 F. Supp., 1212.
[48] Ibid., 1213, 1214–15.

to deny Rutherford his Laetrile. The court informed the FDA that if it were to pursue its battle with proponents of Laetrile, it ought to develop a scientific arsenal.

Forcing the scientific review: 1976 and 1977

Judicial determinations in the Rutherford case

The United States appealed Judge Bohanon's *Rutherford* decision to the court of appeals for the tenth circuit. The court affirmed the decision, but in an opinion that was less self-righteous and angry, and more focused, than the lower court's view.[49] The court did not conclude that Laetrile did (or did not) have medical value; rather, it concluded that "the FDA's record is grossly inadequate and consists merely of a conclusory affidavit of an official of the FDA which in effect declares that [Laetrile] is a new drug because the FDA says it is and thus is subject to all the statutory vagaries of such a designation."[50] The FDA position that only issues that the agency itself thought to be important enough to warrant review could be reviewed in the courts was simply unacceptable to the tenth circuit court of appeals, which concluded that the FDA was obliged to prepare a formal record to justify its determination that Laetrile was not exempted from the new drug approval process under the Act's 1962 grandfather clause. Because the court concluded that the resolution of the underlying problem depended on a scientific inquiry that the FDA was compelled to make, the court of appeals did not reach the constitutional issues suggested in the first district court opinion.[51]

Although the next step in the *Rutherford* odyssey should have been the FDA's development of the Laetrile record required by the court of appeals, the case found its way back into Judge Bohanon's court in March of 1977 on the plaintiff's "Application to Clarify Plaintiff Class."[52] Rutherford claimed to bring the action on behalf of a class composed of all terminally ill cancer patients. Because of the class action organization of the lawsuit, it could continue even if Rutherford, the "named plaintiff," were to die. The United States had opposed the prosecution of the litigation in class action form because, it argued, "the class plaintiffs purport to represent is too ill defined and ephemeral in make up to render its members capable of definite identification."[53] The class was formally certified, and the action was

[49] 542 F. 2d 1137 (10th Cir. 1976).
[50] Ibid., 1140.
[51] Ibid., 1144.
[52] 429 F. Supp. 506 (W.D. Ok. 1977).
[53] Ibid., 508.

prosecuted as a class action because "in cases such as this, where the ultimate effectiveness of a federal remedy may depend in large measure on the applicability of the class action device, all judicial discretion should be directed toward allowing the class action."[54]

In this same opinion the court undertook to determine whether Laetrile was a "new drug," an issue apparently raised by the FDA once again in the hope that the court would not continue its injunction against the enforcement of the Food, Drug and Cosmetic Act against Laetrile importers, pending the FDA's development of the administrative record. This time the court made it very clear that the injunction did not rest upon any judicial finding as to the scientific merit of the Laetrile; the decision rested entirely on the failure of the defendant to do a proper and scientific evaluation. As the court pointed out,

> Defendants have introduced evidence tending to establish the general opposition of medical authority in this country to the use of laetrile. Contrarily, the court is aware of instances of patients and physicians in various parts of the country emphasizing personal experience with laetrile's ability to counteract aspects of the disease's manifestations and discomforts. Regardless, such issue is not before the court, and the court is cognizant that it possesses "neither the facilities nor the expertise" to independently determine the drug's therapeutic value.[55]

The court explained:

> [T]he issue of the efficacy of laetrile is, at most, of secondary importance in this case. The legality of FDA's ban on laetrile is under attack on the theory that FDA arbitrarily and without sufficient basis in fact characterized laetrile as a "new drug;" so far FDA has presented little, if any, evidence to combat that allegation.[56]

[54] Ibid., 509.

[55] Ibid., 510.

[56] Ibid., 511. Judge Bohanon was not the only judge to be concerned by arbitrary and scientifically unjustifiable decisions to ban Laetrile. About the same time, the California Court of Appeals overturned the conviction of Dr. James Privitera for conspiring to prescribe Laetrile in violation of a statute making it a crime to prescribe an unproved cancer treatment. There the court pointed out that "concerning the efficacy of amygdalin, this court, this opinion, does not enter that fray. The effectiveness of amygdalin as a cure for cancer or as a nutritional aid with general health-giving benefits is not, as a matter of law [at issue here]." The California court was straightforward in warning that "to require the doctor to use only orthodox 'state sanctioned' methods of treatment under threat of criminal penalty for variance is to invite a repetition in California of the Soviet experience with 'Lysenkoism.'"

In July 1977, the FDA released the formal report it had been preparing in response to the earlier tenth circuit court opinion.[57] Despite the conflicting evidence that the FDA had received on the efficacy and toxicity of Laetrile, and despite the inconsistent evidence the agency had received on Laetrile's status under the 1962 grandfather clause, the FDA had no trouble reaching exactly the conclusions it had reached regularly since 1970 and exactly the conclusions that it had formally predicted it would reach in this inquiry. The patently closed-minded and passionately anti-Laetrile report read more like a brief in support of the FDA position than a real inquiry into the substantive issues.

At about the same time this report was issued, the FDA intensified its public relations campaign against the use of Laetrile. The November-December 1977 *FDA Drug Bulletin* included an article on the toxicity of Laetrile that was reprinted and widely distributed, and it included an oversized poster containing a "Laetrile warning."[58] An article entitled "Laetrile, the Making of a Myth," which appeared in the December 1976/January 1977 FDA *Consumer* was also reprinted in slick, glossy form, complete with a photograph of "a Laetrile victim" apparently being abused by what appeared to be a greedy and fraudulent snake oil salesman lurking in the shadows.

Scientific research and medical publications

Although little scientific research on Laetrile was published during 1976 and 1977, many of the major medical journals began to carry editorials condemning Laetrile use. Those who opposed the use of Laetrile were also beginning to realize that scientific evidence would be necessary to support their position. Since it was finally becoming apparent that Laetrile might qualify under the 1962 grandfather exemption from the Food, Drug and Cosmetic Act, and because toxicity of Laetrile would have to be demonstrated in order to keep it off the market if it did qualify under that clause, articles with anecdotal evidence of Laetrile toxicity began metastasizing throughout the medical literature. Suddenly, a drug that had created very little interest and that had been presumed to be nontoxic was the

[57] U.S. Food and Drug Administration, "Laetrile," 42 Fed. Reg. 39767 (1977).

[58] "Toxicity of Laetrile," *FDA Drug Bulletin* (November–December 1977): 26–32. The insert contained this message for physicians: "PLEASE POST. The centerfold of this *Drug Bulletin* contains a poster being issued by the Food and Drug Administration to warn about the dangers of laetrile. The Centerfold can be easily removed from this *Bulletin*. We encourage health professionals to post it where it can be seen by cancer victims, their families, and the general public."

subject of an extraordinary amount of toxicity research.[59] The fact that much of this work was done by those who were also taking editorial positions opposing Laetrile, and the transparent legal purpose of this hastened research, cast doubt on its scientific value.

Almost immediately upon the FDA commissioner's release of his report and the administrative record that he had compiled to support it, Rutherford was back before Judge Bohanon to challenge the report's conclusion. The court issued an injunction directing those responsible for enforcement of the Food, Drug and Cosmetic Act not to act against Laetrile or its users. Judge Bohanon was simply unwilling to allow the Laetrile controversy, based on serious questions of science and policy, to be resolved on the basis of what he regarded as a clearly inadequate and partial scientific review. In concluding that the court should intervene only when "the agency really has not taken a single 'hard look' at the salient problems, and has not *genuinely* engaged in a reasoned decision-making,"[60] the court did not attempt to evaluate what scientific results the agency should have reached. Instead, the court merely evaluated the objectivity of the agency's scientific process:

> Considerable evidence called into question FDA's sense of objectivity in this case . . . When this suit was initiated, FDA had declared laetrile a "new drug" without ever having constructed an administrative record in support of such designation . . . Ideally, agency decisions and conclusions should flow from a probing and objective analysis of a carefully amassed and encompassing factual record. When ordered on remand to conduct an appropriate investigation, FDA begrudgingly announced its intention to do so, and then previous to ever having received the evidence on which its conclusions are ostensibly based, FDA reaffirmed its same, entrenched positions on the salient issues in the case . . . Understandably many contributors to the administrative record expressed skepticism concerning the proceeding's fairness.[61]

The court's continuing frustration with the FDA's failure even to permit any neutral scientific review of the evidence appeared as the basis for this third Bohanon opinion. At one point the judge suggested that "the current debate is fierce. The issues appear largely unresolved as to Laetrile's true effectiveness, in large part because

[59] See, e.g., all of the anecdotal evidence from this period collected in Victor Herbert, "Laetrile: The Cult of Cyanide, Promoting Poison for Profit," *American Journal of Clinical Nutrition* 32 (1979):1121–58.

[60] 438 F. Supp. 1287, 1290 (W.D. Ok. 1977).

[61] Ibid., 1290–91, n. 5 (citations omitted).

FDA has prevented adequate testing on humans."[62] Finally, the court concluded that "it is only when a substance is openly used, and its results carefully observed and fully reported that this controversy will be resolved."[63] The court found Laetrile exempted from the operation of the "new drug" regulations under the 1962 grandfather clause because it was apparent that "it is only within the context of FDA's creation of this record that the spectre of Laetrile's toxicity has even been raised."[64] The opinion also included a gratuitous argument that the application of the Food, Drug, and Cosmetic Act to Rutherford to deny him access to Laetrile would constitute a violation of his constitutional right to privacy. That argument was unnecessary to the court's decision, and it has not been the subject of any serious judicial review.[65]

By the end of 1977 it was clear to the court that the FDA was simply not permitting any scientific closure of the Laetrile controversy. Although the court was willing to impose a judicial resolution if there could be no scientific one, the court did all within its legal authority to deflect the issue out of the courtroom and into a setting in which scientists would be able to develop and evaluate reliable data. If the medical community wanted a nonlegal resolution of the Laetrile debate, it would have one – but only if it would provide an honest and rigorous technical evaluation of the scientific questions.

The scientific inquiry begins: 1978

By January of 1978 the scientific community began to take seriously the responsibility imposed by the court in the *Rutherford* case. The January 26, 1978, issue of the *New England Journal of Medicine* maintained a debate on Laetrile that was very different from the earlier debates on how to best combat the "quackery." In "Laetrilomania – Again," Arnold Relman stated that "surely most concerned citizens who are now wondering whether the medical establishment may be stubbornly overlooking a valuable adjunct to cancer therapy ought to be satisfied if Laetrile were tested in a rigorous clinical trial and found worthless. I suspect that there are physicians as well as

[62] Ibid., 1293.
[63] Ibid., 1294.
[64] Ibid., 1298, n. 24.
[65] Eventually, after the case was remanded from the United States Supreme Court, the Tenth Circuit Court of Appeals did formally reject this argument, but with essentially no discussion or analysis. 616 F.2d 455, 457 (10th Cir. 1980), cert. denied 101 S.Ct. 336 (1980).

patients who would like to have any remaining doubt settled in this manner."[66] The "Sounding Board" debate that followed Relman's introduction was a debate on the adequacy of formal clinical testing techniques, rather than on the adequacy of public relations techniques. Guy Newell, deputy director of NCI, explained that "after much discussion, the National Cancer Institute decided that before a decision is made about requesting permission for clinical trial, we should attempt to document by a retrospective review of case records whether or not bona fide responses to Laetrile have occurred."[67] In contrast, Charles Moertel, who conducted a phase II study of Laetrile, announced that clinical studies of Laetrile ought to be begun immediately. He stated, "The simple fact is that Laetrile has never been properly studied in the hands of those competent to make such a judgment" and that "an even more compelling reason for a clinical trial is the lingering doubt, which must be harbored by any scientific mind, that perhaps the overwhelming public acceptance of this therapy could reflect some element of therapeutic effectiveness."[68] It is noteworthy that both Relman and Moertel pointed to the *Rutherford* case as a significant factor encouraging the development of a formal clinical trial of Laetrile.

By coincidence, the FDA's appeal of the *Rutherford* case to the tenth circuit court of appeals was argued on the same day that the *New England Journal of Medicine* published the Relman-Newell-Moertel discussion. By the time the court of appeals had rendered its determination in July, the major substantive issue had been successfully deflected back into the laboratories. The court of appeal's unanimous opinion affirmed the decision of the district court, but on very narrow grounds.[69] The court of appeals refused to consider either the adequacy of the administrative record or the claim that an underlying constitutional right of privacy encompassed a patient's right to use an drug of unproved efficacy. Instead the court concluded that the "safety" and "effectiveness" requirements of the Act were senseless in the case of a terminally ill patient, for whom, by definition, nothing that has been medically proven is also effective.[70] The opinion was very narrowly drawn; the principle on

[66] Arnold S. Relman, "Laetrilomania – Again," *New England Journal of Medicine* 298 (1978): 215–16

[67] Guy R. Newell, "Why the National Cancer Institute Chooses a Case-Record Review of Laetrile,"*New England Journal of Medicine* 298 (1978): 218–19.

[68] "A Trial of Laetrile Now," *New England Journal of Medicine* 218 (1978): 298.

[69] 582 F.2d 1234 (10th Cir. 1978).

[70] Ibid.

which it was based would be applicable in only a few cases. It was a significant opinion primarily because it left the scientific burden on the medical community.

By the end of 1978 the medical community was engaged in a formal scientific evaluation of the toxicity and effectiveness of Laetrile. On September 7, 1978, NCI's retrospective review of Laetrile treatments was completed, and its results – which were far more favorable to those supporting the use of Laetrile than its sponsors would probably have predicted – were published in a special report in the *New England Journal of Medicine*.[71] By the beginning of 1979 the institute had applied to the FDA for an IND to begin formal clinical testing of Laetrile. The FDA reacted to this request more positively than it had reacted to the MacNaughton request eight years previously, and a phase I (pharmacological and toxicological) study of Laetrile commenced shortly thereafter.

Judicial withdrawal in favor of scientific study (1979–1981)

By the time the United States Supreme Court made its only review of the *Rutherford* case, the issue was well on its way to being resolved in the laboratory. Apparently unwilling to become deeply involved in a mixed issue of science and policy that already was being formally resolved by scientists, the United States Supreme Court restricted its evaluation of the case to the narrow issue that constituted the basis of the court of appeal's determination.[72] The Supreme Court chose not even to mention the underlying constitutional and administrative issues in its decision. Instead, the Court unanimously found that the court of appeals had erred in concluding that the "safety" and "effectiveness" provisions of the Act did not apply to the terminally ill and remanded the case to the court of appeals for reconsideration. On reconsidering the case, the court of appeals wrote a very short opinion rejecting the legal argument proffered by *Rutherford*, without offering any analysis.[73] The Supreme Court refused to review this last determination of the *Rutherford* case by the appeals court, and thus that opinion became the final opinion in the case. There remains, however, no binding Supreme Court precedent on the underlying constitutional and administrative issues. The Supreme Court simply avoided those issues when it had to hear the case, and it avoided hearing the case when it otherwise would have had to decide those issues.

[71] Ellison et al., "Special Report on Laetrile."
[72] 99 S. Ct. 2470 (1979).
[73] 616 F. 2d 455 (19th Cir. 1980), cert. denied 101 S. Ct. 336 (1980).

Five years and seven formal opinions after the institution of the *Rutherford* case, that lawsuit had been resolved without the development of any substantial body of nationally binding precedent. On the other hand, the judiciary successfully deflected a serious controversy with both scientific and policy components into the medical laboratories that had refused to entertain the controversy before the lawsuit had commenced. Early in 1981 the Mayo Clinic reported that its phase I study of Laetrile demonstrated that "the administration of amygdalin according to the dosages and schedules that we employed seems to be free of significant side effects."[74] Four nationally respected cancer centers, with an NCI grant and an IND issued by the FDA, are now proceeding to phase II clinical studies, in the "hope that the ongoing studies will provide definitive information regarding possible anti-neoplastic activity, symptomatic benefit, or both, associated with amygdalin therapy."[75] The judiciary has now successfully completed transforming the Laetrile controversy from a legal controversy to be determined in the courts back into a medical controversy that ought to be – and will be – determined in the laboratory.

Conclusion

The study of judicial intervention into the Laetrile controversy does not reveal a legal system anxious to wrest a scientific issue from the hands of laboratory physicians and resolve it in some alternative nontechnical way. At the same time, this study does not reveal a legal system that necessarily withdraws from participation in a public policy controversy whenever technical experts define the controversy as a scientific one. Rather, the legal battle over access to Laetrile demonstrates that the courts may participate in the resolution of a dispute that includes scientific and public policy elements and that the judiciary may regard the process by which the scientists resolve the scientific part of such a controversy as being itself an issue of public policy.

The Laetrile controversy presents an interesting example of attempts at a variety of kinds of closure, because the controversy includes precisely distinguished public policy and scientific questions. First, the antagonists dispute Laetrile's medical value – its safety and efficacy in treating any form of cancer. That is essentially a scientific controversy, although the method by which it is resolved by scientists may have public policy implications. The antagonists also dispute, – at least implicitly, whether any finding as to safety and

[74] Moertel et al., "Amygdalin," p. 594.
[75] Ibid.

efficacy is relevant to the determination of whether the substance ought to be available for the treatment of terminally ill cancer patients. That is essentially a public policy controversy.

Although the "Laetrile question" includes serious scientific and policy controversies, the strongest debaters on each side originally cast their arguments only in scientific terms. Because the traditional scientific community – those who generally have access to the more respected journals and to government agencies – refused to consider seriously the unorthodox arguments advanced in favor of Laetrile therapy, much of the scientific community at first refused to recognize the Laetrile dispute as a legitimate scientific controversy at all. Those who believed that the supporters of Laetrile were unscientific medical charlatans, incompetent to understand the underlying issues, refused to admit their adversaries to the scientific debate, and, because they perceived the issue to be entirely one of science, refused to recognize the very existence of controversy. To most of the medical establishment in 1975, the proponents of Laetrile were patently wrong, and the controversy, – to the extent that one existed – had already been the subject of what Tom Beauchamp describes as "sound argument closure" – that is, consensus acceptance of clearly unassailable reasoning that requires a particular resolution (see Chap. 1, this volume).[76]

This presumed closure was necessarily ephemeral because those engaged in the debate had incorrectly evaluated the credibility of the stakeholders. Those who chose to disbelieve or ignore the conclusions accepted by the traditional medical community were themselves ignored as irrelevant by that community. But the issues were ultimately much broader than the traditional scientific ones, and the class of bona fide participants in the controversy, and thus of real stakeholders in its ultimate resolution, included those who did not participate in the traditional medical dialogue.

The agency to which primary authority for resolving controversies like the one that has developed around Laetrile is delegated is the FDA. The FDA was established by Congress to provide adequate scientific resolutions to what Congress found to be essentially scientific questions. It was established to evaluate medical data generated by trials of proposed new drugs.

The agency's fairness and neutrality in performing these evalua-

[76] Although Beauchamp applies his classification scheme only to policy and value issues, his scheme applies equally well to scientific issues. In fact, his description of "sound argument closure" describes a phenomenon much more common in scientific debate than in value controversy. I use his term here to refer to the part of the Laetrile controversy that includes only scientific questions.

tions were to be guaranteed by an established and respected formal procedure. In reaching their goal, the FDA is required by statute to evaluate scientific data to determine whether a new drug is "*generally recognized* as safe and effective."[77] The charge to the FDA is not to determine whether the drug is safe and effective but to determine whether it is generally recognized as such – that is, whether there is a consensus among traditional scientists that it is safe and effective. In essence, the charge to the FDA is to determine whether there has been "sound argument closure" of the scientific issue. Although the FDA views its drug approval process as an entirely technical endeavor, the federal agency is obliged by the public policy behind its creation as well as by the due process clause of the Fifth Amendment, to provide a fair and evenhanded process for making these technical determinations. In this sense, the charge to the FDA is to provide a forum for what Beauchamp calls "procedural closure" of the underlying issues – that is, to provide a process that will certainly and necessarily impose a resolution on the controversy, at least for some carefully defined purposes. To satisfy its legislative mandate, the FDA has established a procedure for determining whether there has been sound argument closure of essentially scientific disputes. This strange hybrid of procedural closure and sound argument closure, this bastard of science and law, was ineffective in finally imposing closure on the scientific debate surrounding Laetrile because, like the traditional scientists engaged in the debate, it failed to recognize all of the issues at stake.

The function of Rutherford's lawsuit against the FDA was not to impose "procedural closure" on the scientific debate. Like the FDA, the judiciary is simply not equipped to impose a decision that would terminate the sustained discussion that has come to characterize the controversy. The scientific aspects of the debate will have to be the subject of an internal scientific resolution, because no externally imposed solution will satisfy the antagonists. Throughout the *Rutherford* litigation, the courts recognized the futility of imposing an apparently scientific solution external to the scientific basis of the controversy. Of course, the courts could have resolved the controversy by formally addressing the pure policy issue and declaring the constitutional right of privacy to encompass the use of Laetrile, whatever its scientific merit. Such a decision, whatever its legal merit, would have caused the underlying medical issue to remain unre-

[77] Sec. 21 U.S.C. Sci. 321 (p)(1) (1976), emphasis added. For a summary of the statutory scheme applied by the FDA in evaluating new drugs, see Sec. II (a), *supra*.

solved, and it would have left all of the participants in the controversy unsatisfied. The scientific issue is sufficiently important to warrant the courts' decisions to avoid the constitutional issues as long as a scientific resolution remains a real possibility.

The courts have taken the legal and scientific issues seriously and evaluated them rigorously, and they have thus encouraged scientists with appropriate expertise to seek to resolve the scientific question underlying the Laetrile controversy. When it became clear that the FDA and NCI were undertaking studies that would be likely to lead to sound argument closure of the general issues in the Laetrile debate, the judiciary simply withdrew. The courts have never addressed the scientific aspects of the controversy – the question of the safety and effectiveness of Laetrile; they have merely addressed the policy dispute over whether there ought to be a scientific investigation of the drug. By affording both the proponents and opponents of Laetrile use a hearing, the court's action has resulted in something akin to "negotiation closure," as Beauchamp describes it, of that preliminary question. The scientific evaluation of the drug is being conducted, although in a fashion not ideal from either perspective: The proponents and opponents of Laetrile are watching the very clinical trial requested by Laetrile advocates being conducted in a forum and by researchers respected by its detractors.[78]

Although some view the eventual outcome of the *Rutherford* case as a setback for the advocates of Laertrile,[79] those who seek a lasting and scientific resolution of the problem see the judicial deflection of the issue into the laboratories as a major step forward. The consequence of legal intervention was to require scientists, unwilling or politically unable to permit a scientific inquiry, to do the research necessary to bring forward the sound arguments needed as the foundation for meaningful closure. Of course, other social policy issues remain – but they may be rendered moot by an accepted resolution of the primary issues. There will be little medical opposition to a cancer treatment that has been proved effective, and little public demand for one proved ineffective. If the scientific issue is the subject of sound

[78] In 1981 Moertel told the American Society of Clinical Oncology that the NCI study has revealed that Laetrile "is not effective." See "Study Says Laetrile Is Not Effective as Cancer Cure," *New York Times* (May 1, 1981), p. A26. The study report has not yet been published. It is too early to assess whether this study will ultimately bring closure to the national Laetrile debate. Moertel believes that it will; Laetrile advocates hope that it will not. See "Laetrile Study Called 'Set Up,'" *New York Times* (May 1, 1981), p. A26.

[79] See, e.g., Jonathan Brant and John Graceffa, "*Rutherford*, *Privitera*, and *Chad Green*: Laetrile's Setbacks in the Courts," *American Journal of Law and Medicine* 6 (1980): 156–7.

argument closure, there will simply be no scientific or political interest in discussing the unresolved policy issues – at least by those who remain stakeholders only in the Laetrile controversy.

Although the Laetrile controversy ultimately will be resolved by scientists, its resolution has been shaped and directed by legislative action – Congress's "negotiation closure" of some of the policy issues through the enactment of the Food, Drug and Cosmetic Act and the creation of the FDA – and by judicial action, that is, the court's effective imposition of procedural requirements on the laboratory process. Just as the legislature is the paradigm example of a forum established to resolve controversies through "negotiation closure" and the judiciary is the paradigm example of a forum established to resolve controversies through "procedural closure," the laboratory – and its accompanying academic activity – is the paradigm example of a forum that acts to resolve controversies through "sound argument closure." An evaluation of the Laetrile litigation, however, reveals that each of these forums may participate in the resolution of complex controversies of science and public policy in more than one way and that each may act as an appropriate check-up on the others.

16

Politics, science, and the problem of psychiatric nomenclature: a case study of the American Psychiatric Association referendum on homosexuality

RONALD BAYER

In 1973, after several years of bitter dispute, the board of trustees of the American Psychiatric Association (APA) decided to remove homosexuality from the *Diagnostic and Statistical Manual of Psychiatric Disorders*, its official list of mental diseases.[1]

Infuriated by that action, dissident psychiatrists charged the leadership of their association with an unseemly capitulation to the threats and pressures of gay liberation groups and forced the board to submit its decision to a referendum of the full APA membership. And so America's psychiatrists were called to vote upon the question of whether homosexuality ought to be considered a mental disease. The entire process, from the first confrontations organized by gay demonstrators at psychiatric conventions to the referendum demanded by orthodox psychiatrists, seemed to violate the most basic expectations about how questions of science should be resolved. Instead of being engaged in a sober consideration of data, psychiatrists were swept up in a political controversy. The APA, according to its critics, had fallen victim to the disorder of a tumultuous era in which disruptive conflicts threatened to politicize every aspect of American social life. A furious egalitarianism, which challenged every form of authority, had compelled psychiatric experts to negotiate the pathological status of homosexuality with homosexuals themselves. The result was not a conclusion based on an approximation of the scientific truth as dictated by reason but an action demanded by the ideological temper of the times.

To those who viewed the 1973 decision sympathetically, psychiatry

[1] For a fuller description of these events, see Ronald Bayer, *Homosexuality and American Psychiatry: The Politics of Diagnosis* (New York: Basic Books, 1981).

had displayed a remarkable capacity to acknowledge the significance of new research findings and to rethink its approach to sexuality. Psychiatry had not capitulated to the pressure of gay liberation groups but rather revealed an admirable flexibility. Unlike those who were unyieldingly committed to antihomosexual values rooted in the Judeo-Christian past, the leadership of the APA had demonstrated wisdom, insight, and the strength to break with conventional but scientifically unwarranted beliefs.

Both those psychiatrists who fought to preserve the status of homosexuality as a pathology and those who, in alliance with gay liberation groups, wished to remove it from the list of psychiatric disorders understood the profound significance of the battle that had been joined. Each side mobilized the full range of resources it would need to prevail, limited only by the standards of professional decorum. But despite the tactical maneuvers, both sides recognized the very deep and fundamental questions involved: What is normal sexuality? What is the role of sexuality in human existence? Do the brute requirements of species' survival compel an answer to the question of whether homosexuality is a disorder? How should social values influence psychiatry and help to define the concept of mental illness? What is the appropriate scope of a nosology of psychiatric disorders? How should conflicts over such issues be resolved? How should the opposing principles of democracy and authority be brought to bear in such matters? Each side sought to respond to these issues with intellectual rigor consistent with what it considered the standards of "science."

What factors can account for the emergence of this bitter controversy within psychiatry? What factors can account for the resolution that was achieved?

Psychiatry and the pathology of homosexuality: the emergence of a scientific orthodoxy

Homosexuality, despite periods of greater tolerance, has been considered an abomination in the West for much of the past two thousand years. The very nature of human anatomy seemed to reveal a divine plan for the morally acceptable use of the sexual organs. With life short, and human strength virtually the only source of power available for the domestication of nature, the sexual desire felt by men for women seemed a miraculous force whose intended end was procreation. Nonprocreative sexuality represented not only a violation of God's nature but a dangerous diversion of energy from the task of human survival. It is not surprising, then, that homosexuality was the target of repression. Even when the political authorities lacked the will or the power to persecute those who engaged in homosexual practices, the religious authorities condemned them with the moral fury usually reserved for religious heretics. Indeed, in time the act of buggery came to be regarded as a form of religious heresy.

In the early decades of the nineteenth century, what medical discussion of homosexuality did take place clearly bore the mark of the still powerful religious tradition. Although it was acknowledged that in some instances such behavior could be the result of insanity, in most instances it was considered freely willed and therefore a vice.

Only in the last half of the century did homosexuality become the subject of concerted scientific investigation. The early research is of interest primarily because it reflects the extent to which prevailing social values framed and conditioned the scientific effort to wrest from the religious tradition the warrant to speak on issues of sexuality. It was, however, only with the birth of psychoanalysis that a perspective emerged that was to profoundly affect the psychiatric understanding of homosexuality.

For Freud, as for most of those who undertook the scientific study of sexuality in the last years of the nineteenth century and the first years of the twentieth, there was no question but that heterosexuality represented the normal end of psychosexual development. Despite the complex and uncertain process of maturation, "One of the tasks implicit in object choice is that it should find its way to the opposite sex."[2] Here Freud saw no conflict between the demands of convention and nature's course.

In his first effort to account for what he termed sexual inversion, Freud set himself in sharp opposition to those scientists who claimed that homosexuality was an indication of degeneracy. In his *Three Essays on the Theory of Sexuality* he asserted that such a diagnosis could be justified only if homosexuals typically exhibited a number of serious deviations from normal behavior and if their capacity for survival and "efficient functioning" was severely impaired. Since Freud believed that homosexuality was found in men and women who exhibited no other deviations, whose efficiency was not impaired, and who were "indeed distinguished by specially high intellectual development and ethical culture," it made little sense to him to employ the classification "degenerate" for inverts.[3]

Freud's efforts to identify the roots of homosexuality involved a

[2] Sigmund Freud, *Three Essays on the Theory of Sexuality* ([1905]; New York: Avon Books, 1962), p. 133.

[3] Ibid., p. 25.

series of formulations that combined "constitutional" factors and environmental or "accidental" influences. He strove to find a middle ground in the debate between those who asserted that either biology or conditioning forces were exclusively responsible for a homosexual outcome. Although acknowledging in both his case histories and his theoretical work the presence of accidental determinants in many instances of homosexuality, he could not accept an exclusive reliance upon environment. The fact that not everyone subjected to similar influences became homosexual suggested an important role for biological forces. Confronted by an extraordinary richness of detail in his case studies, Freud remarked that he had uncovered a "continual mingling and blending" of what in theory "we should try to separate into a pair of opposites – namely inherited and acquired factors."[4]

Always critical of those whom he termed "therapeutic enthusiasts," Freud was especially pessimistic about the prospects for the psychoanalytic cure of homosexuality: "One must remember that normal sexuality also depends upon a restriction in the choice of object; in general to undertake to convert a fully developed homosexual into a heterosexual is not much more promising than to do the reverse, only that for good practical reasons the latter is never attempted."[5]

It was his therapeutic pessimism as well as his acknowledgment that many homosexuals, though arrested in their development, could derive pleasure from both love and work that sets Frend apart from those who were to emerge as the leading psychiatric and psychoanalytic advocates of the pathological perspective on homosexuality in the last half of the twentieth century. Thus, two of the most prominent figures in the debate on homosexuality within American psychiatry in the 1970s, Irving Bieber and Charles Socarides, were to assert that although same-sex object choice represented a symptom of profound and pervasive psychopathology, it was, with the appropriate form of therapeutic intervention, reversible in many if not a majority of cases.

Though not all psychiatrists shared the more extreme views of Bieber and Socarides, it is clear that the pathological perspective represented the dominant orthodoxy for American psychiatry into the 1960s. And so there was never any doubt about the appropriateness of including homosexuality in the first two official classifications of psychiatric disorders developed by the APA (*Diagnostic and Statis-*

[4] Sigmund Freud, "Psychogenesis of a Case of Homosexuality in a Woman," in *Sexuality and the Psychology of Love* (New York: Collier Books, 1970), p. 157.

[5] Ibid.

tical Manual, Mental Disorders published in 1952, and the revised edition, known as *DSM-II*, published in 1968).

Dissenting views

Early in the twentieth century, under the leadership of sex reformers such as Havelock Ellis and Magnus Hirschfeld, a scientific perspective emerged within which homosexuality was viewed as a normal variant of human sexuality. As the psychiatric interpretation of homosexuality achieved cultural dominance, it eclipsed that viewpoint. Only in the period following World War II was the heterodox perspective to reemerge with renewed vigor. With the pathological status of homosexuality a matter of broad professional and lay consensus, research tended to focus upon the question of etiology, with contending psychodynamic hypotheses vying for recognition. The second issue of interest to investigators concerned with homosexuality was the extent to which therapeutic intervention could be expected to restore normal heterosexual functioning. In both instances it was clinical populations that provided the data. Since it was assumed that all homosexuals suffered from a pathological condition, there was no question about the methodological soundness of relying upon patients for a more general understanding of the disorder.

Among the most important of the postwar critics of the psychiatric orthodoxy were Alfred Kinsey, whose pioneering studies of human sexual behavior challenged the assumption that homosexual activity was restricted to a small number of pathological individuals; Evelyn Hooker, a psychologist whose psychometric studies of nonpatient populations suggested that assumptions regarding the pervasive pathology of homosexuals was unsupportable; Thomas Szasz, who challenged the underlying ideological assumptions of psychiatry and who charged that the very concept of mental illness was a myth designed to advance the power-aggrandizing goals of psychiatrists; and Judd Marmor, the first leading psychoanalytic figure to question the assumptions of the pathological perspective on homosexuality.

In part, the challenge to psychiatry can be explained in terms of the impact of the relativist standpoint of cultural anthropology. More important, however, was the new wave of homosexual activism. Between the homophile movement and the critical line of research, there developed a complex, reciprocal relationship. The existence of the movement had a subtle but nonetheless crucial impact upon the social context within which such research was undertaken. The findings of the research were, in turn, vitally important to the early leaders of the homophile movement, encouraging them in their early, tentative efforts at organizational and ideological development. Finally, it was the struggle for homosexual rights that ultimately transformed this research from an interesting methodological critique of orthodox psychiatric theory and practice into a weapon in the assault on the power of psychiatry.

Gay liberation and the struggle against psychiatry

In its early phases, the modern homophile movement that emerged after World War II was marked by a defensive posture and was chiefly concerned with the dangers that beset homosexuals in their efforts to live anonymously in a society committed to the repressive use of vice squads and the law. Only gradually did those with the audacity to identify themselves as homosexuals begin to challenge the primacy of heterosexual standards. By the late 1960s the tentative thrusts of the early leaders of the movement had become a full-blown attack, with homosexuality presented as an "alternative life-style" worthy of social acceptance on a par with heterosexuality. Mere tolerance was no longer the goal; the demand was for social legitimation.

The struggle for gay liberation was influenced profoundly by the civil rights and feminist movements of the mid- and late 1960s. Like blacks, homosexuals began to see themselves as an oppressed minority injured not only by the arrangement of social institutions but also by deeply entrenched ideological standards that, in ways both subtle and blatant, denied them dignity. Like racism, antihomosexuality required both a fully developed sociocultural critique and a political assault. And homosexuals, like women, began to challenge the dominant standards of sexuality. Sexism was thus perceived as the ideological reflection of the power of male heterosexuals incapable of acknowledging the erotic desires of women or of homosexuals. Like so many other client populations, homosexuals turned on those formerly perceived as protectors, their sense of self-confidence enhanced by an awareness that they were part of an upsurge of protest directed at every social institution in America. Thus American psychiatry emerged as a primary target of their radical disenchantment.

For much of the first half of this century many homosexuals who were willing to express themselves publicly welcomed the psychiatric effort to wrest control of the social definition of their lives from moral and religious authorities. Better "sick" than criminal; better the focus of therapeutic concern than the target of the brutal law. By the late 1960s, however, homosexual activists had discarded whatever lingering gratitude remained toward their former protectors and in a mood of militancy rose up to challenge what they considered the unwarranted, burdensome, and humiliating domination of psychiatry.

The label of mental illness was experienced as painful, itself the source of profound distress. Moreover, it had become clear that the pathological perspective on homosexuality served as a primary prop for the entire system of social and legal disabilities suffered by gay men and lesbians in America. Armed with the techniques of social protest, they subjected American psychiatry to a striking series of jolts.

Picket lines began to appear at lectures given by those who defended the standard psychiatric position on homosexuality. In 1968 the convention of the American Medical Association in San Francisco was leafleted by homosexual activists who took the opportunity of a lecture by Charles Socarides to demand that those who opposed the pathological view of homosexuality be represented at future conventions. They demanded in addition that representatives of homophile groups be invited to participate in such discussions. Finally, they argued for a redirection of scientific research, with the antihomosexual bias of the psychoanalytic perspective being replaced by a value-neutral search for "facts."[6]

A similar protest took place at Columbia University's College of Physicians and Surgeons the same year. Protesting the composition of a panel on homosexuality led by Lawrence Kolb, director of the New York State Psychiatric Institute, the demonstrators demanded "to be participants in considerations of our condition and in the disposition of our fate. It is time that talk stopped being *about* us and started being with us."[7]

Although the struggle on the part of homosexuals had, by the end of the 1960s, taken on the features of a broad social movement, it remained, like other expressions of discontent in this period, radically decentralized. Local groups inspired by the ideological tone of the movement would seize, often in an ad hoc manner, whatever opportunities presented themselves to demonstrate their demands for change. Since the targets of such protest – the media, government, economic institutions, and professional meetings where unacceptable views were being expressed – were so numerous, there tended to be almost no continuity from one effort to the next.

Thus when the APA became the target of homosexual attack in

6 "The Homophile Community v. Dr. Charles Socarides," *Ladder* (September 1968): 29–30.

7 Columbia University Homophile League, "We Protest the Kolb Panel," April 23, 1968, mimeographed (on file, Institute for Sex Research, Indiana).

1970, it was because gay activists in San Francisco saw in the presence of the APA convention in their city yet one more opportunity to challenge the psychiatric profession. That decision was no different from the many others that had preceded earlier challenges to psychiatry. It was the status of the association that gave the decision its significance. With the APA designated as a target, gay groups throughout the country could direct their wrath against a common organizational foe. Furthermore, the generalized antagonism toward psychiatry as a social institution could be transformed into a focused assault upon the psychiatric profession. Most important, the outrage against the view that homosexuality was a mental illness could be translated into a demand for the deletion of homosexuality from the APA's official *Diagnostic and Statistical Manual of Psychiatric Disorders.*

The disruption of the San Francisco convention startled America's psychiatrists, destroying whatever illusions remained about the prospect of conducting the discussion about homosexuality in isolation from the wrathful attention of the gay liberation movement. Even the effort to prevent a repetition of turmoil at the 1971 convention in Washington, D.C., through acquiescence to a demand for an officially sponsored panel on homosexuality – at which the views of gay activists would be heard – failed. At the Washington meeting, the prestigious Convocation of Fellows became the target of attack. Gay leaders declared, "Psychiatry is the enemy incarnate ... You may take this as a declaration of war." Furious psychiatrists shook their fists, comparing the intruders to Nazi storm troopers.

It is within this context, marked by the existence of extraordinary tension, that the emergence of increasing dissatisfaction with the pathological perspective on homosexuality among American psychiatrists must be viewed. Judd Marmor was no longer alone in questioning the assumptions of the orthodoxy.

With psychiatrists as well as workers in the allied mental health professions beginning to doubt the merits of classifying homosexuality as a disease, and with the gay movement increasingly sophisticated in the use of tactics designed to create disorder, the stage was set by the end of 1972 for a full-scale effort to demand the amendment of *DSM-II.* What was now required was an appropriate triggering event that would set in motion the intellectual, professional, social, and political forces that had been generated during the preceding years of protest. That event occurred in October 1972 with a disruptive demonstration at a meeting of behavior therapists in New York City. It was at that session that Robert Spitzer, a Columbia University psychiatrist and a member of the APA's Committee on Nomenclature, was confronted with a demand on the part of gay leaders that they be permitted to present their case against the classification of homosexuality as a disease before his committee. He acceded to that request, and on February 8, 1973, the session was held.

Charles Silverstein, a gay psychologist, presented the argument, stressing the absence of an empirical basis for classifying homosexuality as a disorder. In contrast to the rich scientific literature, psychoanalytic theory was depicted as "subjective," "unsubstantiated," a series of "adult 'fairy' tales." Pointing both to the work of social scientists and to the actions taken by mental health groups that had already rejected the pathological view, Silverstein concluded: "I suppose what we're saying is that you must choose between the undocumented theories that have unjustly harmed a great number of people and continue to harm them and ... controlled scientific studies ... It is no sin to have made an error in the past, but surely you will mock the principles of scientific research upon which the diagnostic system is based if you turn your backs on the only objective evidence we have."[8]

Nothing impressed the members of the Committee on Nomenclature more than the sober and professional manner in which the homosexual case was presented to them. After several years of impassioned denunciations and disruptions, here, at last, was a statement that could be assimilated, analyzed, and discussed in a scientific context. Since none of the committee members was an expert on homosexuality, there was considerable interest in the data that had been presented, much of which was new to those who would have to evaluate the issues raised by the call for a revised nomenclature (telephone interview with Heinz Lehmann, June 23, 1978, and personal interview with Paul Wilson, Bethesda, Maryland, May 16, 1978, members of the committee). That the Silverstein presentation and the discussion that followed it produced such a reaction was remarkable, given the absence on the committee of any psychiatrist who had publicly expressed uncertainty about the diagnostic status of homosexuality.

Within the Nomenclature Committee itself, discussions following the presentation of the gay case in February were affected by pressures to act quickly. Though he was not chairman of the committee, Robert Spitzer, who was committed to an expeditious resolution of the controversy, zealously assumed a central role in directing its considerations, suggesting appropriate clinical and research literature to his colleagues for study. Because of his sense of

[8] "Statement by Charles Silverstein to the Nomenclature Committee of the American Psychiatric Association," February 8, 1973, mimeographed.

mission, he was, despite his unformed views, able to dominate both the pace and the direction of the committee's work. In fact, it was Spitzer's own conceptual struggle with the issue of homosexuality that framed the committee's considerations.

By mid-1973, Spitzer's views had moved quite far. The justification for including homosexuality per se among the psychiatric disorders had become increasingly inconsistent with his understanding of the appropriate focus of a nosological system (interview with author, New York, May 16, 1978). His attention had been drawn to critical analyses of standard psychoanalytic works such as Bieber's and to empirical studies indicating that homosexuals were quite capable of satisfactory adjustment to the demands of everyday life.[9] Contact with gay activists had made it clear that many homosexuals were fully satisfied with their sexual orientations. It began to seem to him that the inclusion of homosexuality in *DSM-II* constituted an unjustifiable extension of the concept of a "psychiatric disorder." Furthermore, as the issues unfolded Spitzer was forced to reconsider the foundations of the psychiatric nosology itself. In rethinking the basis for the classification of aberrant behavior, he concluded that an important distinction existed between what was suboptimal and what could appropriately be considered a psychiatric disorder. Wrestling with the implications of this conceptualization, he began to recognize how great a distance he had traveled from his own psychiatric and psychoanalytic training.

Spitzer's first attempt at formulating a position on homosexuality reflected an effort to steer a middle course between those like Marmor, who saw homosexuality as a normal variant of sexuality, and those like Bieber and Socarides, who characterized it as a psychopathology. Spitzer's strategy entailed the development of a restricted definition of psychiatric disorders that excluded homosexuality while avoiding the implication that it was no different from heterosexuality – in his view the preferred form of human sexuality. His restricted definition of mental disorders, articulated *after he had decided that homosexuality had been inappropriately classified*, entailed two elements: subjective distress and/or "some generalized impairment in social effectiveness or functioning." With the exception of homosexuality and some of the other sexual deviations, Spitzer argued, all other entries in *DSM-II* conformed to this definition of disorder.

For Spitzer there were a number of significant advantages in this

[9] Among the most important studies for Spitzer was Marcel Saghir and Eli Robins, *Male and Female Homosexuality: A Comparative Investigation* (Baltimore: Williams & Wilkins, 1973).

reformulation. Psychiatry would no longer be in the position of claiming that homosexuals who insisted on their own well-being and who were clearly able to function socially were nevertheless sick. Furthermore, removing the label of mental illness from homosexuals would eliminate a major justification for the denial of their civil rights. Finally, such a definition would protect the professional standing of clinicians who sought, through therapeutic intervention, to assist homosexuals seeking a heterosexual adjustment.

The proposed new diagnostic category thus attempted to provide a common ground for those who had been locked in combat for the preceding three years. To homosexual activists it granted the removal of homosexuality from the *Diagnostic Manual*, allowing them to claim a stunning victory. To psychoanalytically oriented psychiatrists, it stated that their own view of homosexuality as suboptimal was not being challenged but rather was not central to the restricted concept of psychiatric disorder. To those seeking an end to the pattern of disruptions that had beset psychiatric meetings, the new classification provided a formula that could remove the APA from the center of controversy. Finally, for psychiatrists concerned with the extent to which the psychiatric nosology had become a tool in the hands of government officials attempting to deprive homosexuals of their rights, the proposed shift promised to put an end to such unwanted collaboration. That all of this could take the form of a theoretical refinement rather than a political accommodation made the proposal more attractive to those willing to yield the polar positions defined in the course of conflict.

Although the Nomenclature Committee never formally voted on Spitzer's proposal to remove homosexuality per se from *DSM-II*, the proposal was passed on to the appropriate decision-making councils of the APA. At each stage it met with unanimous approval, despite the repeated calls by psychoanalytic groups to resist the pressure of the gay movement. By December of 1973, just eleven months after the first presentation before the Nomenclature Committee, the board of trustees was ready to act. On December 15 the board voted 13 to 0, with two abstentions, to delete homosexuality from *DSM-II*, replacing it with the classification "sexual orientation disturbance." "This category," the board stated, "is for individuals whose sexual interests are directed primarily towards people of the same sex and who are either disturbed by, in conflict with, or wish to change their sexual orientation. This diagnostic category is distinguished from homosexuality, which by itself does not necessarily constitute a psychiatric disorder."[10]

[10] American Psychiatric Association, press release, December 15, 1973.

In the press release that followed the December 15 decision, Alfred Freedman, the APA's president, underscored the limited scientific meaning of the vote to delete homosexuality from *DSM-II* but also emphasized its enormous social significance. Closely following Spitzer's line of reasoning about the classification of psychiatric disorders, he asserted that the board had declared neither that homosexuality was "normal" nor that it was as desirable as heterosexuality.

Across the country, newspapers headlined the APA's decision. Washington's two major dailies, the *Post* and the *Star*, reported "Doctors Rule Homosexuals Not Abnormal" and "Victory for Homosexuals." The *New York Times* noted in a front-page story: "Psychiatrists in a Shift Declare Homosexuality No Mental Illness." The gay press exultantly announced the decision in the *Advocate*, declaring, "Gays Leave Psychiatric Sick List," and "Sick No More."[11]

Whereas discussions within the APA's committees and councils had been conducted in an atmosphere of striking consensus, the board's decision provoked a response by many psychiatrists that revealed how profoundly divided American psychiatry was on the issue of homosexuality. The public reaction to the deletion of the older diagnostic category was largely, though certainly not exclusively, expressed in language compatible with the standards of professional decorum. But just beneath the surface, always fueling the intensity of the debate and sometimes breaking through in startling expletives, was a deep bitterness. Denunciations and vilifications, most often muttered in private discussions, characterized the politicized dispute among psychiatrists over their association's new position.

Stung by the significance of the ideological rebuff they had suffered, those who continued to view homosexuality as pathological perceived themselves as having been expelled from the center of psychiatric authority. The liberal, socially oriented leadership of their association had usurped the mantle of science; the APA's councils had fallen victim to a Babylonian captivity. In letters to the Washington headquarters of the association and to *Psychiatric News*, the APA's official publication, they expressed dismay and outrage.

Given the existence of such a sharp outcry against the December 15 vote, it is not surprising that the Ad Hoc Committee against the Deletion of Homosexuality from *DSM-II* attempted to mobilize the forces of dissent in an effort to reverse the decision.[12] Seizing upon by-law provisions designed to provide for a measure of democratic

[11] *Washington Post*, December 16, 1973, p. 1; *Washington Star*, December 16, 1973, p. 1; *New York Times*, December 16, 1973, p. 1; *Advocate*, January 2, 1973, p. 1, and January 16, 1973, p. 1.

[12] Other members of the committee were Lotham Gideo-Frank, Robert J. McDevitt,

professional control over the APA's corporate life, the committee circulated a petition demanding a referendum of the association's membership. Since little more than two weeks remained before such requests had to be filed, the Ad Hoc Committee was forced to move quickly. On December 16 a text was drafted and brought to the annual meeting of the American Psychoanalytic Association in New York City, where it met with an enthusiastic response. More than two hundred signatures were gathered with ease. Armed with these names, the committee sought and gained approval for a referendum to be held in conjunction with the upcoming general election of the APA's officers.

That a decision presented as being based upon the scientific examination of the standards that should apply to the classification of psychiatric disorders would be subject to ratification in a democratic vote of America's psychiatrists astonished many observers. It suggested that psychiatry's claim that it constituted a clinical science like other branches of medicine was at best a self-deception. It is thus remarkable that the same psychiatrists who had charged the APA's board with an unscientific and unseemly capitulation to political pressure now invoked the referendum procedure. When the APA's constitution had been amended to permit such votes, it was to guarantee psychiatrists a voice in the "extrascientific" policy of the association. Certainly there had never been an expectation that diagnostic matters would be opened to a vote.

In defending the decision to employ a referendum on this issue, Charles Socarides – a leader of the dissenting psychiatrists – argued that it was "a wonderfully democratic, vital tool."[13] Irving Bieber attempted to justify the Ad Hoc Committee's strategy by stating that although he was unalterably opposed to democratic decision making in matters of science, it was the board of trustees that had violated the standards of scientific inquiry by voting on the classification of homosexuality. Since that narrowly constituted group had demonstrated a disregard for scientific authority in both its procedures and its decision to remove homosexuality from *DSM-II*, it seemed only appropriate to subject the December 15 statement to a "complete vote" (interview with author, New York City, May 10, 1978). This vote would at least reflect the collective scientific wisdom of the profession, in contrast to the more political orientation of the association's leadership.

Those within the APA's decision-making bodies who had been

Burton L. Nackerson, Armand M. Nicholi, Nathaniel Ross, Vamik D. Volkan, and Harold M. Voth.

[13] Quoted in Charles Hite, "APA Task Force to Study Objections to Science Referenda," *Psychiatric News* (April 3, 1974): 15.

intimately involved in the discussions of the status of homosexuality saw in the referendum a dangerous assault on the principles that ought to govern the resolution of scientific disputes. Although he recognized the importance of eliciting the opinions of America's psychiatrists on the issue, Robert Spitzer nevertheless expressed "severe discomfort" over the recourse to a referendum on this "presumably scientific" matter.[14] John Spiegel, the APA's president-elect at the time, charged that a vote of the association's membership would make a "popularity contest" out of what had been a soberly considered question (interview with author, Waltham, Massachusetts, June 20, 1978).

Despite these objections the decision was made to permit the referendum; to do otherwise would have been politically untenable, considering the intensity of the opposition aroused by the deletion of homosexuality from *DSM-II* and the existence of an apparently unrestricted right to demand such a vote. The leadership of the APA calculated that it would be worth the risk of reversal, profound embarrassment, and violation of scientific principle to avoid provoking a fullscale revolt by a grass-roots movement that would merge the forces of those ideologically opposed to the substance of the decision on homosexuality and those who would consider rejection of the petition a high-handed infringement of their democratic professional rights.

For gay activists the possibility of a reversal after such a hard-won victory was alarming. They sought therefore to map an aggressive strategy that would enhance the political strength of their allies within the APA. The National Gay Task Force assumed the role of orchestrating the campaign, the central aspect of which involved the circulation of a letter signed by leading psychiatrists urging support for the deletion of homosexuality from *DSM-II*. The task force raised the necessary funds for the letter and was responsible for its mailing, although it sought to conceal its role. When this unusual collaborative relationship was made public, it became the source of yet one more acrimonious dispute within the APA.

The referendum was held in April 1974, and just over ten thousand psychiatrists took part in the vote. The results were a clear although not overwhelming expression of support for the nomenclature change (see Table 16.1).

With the referendum having demonstrated a profound division within APA, it is not surprising that Socarides's committee and its allies attempted once more to prevent what it considered premature

[14] Hite, "APA Task Force," p. 17.

Table 16.1 *Results of the referendum on homosexuality*

	Number	Percent
Favoring the board's decision	5,854	58
Opposing the board's decision	3,810	37
Abstaining	367	3
Invalid votes	9	
Not voting on this issue	51	
Total casting ballots	10,091	

Note: Results for "invalid votes" and for "not voting on this issue" were < 1%.
Source: American Psychiatric Association.

closure of the debate on homosexuality. The strategy adopted was to force exposure of the relationship between the National Gay Task Force and the signers of the letter supporting the board of trustees, in the hope of demonstrating that the outcome of the vote had been affected by an act of fraud. At least one effort was made to compel the APA's newspaper, *Psychiatric News*, to reveal the nature of the "collusion."[15] The matter was given broad publicity when Mike Royko, the syndicated columnist, wrote an exposé that appeared in the *Chicago Daily News* on May 17, 1974.[16] It is clear that the opponents of declassification cooperated with Royko in the preparation of this article with the intention of generating the pressure for action that they believed internal APA maneuvers could not possibly achieve.

In September 1974, a special committee was constituted by the APA to study the charges of "collusion" as well as the entire question of the referendum on homosexuality. Although this committee questioned the wisdom of the nonpublic relationship between the National Gay Task Force and the APA leadership, it rejected the charges of unethical conduct. More important, the committee denounced not only the use of a referendum to decide on the status of homosexuality but the procedures of the board of trustees itself in handling the scientific issue:

> The Ad Hoc Committee is opposed to the use of referenda to decide on scientific issues. This is the principal reason why the committee does not recommend another referendum. It also

15 Nathaniel Ross, "Letter," *Psychiatric News* (April 29, 1974).

16 Mike Royko, "Propaganda, First Class," *Chicago Daily News*, May 17, 1974, p. 3. The article also appeared in the *Philadelphia Sunday Bulletin*, May 19, 1974, sec. 4, p. 3.

> does not recommend to declare the referendum invalid. This would cause further confusion without any benefit to anyone. Referenda on facts of science make no sense . . . Scientific matters should be discussed by a broad and informed panel of experts. The Board of Trustees may or may not choose to accept the statement of experts; but neither they nor the membership in its entirety should be put into the position of deciding scientific questions by vote.[17]

Here, in the face of a wrenching political dispute, the ideology of science was invoked in its purest form as the only acceptable standard for resolving conflicts in which psychiatrists held such disparate views.

Socarides's group responded to the report with derision.[18] The committee had failed in its responsibility to call before it those charged with misconduct. It had acknowledged the accuracy of the assertion that the undisclosed role of the National Gay Task Force raised serious questions about the referendum, but in response had only proposed guidelines for the future. Last, it had endorsed an antidemocratic principle to defend the ideological authority of the dominant forces in the APA leadership. Since their perspective on homosexuality was incompatible with that of the leaders of the association, Socarides and his supporters invoked the principle of democracy in their struggle to win for themselves the warrant to speak in the name of science.

At its December 1974 meeting, the board of trustees accepted the committee's findings on the conduct of those accused of unethical behavior. It also approved the recommendations on the principle of public disclosure for all material designed to influence the membership of the association. The trustees sidestepped, however, the difficult question of the appropriate role of referenda in "scientific" and "nonscientific" disputes.[19] Having experienced the consequences of their action on homosexuality, they sought to avoid a potentially disruptive debate on this fundamental question touching on the scientific status of psychiatry.

What can account for the speed with which so many psychiatrists, the APA, and so many other allied mental health professionals had altered their thinking on homosexuality? What made the arguments of gay activists appear so credible when the work of Hooker, Kinsey, and others had been dismissed so readily for more than two decades? Those who have denounced the deletion of homosexuality from

[17] Report, "Ad Hoc Committee to Investigate the Conduct and the Referendum," Recommendation 4.

[18] Charles Socarides to APA headquarters.

[19] "Minutes of the Board of Trustees Meeting," December 13–14, 1974.

DSM-II claim that the APA was intimidated into taking its action and that despite the association's official posture, a vast majority of psychiatrists continue to view homosexuality as a pathological condition. That the APA responded to the concerted pressure of an angry, militant movement that had made full use of coercive and intimidating tactics is undeniable. To assert, however, that the decision of December 1973 represented nothing more than a capitulation in the face of force involves a great distortion. Although it is difficult to determine that precise proportion of psychiatrists who have adopted the nonpathological view, it is clear that the numbers are substantial.

Those who assumed the most central role in pressing for change within the association have tended to minimize the importance to this process of the disruptive challenge by gay activists. They have preferred to characterize their motivations and concerns as preeminently professional. At best, they have been willing to acknowledge that the homosexual attack accelerated a process of rethinking that had begun *within* psychiatry. Robert Spitzer was unusual in this regard, acknowledging quite openly the importance of gay pressure. When publicly asked by Irving Bieber whether he would consider removing fetishism and voyeurism from the psychiatric nomenclature, he responded, "I haven't given much thought to (these problems) and perhaps that is because the voyeurs and the fetishists have not yet organized themselves and forced us to do that."[20] More typical of those who had pressed for change was John Spiegel's comment that "while the agitation of the gay movement quickened our sympathetic awareness of the gay concerns, the action taken was not a response to gay demands as such. It was a scientifically based decision."[21]

For those psychiatrists who were ultimately to side with the demands of the gay liberation groups, whatever initial opposition was provoked by the indecorous use of disruptive tactics was quickly overcome. Politically more liberal than other physicians or the general population,[22] many psychiatrists understood that gay activists had merely relied upon the forms of social protest then being used by the disenfranchised throughout the United States. To support demonstrations, sit-ins, and disruptions by students, racial minorities, welfare mothers, and antiwar activists while opposing them when used against the psychiatric establishment would have been self-serving hypocrisy. But more important than this willingness to tolerate the discomfort produced by being the target of protest was

[20] *New York Times*, December 23, 1973, sec. 4, p. 5.

[21] Unpublished letter of John Spiegel to *Time*, September 15, 1975.

[22] Arnold Rogow, *The Psychiatrists* (New York: Delta Books, 1970), pp. 118–50.

the almost visceral recognition on the part of many psychiatrists that the list of gay grievances had substantive merit.

Psychiatrists responded with great concern to the charge that their diagnostic standpoint had become a major prop for social repression; that the stigmatization brought about by psychiatric classification was especially virulent; that rather than a source of melioration psychiatry had become the source of great pain and suffering. Perhaps more than any other group, homosexuals were the victims of what many academic sociologists had claimed was the inevitable consequence of "labeling" deviant behavior. In the social climate of the early 1970s, it would have been difficult for liberal psychiatrists to ignore the parallels between the discontent of racial minorities, the poor, opponents of the war in Vietnam, and the gay liberation movement, which had been linked by its leaders to the other movements of protest. With pressure mounting within psychiatry for the mobilization of its professional resources on behalf of social change, the demands of gay activists had an unusual force. The capacity of the orthodox psychiatric perspective on homosexuality to command allegiance began to wane.

Political liberalism alone cannot, however, explain the dramatic shift in outlook on the part of psychiatrists. Equally important was the growing confusion about the scope of the profession's concerns and the concomitant interest in the development of a narrower, less inclusive definition of mental illness. With the theoretical foundations for classifying homosexuality as a psychiatric disorder uncertain, it was possible for "extraprofessional" values to assume greater salience than otherwise might have been the case.

Finally, and framing this dispute, was the profound transformation that had occurred in attitudes towards sexuality during the first seven decades of the twentieth century. No longer were procreative ends hegemonic. With sexual pleasure acknowledged as an end itself, a central element in the sociocultural antagonism to homosexuality had been undercut. It was under these conditions that psychiatry, so bound to prevailing cultural values, could exercise its relative autonomy to determine the appropriateness of labeling homosexuality a disease. It was under these conditions that psychiatry, internally confused about the scope of the concept of mental disease, could be so open to the pressures of gay liberation.

Conclusion

From the outset in the nineteenth century the dispute over the pathological status of homosexuality was suffused with explicitly sociopolitical features. The precocious critics of the disease of homosexuality could not help but realize that their methodological critique of the emergent psychiatric perspective entailed a confrontation with a society that viewed homosexuality with a mixture of abhorrence and dread. Those who advanced the psychiatric view realized that they posed a challenge to the social order's belief that homosexuality constituted a profound evil rather than a disease. Under such conditions, both proponents and opponents of the pathological perspective could not have approached their subject with "scientific disinterest." Thus, there was never a period in which the dispute could be perceived solely as an asocial, apolitical conflict among medical scientists.

But that festering dispute was fundamentally transformed by the upsurge of the gay liberation movement in the United States during the late 1960s and early 1970s. Sociopolitical factors were pressed to the foreground of the discussion. Indeed, the gay agenda may be understood as having centered upon the effort to reveal the political nature of psychiatric diagnosis. It sought to politicize that which had been shrouded in the language of scientific discourse. This, in brief, has been the strategy of insurgencies against scientific orthodoxy.

Confronted with the political challenge of lesbians and gays, America's psychiatrists were compelled to translate the demands being made upon them into claims about the scientific soundness of the psychiatric orthodoxy. And so the political debate became an intraprofessional scientific encounter. Each side sought to succeed at first with sound arguments – marshaling the relevant empirical material. Ultimately the disputants were required to acknowledge that very different understandings of the concept of mental illness lay at the heart of their disagreement.

The debate within the various councils of the APA followed well-established procedures designed to resolve conflicts over issues of scientific merit. Only when the APA was forced to place the issue before a vote of its full membership did the process established for the resolution of scientific controversies falter. With that vote, however, procedural closure had been achieved, and homosexuality was removed from *DSM-II*. But it is important to underscore that that closure did not put an end to the controversy; almost 40 percent of America's psychiatrists remain committed to the pathological perspective (see Table 16.1). Indicative of the unsettled status of the argument was the controversy that flared between 1976 and 1978 over the status of homosexuality in *DSM-III*.

For the time being, the more ferocious aspects of the controversy appear to have disappeared. Those who hold to the pathological perspective quietly pursue the implications of their orientation. The victors of 1973 hold on to their dominant outlook. Thus, while there is no controversy, neither is there resolution. As long as there is no social consensus about the status of homosexuality, there is little prospect for a resolution of the conflict within psychiatry.

This dispute highlights a number of themes in the nature of controversies in science: how they arise, run their course, and how attempts to achieve closure are made. Yet because of the very special nature of psychiatry as a social institution, elements that were so crucial in this dispute may not be as salient in other conflicts in which science becomes embroiled. Therefore great caution is necessary in using the data presented here to derive conclusions about scientific controversies more generally. The preeminence of social values in the controversy about homosexuality may, in fact, indicate that when norms of human behavior become the focus of scientific concern, the patterns of dispute to be expected will differ not only in degree but in kind from those that one may expect to find when, for example, geophysical matters are at issue. The extent to which special, overtly political interests are at stake when clinical scientists struggle over the appropriateness of restoring to health those whose behavior is socially deviant may differ fundamentally from those situations in which decisions must be made regarding theories of continental drift. Since "health" is both a cultural and a biological standard, it cannot be otherwise.

17

The diagnostic status of homosexuality in *DSM-III*: a reformulation of the issues

ROBERT L. SPITZER

In 1973 the American Psychiatric Association Board of Trustees passed a resolution that removed homosexuality per se from the list of mental disorders in the second edition of the association's *Diagnostic Manual of Mental Disorders* (*DSM-II*).[1] This decision was made after an acrimonious debate within the association initiated by lobbying efforts by gay activists. In the place of the category of homosexuality was substituted a category called "sexual orientation disturbance" that was for "individuals whose sexual interests are directed primarily toward people of the same sex and who are either disturbed by, in conflict with, or wish to change their sexual orientation." In a statement supporting the board of trustees' action, I maintained that the revision was "scientifically sound and a reasonable compromise between two extreme viewpoints about homosexuality held by many members of our profession. It avoids commitment to either the notion that homosexuality is invariably a mental disorder or that it is merely a normal variant of heterosexuality." However, those who opposed the resolution and argued that homosexuality always represents psychopathology regarded the decision as a triumph of politics over science.

It is not surprising that those who had opposed the 1973 decision made no attempt to reopen the question of the classification of homosexuality as a mental disorder when, in 1974, work began on the third edition of the manual (*DSM-III*)[2] under my leadership as newly appointed chairperson of the APA Task Force on Nomenclature and

1 American Psychiatric Association, *American Psychiatric Association Diagnostic Manual of Mental Disorders*, 2d ed. (Washington, D.C.: American Psychiatric Association, 1968).

2 American Psychiatric Association, *American Psychiatric Association Diagnostic Manual of Mental Disorders*, 3d ed. (Washington, D.C.: American Psychiatric Association, 1980).

Statistics. First, these individuals may have assumed that in view of the central role that I played in the 1973 decision, their views would hardly be given a favorable hearing. Furthermore, no one wished to reenact the bloody battles of 1973. Instead, many of the same people who had argued for the elimination of homosexuality from the classification of mental disorders in 1973 now focused their attention on the developing *DSM-III* classification, which included a category that was an elaboration and refinement of the 1973 category of sexual orientation disturbance.

This *DSM-III* category underwent several changes in both name and substance. Initially baptized *homodysphilia*, in an effort to improve upon the lack of specificity in the term *sexual orientation disturbance*, the name of the category was quickly changed to *dyshomophilia* – later to *homosexual conflict disorder* – and finally to *ego-dystonic homosexuality*. In response to a barrage of criticism that it made no sense to single out homosexuals and to define the disorder on the basis of distress alone (since both homosexuals and heterosexuals can be distressed by their sexual impulses), the defining features were modified to emphasize, first, a persistent pattern of absent or weak heterosexual arousal that "significantly interferes with initiating or maintaining wanted heterosexual relationships" (*DSM-II*, p. 282). Only secondarily is there "a sustained pattern of homosexual arousal that the individual explicitly states has been unwanted and a persistent source of distress" (ibid.). Although not stated, the notion that for some homosexuals the absence of heterosexual arousal may represent psychopathology is implicit.

Many of those who championed the 1973 decision to remove homosexuality as a disorder only reluctantly went along with the DSM-II category of sexual orientation disturbance, viewing it as a politically necessary compromise that did little harm, since in its vagueness there was little or no implication that for some homosexuals the absence of heterosexual arousal might represent psychopathology. (As Frank Kameny, a gay activist, remarked to me in 1973, he had no objection to the category of sexual orientation disturbance, since any homosexual who was distressed at being homosexual was clearly "crazy" and in need of treatment by a gay counselor to get rid of societally induced homophobia.) Whereas in 1973 I was viewed by many as a radical who had joined forces with "gay lib," now, because of my defense of the *DSM-III* category I had become a reactionary betraying my own homophobia. Once again, it was charged (although now by those who regarded homosexuality as merely a normal variant) that politics had triumphed over science.

Requirements for participation in a discussion of the status of homosexuality in a classification of mental disorders

I suggest that a productive discussion of the status of homosexuality in a classification of mental disorders requires (in addition to the normal rules of etiquette – often suspended in nosological debates) the following conditions:

1. We must specify the reasons for identifying certain conditions in nature as pathological, that is, as "disorders."
2. We must define mental or physical disorder – or at least specify the key concepts or criteria (however vague or difficult to operationalize).
3. We must apply these concepts not only to homosexuality but to other conditions whose status as disorders is questionable, such as persistent lack of sexual desire, voyeurism, and fetishism, to ensure that the same standards are applied to all conditions that are considered "disorders."
4. If, after these inquiries are made, the status of homosexuality as a disorder still is not clear, given present knowledge, we must determine what additional facts are needed to clarify the situation.

Failure to comply with these requirements leads to the confusion, acrimony, and endless, hollow debate that has characterized so much of this controversy. With the wisdom of hindsight, I must say that my own small contribution to this area of debate has not been totally exemplary.

I will now try to live up to my own strictures.

Reasons for identifying certain conditions in nature as mental or physical disorders

Nature doesn't know about disorders. Over the course of time, all cultures have evolved concepts of illness or disease, in order to identify certain conditions that, because of their negative consequences, implicitly have a call to action to a special group of caretakers (in our society, the health professions, to provide treatment); to the person with the condition (to assume the sick or patient role); and to society (to provide a means for delivery of health care and in some instances to exempt the sick individual from certain responsibilities). The advantage the identifying such conditions is that it facilitates individuals with those conditions receiving care that may be helpful to them.

When the reasons for identifying certain conditions as mental or physical disorders are understood, it will be apparent that the question "Is condition A (whether it be homosexuality, schizophrenia, left-handedness, or illiteracy) a disorder?" is more precisely stated as "Is it useful to conceptualize condition A as a disorder?" or "What are the consequences (to society, the individual with the condition, and the health professions) of conceptualizing condition A as a disorder?" Usually what are at stake in discussions of this kind, are not the facts about condition A but rather the definition or concept of what a mental disorder is.

Definition of mental and physical disorder

When I first was given the job of considering the claims of the gay activists that homosexuality should not be regarded as a mental disorder, I was confronted with the absence of any generally accepted definition of mental disorder. I therefore reviewed the characteristics of the various mental disorders and concluded that, with the exception of homosexuality and perhaps some of the other "sexual deviations," they all regularly cause subjective distress or were associated with generalized impairment in social effectiveness or functioning. It became clear to me that the *consequences* of a condition, and not its *etiology*, determined whether or not the condition should be considered a disorder. Therefore it seemed irrelevant to me then (and still does now) whether or not a condition is the result of childhood conflicts and intrapsychic anxieties, since many desirable conditions that no one would suggest are disorders, such as ambition and self-discipline, may also result from conflict. (Robert Stoller even asserts that both heterosexuality and homosexuality do not arise as "a result of a benign, relatively conflictless infancy, childhood and adolescence but rather ... are compromise formations erected to protect one's capacity for erotic pleasure against noxious elements" [letter to the author, September 30, 1977]). I therefore proposed that the criterion for a mental disorder was either subjective distress or generalized impairment in social effectiveness. Using these criteria, neither homosexuality nor many conditions associated with heterosexuality, such as inhibited sexual desire, inhibited sexual excitement (frigidity), premature ejaculation, and Don Juanism, qualify as mental disorders. It seemed to me inconsistent for homosexuality (and not any of these other conditions) to be singled out as a mental disorder merely because of negative societal attitudes.

When I became involved in the development of *DSM-III*, I was forced to consider the issue of these and other sexual conditions vis-à-vis the definition of a mental disorder. I also became preoccupied with the question of what determines the boundaries of the concept of medical disorder. It became clear to me that although the definition of mental disorder proposed at the time of the controversy regarding homosexuality was suitable for almost all of the conditions generally regarded as mental disorders, a broader definition was necessary.

To make a long story short, with the help of Jean Endicott and other colleagues on the Task Force on Nomenclature and Statistics, I evolved a very complicated definition of medical and mental disorder, replete with specified criteria.[3] At the time I hoped that this complicated formulation might provide a relatively clear boundary between disorder and nondisorder and that the definition, with only slight modification, might be suitable for inclusion in *DSM-III*. However, my colleagues on the task force were not as impressed, and regarded the effort as a more benign manifestation of my compulsive personality disorder. They finally convinced me that no precise definition of disorder (physical or mental) was possible or even useful. However, after endless rumination and discussion with various colleagues, certain key concepts do appear useful and were helpful in providing a rationale for decisions as to which conditions should be included or excluded from the *DSM-III* classification of mental disorders and as guides in defining the boundaries of the various mental disorders. These concepts were eventually incorporated in a short statement that is included in the introduction to *DSM-III:*

> In DSM-III each of the mental disorders is conceptualized as a clinically significant behavioral or psychological syndrome or pattern that occurs in an individual and that is typically associated with either a painful symptom (distress) or impairment in one or more important areas of functioning (disability). In addition, there is an inference that there is a behavioral, psychologic or biological dysfunction, and that the disturbance is not only in the relationship between the individual and society. (When the disturbance is *limited* to a conflict between an individual and society, this may represent social deviance, which may or may not be commendable, but is not by itself a mental disorder.) (p. 6)

It should be noted that this definition makes no attempt to provide precise definitions for several of the key concepts referred to in

[3] Robert Spitzer and J. Endicott, "Medical and Mental Disorder: Proposed Definition and Criteria," in *Critical Issues in Psychiatric Diagnosis*, R. L. Spitzer and D.F. Klein, eds. (New York: Raven Press, 1978), pp. 15–39.

phrases such as "*impairment* in one or more *important* areas of functioning" and "behavioral, psychologic or biologic *dysfunction*."

Application of these concepts to various conditions

Let us start with the easy ones first. There is no difficulty in demonstrating that the major mental disorders such as schizophrenia, organic brain syndromes, anxiety disorders, and personality disorders meet the criteria. They all involve either distress or obvious impairment in important areas of functioning.

What about such conditions as pedophilia, fetishism, voyeurism, and masochism, which in *DSM-III* are grouped together in the class of paraphilias? The essential feature of disorders in this group is that unusual or bizarre imagery or acts are necessary for sexual excitement. As is well known, frequently people with these conditions do not claim distress, or they claim that any distress that they have is because of societal sanctions. After agonizing over whether or not these conditions should be considered disorders only if the person was distressed by the symptom (as many expected, given the 1973 decision), we decided that even in those cases where there was no distress, the behavior represented *impairment in an important area of functioning*. That is, the necessity for sexual arousal of the unusual or bizarre imagery or acts was regarded as impairment in the important area of *sexual* functioning that justified the inference of a behavioral or psychological dysfunction.

What about the psychosexual dysfunctions, such as impotence, absent sexual desire, and inhibited female or male orgasm? Again, there is the assumption that sexuality is an important area of functioning, and there is the normative notion of the "complete sexual response cycle" such that the absence of any of the component phases justifies the inference of a dysfunction, whether or not the individual is distressed by the symptom. However, in the case of the diagnosis of inhibited female orgasm, *DSM-III* acknowledges lack of certainty in the boundary between "normal variation" and "pathological inhibition" by stating that

> Some women are able to experience orgasm during noncoital clitoral stimulation, but are unable to experience it during coitus in the absence of manual clitoral stimulation. There is evidence to suggest that in some instances this represents a pathological inhibition that justifies this diagnosis whereas in other instances it represents a normal variation of the female sexual response. (p. 279)

Let us now apply these concepts of a mental disorder to homosexuality. The reader should realize that for the remainder of this paper the term *homosexuality* refers not only to a persistent pattern of homosexual arousal but also to a persistent pattern of absent or weak heterosexual arousal. This is sometimes referred to as "exclusive homosexuality."

It is now clear that homosexuality often is not associated with distress. Furthermore, many claim that if there were no societal discrimination against homosexuality and no societal expectation of heterosexuality, homosexuals would never be distressed by their sexual orientation. (Although this argument seems reasonable, others argue that regardless of society, in at least some homosexuals there would always be a sense of inadequacy at the inability to function heterosexually.) In any case, using the criterion of distress, there is no basis for considering homosexuality per se a mental disorder.

Does homosexuality represent "impairment in one or more important areas of functioning?" The gay activists and their supporters argue that the answer is clearly no, since there is no impairment in sexuality. They refuse to accept heterosexual functioning as the norm. Furthermore, if pressed, they argue that if some homosexuals cannot function heterosexually, far more heterosexuals cannot function homosexually, yet no one in a position of authority suggests classifying exclusive heterosexuality as psychopathology.

In my judgment the question of whether or not heterosexual functioning should be used as the norm – so that inability to function heterosexually is impairment in a major area of functioning – is a value judgment and not a factual matter. It should be understood that there is always a value judgment in deciding that a particular area of functioning is "important." The only difference is that no one would argue that the ability to function socially and occupationally is unimportant, and, as already noted, it is also possible to have a consensus (at least among mental health professionals) that an important area of functioning is sexuality, justifying the categories listed under "Psychosexual Dysfunction." However, it is not possible at the present time for mental health professionals to agree on whether or not heterosexual functioning represents an "important" area of functioning.

What guidelines might be available for determining what areas of functioning are "important"? I suggest that the concept of *inherent disadvantage*, relative to other individuals, that results from not being able to function in a particular area, is useful. For example, an individual who is not able to test reality because of delusions or hallucinations, or an individual who is not able to function occupationally because of depression clearly is at an inherent disadvantage in efforts to satisfy basic biological and psychological needs.

Although homosexuality in our culture is a disadvantage because of our society's interference with the pursuit by homosexuals of basic needs, it is not at all clear whether or not homosexuality is a disadvantage in all cultures or subcultures.[4] That is, it is not at all clear whether or not homosexuality entails inherent disadvantage.

Various arguments have been presented on both sides of the issue of the pathological status of homosexuality that attempt to present the issue as a factual matter rather than one involving a value judgment. For example, I. Bieber and D. Klein refer to the evolutionary significance of genetically controlled built-in mechanisms for ensuring heterosexuality.[5] They assume that if these mechanisms do not function, this is evidence of pathology. Although I believe that they are correct in postulating the existence of these mechanisms, I do not necessarily assume that failure of these mechanisms to function indicates pathology. For example, if it could be shown that there were a built-in genetic mechanism for expressing aggression (perhaps useful only a hundred thousand years ago) that does not function now in most individuals because of society's need to control aggression, no one would argue that this inhibition of aggression represented psychopathology.

It has also been argued, for example by John Meyer, that homosexuality clearly represents psychopathology because it interferes with the ability to reproduce (letter to the author, October 11, 1977). Again, there is an implicit value judgment that the ability to reproduce is desirable and is an important area of functioning.

Richard Green, who champions the view that homosexuality is merely a normal sexual variant, gets himself into trouble in dealing with the argument that homosexuality "runs counter to an evolution bent on survival of the species and therefore is contrary to the ultimate design of nature." He asserts that "the very fact that the species homo sapiens has survived, in spite of the continued existence of homosexual individuals over the eons, disproves this tenet, as it shows the compatible coexistence of heterosexuality and homosexuality enabling a species to survive."[6] This is analogous to arguing that the coexistence of a disease such as cancer, and health in the

4 See C. S. Ford and F. A. Beach, *Patterns of Sexual Behavior* (New York: Harper & Row, 1970), and W. Davenport, "Sexual Patterns and Their Regulation in a Society of the Southwest Pacific," in *Sex and Behavior*, F. A. Beach, ed. (New York: Wiley, 1965), pp. 164–207.

5 Irving Bieber, "On Arriving at the APA Decision," in *Homosexuality: A Case Study in Scientific Controversies* (New York: Plenum, 1982); D. Klein, "A Proposed Definition of Mental Illness," in *Critical Issues in Psychiatric Diagnosis*, R. L. Spitzer and D. F. Klein, eds. (New York: Raven Press, 1978), pp. 41–73.

6 R. Green, "Psychiatry, Homosexuality and Mental Illness: Non-closure in a Non-scientific Discipline," in *Homosexuality*.

species as a whole over the eons, indicates that the disease in question has no negative consequences for reproduction and survival of the species. A more convincing defense of homosexuality as a normal variant would be to argue that reproduction and survival value (apparently negative for homosexuality) should not be a consideration in the definition of mental disorder.

Additional facts that would clarify the status of homosexuality as a mental disorder

If, as I believe, there is some reason to be unsure whether or not homosexuality should be regarded as a mental disorder, since the fundamental issue is whether or not homosexuality itself results in an inherent disadvantage, are there facts about the condition that would help resolve the issue?

An important factual matter that used to be regarded as relevant was the question of whether or not homosexuality was always associated with "other" signs of psychopathology. I accept the evidence from recent studies that indicates that the range of psychopathology seen in homosexuals differs little if any from that seen in heterosexuals.[7] If future studies, perhaps with more sensitive measures, reveal significant differences in associated psychopathology (such as difficulties in interpersonal relationships), then that might weigh in favor of classifying homosexuality as a mental disorder, although an alternate interpretation has always been that any associated psychopathology is merely a reaction to societal oppression.

Often in discussions of this kind a hope is expressed that some biological "abnormality" in homosexuals will be discovered, such as an endocrine or genetic disturbance, that will resolve the issue once and for all. I fail to see how this would answer the question any more than would knowledge of the biological cause or antecedents of left-handedness (which surely must have a cause) would itself indicate whether that condition should be regarded as a normal variant or pathology. Similarly, hard evidence (there already is some evidence) that certain kinds of family relationships predispose to the development of homosexuality, or any other evidence regarding etiological factors, will not resolve the issue.

Ordinarily, the availability of a procedure (treatment) for modifying

[7] A. P. Bell and M. S. Weinberg, *Homosexualities: A Study of Diversity among Men and Women* (New York: Simon & Schuster, 1978); M. Siegelman, "Adjustment of Homosexual and Heterosexual Women," *British Journal of Psychiatry* 120 (1972): 477–81, and "Adjustment of Male Homosexuals and Heterosexuals," *Archives of Sexual Behavior* 2 (1972): 9–25.

a condition does not bear on a judgment as to whether or not the condition itself is pathological. Thus, we judge tuberculosis or schizophrenia to be disorders whether or not we have effective treatments for these conditions, since there is no ambiguity as to whether or not such conditions result in distress, disability, and disadvantage. On the other hand, it is clear that a large part of the motivation of gay activists to declassify homosexuality as a disorder was a result of their conviction (misguided or not) that there was no effective treatment and therefore no possible benefit to be derived from adopting the patient role.

If there were a "treatment" for homosexuality (I use quotation marks because the term presupposes pathology) that was available and effective in most cases, I very much doubt that there would be much objection to classifying it as a disorder. (Let me hasten to add that I would make the same prediction regarding left-handedness, although a far stronger case can be made for the inherent disadvantage of homosexuality than can be made for left-handedness.)

Critique of the *DSM-III* diagnosis of ego-dystonic homosexuality

The *DSM-III* diagnosis of ego-dystonic homosexuality has been criticized, and even mocked as an expression of irrational prejudice and latent homophobia. Let us examine the arguments.

The absence of a comparable category for heterosexuals

It has been asserted that if there is a need for a separate category for homosexuals distressed with their sexual orientation, there should be a comparable category for heterosexuals. There are several problems with this argument. Certainly heterosexuals are frequently distressed by some aspects of their sexuality, such as impulses to engage in "perverse" acts, but there is not a single case in the scientific literature that describes an individual with a sustained pattern of heterosexual arousal who was distressed by being heterosexually aroused and wished instead to acquire homosexual arousal in order to initiate or maintain homosexual relationships. (Note that this wording is taken from the essential features of ego-dystonic homosexuality, with "heterosexual" and "homosexual" interchanged.) It is true that one hears of people, primarily women, who as part of a commitment to feminist ideology have expressed a wish to expand their sexual responsiveness to include women. In such cases, the wish for acquiring homosexual arousal is clearly best understood in ideological terms, and the developmental origins and course of this "condition" bear little resemblance in any meaningful way to ego-dystonic homosexuality. (The reader will be interested to know that early in the deliberations of the *DSM-III* Advisory Committee on Psychosexual Disorders, a serious proposal was made for a category that could be applied to "patients" who sought help in overcoming inhibitions that prevented them from enjoying bisexuality. It was asserted that there actually were such individuals, from the western parts of this country, who actually sought professional help with this problem. However, it was pointed out that a distinction had to be made between inability to perform sexually in a normative manner, such as heterosexually, and inability to perform in a deviant or extraordinary manner, such as homosexually or bisexually. Society is prepared to assign the patient role to individuals with inability to perform normatively, but certainly not to individuals interested in functioning in nonnormative ways.)

The underlying premise of the argument that there should be a comparable category for heterosexuals is the assumption that homosexuality and heterosexuality are essentially comparable conditions, differing only in prevalence. Thus, what is the difference between a homosexually oriented person unable to function heterosexually (a potential candidate for the category of ego-dystonic homosexuality) and a heterosexually oriented person unable to function homosexually (a description that applies to the vast majority of heterosexuals)? Although I must admit that I used this argument myself in 1973, I now see that with the concept of inherent disadvantage, the argument falls apart. The inability of most heterosexuals to function homosexually puts them at no disadvantage except in extraordinary situations, such as when they are in jail. On the other hand, inability to function heterosexually has the built-in consequence of preventing, or at least interfering with, the ability to procreate – a matter that some (including some patients with ego-dystonic homosexuality) judge to be not inconsequential.

The suitability of already existing categories

One of the major arguments of those opposed to the inclusion of ego-dystonic homosexuality in *DSM-III* has been that there is no need for the category, since already existing categories can subsume the clinical phenomena. For example, if a homosexual is depressed that he or she is unable to function heterosexually, why not make a diagnosis of depression? If a homosexual is distressed by the compulsive quality of his or her sexual impulses, why not diagnose obsessive-compulsive disorder?

This argument makes sense only if one believes that in no case is the absence of the heterosexual arousal pattern and the presence of the homosexual arousal pattern the pathology. It is significant that in all of the critiques of the need for this category, there has been no discussion of the significance of the absent heterosexual arousal pattern. Clearly, this is because the lack of a heterosexual arousal pattern is not viewed by these critics as ever indicative of pathology, as long as there is a homosexual arousal pattern. This gets us right back to the clincher: the value judgment that an important area of functioning is heterosexuality, and not just sexuality.

DSM-III takes the position that the mental health profession cannot now arrive at a consensus as to the positive value of heterosexuality. However, if an individual patient asserts that his or her problem is absent or inadequate heterosexuality and a sustained homosexual arousal pattern, the legitimacy of the focus of the complaint – the absent heterosexual functioning – and not merely the associated distress, is accepted.

It would seem that the critics of the category of ego-dystonic homosexuality are united in believing that when someone complains of an absent heterosexual arousal pattern and the presence of a homosexual arousal pattern, the only appropriate therapeutic activity is to help the person become more comfortable with his or her homosexuality. The ethics of this approach, in my judgment, are highly questionable. By having this category, *DSM-III* acknowledges that in at least some cases an appropriate therapeutic activity is to help the individual develop a normative sexual arousal pattern.

How should decisions regarding the classification of mental disorders be made?

Diagnoses are man-made, and ultimately someone has to decide what conditions should or should not be included in a classification of mental disorders. The issues, then, are who should decide and what procedures should be followed in making such decisions.

In the case of the category that eventually became known in *DSM-III* as ego-dystonic homosexuality, the topic was initially debated at great length within the Advisory Committee on Psychosexual Disorders. Several of the members opposed its inclusion and instead proposed a category of "amorous relationship disorder" that could be used for both heterosexuals and homosexuals in some way distressed by their inability to sustain desired relationships. (Eventually the Task Force on Nomenclature and Statistics concluded that such a category lacked specificity.) As a result of criticisms of initial drafts of the category, the category of ego-dystonic homosexuality was extensively revised numerous times. Eventually it became clear that a consensus could not be reached by the advisory committee even though a majority favored the inclusion of the ego-dystonic homosexuality category. For this reason, and because one member of the advisory committee threatened to take the issue to other components of the APA, the matter went to the task force, whose job it was to oversee the development of *DSM-III* and to ensure consistency in the application of basic concepts throughout the classification.

Although Judd Marmor, a past president of the APA and an initial critic of the category, argued against the inclusion of the category at a meeting of the Council on Research and Development (the component of the APA to which the Task Force on Nomenclature and Statistics reported), that council declined to become involved in the issue. Instead the council insisted that the issue be decided by the Task Force on Nomenclature.

In my judgment, a familiarity with the over 180 pages of memoranda and letters concerning this category leads to the following conclusions:[8]

1. A serious effort was made to consider all views on this matter.
2. The issue was debated with consideration given to applying uniform standards for a condition to be regarded as a mental disorder, taking into account the potential advantages and disadvantages to the patient, to society, and to the mental health profession, of having the category included in the classification.
3. In response to criticisms of initial drafts of the category, basic concepts were modified.
4. A decision was eventually made by the task force when a large majority sustained the majority decision of the Advisory Committee on Psychosexual Disorders.
5. The decision-making process was entirely left to psychiatrists with presumed expertise in the relevant areas (sexuality, in the case of the Advisory Committee on Psychosexual Disorders, and nosology in general in the case of the Task Force on Nomenclature and Statistics).

A comparison between the decision-making process regarding the inclusion of ego-dystonic homosexuality in *DSM-III* and the removal of the category of homosexuality from *DSM-II* in 1973 indicates that progress has been made toward finding appropriate ways of resolving difficult diagnostic issues.

[8] Ronald Bayer and Robert Spitzer, "Annotated Selected Correspondence on the Diagnostic Status of Homosexuality in DSM-III," in *Homosexuality*.

A bit of whimsy: Can the diagnosis of sexual disorder be value free?

Imagine the following dialogue between a nosologically sophisticated patient (P) and his doctor (D):

D: Let me summarize what I think is your problem. You have been very depressed and upset ever since you lost your job. We call that an "adjustment disorder with depressed mood."

P: I certainly agree with you there.

D: In addition, you have told me that ever since you were an adolescent, the only way you can be sexually aroused is by women's shoes. We call that "fetishism."

P: Wait a minute. That doesn't bother me.

D: Well, as a matter of fact, whether you are bothered by it or not, according to the new diagnostic and statistical manual of the American Psychiatric Association, *DSM-III*, you have a mental disorder.

P: Wait a minute! I have done some reading in this area, and I find this very confusing. Why is my preference for women's shoes a mental disorder?

D: You must admit that your behavior is extremely atypical. Hardly anyone needs women's shoes to be turned on.

P: Certainly you are not using a statistical concept of normality; otherwise, you would have to classify genius as pathology.

D: You have a point there. However, from the evolutionary point of view, if everyone had your condition the human race would die out!

P: But I understand that your APA has decided that homosexuality itself is not a mental disorder, so certainly you cannot use that argument. And if everyone were a psychiatrist, we would also be in big trouble.

D: Well, maybe it comes down to our view that certain critical life experiences and inner conflicts explain your behavior.

P: I am sure that is correct, but I understand that all behavior, including your sexual preferences, whatever they are, are also determined by your life experiences and the way you have resolved your inner conflicts. I am beginning to think that there is something about my preference itself that your profession doesn't care for.

D: Well, I think you may have hit on something there. We do believe that optimal sexual functioning involves two human beings (at least), and not exclusively or preferentially inanimate objects.

P: Why do you believe that?

D: I guess we believe that if you are unable to be sexually aroused by another human being, that you are at a disadvantage.

P: Why is it a disadvantage? Shoes are easy to get.

D: I guess that deep in our bones we must believe that sex is more fulfilling when it is between human beings.

P: That doesn't sound like a scientific fact but only your value judgment . . .

To sum up the argument of this chapter: the concept of "disorder" always involves a value judgment. In the case of homosexuality, it is not possible to reach a consensus at the present time as to whether or not heterosexual functioning is an important area of functioning such that inability to function heterosexually because of an exclusive homosexual orientation should be conceptualized as a mental disorder. Given this state of uncertainty, it seems reasonable to adopt a middle position, as does *DSM-III,* that in some cases homosexuality is usefully conceptualized as a mental disorder.

18

On arriving at the American Psychiatric Association decision on homosexuality

IRVING BIEBER

The deletion of the term *homosexuality* from the American Psychiatric Association's revised diagnostic and statistical manual was not simply arrived at after carefully considered judgment by a group of psychiatrists. It was the climax of a sociopolitical struggle involving what were deemed to be the rights of homosexuals.

It is my aim here to separate out the psychiatric and conceptual issues from the sociopolitical issues, to document my own theoretical and clinical position, and to describe the events that I participated in and observed – all of which I trust will bring into focus the elements that went into the American Psychiatric Association's decision of 1974.

The complexity of homosexuality as a category of human adaptation has stimulated argument and controversy among lawmakers, the clergy, and behavioral, social, and biological scientists. Is it a sin, a crime, a deviation? Is it a dislocation of sexual development or an illness? Is it a constitutional disorder, a genetic misprint, a habit? The question of whether homosexuality is or is not an illness played an important role in the APA decision. Coming from another direction was the influence of the gay activist groups who believed that prejudice against homosexuals could be extinguished only if, as homosexuals, they are accepted as normal. They claimed that homosexuality is a preference, an orientation, a propensity; that it is neither a defect, a disturbance, a sickness, nor a malfunction of any sort. Therefore, homophile leaders and their followers consistently impugned the motives and ridiculed the work of those psychiatrists who asserted that homosexuality is other than normal. The long-term research that has engaged my attention for many years has demonstrated that homosexuality is other than a normal sexual adaptation.

In 1962, the research team that I had led, consisting of nine practicing psychoanalysts and two psychoanalytically trained psychologists, published the findings of a nine-year study of male homosexuals.[1] The team psychiatrists and 77 respondents to a 500-item questionnaire were members of the Society of Medical Psychoanalysts, whose roster consisted of faculty and graduates of the Psychoanalytic Division of the Department of Psychiatry of New York Medical College. The research sample consisted of 106 male homosexuals and a comparison group of 100 male heterosexuals, all in psychoanalytic treatment with members of the society. The data obtained were analyzed statistically in consultation with statistical experts,[2] and the clinical implications were carefully analyzed and evaluated.

When the volume first appeared, critics questioned the methodology on two major points: First, how reliable were data obtained from analysts, rather than directly from the patients, whose information concerned not only themselves but their families, whom the analysts had never seen? Second, could the findings obtained from a white, middle- and upper-class population be generalized to the homosexual population at large? I am now in a position to address those issues.

Research findings

In the many years since our volume was published, I have interviewed more than 1,000 male homosexuals in psychoanalytically focused psychiatric interviews. I have also examined about 75 pairs of parents of patients. In all regards, the data were in accord with the findings reported in our volume. Most subjects of this large sample were interviewed at a city hospital, came from a lower socioeconomic strata, and belonged to one of three ethnic groups: black, Puerto Rican, or white, distributed about evenly.

In 1960, Westwood had published a study in England of 127 working-class male homosexuals, only 5 percent of whom had ever been in psychiatric treatment.[3] In those areas where our study tapped similar items, the findings were similar. A study of nonpatient homosexuals by Snortum and his coworkers, and another by Evans, each using our questionnaire, reported findings similar to ours, although there were differences in interpretation.[4] The consistency of

[1] I. Bieber, H. Dain, P. Dince, *Homosexuality: A Psychoanalytic Study of Male Homosexuals* (New York: Basic Books, 1962).

[2] Richard Christie, Professor of Social Psychology, Columbia University; Robert Cohen, Professor of Social Psychology, New York University; Robert Lee, research psychologist, IBM.

[3] G. Westwood, *A Minority: A Report on the Life of the Male Homosexual in Great Britain* (London: Longman Group, 1960).

[4] J. R. Snortum, "Family Dynamics and Homosexuality," *Psychological Reports* 24

findings on nonpatients coming from lower social classes supports the appropriateness of generalizing from a patient sample to the homosexual population at large.

Our study contained questions that tapped the following areas: interparental relationship; mother-son relationship; father-son relationship; siblings; the triangular family system; developmental aspects of the prehomosexual child; homosexuality in adolescence; sexual adaptation of the male homosexual; latent homosexuality; and the results of treatment. Space will permit only a brief review restricted to the relationship of mother and son, father and son, and aspects of the development and socialization of the child. Statistics describing the homosexual and heterosexual sample will be given to emphasize the differences between them, bearing in mind that psychological problems brought each group into therapy.

The items listed in Table 18.1 significantly differentiated the homosexual from the heterosexual sample at levels of confidence varying from .05 to .001. For about 80 percent of the mothers in the homosexual sample, a picture emerged of a woman who was overly close to this particular son, spent a great deal of time with him, and preferred him to his siblings. More often than not, she openly preferred him to his father. The son became her confidant, and some mothers even confided to him the details of their sexual life with the father or lovers. Such a mother carried on a nongenital yet sexually oriented, romantic relationship with a son who replaced the father in this role. Not only was there an exaggerated concern about the son's health and possible injury, but other salient findings revealed that these mothers interfered with the son's assertiveness, and they tended to dislocate his relationship with the father, with siblings, and with peers. We called this type of mother "close-binding-intimate" (CBI). Most mothers of homosexuals whom, at a later time, we were able to interview also conformed to this profile. But not every mother of a homosexual had these characteristics. In the study, six were found to be detached and uninterested; in a few cases, the sons had been boarded out to institutions and foster homes. We found that homosexuality can develop without the frequently occurring CBI mother-son bond.

The items listed in Table 18.2 distinguished the fathers of the homosexual sample from the fathers of the heterosexuals. The father-son relationship was almost the diametrical opposite of that between mother and son. The paternal portrait was one of a father who was either detached or covertly or overtly hostile. Detachment of

(1969): 763 – 70; R. B. Evans, "Childhood Parental Relations of Homosexual Men," *Journal of Consulting and Clinical Psychology* 33 (1969):129–35.

Table 18.1 *The mother–son relationship*

Questionnaire findings	Level of confidence
The patient was mother's favorite.	.050
The mother demanded to be the center of patient's attention.	.001
Mother was dominating.	.050
Mother was seductive.	.050
She spent a great deal of time with patient.	.001
She did not encourage masculine attitudes and activities.	.001
She discouraged masculine attitudes and activities.	.010
She encouraged feminine attitudes and activities.	.001
She was puritanical.	.050
She was considered to be sexually frigid by her son.	.001
She allied with her son against her husband.	.010
She openly preferred son to husband.	.010
She was more intimate with the patient than with his siblings.	.050
She interfered with heterosexual activities.	.010
The patient was the mother's confidant.	.050
The mother was the patient's confidant.	.050
The mother was unduly concerned with protecting the patient from physical injury.	.050
The mother's concern about health and injury caused her to interfere with or restrict his play, socializing, and other activities.	.001
The patient considered his mother to be overprotective.	.010
The patient was excessively dependent on his mother for advice in making decisions.	.010
The mother babied the patient.	.010
She administered frequent enemas.	.050
The patient could cope with the mother more easily than with the father.	.010

a parent from a child is in itself an oblique manifestation of hostility, and perceptive children easily catch subtle attitiudes toward them. But even where hostility may not be present, children are apt to interpret the detachment as a negative attitude of some sort. Although we found that most mothers were CBI, there was also, as pointed out, variance in the mother-son pattern. Some mothers were not remarkable; some were detached; a few were hostile. The father–son relationship, however, revealed uniformly an absence of loving, warm, constructive paternal attitudes and behavior. In my

Table 18.2 *The father–son relationship*

Questionnaire findings	Level of confidence
The patient was not his father's favorite.	.001
Another sibling was the father's favorite.	.010
The patient was the child that the father least favored.	.010
The patient did not feel accepted by father.	.010
The father spent very little time with patient or was absent.	.001
The father did not encourage masculine activities.	.050
The patient consciously hated his father.	.001
He hated and feared his father.	.001
He did not respect his father.	.010
He did not accept his father.	.001
The father did not express affection to the patient.	.010
The father had less respect for the patient than for other sons.	.050
The patient did not side with his father in parental arguments.	.010
He did not cope with his father more easily than with his mother.	.010
He feared that his assertiveness would anger his father.	.010
He feared that his father would hurt him.	.050
He felt his father did not consider his needs.	.010
He did not feel currently respected by his father.	.010
He did not admire his father.	.001

long experience, I have not found a single case where, in the developing years, a father had a kind, affectionate, and constructive relationship with the son who becomes homosexual. This has been an unvarying finding. It is my view, and I have so stated and written, that if a father has a kind, affectionate, and constructive relationship with his son, he will not produce a homosexual son, no matter what the mother is like. It turns out, contrary to popular thought and some psychoanalytic theories, that the crucial and determining relationship in the evolution of male homosexuality is usually not with the mother but instead with the father. When a homosexual is questioned about his childhood with his father, frequent answers are of this type: "He was not there"; "I don't remember"; "He played no part"; "He was not interested"; "He was hostile, mean"; "He hit me"; "I hated him"; and so on. The son leaves his childhood years with, on the one hand, a profound fear and conscious or unconscious hatred of his father, and on the other a deep yearning for his acceptance and affection.

These elements of the father–son relationship dominate the psychopathology of the adult male homosexual.

Table 18.3 lists the developmental aspects for which we found statistically significant differences between the heterosexuals and the homosexuals during their childhood years. We noted a continuity of traumatic experiences with males, starting with the father. Brothers were also usually feared and hated, and the prehomosexual child had difficulties with same-sex peer groups, until adolescence. The consistent history of unremitting fear of and hostility to other males throughout childhood has led me to conclude that male homosexuality is basically an adaptation to a disorder of a man's relationship with other men. In childhood and beyond, affection and trust are felt toward the mother, perhaps women relatives, sometimes a sister, and not infrequently girl friends. At adolescence there tends to be a change, and social life improves. With the opening of this era there is an abatement of the combative play of preadolescent boys, making it easier for the homosexual teenager to come into contact with peer mates.

The psychoanalyses of adult homosexuals reveal that they perceive other men in two ways: as aggressors who are feared and identified with the father, successful brothers, and combative peers; and as nonaggressors and as homosexuals like themselves who are not feared. This division enables them to come into a relationship with men, men who need not be defended against. Relationships with women in adult life are more trustful and positive, as long as the possibility of sex and romance is excluded. The fear is that a heterosexual attempt will elicit a dangerous, even lethal attack by combative men. A frequently reported dream tells the story. The opening segment depicts the patient with an attractive young woman. Then there is a threat of attack or an actual attack by an aggressive male. In the third sequence, heterosexuality is abandoned, and the dreamer is involved in substitutive homosexual activity.

The cues that "turn on" a homosexual illuminate the operant psychopathology. Elsewhere I have delineated three categories,[5] but undoubtedly there are more. The first relates to the displacement of a heterosexual stimulus to a homosexual object; that is, a particular feature of a woman significant in the patient's life is perceived in a man. It becomes an arousing stimulus, although there is no awareness of it. One patient became aware that he was attracted to men who had eyebrows like his mother's. Actually he was reacting to a

[5] I. Bieber and T. Bieber, "Male Homosexuality," *Canadian Journal of Psychiatry* 24 (1979):409–22.

Table 18.3. *Developmental aspects*

Questionnaire findings	Level of confidence
The patient was excessively fearful of physical injury.	.001
He avoided physical fights.	.001
Play activity before puberty was predominantly with girls.	.001
He was a lone wolf in childhood.	.001
He did not participate in competitive games.	.001
He did not play baseball.	.001
He was a clinging child.	.010
He was reluctant to start school.	.010

heterosexual stimulus, a feature associated with a woman, his mother, but the incestuous aspect of the arousal had to be disguised by displacement to a male. Other physical features also serve as arousal cues, particularly eyes and skin texture. Some homosexuals are attracted only to men with very smooth skin or who are hairless, or men whose fat distribution suggests femininity or whose attitudes and behavior may suggest femininity. Thus, feminine traits are not infrequently sought for their ability to stimulate sexual arousal, although such motivations may be completely unconscious.

The second category includes the eroticizing of the feared stimulus. For example, a bisexual patient became aware that he was aroused by men who had hands like his father's. His father was a strong, burly workman with large, thick hands. He had often beaten the patient with his hands. Yet, as an adult, hands like the father's excited the patient sexually. During an interview, an exclusively homosexual young man told me that he was attracted to hairy men. I obtained the following dream from him: In the opening scene he is in bed with a young woman. It is obviously a heterosexual setting. Next, there appears from under the bed a big, brawny, very hairy arm with knife in hand, threatening to kill him. Again, the dream sequence starts with a heterosexual segment; then there is a lethal attack by a man whose arm is covered with hair, the stimulus that is homosexually arousing.

Another common arousal phenomenon is the sight of a large penis. Only the very large penis is sexually exciting to certain homosexuals. It is the symbol of ultimate masculinity. The dynamics of the compulsive urge to possess this symbol of masculinity, which is also feared, is an attempt to neutralize fear through eroticizing the feared stimulus. Unconscious attitudes toward the large penis vary. One patient, the night after he had fellated a large penis, had a dream that he had blood on his teeth, a depiction of his wish to castrate the penis that had consciously excited him.

A third category consists of interpersonal reparative attempts. Many homosexuals become sexually excited when a man shows them warmth and acceptance. They are turned on by it even though the partner may not conform to the physical type who is usually arousing. This is a central element in the dynamics of homosexuality. As previously stated, the homosexual leaves his childhood relationship with his father and other males with profound fear and hatred but also with envy, admiration, and the wish to be accepted and loved by them. The homosexual response very often is based on a need to redeem masculine affection.

Results of treatment

Of the 106 homosexuals who started psychoanalytic therapy, 29 were exclusively heterosexual at the time the volume was published. This represented 27 percent of the total sample. Fourteen of these 29 had been exclusively homosexual when they began treatment; 15 were bisexual. In 1965, in a follow-up study of the 29, I was able to reclaim data on 15 of the 29.[6] Of these 15 men, 12 had remained exclusively heterosexual; the other 3 were predominantly heterosexual, had married, but had occasional episodes of homosexuality when under severe stress. Of the 12 who had remained consistently heterosexual, 7 had been among the 14 who had been exclusively homosexual when they started treatment. Thus, 7 men who started treatment exclusively homosexual had been exclusively heterosexual for at least six or seven years.

Heterosexual development and the concept of normalcy

Before going on to the events that led to the American Psychiatric Association (APA) decision, I shall discuss the role of heterosexual development and the concept of normalcy. Neither Freud nor Kinsey believed that the choice of mate – at least the sex of the mate – was biologically determined. This view has always seemed odd to me, since Freud was a fine physician, a trained neurologist, and Kinsey was a biologist. Every species whose reproduction depends on heterosexual mating is provided with inborn mechanisms to guaran-

[6] I. Bieber, "Sexual Deviations," in *Comprehensive Textbook of Psychiatry*, ed. A. Freedman and H. Kaplan (Baltimore: Williams & Wilkins, 1967), pp. 959–76.

tee heterosexual arousal. That man is the only species of whom this is not true is an untenable idea. In primates, olfaction is very clearly the guaranteeing mechanism, and this is also the case for dogs, cats, horses, and other animals. In 1959, I published a paper on the role of olfaction in human sexuality.[7] The data were largely, though not entirely, derived from dreams and clearly indicated that olfaction was an important modality in early childhood sexuality. I formulated the theory that olfaction in man, as in other animals, was the sensory modality that initiated heterosexual arousal.

A male dog at the age of three months will not respond with sexual excitement to an estrous bitch's odor. His olfactory system is not impaired, but his sexual response has not yet developed. Arousal apparently involves neuroendocrine organization and development, and at nine months a male dog will be sexually excited by a bitch's odor. In humans, at somewhere between the ages of two and three, but closer to two, the neuroendocrine basis for olfactory response to sexual stimuli becomes established, and the child begins to respond with excitement to individuals of the opposite sex. Kalogerakis, a psychoanalyst, followed a child through this period and had the opportunity to observe the beginnings of a foot fetish.[8] The central dynamic of a foot fetish is olfactory. He was able to observe how stimulation in one sensory modality spread and became integrated with other sensory modalities. At first, this three-year-old boy was stimulated by the foot odor of his young aunt who lived with the family. He enjoyed playing with her feet when she kicked off her shoes, and he would comment about the smell. When his mother's friends came to visit and they sat at table, he would sit under it gazing at their feet. Several months after this behavior was observed, he took to looking with avid interest at newspaper advertisements that depicted barefooted little girls, indicating an integration between the visual and olfactory modalities. I think that normal human heterosexual development is initiated through olfaction and rapidly becomes integrated with other modalities, particularly vision, as well as hearing and touch. In the adult, olfaction does not remain as a salient stimulus modality, although it undoubtedly plays a greater part than most realize. Thus humans are biologically programmed for heterosexuality, and all go through an early phase of heterosexual development and arousal capability before some veer off into a substitutive homosexual adaptation. Homosexuality develops only after early

[7] I. Bieber, "Olfaction in Sexual Development and Adult Sexual Organization," *American Journal of Psychotherapy* 13 (1959):851.

[8] M. Kalogerakis, "The Role of Olfaction in Sexual Development," *Psychosomatic Medicine* 25 (1963):420.

heterosexuality has been dislocated by fears related to experiential factors such as a disturbed parent–child relationship, disturbed same-sex peer group relations, and other traumatic situations that inhibit heterosexuality.

Now to turn to the issue of normalcy: It has been the position of the gay activists that homosexuality is within the normal range of human sexual behavior. They claim that the only reason psychiatrists and others designate homosexuality as abnormal is because society does not look upon it as socially acceptable. They also claim that homophobic psychiatrists have used prejudicial social criteria to designate homosexuality as other than normal. This, of course, is not the case. Any phenomenon that is statistically normal refers to the average, the mean, or near the average. Normalcy also refers to health as opposed to illness or pathology. The common cold is statistically normal because most people catch cold, but having a cold is not a normal condition. A cold is a viral infection, and there are demonstrable pathophysiological signs and symptoms that make the cold abnormal. In designating any behavior or condition as other than normal, psychiatrists use neither frequency distribution nor standards of social acceptability as criteria. Behavior is psychologically abnormal when it is based on irrational or unrealistic fears. In the case of homosexuality, as I have emphasized, the fears are of hostile responses by other men, should heterosexual, romantic wishes be fulfilled. Such fears may have been realistic during early life when other males were in fact hostile, as in cases where a father was specifically hostile when observing the son's closeness with the mother. But in adult life, the fear of injury from other men for heterosexual activity is not rational. Homosexuality first develops as a consequence of such fear and is maintained in adult life, usually unconsciously, by the continuity of fear.

The APA decision

The issue under consideration in the decision was whether the designation *homosexuality* should be removed from *DSM-II*.[9] As it stood, there was a major category in the manual entitled "Sexual Deviation," which included homosexuality without naming it as such:

> This category is for individuals whose sexual interests are directed primarily toward objects other than people of the

[9] American Psychiatric Association, *American Psychiatric Association Diagnostic Manual of Mental Disorders*, 2d ed. (Washington, D.C.: American Psychiatric Association, 1968), p. 33.

> opposite sex, toward sexual acts not usually associated with coitus or toward coitus performed under bizarre circumstances as in necrophilia, pedophilia, sexual sadism and fetishism.

Subsumed under a subcategory "*Sexual Orientation Disturbances (Homosexuality)*," homosexuality was defined as follows:

> This is for individuals whose sexual interests are directed primarily toward people of the same sex and who are either disturbed by, in conflict with or wish to change their sexual orientation. This diagnostic category is distinguished from *Homosexuality* which by itself does not constitute a psychiatric disorder. Homosexuality *per se* is one form of sexual behavior, and with other forms of sexual behavior which are not by themselves psychiatric disorders are not listed in this nomenclature.

On the one hand, the manual explicitly states that homosexuality is not a psychiatric disorder and is not listed as such; on the other hand, homosexuality appears as a subcategory under the general classification of sexual deviation, which suggests that homosexuality is included. This ambiguity is commented upon in a footnote: "This term (*Sexual Deviation*) and its definition are inconsistent with the change in thinking that led to the substitution of *Sexual Orientation Disturbances* in the list below. However, since no specific recommendations were made for changing this category or its definition, this category remains unchanged for the time being."

The question at issue centered not only upon whether or not *homosexuality* should be unambiguously excluded from the psychiatric diagnostic manual, but if included, should it appear under another heading, such as "Sexual Dysfunctions" (which would also list other conditions such as sexual frigidity and sexual impotence, psychopathologies that had not appeared in *DSM-II*).

Considerations that were extraneous to the decision became entangled in the controversy. An appropriate course would have been to adhere to clinical and other data in deciding whether homosexuality was an adaptation that fell within a wide spectrum of normal sexual behavior, or whether it was a manifestation of abnormal sexual development consequent to disturbances in childhood and adolescence. If homosexuality were a variant of normal sexual behavior, obviously it did not belong in a diagnostic manual. If it were other than normal, it did belong in the manual. The questions that were raised at the outset in committees – that is, Is homosexuality a mental disease? Are homosexuals as well adjusted as heterosexuals? – only obfuscated the basic issue. Mental disease in the *DSM-II* applied only to psychotic illness, and no psychiatrist with expertise in the subject has ever considered homosexuality as a mental disease. Factors such as excellent occupational performance and good social adjustment were cited as evidence of the normalcy of homosexuals; such factors do not, however, exclude the presence of psychopathology. Although psychopathological disorders frequently impair occupational and social adjustment, the converse is not necessarily so. Men with potency problems or women who are sexually frigid may function well at their jobs and in their social circles, yet, excluding the infrequent occurrence of physical illness, these conditions are viewed as symptomatic of sexual psychopathology.

A number of important events and circumstances preceded the voting by the Board of Trustees of the APA in December, 1973, and the vote of the general membership in 1974. On September 28, 1967, the United States Department of Health, Education and Welfare of the National Institute of Mental Health issued the following press release:

> Formation of a task force to assist the National Institute of Mental Health in developing a research program in the area of human sexuality, especially homosexuality, was announced today by Dr. Stanley F. Yalles, Director of the National Institute of Mental Health, United States Public Health Service. A twelve member inter-disciplinary task force was drawn from psychiatry, sociology, the clergy, law, psychology, and anthropology. The chairman is Evelyn Hooker, Ph.D., a noted research psychologist, from the University of California at Los Angeles. Besides encouraging research into the problem of homosexuality, the task force will look for better ways to enable homosexuals to lead more satisfactory lives in the society . . .
>
> Dr. Yalles said that one of the purposes of appointing a task force of eminent persons was to encourage outstanding scientists to conduct research on homosexuality . . . Dr. Hooker added that there was a need for basic and applied research in the field.

The study that I had conducted in 1962 was by then widely known. In 1964 the APA had bestowed upon my group the Hofheimer Research Award, Honorable Mention, yet, neither I nor any member of the research team was invited to join the task force. Other colleagues who had published important contributions were also overlooked: Lionel Ovesey, Charles Socarides, and Lawrence Hatterer.[10] Common to our work was the conclusion that homosexuality

[10] L. Ovesey, *Homosexuality and Pseudohomosexuality* (New York: Science House, 1968); C. Socarides, *The Overt Homosexual* (New York: Grune & Stratton, 1968); L. J. Hatterer, *Changing Homosexuality in the Male* (New York: McGraw-Hill, 1970).

was not a normal sexual varient. Not a single psychiatrist who held the view that homosexuality was anything other than an adaptation within the range of normal sexual organization was invited to participate. Evelyn Hooker had written two papers based on a study of the adjustment of homosexual men, members of a homosexual association known as the Mattachine Society.[11] She found that their adjustment was in the normal range, in some ways even superior to that of heterosexuals.

The only other member of the task force who had written on the subject was Judd Marmor,[12] a well-known psychiatrist, psychoanalyst, and a former president of the APA. In his view, homosexuality is a normal variant but has been considered pathological because many psychiatrists had not freed themselves from the long-standing prejudices against homosexuality.

The task force issued its final report on October 10, 1969. First, they recommended the establishment of a center for the study of sexual behavior to include research, training, education, prevention and, treatment. A second category encompassed considerations of social policy in regard to legal and societal prejudices, emphasizing the adverse effects of social discrimination. "Homosexuality presents a major problem for our society largely because of the amount of injustice and suffering entailed in it not only for homosexuals but also for those concerned about them."

The gay liberation movement

The political, intellectual, emotional, and psychiatric thrust of this movement is well demonstrated by the writings of Franklin E. Kameny, a leading spokesman.[13] In a statement describing the forces that gave rise to the gay liberation movement, he wrote:

> Gay liberation as a formal entity had its birth in a riot by homosexuals in late June, 1969, at a bar called the Stone Wall on Christopher Street in Greenwich Village, New York City. The details of the riots and the incident which precipitated it or its merits are not important . . . The time was ripe, the homosexual community was ready and if the riot had not occurred at the Stone Wall at that time it would have soon occurred elsewhere. What is important is the message conveyed by the riot, probably

[11] E. Hooker, "The Adjustment of Male Homosexuals," *Journal of Projective Techniques* 22 (1957):17–31; and "Male Homosexuality in the Rorschach," *Journal of Projective Techniques* 22 (1958):33–54.

[12] J. Marmor, *Sexual Inversion* (New York: Basic Books, 1965).

[13] F. Kameny, "Gay Liberation and Psychiatry," *Psychiatric Opinion* 8 (1971):18–27.

> the first ever staged by homosexuals. The message was we have been shoved around for some three thousand years. We are fed up with it and we are starting to shove back. If we don't get our rights and the decent treatment as full human beings which we deserve and get them now, there's going to be a lot more shoving back.

In another statement, psychiatry is taken strongly to task as not only an adversary but as "the archenemy":

> The unproven allegation that homosexuality is pathological (or any of its semantic guises – a disturbance, a disorder, a neurosis, etc. or as a symptom of any of these) is recognized as not only destructive to the self-respect, self-esteem, self-confidence and self-image of the homosexuals (thereby in self-fulfilling fashion helping to produce some of the very symptoms which psychiatrists claim to find in their homosexual patients and in all homosexuals) but as perhaps the major supportive fact that is currently (now that organized religion is coming around) behind the negative attitudes of society at large. These attitudes inculcate to the homosexual (as it does to blacks in ways which are different in detail but identical in essence and results) a feeling of inferiority and of second-ratedness. Thus, psychiatry wreaks major violence upon the psyche of the homosexual both first-hand and through the aid which it offers to the maintenance of social bigotry.

The gay activists thus explicitly targeted psychiatry as its main enemy. Among their major activities was the disruption of psychiatric meetings. My first direct contact with the Gay Activist Alliance occurred during the 1970 annual meetings of the APA in San Francisco. I was a member of a panel on "Transsexuals and Homosexuals." As we were preparing to start, a number of gays dressed in fantastic garb entered the meeting hall, distributed literature, behaved as if they intended to disrupt the meeting, and, in fact, did. We finally got under way when arrangements were made for gay representatives to remain and be given the opportunity to speak.

My next direct contact with disruptive tactics occurred in 1972 at the APA annual meeting in Dallas. I was to present a paper entitled "Homosexual Dynamics in Psychiatric Crisis."[14] When I learned from an informed source that the gay activists intended to disrupt the meeting, I conferred with several colleagues who were in charge of arrangements. They worked out an agreement with the gays to

[14] I. Bieber, "Homosexual Dynamics in Psychiatric Crisis," *American Journal of Psychiatry* 128 (1972):1268–72.

deliver their remarks following my presentation. Frank Kameny was their major speaker.

The gay activists had from the beginning of their social protest action blamed the psychiatrists for perpetuating discriminatory practices against homosexuals, and although homosexuality was listed in the *Diagnostic and Statistical Manual* with the limitations previously described, the homosexual community nevertheless considered its inclusion to be damaging. Many psychiatrists shared the gay point of view, especially the younger colleagues. In June 1972, the official newspaper of the APA, *Psychiatric News*, carried an article headed "Psychiatrists Blast Colleagues." It discussed Marmor's sharp criticism of psychiatrists who he said were mistakenly treating homosexuality as a psychiatric disorder. On February 9, 1973, the *New York Times* reported that at a meeting of the Nomenclature Committee of the APA, held the previous day in New York, the gay activists had presented a group of studies supporting the view that many homosexuals are indistinguishable from heterosexuals in social adjustment. Marmor was quoted as saying, "Homosexuality merely represents a variant sexual preference which society disapproves of but it does not constitute a mental disease." The inference is that those psychiatrists who viewed homosexuality as other than normal considered it to be a mental disease. Psychoanalysts have never considered it to be a mental disease, any more than sexual impotence and frigidity would be diagnosed as a mental disease. Freud's well-known letter to a mother of a homosexual, written in 1935, stated, "I gather from your letter that your son is a homosexual. Homosexuality is surely no advantage but is nothing to be ashamed of, no vice, no degradation. It cannot be classified as an illness; we consider it to be a variation of the sexual function produced by a certain arrest of sexual development."[15]

Soon after the *Times* article appeared, Socarides and I decided to constitute an ad hoc committee for those APA members who believed that the term *homosexuality* should not be removed from the diagnostic manual. We cochaired this meeting and composed a letter stating our position and sent it to all members of the Board of Trustees of the APA in April 1973. Our major points were that our studies indicated that homosexuality was not a normal variant and did belong in the statistical and diagnostic nomenclature. Moreover, deletion threatened the prophylactic treatment of children who, in preadolescence, constitute a population at risk for becoming homosexual. Such children, particularly boys, are easily identified, not only by psychia-

[15] S. Freud, in *American Journal of Psychiatry* 107 (1951):786.

trists but by teachers and peer-mates. If treated, many may not become homosexual. The question remained: What will the APA position be on the prophylactic treatment of this population of preadolescent boys? We requested the trustees of the APA to constitute a task force whose views would include the ones we represented, unlike the homogeneous group that had been appointed by the National Institute of Mental Health.

In the fall of 1970, the New York District Branch of the APA appointed Socarides as chairman of a task force to investigate the problem of sexual deviation. I became a member of this group, along with ten other psychoanalysts who represented a broad spectrum of theoretical orientations. The members who participated came from analytic societies affiliated with the Columbia Psychoanalytic Society, the Horney Institute, the New York Psychoanalytic Society, and the Society of Medical Psychoanalysts. Since we were all especially interested in the problem of homosexuality and since it was the topic of the moment, it became our first subject. After working together for about two years, the task force submitted a detailed report of its findings and conclusions. We unanimously agreed that homosexuality was not a normal sexual variant but was a manifestation of psychopathology; that it was experientially not organically derived, and that a significant number of homosexuals could shift to exclusive heterosexuality if they had a psychoanalytic type of psychotherapy. The council of the New York District Branch refused to accept our report on the ground that the issue of homosexual normalcy or pathology was controversial and that the society could not take a stand on controversial scientific issues. Yet this same council did indeed take a stand on a controversial issue when it directed its delegate to the assembly at the district branches to vote to remove homosexuality from *DSM-II*. Further, they took this stand without consultation with the general membership of our district branch.

In the fall of 1972, I was appointed chairman of this task force, but shortly thereafter we were instructed by the council to disband, on the ground that since they had submitted our report on homosexuality there was no reason for our committee to continue its work, even though we had been appointed as a task force to study the broader topic of sexual deviation.

Shortly thereafter, Robert Spitzer, a leading member of the Nomenclature Committee, invited me to participate on a panel of which he was chairman, scheduled for the 1973 APA meetings in Honolulu. The subject again was whether homosexuality should remain in *DSM-II*. My fellow panelists were Ronald Gold, a representative of the Gay Activist Alliance; Richard Green; Judd Marmor;

Charles Socarides; and Robert Stoller. There was an audience of about 2,000, and it was evident from the mood and response that the majority supported the Gay Alliance position. The first issue was clear: Was homosexuality a normal sexual variant or an expression of psychopathology? The second issue was sociopolitical. Did the inclusion of homosexuality in *DSM-II* significantly contribute to the continued prejudice against homosexuals, and, if so, was the solution one of removing the term from the manual, even if homosexuality was deemed to be pathological? Instead of keeping to the issue, clarity was lost by introducing the concept of mental illness and by discussing a new set of criteria for diagnosing psychiatric conditions.

Spitzer was appointed chairman of a subcommittee of the Committee on Nomenclature and Statistics to investigate the broad problem of homosexuality and to determine whether or not it belonged in the diagnostic manual. He introduced two criteria for determining which psychiatric conditions should be listed in *DSM-II*. The condition must (1) regularly cause distress, or (2) interfere with social effectiveness. In a position paper published in *Psychiatric News*, I stated that psychopathology can be ego-syntonic and not cause distress; that social effectiveness – that is, the ability to maintain positive social relations and perform work effectively – may coexist with psychopathology, in some cases even of a psychotic order.

In a dialogue with Spitzer, reported in the *New York Times* December 23, 1973, I pointed out that there were several conditions in the *DSM-II* that did not fulfill his criteria: voyeurism, fetishism, sexual sadism, and masochism. He replied that these conditions should perhaps also be removed from the *DSM-II* and that if the group so affected were to organize as did the gay activists, they, too, might find that their conditions would be removed as a diagnostic entity from *DSM-II*.

On December 14, 1973, the Board of Trustees of the APA was to meet in Washington, D.C., to vote on whether or not to remove the term from the manual. Socarides and I agreed that our viewpoint should be represented, and we decided to go to Washington. Robert McDevett, a fellow psychiatrist, joined us there, and we each addressed the committee, offering our considered opinions as to why the term should not be deleted. I stated that apart from scientific error, if they voted for removal their decision would be interpreted by the gay community and the public as an APA declaration that homosexuality is normal. My major concern was and continues to be the effects of their decision on prophylaxis and the treatment of children and adolescents who show clear-cut signs of developing homosexuality. I emphasized that scientific issues must be settled by research, not by vote, and that a task force of colleagues holding varying opinions should be constituted to study the problem further. The Executive Council, with two absentions, voted unanimously to remove the term *homosexuality* from the *DSM-II.*

Now that a vote had been used to settle a scientific issue, it was Socarides's thought that the entire membership should therefore have a voice and that a petition for referendum should be initiated. The required 200 signatures were soon obtained, and Socarides and I became cochairmen of an ad hoc committee for the referendum.

Before the vote, two letters were circulated to the entire APA membership. One was signed by the president-elect, vice-president, and the three candidates for president-elect of the APA; the other letter was written by a group calling themselves the Committee of Concerned Psychiatrists. Each of the letters asked the membership to support the Board of Trustees and vote against the referendum. Ballots were returned by 9,644 members, roughly 37 percent of the membership; 5,834 backed the board and 3,810 voted against the decision we supported. We later learned that there was an overrepresentation of younger colleagues who supported the board, though it probably did not affect the outcome.

Shortly after the referendum was completed and action was taken to remove the term *homosexuality* from the *DSM-II,* the following circumstances came to light: The letter that had been sent to the entire membership under the signature of the two officers and three candidates for office had been written by a gay activist group and had been financially supported and distributed by them. The failure to identify gay sponsorship prompted demands that the APA investigate this seeming impropriety. It was the National Gay Task Force who had written the letter, solicited the signatures, purchased a membership list from the APA for $360, and mailed the letter to 17,900 psychiatrists, urging them to vote in the referendum and to uphold the APA trustee's decision to eliminate the term *homosexuality* from the list of mental disorders. A number of APA members questioned the propriety of sending a letter to the membership before the vote on the referendum without noting that the letter had been financed by the National Gay Task Force. The ethics committee of the APA investigated the matter and came to the conclusion that although the actions had been unwise, there had been no impropriety.

After the decision

On April 6, 1976, more than two years after *homosexuality* had been removed from the *DSM-II,* a meeting that was to be devoted to a discussion of male homosexuality was scheduled to be held at the New York Academy of Medicine under the auspices of the Columbia Psychoanalytic Society. The three main speakers were to be Socarides, Ovesey, and myself. The meeting was completely disrupted by the gay activists, and the papers were not given. Thus, several years after winning their position, the gays were still breaking up psychiatric meetings.

The November 1977 issue of *Medical Aspects of Human Sexuality*, a magazine widely circulated among physicians in the United States, published the results of an analysis of the first 2,500 replies to a questionnaire it had sent out to 10,000 psychiatrists. The questions and answers follow:

Q. Is homosexuality usually a pathological adaptation (as opposed to a normal variation)?
A. Yes, 69 percent; no, 18 percent; uncertain, 13 percent

Q. Can homosexuals become heterosexual via therapy?
A. In most cases, 3 percent; fairly often, 37 percent; almost never, 58 percent

Q. Are homosexual men generally less happy than others?
A. Yes, 72 percent; no, 26 percent

Q. Are homosexual men generally less capable than heterosexual men in mature loving relationships?
A. Yes, 60 percent; no, 39 percent

Q. Are lesbian women less capable than heterosexuals of mature loving relationships?
A. Yes, 55 percent; no, 43 percent

Q. Are homosexuals' problems in living a result of personal conflicts more than of social stigmatization?
A. Yes, 70 percent; no, 28 percent

Q. Can bisexuals have successful heterosexual marriages?
A. Usually, 21 percent; occasionally, 65 percent; almost never, 12 percent

Q. Are homosexuals generally more creative than heterosexuals?
A. Yes, 22 percent; no, 74 percent

Q. Are homosexuals generally a greater risk than heterosexuals to hold position of great responsibility?
A. Yes, 42 percent; no, 54 percent

Harold I. Lief contributed the following interpretative remarks in the article: "The present findings suggest three possibilities: 1. The APA vote was influenced by political and social considerations in that the vote was perceived as a step toward stopping the denial of rights to homosexuals. 2. Those who were more inclined to answer this survey had stronger feelings about their point of view than those who did not respond. 3. Psychiatrists' opinions on the matter had changed since 1972."

Summary

The factors that seemingly determined the decision of the APA to delete *homosexuality* from *DSM-II* may be summarized as follows: (1) The gay activists had a profound influence on psychiatric thinking. (2) A sincere belief was held by liberal-minded and compassionate psychiatrists that listing homosexuality as a psychiatric disorder supported and reinforced prejudice against homosexuals. Removal of the term from the diagnostic manual was viewed as a humane, progressive act. (3) There was an acceptance of an altered concept of psychiatric conditions. Only those disorders that caused a patient to suffer or that resulted in adjustment problems were thought to be appropriate for inclusion in the *Diagnostic and Statistical Manual.*

The way the APA decision was arrived at was unique in that never before had a scientific controversy been settled by vote of the members of a large professional society. There was no precedent for this procedure in the APA and probably not in any other scientific organization.

The initial decision of the board of trustees was also arrived at by a vote – unanimous, with two absentions – and had the subsequent referendum not taken place the board's pronouncement would likely have stood without protest. Yet their decision by no means reflected broad agreement among the membership. A substantial minority, 39.5 percent, voted against deleting *homosexuality* from *DSM-II.*

In the final analysis, scientific controversies are settled in time, when the overwhelming weight of evidence makes the continuation of controversy irrelevant. Long before convincing evidence is in, however, there usually tends to be a polarization of opinion, with one side attaining decision-making power and influence. It seems obvious enough that scientific differences should be settled by scientific methods, not by vote nor by power politics, but given a choice between a small group decision and a democratic vote, I do not regret that the APA had a mechanism that permitted the membership to have a voice in the outcome.

19

Values in the debate over workplace safety and health: the rancorous rhetoric about regulation

GILBERT S. OMENN

The Occupational Safety and Health Administration (OSHA) stands in the eye of the storm about government regulation. Mere mention of OSHA in many quarters triggers rancorous rhetoric, revealing underlying conflicts about the power relationships of labor and management, about social responsibility of corporations, about "internalizing" the costs of products and production, and about concepts of freedom in our society.

One of the more amusing political moments of 1980 was the Kissingerian negotiation at the Republican National Convention aimed at placing former President Ford on the ticket with Mr. Reagan. One wag, impressed with Kissinger's negotiating skills and with his chutzpah in agreeing with the television interviewers that Ford would be "co-President," warned Reagan that by the time the areas of responsibility had been distributed, he – Reagan – might be left with "nothing but OSHA"! Dignified businessmen, upon joining a discussion about OSHA, are prone to begin talking about heights of toilet seats and spacing of rungs on ladders – as examples of the weighty matters that they think preoccupy OSHA inspectors. And the sagebrush conservatives who moved to Washington in 1981 love to describe the big poster of "The Cowboy after OSHA," wearing goggles and helmet; a net surrounding his horse; and a box safely dangling beneath the horse's derrière as an "antipollution device."

In the Closure Project meetings about occupational safety and health and about certain specific standards promulgated by OSHA, it is essential to recognize the serious conflicts in values and high emotion that set the context for any specific debates.

Employers with good safety records complain that OSHA cites them for violations that are irrelevant to good safety practices. They also decry OSHA's total reliance upon engineering controls and OSHA's general refusal to recommend personal protective devices, whether earmuffs against noise or respirators against inhaled dusts. Small businessmen especially resent the OSHA inspectors and their supposedly picayune approach; for years, they have sought congressional relief from OSHA access.

Labor unions are no happier. They criticize OSHA for relying upon industry for health, engineering, and cost information and for utilizing cost estimates at all. They regard penalties as too small, inspectors as too few, and the willingness of OSHA to let the states carry on nonfederal programs as irresponsible.

Labor leaders, such as Anthony Mazzocchi of the Oil, Chemical, and Atomic Workers Union, consider discussions of scientific and ethical issues in occupational health sterile and disassociated from reality. Mazzocchi regularly points out that "work is terrible," that he worked in mines and in fume-filled factories and he didn't like it one bit. In contrast, he points out, he's now just as much a bureaucrat as government officials and nearly as far away from the terrors of many work environments as the corporate officials and professors. He makes clear that his sympathies remain with the workers, and for good reasons.

I found excellent preparation for this discussion in reading Studs Terkel's book, *Working* (1972). He pointed out that

> this book, being about work, is, by its very nature, about violence – to the spirit as well as the body. It is about ulcers, as well as accidents, about shouting matches as well as fistfights, about nervous breakdowns as well as kicking the dog around. To survive the day is triumph enough for the walking wounded among the great many of us . . . Lately there has been a questioning of (the) "work ethic," by the young. Strangely enough, it has touched off profound grievances in others, hitherto devout, silent and anonymous. Unexpected precincts are being heard from in a show of discontent. . . . The drones are no longer invisible or mute . . . Nor are they exclusively of one class. Markham's Man with the Hoe may be Ma Bell's girl with the headset . . . Scrooge has been replaced by the conglomerate. Dickens' people have been replaced with Beckett's. (Introduction, p. xi)

Terkel's approach of listening to working people and reporting their frustration, anger, joys, and accommodations actually builds upon a similar effort more than a century earlier. Henry Mayhew, in 1850, astonished and horrified readers of the London *Morning Chronicle* by digging deep into the lives of London's laborers and evoking from those invisible people themselves the wretched conditions in which they lived. His letters ran six full columns and averaged more than ten thousand words.

One might ask, Where did OSHA go wrong? Such a question is stimulated particularly by statements in the much-cited monograph of Robert Smith of the American Enterprise Institute. Smith described the passage of the Occupational Safety and Health (OSH) Act in 1970 in terms that we here would consider to represent "closure" on the overall subject of government protection of workplace safety and health. According to Smith, every witness before the House subcommittee holding hearings on the bill agreed that there was a need for the legislation, and the initial House and Senate versions of the act passed by margins of 383 to 5 and 83 to 3, respectively. I have reviewed the circumstances of the passage of the original act to determine whether closure was indeed achieved on the need for legislation that created a regulatory federal role.

Lack of closure on the intent of the legislation

The Occupational Safety and Health Act, signed by President Nixon on December 20, 1970, gives the secretary of labor strong powers, to be exercised through the assistant secretary who is administrator of OSHA, both to set standards and to enforce standards to improve and protect the safety and health of workers in this country.

The combining of the standard-setting and enforcement functions was a highly controversial issue. The Nixon administration wanted to restrict the secretary of labor to enforcement of standards that would be issued by an independent board of presidential appointees with substantial professional competence. The relevant committees in the Congress were at loggerheads on this matter and on the underlying philosophies. A minority report of the House Committee on Education and Labor supported the administration, but the majority report (by the Democratic members) opposed this separation of functions. The legislative history recorded explicit fears that employer perspectives would not be likely to be given a fair or adequate hearing within a standard-setting process carried out entirely by the Department of Labor. The majority considered the separation of functions unworkable and unnecessary to ensure access to expert advisers. Nevertheless, the full House voted overwhelmingly for the minority recommendation, President Nixon's approach.

Meanwhile, the Senate Labor Committee and the full Senate rejected the concept of an independent board, in the belief that a more efficient and sounder program would result from putting full responsibility for both functions in one department. This major difference and a number of others were resolved in the subsequent House-Senate Conference committee; the House acceded to Senate views on this point.

Despite formal closure in the form of passage of the OSH Act, lack of agreement on values, approaches, and means underlies the severe difficulties experienced in staffing the agency, adopting and setting safety and health standards, developing enforcement strategies, and gaining public confidence. The case studies presented in Chapters 20 to 22 by others provide excellent examples of the difficulties that bedevil the agency, I believe. This overview essay is intended to lay some groundwork for tying together the conceptual and ethical issues and political and economic conflict manifest in the case studies.

Placing more emphasis upon health standards

OSHA has been criticized severely for its preoccupation with safety standards, especially standards based upon highly prescriptive specifications. Thousands of standards were adopted in the very early days of OSHA, whereas barely a dozen health standards have been issued (see Table 19.1). OSHA was required by the Act to issue or adopt federal or consensus safety standards within two years. The only practical means was to adopt existing standards from the American National Standards Institute (ANSI), an industry-supported consensus organization; from the national Fire Protection Association; and from federal requirements for coal miners, longshoremen, and a broad array of federal contractors under the Walsh-Healy Act. The result was an astonishing array of thousands of truly picayune specifications. More than any other problem, I believe, this approach to safety, and inspections based upon these specifications, has established the widespread negative image of OSHA. Under the Nixon administration there was a stubborn refusal to improve the situation. Morton Corn, in the Ford administration, began to criticize the preoccupation with safety specifications, and then President Carter and Eula Bingham made political hay by eliminating some eleven hundred regulations in one protracted swoop.

There is broad agreement that OSHA should place increasing emphasis on occupational diseases and their prevention. Injuries are well recognized by workers, and employers have at least some incentives to reduce injuries, because workmen's compensation premiums are based upon experience ratings and because wage premiums often must be paid to entice workers to accept obvious risks. Nevertheless, the efficacy of OSHA's safety standards has been challenged on the basis that employers could take steps that would be more effective at less cost.

The difficulty in getting moving on health hazards springs from many sources. Their effects are not obvious; health problems develop slowly from cumulative exposures; they are of many kinds, affecting various bodily functions; and they often are the result of complex interactions with other factors, especially individual behaviors, the most important of which is cigarette smoking. OSHA, responding to public and worker fears, has concentrated nearly all of its standard-setting effort on workplace carcinogens. (Coke oven exposures, incidentally, were a significant factor in the original legislative push from I. W. Abel, president of the United Steel Workers Union and newly installed (after Walter Reuther's death) as head of the American Federation of Labor–Congress of Industrial Organizations (AFL-CIO) Industrial Union Division). In general, scientific data come from animal studies, plus clusters of clinical cases and epidemiological investigations, which are often plagued by limitations in sample size or by confounding variables. Extrapolations based on animal studies involving high-dose exposures are controversial. The debates about interpretation of the scientific data on chemical hazards should come as no surprise to those who have followed the decades of debate about the low-dose risks of radiation exposures, a much simpler problem.

The focus on workplace carcinogens, moreover, has forced OSHA into a confrontation on the question of costs of compliance, since most scientists hold the view, at least for populations, that there is no recognizable threshold for carcinogenic effects. Thus, there is no tolerable amount of the compound that can be associated with "safety" if the goal is zero-risk.

Achieving closure in the face of scientific uncertainty

Scientific uncertainty leads to polarized responses. In general, management adopts a "Prove it is dangerous" attitude, whereas labor insists that "It is better to err on the side of safety" and that protection of the worker is entirely the responsibility of management. Both embrace the term *conservative* to describe their positions, management claiming to be conservative in jumping to conclusions and labor claiming to be conservative in protecting the worker. Scientists are readily caught up in the polarization. Those who point out flaws in the studies or resist concluding that hazards are real and must be reduced urgently are lumped with the interests of management. Those who accept positive evidence of hazard yet disregard better studies that fail to show any effects are generally thought to be aligned with labor. There is little tolerance on either side for scientists in the middle, struggling to review the evidence objectively and to apply rules of evidence, with full recognition of the limitations of the

Table 19.1. *A chronology of health standards promulgated by OSHA: 1972–1981*

Health hazard	Proposal published	Final standard published	Court case	Present standing
Asbestos	1/12/72	6/2/72	4/7/74 (affirmed)	No major actions planned.
13 carcinogens:	7/16/73	1/29/74	3/17/75, 12/17/74 (affirmed)	No major actions planned.
4-nitrobiphenyl; α-naphthylamine; methylchloromethyl ether; 3,3′-dichlorobenzidine; bischloromethyl ether; β-naphthylamine; benzidine; 4-aminodiphenyl; ethyleneimine; β-propiolactone; 2-acetylaminofluorene; 4-dimethylaminoazobenzene; *N*-nitrosodimethylamine; and MBOCA (vacated 12/17/74, reproprosed 2/3/75, no further action planned)				
Vinyl chloride	5/10/74	10/4/74	*Society of the Plastic Industry* v. *OSHA*, 1/21/75 (affirmed)	OSHA may look at PVC dust.
Coke ovens	7/24/75	10/22/76	*Am. Iron and Steel Institute* v. *OSHA*, Third Circuit, 1978 (affirmed)	Industry proposal for Supreme Court review withdrawn.
Benzene	5/27/77	2/10/78	*Am. Petroleum Institute* v. *OSHA*, Supreme Court, 1980 (vacated)	Only original consensus standard remains in place.
Cotton dust	12/28/76	6/23/78	D.C. Circuit, 1979 (affirmed)	Pending before Supreme Court; OSHA's Auchter asks court to allow further review for C/B analysis.
Cotton dust in gins	12/28/76	6/23/78	Fifth Circuit, 1980 (vacated)	No plans to repropose.
DBCP (dibromochloropropane)	11/1/77	3/17/78	None	No plans to reopen.

(inorganic)			filed 5/5/78	standard to OSHA for reconsideration in light of benzene decision (but standard remains in effect).
Acrylonitrile	1/17/78	10/10/78	None	No plans.
Lead	10/3/75	11/14/78	D.C. Circuit, 1980 (affirmed, but for selected industries)	Pending before Supreme Court. OSHA asks to reconsider whole standard.
Carcinogens policy	10/14/77	1/22/80 (amended 1/19/81 and 1/23/81)	None yet (no standard involved)	Candidates list issued 8/12/80; working on priorities list.
Access to medical records		5/23/80 8/21/80, effective 4/28/81, stayed for construction	*AFL-CIO* v. *OSHA* *Louisiana Chem. Assn.* v. *OSHA* (hearings scheduled July 81, Appeals Court)	
Noise (hearing conservation)	10/24/74	1/16/81		Temporarily stayed (thru 6/1/81), pending further OSHA study
Chemical labeling	1/16/81; with-drawn 2/12/81			
Beryllium	10/17/75			Still proposed. Benzene ruling delays it. Risk data poor. Consensus standard remains in effect. Industry may challenge consensus standard on sampling technique (personal monitoring devices).

Source: Prepared by Susan O'Hara, Labor Branch, U.S. Office of Management and Budget, Washington, D.C.

methods. There is also a striking impatience, based on the view that if the exposure is hazardous, then surely workers have suffered long enough. The result, all too often, is that tremendous efforts are expended and disputes are exaggerated because of inordinate attention to what may be marginal risks, while more serious, well-recognized hazards on which closure has been prevented by litigation or economic arguments are left unresolved.

I believe, as Mazur, Marino, and Becker (1979) have argued, that scientists have important contributions to make as experts and that it is both useful and desirable to separate disputes about facts from disputes about values. Explicit questions about facts and extrapolations in dispute must be isolated and addressed directly, rather than being lumped together with options for overall decisions. In their rush to influence decisions on what should be done, scientists are tempted to undermine their own special contributions.

Knowledgeable physicians may have a better professional temperament than many scientists for dealing with these situations at the decision-making level. After all, good clinicians are faced frequently with the need to make judgments on whether to do something or whether to wait, whether to seek more information and, if so, what specific kinds of information to seek.

Through what Tom Beauchamp calls "sound argument closure" and "negotiation closure" (Chap. 1, this volume), scientific and policy officials in the Carter administration finally reached closure on certain rules of evidence for reviewing animal studies and epidemiological studies of health effects of potential human carcinogens. The separate guidelines for risk assessment published by the Environmental Protection Agency (EPA), the Consumer Product Safety Commission (CPSC), the Food and Drug Administration (FDA), and OSHA were brought together under the aegis of the Interagency Regulatory Liaison Group's working group on risk assessment, with a significant assist from the White House Office of Science and Technology Policy. The working group published an extensive paper presenting closure (compromises) on such knotty and controversial issues as how to deal with tumors having benign histology, how to handle evidence both positive and negative for a given substance, what statistical requirements should be met, and how many studies or substudies constituted adequate confirmation. There had been recognition at the policy level that inconsistent agency positions on rules of evidence damaged the entire process for all agencies. Scientists and agency representatives were forced to close their differences through sound argument and negotiation. The result showed a clear tendency toward erring on the side of declaring substances potentially hazardous to man. Such a declaration represents only a first step, to be followed by characterization of the relative potency of such substances, the kinds and sources of exposure, and the types of human subpopulations most susceptible (Calkins, Dixon, Gerber, et al., 1980). A combination of qualitative and quantitative information with a clear description of the extent and bases of uncertainty seems to me to provide a substantial improvement over routinized quantification of risk, in which data are plugged into a black box called a "model."

Finally, as outlined by Calkins and his coworkers (discussed more fully later in this chapter), responsible government agencies and interested parties in labor and management must take the scientific information and the characterization of risk and decide which exposures warrant urgent attention and then which actions would be most effective in reducing those exposures. Given our political system, it is likely that the Carter and Reagan administrations would have OSHA administrators drawing the line on action at different places. However, it is quite reasonable, in my opinion, to expect each administration to accept the same rules of evidence, the same detailed scientific review, and similar characterization of risk, achieving a form of closure on a scientific approach to uncertainties in the scientific data. A major report from the National Research Council has elaborated my point of view (NRC, 1983).

Determining an acceptable objective in setting health standards

The OSH Act juxtaposes absolute criteria for safety with a conflicting directive to respect "feasibility." Specifically, the Act, Sec. 6(b) (5), orders "the Secretary, in promulgating standards . . . , (to) set the standard which most adequately assures, *to the extent feasible,* on the basis of the best available evidence, that *no employee will suffer* material impairment of health or functional capacity even if such employee has regular exposure to the hazard dealt with by such standard for the period of this working life" (emphasis added).

It is difficult enough to establish a standard that will ensure that no worker, no matter how predisposed by genetic traits or illness or personal habits, will suffer impairment of health *or* function, unless that standard is zero exposure. But it would truly require Solomonic wisdom to accomplish that objective while respecting feasibility. Such is the nature of closure in an Act of Congress.

The legislative history gives very little instruction on how to meet the phrase "to the extent feasible." The phrase was inserted late in the process by Senate conferees. The courts, especially the U.S.

District of Columbia Circuit Court, have rendered a series of opinions based upon comparison with the explicit instructions to accommodate economic feasibility in the statutory language of certain sections of the Clean Air Act and other acts contemporary with the OSH Act. In general, the courts have directed that OSHA *may* analyze or consider evidence about costs, in order to accomplish the health objective most expeditiously. Standards that would force many individual plants to close are reckoned feasible, whereas a standard that would force the closing of an entire industry might be ruled infeasible. These issues will be discussed in detail in the specific case papers.

Assuring protection of every employee

The matter of differential susceptibility among persons exposed has received little attention. Upon review of the Clean Air Act and the OSH Act, Friedman (1979) and I concluded that individual differences in susceptibility have been recognized in the statutes, legislative history, and regulatory standard-setting processes. There can be no doubt in either case that Congress intends that the agencies seek to identify and protect the most susceptible individuals. The legislative history of the Clean Air Act even draws a line between those on life-sustaining equipment and all others. EPA has made specific efforts to identify high-risk groups in the population in setting standards for photochemical oxidants (ozone) and for airborne lead (see Omenn, 1979). OSHA has made little explicit effort, either in setting standards or in monitoring and medical surveillance requirements. However, OSHA does require, in the cases of lead and arsenic, that workers be tested periodically and that workers found to have high blood levels of these substances be removed from exposure until their individual levels fall within a tolerable range. OSHA does not recommend any investigation to assess possible differential susceptibility of the workers. In fact, many advocates of tougher OSHA enforcement and tighter OSHA standards argue that worker differences should be dismissed out of hand and all workers considered alike, speculating that such differences could account for only a small part of the risks anyway (see Ashford, 1976).

Preemployment screening, especially for genetic traits, has been much discussed of late. It is important to note that there is no basis at the present time for routine preemployment screening or medical surveillance for genetic predispositions. There is ample reason to suspect that common genetic traits may predispose groups of people to higher risk in various work settings, but no one should recommend moving from such suspicions to routine screening. In order to assess the significance of such potential risk factors and to offer information that those people may feel entitled to have for their own use, I do believe it is desirable to initiate a research program to evaluate such predispositions, – which range from glucose-6-phosphate deficiency and methemoglobin reductase deficiency in red blood cells, which make the red cells susceptible to breakdown by oxidizing chemicals, to certain allergic responses to dyes and industrial chemicals. A responsible research program would begin with careful negotiations with workers and their unions and with management to define the objectives, the access to resulting data, the experimental nature of the observations, and the steps that each side would want to consider as follow-up should positive and useful results be found (see Omenn, 1982).

I must emphasize that these same criteria should apply to any preemployment screening or surveillance scheme on the job, whether genetic traits are involved or not. One of the more persistent examples of inappropriate screening is the use of the low-back X-ray examination to classify workers or job applicants as having back problems that exclude them from work in the railroad industry (Rockey, Fantel, and Omenn, 1979). The X-ray examination is a notoriously poor basis for prediction of back problems, yet large numbers of workers are denied jobs and "branded" as having a potentially disabling health problem.

Finally, I believe it is reassuring to note that good relations between unions, management, and academic researchers can be achieved. A model is the cooperative investigation by the Harvard School of Public Health, the United Rubber, Cork, Linoleum and Plastic Workers of America (URW), and the B. F. Goodrich, Armstrong Rubber, and Mansfield Tire & Rubber companies. A formal agreement was signed in 1971 (Peters, Monson, Burgess, et al., 1976). The first component was the Occupational Health Committee, composed of three management and three union representatives. It reviewed and approved the research plans proposed by the university research team, made recommendations for implementation of the findings, and reviewed occupational health problems referred to it by local health and safety committees for possible referral to the university. The second component was an Occupational Research Study Group at Harvard. This research team conducted epidemiological studies and in-plant investigations to define potentially hazardous conditions. The group also assisted in the development of safe standards for occupational environments. The group included members from the departments of environmental health sciences, epidemiology, and physiology. The program included studies of work and type of exposures. Another example of cooperative activity can be found at the University of Washington. The industrial hygiene program of the Department of Environmental Health is supported by the state Department of Labor and Industries, with an oversight committee that includes representatives of the state Labor Council and the Association of Washington Business.

Thus, I am persuaded that arrangements can be made to permit research investigations. It is important to seek workers' consent for studies of safety and health risks, even though individual workers may not feel they can actually have much involvement or influence (see Beauchamp, Chap. 1, this volume).

OSHA's process for obtaining expert advice and review

As has been mentioned earlier, the issue of how to assure the independence and objectivity of the scientific and technical input for the OSHA standard-setting process arose during the passage of the OSH Act. This issue has not disappeared. For example, the American Industrial Health Council (AIHC) is actively campaigning for an independent board or agency to review data before and after regulatory agency actions. There was wide interest in our policy paper from the Office of Science and Technology Policy entitled "Identification, Characterization and Control of Potential Human Carcinogens: A Framework for Federal Decision Making" (Calkins, Dixon, Gerber, et al., 1980). As noted earlier, I emphasize the choice of the word *characterization* of risk, rather than the usual term *quantitation*. Too often quantitation of risk estimates is achieved by routinized use of a "model," into which highly aggregated data are plugged, in which assumptions are buried, and from which a number or a number with confidence limits is generated. In contrast, characterization of risk deliberately juxtaposes qualitative and quantitative information. The three components of characterization of risk are

1. Description of the nature of the exposures, the sources of the exposures, and the numbers of people subjected to various exposures
2. Description of the health effects by type, reversibility, kinds of disability or illness caused, and quantitation of relative potency of the agent in one or more standard test systems
3. Review of evidence of variation in susceptibility to the agent among individuals or subgroups of the population

Characterization involves comparative ranking of the probability and severity of health effects with other risks faced by similar groups or caused by similar sources.

In that paper, we recommended a clear, two-stage process in which review of the scientific data and characterization of the results are separated from the agency rule making. We did not insist that the first stage be carried out by a different agency, but we emphasized that the stages of risk assessment and risk management must at least be separated, as EPA has accomplished, within an agency. On the other hand, we opposed removing it from government altogether, since there must be communication between the parties in the two stages and since the government, in the end, must be responsible for the process and the decisions.

The AIHC convinced Senator Eagleton, a good friend of labor, to insert language in the fiscal 1981 appropriations bill for the FDA directing that a study be supported, presumably at the National Academy of Sciences, on the desirability and feasibility of a "science court" or science panel. The science court concept (see Mazur et al., 1979) has considerable appeal. It also carries several limitations and risks. First of all, it invites an adversary procedure in which polarization of views most likely would precede any movement toward closure. Of course, as Kantrowitz emphasized (1978), scientists themselves often argue and generate controversy. Frankel's chapter on continental drift (Chap. 7, this volume) describes such a controversy. Second, there would be pragmatic constraints, such as the tremendous breadth of potential topics for any such court to consider and the likelihood of very great reliance on staff, rather than on the presumably prestigious, objective members of the court. Third, most of the controversies of great public interest do not hinge on scientific disputes but on different values and power positions, so the public may be disappointed and scientists may be frustrated that the court cannot bring closure on the larger issue. Fourth, the public and the government already receive well-informed, reasonably objective expert panel reviews from the National Academy of Sciences and from other professional bodies. When I served in the White House Office of Science and Technology Policy, we turned often to these bodies for intensive review of such matters as the status of and prospects for the use of photovoltaic cells to store solar energy and the rate and potential consequences of accumulation of carbon dioxide in the atmosphere. The congressional Office of Technology Assessment also utilizes expert panels, generally in response to issues raised by members of Congress and their staffs. In the health arena, the National Institutes of Health (NIH) has established a mechanism for state-of-the-art reviews and debate in the form of "technical consensus" exercises. And the health and environmental regulatory and research agencies jointly sponsor a National Toxicology Program aimed at objective testing and evaluation of potentially toxic chemicals.

Thus, on the scientific issue our society already has some effective mechanisms, but these lack the visibility and notoriety that might come in a "sitting" science court. My personal preference is to invest more effort in making existing mechanisms work better, rather than turning to a substitute. I do support experiments with the science court concept such as that attempted for the power line decision in Minnesota (Mazur et al., 1979).

The OSH Act does not require review of OSHA's standard-setting actions, of the data it uses, or of its interpretation of such data by scientifically or technically competent individuals outside the agency. The Act prescribes the composition of advisory committees that the secretary may appoint, to ensure that they primarily represent interests balanced against each other and are only secondarily vehicles for providing scientific and technical advice. OSHA has utilized advisers far less, and far less effectively, than have its counterpart agencies EPA and FDA. A study comparing the utilization of such advisory committees commissioned by the Office of Science and Technology Policy is available (Greenwood, 1984).

For an agency that must rely on its own staff to an extraordinary degree, OSHA has been especially damaged by serious difficulties in recruiting technical staff and in stabilizing the internal structure of the agency. In addition, OSHA must rely entirely upon the National Institute of Occupational Safety and Health (NIOSH) for its relevant research and development work, and this relationship, stormy and poor before 1977, was much improved under Eula Bingham and Anthony Robbins, the heads of OSHA and NIOSH in the Carter administration. Also, OSHA must rely upon the office of the general counsel of the Department of Labor for its legal staff. Since so much of the review of scientific data is based upon how lawyers think the "brief" will read in court, this point further demonstrates the limited resources and autonomy of OSHA.

The extent to which expert advice is sought probably reflects underlying philosophical and political strategies. Lee (1977), for example, has contrasted the intuitionism of the Judeo-Christian tradition, of the external forms and "good" of Plato, and of the "moral duty" of Kant with the utilitarianism of Jeremy Bentham and John Stuart Mill. Finding units for "the greatest benefit for the greatest number" and accounting for the differing values struck Lee as insuperable tasks. Mill himself apparently doubted whether the greatest benefit principle could be scientifically applied on a quantitative basis. Intuitionists tend to become exasperated with scientists and economists who insist on evidence and analytic rigor, especially when such experts often end up disputing each other.

It is likely that ethicists could offer substantial assistance in these debates over values, since so often the values remain implicit and unacknowledged. The capacity to formulate ethical and moral dimensions – not in the jargon of moral philosophy, but in the context of workplace experiences – might well help all parties to grapple more directly with the hard choices that affect worker autonomy, company liability, and the self-esteem of individuals on both sides. There are specific ethical dilemmas for company health professionals, for engineers and chemists who design the production schemes, and for workers who may take risks with adverse consequences to their fellow workers or their families or for consumers.

Documenting the basis for OSHA's standards and decisions

Nearly every significant OSHA decision, including standard setting, has been attacked, often from both sides, in the courts. Thus, preparation for judicial review has become a dominant element in the work of OSHA and in the review by the office of the secretary and the general counsel in the Department of Labor. Beyond the Department of Labor, President Carter's Executive Order 12044 on regulatory reform required agencies to submit proposed regulations to a regulatory analysis review group chaired by a member of the Council of Economic Advisers to perform cost estimates and, then, for actions with potential costs exceeding $100 million, to do cost-effectiveness analyses of alternative actions. Occasionally the several regulatory agencies themselves disagreed. A good example in 1980 was the OSHA defense of the benzene standard, which the solicitor general was preparing to argue before the Supreme Court. OSHA insisted on claiming that it is infeasible to do any kind of quantitative risk assessment, yet the EPA, FDA, and CPSC each had published and utilized quantitative risk assessment methods (see References). EPA had used such methods extensively. It was necessary to narrow the OSHA argument.

An excellent view of the attitudes of federal courts has been provided by Judge David Bazelon, who served for thirty years as chief judge of the U.S. Circuit Court in Washington, D.C. That court's case load predominantly involves challenges to federal administrative actions; in recent years, the caseload has been dominated by issues relating to the frontiers of technology. As Judge Bazelon has written

> The judiciary . . . is not the proper forum for developing relevant information or selecting values, at least in these issues (of environmental, safety and health regulation). What, then, is the Court's role? In brief, I believe that the judicial responsibility is to monitor and scrutinize the administrative process. Our task

> is to assure that the agency's decision-making is thorough and within the bounds of reason. The agency's decisional record must disclose the evidence heard, the policies considered, the agency's precise reasons for resolving conflicts in the evidence. This includes their basis for selecting one scientific view rather than another. This permits quality checks through peer review, legislative oversight, and public attention. Only if decision-makers disclose assumptions, doubts, and moral and political tradeoffs can experts and citizens evaluate administrative action. Only then can professional peer review bring to light new data or challenge faulty assumptions. And only then can Congress and the people understand and debate the value choices implicit in regulatory action. (Bazelon, 1980, p. 5)

Several federal courts have applied this approach in reviewing OSHA health standards. A good contrast exists in the decisions on the benzene standard, which was remanded for further work, and the cotton dust standard, which was strongly supported by the Court. It is not yet clear whether the judiciary can reach closure on the proper role and proper reach of the courts in these matters. The judiciary has a procedural system of precedents and appeals that should bring periodic closure in the form of a Supreme Court decision, at least when those decisions are interpretable on the larger issues, rather than being narrowly focused on technical or specific aspects.

In its consolidated opinion on cotton dust cases, *AFL-CIO* v. *Ray Marshall* (Secretary of Labor), the U.S. Court of Appeals, Washington, D.C. Circuit, declared (617 F. 2d 636, D.C. Cir., 1979):

> Congress has made it clear that OSHA must protect every worker from the risks of material health impairment due to occupational exposure to hazardous materials. OSHA is constrained only by the requirement that its standards be technologically and economically feasible for the regulated industries. In sum, our review leads us to uphold the standard, except for its application to the cottonseed oil industry. In every aspect except that one, OSHA explained the evidence in the record it found persuasive and the policies that guided its judgment when facts proved insufficient. As for the cottonseed oil industry, however, we are unable to discern the basis for the agency's determination. We do not conclude that the standard is infeasible for the industry; rather, we remand that portion for reconsideration or clarification by the agency.

The textile industry appealed the decision to the Supreme Court primarily to gain a review of the matter of cost-benefit analysis, in hopes of gaining a requirement that OSHA do such analyses. The hearing was held on January 21, 1981. Subsequently, the new administration requested that the Supreme Court remand the case for further work, since the new OSHA intended to rereview the cotton dust record and carry out additional cost–benefit calculations. In the meantime, the industry made substantial progress in implementing the new standard. Ironically, OSHA did perform quite detailed cost–benefit analyses during the Regulatory Analysis Review Group process in the Carter administration, but OSHA at that time decided to take the position that such analyses could not be required of the agency.

Finding complementary actions or alternatives to regulation

There are inherent difficulties when the law is the predominant mechanism for the social control of technology. As Ashford (1976) pointed out, the law cannot be implemented successfully so long as one continues to ignore the fundamental conflicts and tensions that exist between groups of people or between institutions in our society. No single or simplistic approach can succeed. It is necessary to seek convergence of multiple approaches: from the law and the courts; from market incentives; from generating, disseminating, and utilizing relevant knowledge; and from development of personnel in unions, management, professions, and government with the requisite knowledge of the technical and policy issues.

Most economists prefer "free-market" approaches to achieve defined social goals, even such broad goals as health and welfare. R. S. Smith, for example, began with the question, "Are we producing 'enough' safety and health?" Since resources are required to enhance safety and health, he wants us to determine whether the people who derive benefits from the "product," especially the employers, are willing to pay for it. Benefits accrue to the employers in the forms of higher productivity, less time lost to injuries and illness, less turnover, less need to replace experienced workers with untrained workers. The quite imperfect workmen's compensation system also provides incentives for the employer by shifting at least part of the cost of injuries from the individual employee to the employer, giving rise to the adage that "safety pays." To be effective, however, a drastic change in workmen's compensation premiums would be required, moving from the industry injury experience to a rate of "injury tax" levied on the basis of the individual firm experience.

Any such program could be combined with a government-sponsored information campaign for workers and employers, allowing employers to respond with flexible, up-to-date, hazard-related programs to reduce the injury rates and the resultant injury tax. Such an approach has the advantages of avoiding the cumbersome and controversial standard-setting process, the difficulties of updating standards to cover technological advances, the failure of any standard to relate to hazards across a broad array of operations, and the reliance on enforcement by citation.

Even Smith insists that a standards approach must be maintained for occupational health, although he would utilize a cost–benefit analysis based upon the question, "Would the beneficiaries be willing to pay the costs of this standard if they were fully informed as to the hazards involved?" By his criterion, the noise standards that OSHA proposed in 1975 were too stringent, the asbestos standard was too lenient, and the seat belt interlock system was too stringent.

A crucial ethical issue is the definition of beneficiaries: One might argue that the employers and stockholders are beneficiaries of the production, just as are the wage earners.

Tax approaches and tort approaches are alternative or additional possibilities. Tort approaches were rejected explicitly in the passage of the OSH Act, and tax approaches, which were very little discussed, would have been rejected, according to Mendeloff (1979). The latter conclusion is based in part on commentary from Kneese and Schultze asserting that congressional staffs generally lack economists and analysts and that the legal background of most congressmen inclines them to make the reallocation of legal rights and duties, rather than economic incentives, the usual tool in seeking changes in behavior. The congressional inclination is strongly reinforced by interest-group lobbying.

The OSH Act did include a provision for making more information available to workers, mostly because of pressure from Ralph Nader's group, which later published a book entitled *Bitter Wages* (Page, 1973). The motivation was not so much to offer workers information upon which they might act individually to protect themselves or remove themselves from undue exposures but to arm workers with information with which they collectively might overcome the inertia of both union and government bureaucrats.

It is expected that the Reagan administration and the Congress will open discussion about alternatives to regulation in the area of occupational safety and health. The Department of Labor, in response to a directive in the 1977 Black Lung Benefits Reform Act, produced a report in 1980 examining various alternatives to improving occupational health. Health protection is much more difficult, as noted earlier, than the safety issue, since direct causal links are few, confounding factors are many, and years or decades may be required for the clinical onset of disease. Thus, it is extremely difficult to fix the costs and responsibilities with individual employers in many cases. The department reviewed suggestions to restructure the worker's compensation systems, to build upon Social Security disability insurance programs, to establish compensation programs for specific diseases (black lung model), to increase product liability compensation, to enhance medical and environmental monitoring, and to increase worker education and training. A rational approach would combine these avenues with high-priority, well-based regulatory actions.

Early actions of OSHA in the Reagan administration

The new administrator of OSHA, Thorne Auchter, a former construction industry official, moved quickly to put his stamp on the agency (Smith, 1981). In general, he relaxed, postponed, or withdrew many of the occupational health requirements promulgated by the previous administration. OSHA withdrew a requirement for labeling toxic chemicals in the workplace, charging that it was too complicated; ordered reconsideration of the lead and cotton dust standards; exempted the construction industry from a requirement to provide medical records to workers; delayed a requirement that the smelting industry comply with stiffer limits on permissible levels of lead in blood samples; and downgraded efforts to develop new health standards for asbestos, formaldehyde, ethylene oxide, and cadmium. The health standard for noise, issued on January 16, 1981, by the outgoing Carter administration, was postponed, with a decision promised by August. On August 20, Auchter announced that the standard would be placed at 85 decibels, as before, and that monitoring would be required but that some flexibility would be offered employers in finding cost-effective means of compliance. A few aspects were left open for further comment, including the question of how much information must be made available to workers about the noise levels at their job sites. Auchter stated that he seeks greater emphasis on workplace safety than on health hazards, many of which are hard to demonstrate. He wants to leave enforcement more to the states, a move that organized labor resists, because of past experience with this approach. Finally, he slashed the budget for educational materials, after ordering the destruction of a booklet on brown lung disease that was written and graphically illustrated for workers in the textile industry.

Workers' responsibilities

A very interesting test of workers' responsibilities in ensuring their own health is the interaction of asbestos exposure and cigarette smoking. It is well established that asbestos exposure can induce chronic lung disease, lung cancer, and cancer of the pleura or peritoneum. However, a large majority of excess cases of cancer occurs in smokers (Hammond and Selikoff, 1973). As a result, a leading producer of asbestos, in addition to making major efforts to reduce ambient exposures, proposed a policy under which no smokers would be hired for certain jobs and current employees who are smokers would be offered the choice of participating in a company-sponsored "stop smoking" campaign or else being transferred to a different job within the company. Workers have taken the company to arbitration and to court, charging discrimination against smokers. Thus far, at least three arbitration panels have upheld the company position (Paul Kotin, letter to author, 1981).

As in so many issues in our society, the courts will have to become involved in addressing the distribution of responsibility for health protection in the workplace. It is understandable that companies wish to reduce their liability for high-risk workers and to take as many measures as feasible to protect the health of their employees; on the other hand, unions, on the basis of much bitter experience, suspect companies of shifting responsibility and cost and of thereby denying certain employees or potential employees some job opportunities.

Distribution of responsibility is not a new issue. Often worker practices offered relatively easy routes to improved health, although sometimes the opportunities were missed for many years. For example, Percivall Pott's report more than two hundred years ago on the occurrence of cancer of the scrotal skin in English chimney sweeps concluded that gross contact with soot was the cause of the cancer. His emphasis on its long latent period stands as a milestone in occupational medicine. However, Pott did not suggest avoidance of contact with soot as a means of prevention. Perhaps he thought it impractical for a chimney sweep, an essential worker in Pott's London, to do so. Nevertheless, as Clemmensen (1951) noted, Pott's 1775 book with its brief paragraphs on "soot wart" apparently inspired a ruling just three years later by the Danish Chimney Sweepers Guild that its members should bathe daily. The value of that ruling was revealed a century later when Butlin (1892) investigated the relative rarity of scrotal skin cancer in chimney sweeps outside of England and found that on the Continent it was apparently prevented by frequent bathing and protective clothing.

J. B. S. Haldane went even further in his book, *Heredity and Politics* (1938): "But while I am sure that our standards of industrial hygiene are shamefully low, it is important to realize that there is a side to this question which so far has been completely ignored. The majority of potters do not die of bronchitis. It is quite possible that if we really understood the causation of this disease we should find that only a fraction of potters are of a constitution which renders them liable to it. If so, we could eliminate potters' bronchitis by regulating entrants into the pottery industry" (p. 179).

As nasty as his prescription may sound, we must recognize that a good deal of screening and denial of job opportunity exists in this country, totally unrelated to genetics, whether based upon sex (now outlawed) or back X-rays (see Rockey, Fantel, and Omenn, 1979) or other attributes, including well-recognized preexisting illness. The ethical and legal ramifications of practical decisions about workers' rights, about protection of workers' health, and about limitation of employers' liability were highlighted in a brief case study in the Hastings Center Report (Stevenson 1980). It concerned a thirty-two-year-old man who took a position as a technician in a chemical plant and was then found to have chronic renal disease. The commentators discussed the autonomy of individual workers, the possibility of false representation, and the claims to a single job opening by different persons on the competing grounds of affirmative action for medically handicapped individuals, of seniority, and of equal opportunity for women. Real-life case studies of this type can offer a useful complementary perspective to the studies presented in this volume, based upon the standard-setting processes for individual chemical hazards.

OSHA compliance strategy

OSHA has a variety of strategies available to enhance occupational safety and health. Enforcement through inspections and penalties has dominated OSHA's approaches thus far, but in the Carter administration increasing attention was given to penalty-free consultations, to technical assistance, and to training programs for employees and employers. However, this balanced approach came into conflict with yet another federal court decision dealing with state benchmarks to be developed by OSHA for inspection forces.

The background of this case again is found in the OSH Act of 1970.

The Act preempted all state workplace health and safety laws, but it allowed states to reenter the programs if the Department of Labor approved their plans. The Act specified eight criteria for approval, most of which require state programs to be "at least as effective as the Federal program." Two criteria specify that the state give satisfactory assurances that a state agency will have qualified personnel necessary for the enforcement of such standards and that the state will devote adequate funds to the administration and enforcement of such standards.

The AFL-CIO, which strongly opposes state-level programs, brought suit. In 1978, a U.S. Court of Appeals finally decided that the criteria in the Act require approved state plans to have "fully effective" inspection forces and directed OSHA to develop a five-year plan with "benchmarks" for numbers of inspectors to make state inspection forces "fully effective." The court left it to the Department of Labor to determine the definition of "fully effective" inspection forces, but the court itself defined fully effective enforcement in terms of inspection personnel.

Unfortunately, OSHA has no good evidence on the relative effectiveness of enforcement vis-à-vis other compliance strategies, and the court decision forced OSHA to rely only on inspections at a time when OSHA itself was modifying the federal strategy to form a balanced array of compliance activities. To produce the benchmark model, OSHA needed information on five variables: number and characteristics of industry establishments by state; anticipated number of inspections and time required for each accident, complaint, and other special inspection; number of inspections each inspector can perform; frequency of general inspections necessary (by establishment type) for "fully effective" enforcement; and time required for a "fully effective" inspection, by industry.

OSHA had fairly good data on the first three variables. However, there is no useful information on frequency or intensity of inspections required to induce compliance with OSHA standards, and these are the two most critical variables in the benchmark model. Nor is there any way to resolve the essential subjective and value judgments of the question "How effective is 'fully effective'?" with scientific and technical data. The court did not allow time to seek new information in any systematic way. So OSHA convened a panel of practitioners and others expert in the field of occupational safety and health to present a consensus recommendation on the frequency and duration of "fully effective" inspections. Rather than recognize the essentially political nature of this process, OSHA sought to clothe the subjective and social characteristics of the decision in technical terms – by employing a panel of experts and a quantitative model. This exercise was a serious distortion of the expert panel approach, since there was a gross lack of relevant data, and the experts did not have experience with the criterion "fully effective" that the court had imposed.

Fortunately the court decision does not bind OSHA to provide the resources necessary to implement the model benchmarks. The numbers of personnel, both federal and state, are simply not feasible within agency resources. In fact, the political process was reactivated late in the budget process for the 1982 federal budget (December 1980), when top officials of the Department of Labor and of the Office of Management and Budget (including the author) reviewed the priorities of OSHA. It was decided to continue the balanced compliance strategy, providing additional resources for technical assistance, rather than to seek implementation as early as 1982 for the benchmark inspection forces.

Conclusion

As we review individual OSHA decisions in standard setting and in enforcement in the case study approach described in this volume, several general issues should be kept in mind:

- What specific information triggered OSHA's interest and determined OSHA's pace: an accident, a sensational news story, a scientific report, a labor dispute, related action by another regulatory agency?
- Is the standard well based in scientific and technical data, or does it represent mostly judgment and intuition about likely risks?
- Was a reasonable effort made to obtain balanced and expert scientific advice? How influential was such advice?
- Does the standard protect only the average worker or ensure protection for the most susceptible?
- Does the OSHA decision to take action on a particular hazard reflect good judgment about priorities vis-à-vis other actions that need to be undertaken, within available resources and within the tolerance of the business community, labor, and the public?
- Is regulation necessary, or would some other OSHA approach be sufficient or preferable?
- What compliance strategy, if any, has been initiated?
- If cost data are available, who prepared the estimates? Are the costs allocated to different affected parties or just lumped together in a way that tends to defy critical analysis? If

benefits are estimated, what is the distribution of benefits, and how does that distribution compare with the distribution of costs?

Has OSHA provided a well-documented and logical record to support its decisions?

In assessing the extent and significance of "closure," we must recognize in the general application of the OSH Act and in the various specific cases that political decisions often represent agreements to disagree, to compromise in ways that leave the conceptual, ethical, and interest-group arguments open for continued debate in the same or other forums.

There have been numerous efforts to revise the OSH Act, with little effect thus far. There have been striking shifts in the approaches of those implementing the Act. And there have been pitched battles in agency hearings and in the federal courts when preliminary or final rules are promulgated. The interested parties are known to have agents positioned so that they can rush to file petitions in "friendly" federal courts immediately after OSHA formally announces a new ruling. Thus, it was no accident that the benzene case began in the Fifth Circuit in Louisiana, producing for the American Petroleum Institute a different result than would have been expected in the District of Columbia circuit. Furthermore, it is no accident that in the benzene case the Supreme Court found a way to issue a decision (close the case) based upon narrow issues rather than confront the larger questions of congressional intent, public interest, and moral values in the still-open issue of whether or not OSHA is required to perform cost effectiveness and cost–benefit analyses.

The broad field of occupational safety and health offers many opportunities for the full range of mechanisms to seek closure. As Rich has emphasized (Chap. 5, this volume), specific issues may arise in a wide array of political and economic institutional settings, each having preferred mechanisms for closing or sustaining controversies.

Sound argument and negotiation together represent the strongest basis for ferreting out factual disputes and seeking mutually acceptable solutions. In everyday working situations, this mode is predominant. As Beauchamp noted (Chap. 1, this volume), many controversies turn more on incomplete factual information than on differences of normative principle. In more controversial situations, labor and management may resort to arbitration, especially if the situation is covered by contract terms that are routinely referred to arbitration. These modes of conflict resolution are most applicable in situations where hazards are recognized as hazards and both sides are willing to work toward reduction of the risks in specific workplaces. In order to gain attention for hazards in certain settings, labor has found it necessary to seek broader arenas, especially government intervention. The inclusion of occupational safety and health issues in collective bargaining and in agreements for arbitration serves as a good indicator that the union is seriously interested in these issues. Unions and their contracts vary significantly in their investment of effort in pursuing these issues, compared with issues involving wages, hours, and benefits. It is rare for a union to strike over health hazards.

Expert scientific panels are more likely to be called in by regulatory agencies to examine evidence of previously unrecognized hazards. They can be valuable in helping to review the evidence but cannot be expected to generate action plans and should not be forced to play policymaker for reducing those hazards on a national or industrywide basis.

Generalizations about the value of each type of closure mechanism are particularly awkward in our discussions of occupational safety and health, since we have never focused the topic on a specific decision or action. Can we reach closure on the notion that the health and safety of workers is important and should be protected? Yes, we have, through the procedural and rhetorical device of an act of Congress and the administrative machinery that the act prescribed. Can we reach closure on the extent of responsibility of the worker and of the employer? I doubt that we can, since the question is far too vague. Can we reach closure on a more specific question – whether workers should be absolved of all personal responsibility for safety and health protection? I think the answer should be no, both on practical grounds and on the moral grounds having to do with the effects of irresponsible actions upon fellow workers and the production line. Many would insist that all responsibility lies with the employer and would cite legal precedents. Can we reach closure on the need to regulate exposures to asbestos? Yes, there should be no doubt. Can we reach closure on the precise level of asbestos to be tolerated? Probably not, unless it is economically quite simple to reach zero exposures. Should preemployment screening examinations be permitted as a basis for excluding applicants from work? Such a question raises too many more questions and cannot be subjected to a brief answer or a closure mechanism until it is made far more specific. In fact, the Hastings Center has launched a new project devoted to formulating the many practical issues and ethical aspects suggested by that question.

From the beginning, our discussion of the occupational safety and health topic has been broad and general, quite in contrast to the issue of whether the FDA should permit Laetrile to be prescribed as a drug or the issue of whether there should be a moratorium on the building of nuclear power plants. Labor–management relations are such an important aspect of our society and such a rich reflection of our societal tensions and values that every type of closure mechanism can be employed. In the pragmatic world of labor–management relations, partial closure by compromise within morally acceptable or morally unobjectionable boundaries will be frequent (Beauchamp's "negotiation closure"). Nonepistemic factors, strongly emphasized by McMullin, (Chap. 2, this volume), have tremendous influence.

Nevertheless, we can be confident that our discussions about closure on this complex and deeply rooted set of problems will not lead to closure. OSHA, more than all of the other regulatory agencies created in 1969–70 in the wave of environmental protection laws, stands in the midst of centuries of conflict between workers and management, a tension that is certain to remain, no matter how thoughtfully or imaginatively moderated by open discussion and common social purpose.

References

Ashford, N. 1976. *Crisis in the Workplace*. Cambridge, Mass., MIT Press.

Bazelon, D. March 1980. "Risk and Public Policy: To Live and . . . Become Bold." Fourth Annual William O. Douglas Award Lecture, March, 1980, p. 5.

Butlin, H. T. 1892. "Cancer of the Scrotum in Chimney-sweeps and Others": II. "Why Foreign Sweeps Do Not Suffer from Scrotal Cancer." *British Medical Journal* 2:1.

Calkins, D. R., Dixon, R. L., Gerber, C. R., et al. 1980. "Identification, Characterization and Control of Potential Human Carcinogens: A Framework for Federal Decision-making." *Journal of the National Cancer Institute* 64:169–175.

Clemmensen, J. 1951. "On the Etiology of Some Human Cancers." *Journal of the National Cancer Institute* 12:1.

Friedman, R. D. 1979. *Individual Susceptibility and Environmental Standards: A Report to the Office of Science and Technology Policy*. Washington, D.C.: Conservation Foundation.

Greenwood, T. 1984. *Knowledge and Discretion in Government Regulation*. New York: Praeger.

Haldane, J. B. S. 1938. *Heredity and Politics*. New York: Norton.

Hammond, E. C., and Selikoff, I. 1973. "Relation of Cigarette Smoking to Risk of Death of Asbestos-associated Disease among Insulation Workers in the United States." In Bogovski, P., et al. eds., *Biological Effects of Asbestos*. International Agency for Research on Cancer, Scientific Publication No. 8. Lyon, France, pp. 312–17.

Hastings Center. Dec. 1980. "Case Study on Health Risks and Equal Opportunity." *Hastings Center Report* 10 (6): 25–6.

Interagency Regulatory Liaison Group Work Group on Risk Assessment. 1979. "Scientific Basis for Identification of Potential Carcinogens and Estimation of Risks." *Journal of the National Cancer Institute* 63:214–48.

Kantrowitz, A. 1978. "A Technologist Looks at Anti-technology." Messenger Lectures, Cornell University, November 6–10, 1978.

Lee, W. R. 1977. "Some Ethical Problems of Hazardous Substances in the Working Environment." *British Journal of Industrial Medicine* 34:274–80.

Mazur, A., Marino, A. A., and Becker, R. O. 1979. "Separating Factual Disputes from Value Disputes in Controversies over Technology. *Technology in Society* 1:229–37.

Mendeloff, J. 1979. *Regulating Safety: An Economic and Political Analysis of Occupational Safety and Health Policy*. Cambridge, Mass.: MIT Press.

National Research Council. 1983. *Risk Assessment in the Federal Government: Managing the Process*. Washington, D.C.: National Academy Press.

Omenn, G. S. 1979. "Genetics and Epidemiology: Medical Interventions and Public Policy." *Social Biology* 26:117–25.

Omenn, G. S. 1982. "Predictive Identification of Hypersusceptible Individuals." *Journal of Occupational Medicine* 24:369–74.

Page, J. A. 1973. *Bitter Wages: Ralph Nader's Study Group Report on Disease and Injury on the Job*. New York: Grossman.

Peters, J. M., Monson, R. R., Burgess, W. A., et al. 1976. "Occupational Disease in the Rubber Industry." *Environmental Health Perspectives* 17: 31–4.

Pott, P. 1963. *Chirurgical Observations Relative to the Cataract, the Polypus of the Nose, the Cancer of the Scrotum, the Different Kinds of Ruptures, and the Mortification of the Toes and Feet*. London, 1775. Reprinted in National Cancer Institute Monograph No. 10, Bethesda, Maryland, pp. 7–13.

Rockey, P. H., Fantel, J., and Omenn, G. S. 1979. "The Legal Implications of the Use of Pre-employment Low Back X-ray Examinations in the Railroad Industry." *American Journal of Law Medicine* 5:197–214.

Smith, R. J. 1981. "OSHA Shifts Directions on Health Standards." *Science* 212:1482–83.

Smith, R. S. 1976. *The Occupational Safety and Health Act: Its Goals and Its Achievements*. Washington, D.C.: American Enterprise Institute for Public Policy Research.

Stevenson, Robert F. 1980. "Health Risks and Equal Opportunity," commentaries by Jerome A. Paulson and Laurence Thomas. *Hastings Center Report* 10:25–6.

Task Force of the Presidential Advisory Group on Anticipated Advances in Science and Technology. 1976. "The Science Court Experiment: An Interim Report." *Science* 193:652–6.

Terkel, S. 1972. *Working: People Talk about What They Do All Day and How They Feel about What They Do*. New York: Random House (Pantheon Books).

U.S. Consumer Product Safety Commission. June 13, 1978. "Classifying,

Evaluating, and Regulating Carcinogens in Consumer Products." *Federal Register* 43:25658.

U.S. Department of Labor. 1980. *An Interim Report to Congress on Occupational Diseases*.

U.S. Environmental Protection Agency. May 25, 1976. "Interim Procedures and Guidelines for Health Risk and Economic Impact Assessments of Suspected Carcinogens." *Federal Register*. 41:21402.

U.S. Food and Drug Administration. March 20, 1979. "Chemical Compounds in Food-producing Animals: Criteria and Procedures for Evaluating Assays for Carcinogenic Residues." *Federal Register* 44:17070.

U.S. Occupational Safety and Health Administration. October 4, 1977. "Identification, Classification and Regulation of Toxic Substances Posing a Potential Occupational Carcinogenic Risk." *Federal Register* 42:54148.

20

The successful experiment that failed

MARVIN S. LEGATOR

Epidemiological studies are of limited value in detecting agents that can induce neoplasms. There are, however, a series of procedures that can be performed in high-risk populations that have the potential to identify carcinogenic and mutagenic chemicals in our environment. Here I will discuss the incidence of cancer in the workplace; the limitations of classical epidemiological studies; available relevant short-term procedures for detecting carcinogens or mutagens in the workplace; the Dow Chemical Company experience with these methods; and the reasons why they are not being routinely used in industry to screen workers for exposure to carcinogenic and mutagenic agents.

Cancer and the workplace

The prevention of cancer and genetic diseases, as opposed to their treatment, is based on three premises: (1) Cancer and genetic diseases are caused, in part, by exogenous factors such as chemicals in the workplace; (2) such factors can be identified; and (3) once identified, these factors can be eliminated, or measures can be taken to minimize exposure.

At the present time the debate is particularly active about how much our life-style causes or contributes to our overall cancer burden – that is, how much our personal habits (including nutrition, smoking, and alcohol consumption), as opposed to factors not under our personal control (such as air pollution, food additives, and the occupational environment), contribute to the development of cancer.

Cancer is a disease caused by many factors and such factors as smoking and occupational hazards can act in an additive or synergistic fashion with other elements in the genesis of a specific neoplasm. It is reasonable to state that both life-style and daily exposure to chemicals in the environment, as they affect genetic composition, will determine whether malignancy is to develop.

Although an exact percentage cannot be assigned to compare the importance of life-style with environmental hazards in the induction of cancer or genetic diseases, there can be little question that a significant factor, especially in the petrochemical industry, is exposure of workers to the materials that they handle. Although only an extremely small percentage of industrial chemicals has been adequately studied, we have already identified arsenic, asbestos, auramine, benzene, benzidine, bischloromethyl ether, cadmium oxide, chrome, hematite, 2-naphthylamine, nickel, *N*, *N*-bis(2-chloromethyl) 2-naphthylamine, vinyl chloride, and soot and tars as known human industrial carcinogens. In fact, as of 1982, of the 21 chemicals or processes that have been shown to be carcinogenic in humans, 15 are present in various workplaces. In addition to the 21 known human carcinogens, approximately 200 additional chemicals have been identified as potential human carcinogens, on the basis of definitive animal studies.[1]

In a monograph on cancer and the worker published by the New York Academy of Sciences,[2] our poor record of protecting the worker was emphasized in the following examples:

> Thousands of coke-oven workers in the steel industry are still inhaling the same kinds of substances that caused cancer among the chimney sweeps. The result: they are dying of lung cancer at a rate 10 times that of other steel workers.
>
> One hundred years ago miners in Central Europe were found to be dying of lung cancer. Fifty years ago, scientists identified radioactivity in the mines as the cause of their disease. Yet in 1971, thousands of American uranium miners were working with radioactive materials under conditions that tripled their chances of dying from lung cancer.
>
> Today, 130 years after scrotal cancer was discovered among copper smelter workers exposed to arsenic, some 1.5 million American workers are inhaling the same substance. These workers are also dying of lung and lymphatic cancer at two to eight times the average national rate.
>
> Eighty years ago scientists discovered that aromatic amines

[1] L. Tomatis, "The Value of Long-Term Testing for the Implementation of Primary Prevention," in H. H. Hiatt, J. D. Watson, and J. A. Weinstein, eds., *Origins of Human Cancer*, Cold Spring Harbor Symposium on Cell Proliferation No. 4 (Cold Spring Harbor, New York, Cold Spring Harbor Laboratory, 1977), pp. 1339–580.

[2] M. Boland, ed., *Cancer and the Worker* (New York: New York Academy of Sciences, 1977).

were causing bladder cancer among German dye workers. Such amines as benzidine and beta-naphthylamine were banned or taken off the market decades ago in the United Kingdom, Switzerland, Japan, Italy, and the Soviet Union. Yet in 1973, thousands of American workers were still literally sloshing around in these chemicals. These workers are now developing – and will continue to develop – bladder tumors at an epidemic rate. As of three years ago, half (27) of the former employees at one benzidine plant had developed bladder cancer.

Seventy-five years ago asbestos was known to cause a fatal lung disease. It was identified 25 years ago as a potent cause of lung cancer. But very recently, workers in dozens of asbestos factories and hundreds of asbestos-related trades were laboring in asbestos dust so thick that it blocked the light. The grim outlook: Of one million current and former asbestos workers in this country 300,000–400,000 can expect to die of cancer, more than double the national rate.

Table 20.1 lists incubation or latent period, risk ratio, and occupation for several chemical carcinogens. This compilation was reported in 1974.[3] In conjunction with this listing, however, it should be emphasized that the vast majority of industrial chemicals have not been evaluated either in animals or in humans for carcinogenic activity.

Although we do not have the information necessary to place an exact figure on the number of cancers due to employee exposure in the workplace, there can be little doubt that it is substantial. The workplace has been characterized as a mystery world. At the present time, there is no way of determining what the 20,000 to 40,000 industrial chemicals introduced as part of manufacturing processes in the past several years may be doing to workers' health. Equally alarming is the fact that what happens to the worker is simply an exaggerated effect of what occurs in the general population, because the worker is more highly exposed than the general public.

Studies carried out with vinyl chloride and asbestos illustrate that effects occur with industrial chemicals far beyond the confines of the plant.[4] Chemicals that have been identified as carcinogens have been

[3] P. H. Cole and M. B. Goldman, "Occupation," in J. P. Fraumeni, ed., *Persons at High Risk of Cancer* (New York: Academic Press, 1975), pp. 167–84,

[4] J. K. Wagoner and P. F. Infante, "A Review of the Methodologic Approaches in the Assessment of an Association between Vinyl Chloride Exposure and Reproductive Hazards," in *Proceedings of a Workshop on Methodology for Assessing Reproductive Hazards in the Workplace*, PHHS (NIOSH) Publication No. 81–100 (Washington, D.C.: U.S. Government Printing Office, 1978), pp. 43–52.

Table 20.1. *Classification of occupational carcinogens*

Agents	Affected organs	Latent period (in years)	Risk ratio	Occupation
Organic agents				
Aromatic hydrocarbons:				
Coal soot Coal tar Other products of coal combustion	Lung, larynx, skin, scrotum, urinary bladder	9–23	2–6	Gashouse workers, stoker, producer of asphalt and coal tar, pitch workers; coke oven workers; mine still cleaners; chimney sweeps
Petroleum Petroleum coke Wax Creosote Anthracene Paraffin Shale Mineral oils	Nasal cavity, larynx, lung, skin, scrotum	12–30	2–4	Contact with lubricating, cooling, paraffin or wax, fuel oils, or coke rubber fillers; retortment; textile weavers; diesel jet testers
Benzene	Bone marrow (leukemia)	6–14	2–3	Explosives, benzene, or rubber cement workers; distillers; dye users, painters; shoemakers
Auramine Benzidine α-Naphthylamine	Urinary bladder	13–30	2–90	Dyestuffs manufacturers and users; rubber workers (pressmen, filtermen,

β-Naphtylamine				laborers); textile dyers; paint manufacturers
Magenta				
4-Aminobiphenyl				
4-Nitrobiphenyl				
Alkylating agents:				
Mustard gas	Larynx, lung, trachea, bronchi	10–25	2–36	Mustard gas workers
Others:				
Isopropyl oil	Nasal cavity	10+	21	Producers
Vinyl chloride	Liver (angiosarcoma), brain	20–30	200 (liver) 4 (brain)	Plastic workers
Bis(chloromethyl) ether Chloromethyl methyl ether	Lung (oat cell carcinoma)	5+	7–45	Chemical workers
Inorganic agents				
Metals:				
Arsenic	Skin, lung, liver	10+	3–8	Miners; smelters; insecticide makers, and sprayers; tanners; chemical workers; oil refiners; vintners
Chromium	Nasal cavity and sinuses, lung, larynx	15–25	3–40	Producers, processors, and users; acetylene and aniline workers; bleachers; glass, potters, and linoleum workers; battery makers
Nickel	Nasal sinuses, lung	3–30	5–10 (lung) 100+ (nasal sinuses)	Nickel smelters, mixers, and roasters; electrolysis workers

Agents	Affected organs	Latent period (in years)	Risk ratio	Occupation
Fibers:				
Asbestos	Lung, pleural, and peritoneal mesothelioma	4–50	1.5–12.0	Miners; millers; textile, insulation, and shipyard workers
Dusts:				
Wood	Nasal cavity and sinuses	30–40		Woodworkers
Leather	Nasal cavity and sinuses, urinary bladder	40–50	50.0 (nasal sinuses), 2.5 (bladder)	Leather and shoe workers
Physical agents				
Nonionizing radiation:				
Ultraviolet rays	Skin	Varies		Farmers; sailors
Ionizing radiation:				
X-rays	Skin, bone marrow (leukemia)	10–25	2–9	Radiologists, medical personnel
Uranium Radon Radium Mesothorium	Skin, lung, bone marrow, (leukemia)	10–15	3–10	Radiologists, miners, radium dial painters, radium chemists

Source: Data from P. H. Cole and M. B. Goldman, "Occupation," in J. P. Fraumeni, ed., *Persons at High Risk of Cancer* (New York: Academic Press, 1975), pp. 167–84.

found, in addition to inducing neoplasms, to cause a spectrum of genetic effects, including chromosomal damage and gene mutations. These effects cannot be detected in classical epidemiological studies, but data from animal studies show that there can be little doubt about the potential for them. Indeed, the spectrum of genetic effects induced by chemical carcinogens, although hidden, may be far more important than the frank induction of cancer.

The human carcinogenic potential of industrial materials should be investigated in definitive animal studies. Ideally these studies should be carried out before such chemicals are released into the marketplace. Most large chemical companies are presently screening new products for carcinogenicity, and with stricter regulatory requirements it is hoped that the majority of cancer-producing chemicals will be eliminated before human exposure occurs. Most chemicals in use at present, however, have either not been screened or have been inadequately screened for carcinogenic activity. Only with the advent of the Toxic Substance Control Act in the late 1970s have steps been initiated to evaluate industrial chemicals.

Shortcomings of conventional epidemiological studies for identifying carcinogens

Several obvious shortcomings may affect epidemiological studies of occupational or environmental cancer. First, it is difficult to assemble a sufficiently large group of workers (a cohort) all suffering the same exposure, to permit one to draw a definitive conclusion concerning the risk of cancer at a particular site. Second, although cancer is a "major killer" in the United States and other developed nations, from a statistical or population point of view, death from a particular type of cancer is actually a "rare" event.

Thus, the incidence (rate of new cases) of lung cancer, for example, in the United States in 1970 was 40 cases per 100,000; statistically speaking, the probability that an individual would contract lung cancer in the year 1970 in the United States was 1 in 2,500.[5] Consequently, because of this low risk and unavoidable statistical variation, detection of a slight excess in incidence of lung cancer would clearly require access to records of tens of thousands of specifically exposed individuals.

A recent, well-designed epidemiological study by O'Berg at Du

[5] U.S. Department of Health, Education and Welfare, *Vital Statistics of the United States, 1979*: Vol. 2, *Mortality* (Washington, D.C.: U.S. Department of Health, Education and Welfare, Public Health Service, 1970), Section 1, General Mortality, pp. 1–6.

Pont and Company compared the rates of cancer of the lung and of cancer in other sites in the body in workers matched for age, and for histories of smoking or not smoking, who were exposed to acrylonitrile. The study showed a significant increase in lung cancer mortality in the exposed workers. Yet the number of exposed workers was quite small.[6] In 1980 this apparent excess mortality from lung cancer in acrylonitrile-exposed workers was questioned because of the possibility of different smoking histories (degree and length of smoking) between exposed and unexposed workers, even though the study had been matched for smokers and nonsmokers.[7] Thus, when the risk is low and confounding effects by another important causative agent (such as smoking) are present, epidemiological studies require very large populations and extremely cautious design.

When a malignant tumor rarely occurs in the absence of exposure to occupational agents such as vinyl chloride or asbestos (tumors related to angiosarcoma of the liver and mesothelioma, respectively, are examples), and when the association of the agent to the disease is strong, it is easier to show an excess risk because of exposure. Otherwise, the amount of data that must be generated to designate a specific agent as a human carcinogen is formidable, and in many instances gathering the data is an impossible task. In addition to the large populations required, the long latency period for cancer induction (ten to forty years) adds a further problem to the epidemiological method: Multiple exposure to other agents during this latency period confounds the study.

The main drawback of an epidemiological study as a primary tool for safeguarding high-risk populations is that the results come too late or after the fact. Consider that in the new cases in which an agent has been identified as a human carcinogen, it has been identified only because a sufficiently large number of individuals has already developed the disease. It may be anticipated that many more will develop the disease after the symptomless latency period. Thus, classical epidemiological studies cannot be considered as a primary tool for identifying carcinogens. Although such studies can provide proof and can corroborate animal and short-term tests, ideal studies are difficult to perform, and the information, when it does surface, comes too late to permit remedial action.

[6] M. O'Berg, "Epidemiologic Study of Workers Exposed to Acrylonitrile," *Journal of Occupational Medicine* 22(4) (1980): 245–52.

[7] U.S. Environmental Protection Agency, *Acrylonitrile*, Health and Environmental Effects Profile No. 7 (Washington, D.C.: Office of Solid Water, April 30, 1980).

Ideal methods of identifying carcinogens in high-risk populations

Given the large number of chemicals that have not been evaluated and the severe limitations of classical epidemiological studies, what is needed are procedures that would allow carcinogens in the workplace to be identified without recourse to the usual epidemiological studies. Ideally we would like to monitor workers continuously by short-term procedures that could identify potentially harmful products and combinations of products long before clinical symptoms appear. The utopian approach to this problem would be to have a number of accurate, economical tests that could be conducted in humans that would not pose any danger to the subject being tested and that would yield results quickly. Further, the results should serve as an advance warning system so that we could take immediate action before any clinical symptoms occur.

In very general terms, our goal is to detect, quickly and accurately, those chemicals in the workplace that are potential carcinogens or mutagens long before we see any neoplasms or adverse birth effects. The fact is that at this very moment such procedures are available. It is entirely conceivable that we could, through utilizing a number of available procedures, and without any additional laboratory work, identify those chemicals in the workplace that are carcinogenic or mutagenic.

Tests for detecting carcinogens in the workplace

Theoretical basis

What are these procedures? They are a series of short-term, noninvasive tests that can be conducted in high-risk populations, using blood, urine, or semen, to determine whether recent exposure to carcinogens or mutagens has occurred. These procedures utilize the assumption that mutations are part of the carcinogenic process and that a high correlation exists between carcinogenic and mutagenic activities of chemicals. The theoretical basis for this association has been reviewed, and the most likely mutation model of carcinogenesis was presented.[8] The mutation model indicated that a series of discrete actions must occur in order for a chemical to induce cancer:

1. Adduct formation with DNA, or other DNA damage

[8] P. H. Cole and M. B. Goldman, "Occupation," in J. P. Fraumeni, ed., *Persons at High Risk to Cancer* (New York: Academic Press, 1975), pp. 167–184.

2. Induction of gene mutations
3. Induction of chromosome aberrations
4. Shift of DNA repair from error-free excision repair to error-prone systems, facilitating mutagenesis
5. Reduced DNA repair, elevating the frequency of chromosome aberrations
6. Suppression of DNA replication and mitosis in normal cells, facilitating clone formation of cells with mutated regulator genes
7. Induction of a lethal necrotic effect, leading to the compensatory growth required to establish clones of cells with mutated regulator genes

For the induction of cancer, this series of events occurs in somatic cells. A similar series of events, including point mutation and chromosomal damage (mutation), occurs in germinal cells, but the expression of the final outcome of the event is different. If a germinal cell mutation is not lethal, a live birth will occur, but the infant will have a genetic defect. If these events are lethal, however, this lethality may show up as early fetal loss, late fetal loss, abnormal sperm, or sperm lethality. All of the short-term procedures that are used to identify potential carcinogens and mutagens in the workplace capitalize on the spectrum of events that are induced in either somatic or germinal cells. The event that is measured can be classified as a mutagenic response. These responses simply identify a phase in the overall process that might result in an abnormal outcome, such as cancer or a genetic defect.

Table 20.2 lists the events and procedures that can be used in monitoring such events in humans, as well as the practicality of using the procedure at present to monitor high-risk populations. From Table 20.2 it can be deduced that the procedures available for monitoring workers include determination of three events: the occurrence of cytogenetic abnormalities, of point mutations, and of the spectrum of events in spermatogenesis, including the induction of an extra YF body.

Tests

The most useful test for evaluating adverse effects of chemicals in a human population is cytogenetic analysis. This technique, when used with a sufficient number of cells and a sufficient number of subjects in both exposed and control groups, is probably the most powerful tool available for evaluating high-risk groups. In a recent presentation, a summary of the assessment of cancer risks by the International

Table 20.2. *Carcinogens and mutagens in high-risk populations*

Event	Procedures	Status
Somatic cells		
1. Adduct formation with DNA (macromolecules)	Alkylation of macromolecules	Advanced developmental stage
2. Induction of gene mutations	Body fluid analysis	Available procedure
	HGPRT variants	Advanced developmental stage
3. Induction of chromosome aberrations	Metaphase analysis	Available procedure
	Sister chromatid exchange	Available procedure
4. Repair induction	Alkaline elution	Developmental stage
	Unscheduled DNA synthesis	Developmental stage
Germinal cells		
1. Early or late fetal loss	Increase in spontaneous abortion (obtained by questionnaire)	Available procedure but difficult to derive significant differences
2. Unequal chromosome segregation	YFF sperm study	Available procedure
3. Abnormal sperm	Test for morphological variants	Available procedure
4. Aspermia or oligospermia	Sperm count	Available procedure

Agency for Research on Cancer (IARC) (1980)[9] was correlated with cytogenetic studies.[10] Of the 442 chemicals evaluated by IARC since it began providing information on chemical carcinogens in 1968, sufficient evidence of carcinogenicity in animals had been accumulated in 112, but only 53 chemicals had been tested in human studies. These 53 chemicals are categorized as chemicals carcinogenic for humans (17, total); chemicals probably carcinogenic for humans (18, total); and chemicals for which data are insufficent (18, total).

Of the 17 chemicals or chemical processes classified as carcinogenic for humans, 4 have been tested for cytogenetic aberrations in humans and were found to be positive: arsenic compounds,[11] benzene,[12] bis (chloromethyl) ether and technical grade (chloromethyl) methyl ether,[13] and vinyl chloride.[14] Of the 18 chemicals or chemical processes classified as probably carcinogenic for humans, 3 have been tested for production of cytogenetic aberrations in humans, and results were positive: cyclophosphamide,[15] epichlorohydrin,[16] and tris(1-aziridinyl)phosphine sulfide (thio-TEPA).[17]

[9] International Agency for Research on Cancer, IARC Monographs on the Carcinogenic Risk of Chemicals to Humans, Volumes 1 to 20 (Lyon, France: World Health Organization).

[10] H. F. Stich and A. B. Acton, "Can Mutation Theories of Carcinogenesis Set Priorities for Carcinogen Testing Programs?" *Canadian Journal of Genetics and Cytology* 21 (1979): 155–77.

[11] G. Beckman, L. Beckman, and I. Nordenson, et al., "Chromosomal Aberrations in Workers Exposed to Arsenic," in Kare Berg, ed., *Genetic Damage in Man Caused by Environmental Agents* (New York: Academic Press, 1979), pp. 205–21; I. Nordenson, G. Beckman, L. Beckman, et al., "Occupational and Environmental Risks in and Around Smelters in Northern Sweden. II: Chromosomal Aberrations in Workers Exposed to Arsenic," *Hereditas* 88 (1978):47–50.

[12] I. M. Tough, P. G. Smith, B. Court, et al., "Chromosomal Studies in Workers Exposed to Atmospheric Benzene: The Possible Influence of Age," *European Journal of Cancer* 19 (1970): 49–55; D. Picciano, "Cytogenetic Study of Workers Exposed to Benzene," *Environmental Research* 19 (1979):33–8.

[13] Z. Zudova and K. Landa, "Genetic Risk of Occupational Exposures to Haloethers," *Mutation Research* 46 (1977):242–3.

[14] F. Funes-Cravioto, B. Lambert, J. Lindsten, et al., "Chromosome Aberrations in Workers Exposed to Vinyl Chloride," *Lancet* 1 (1975):459; A. Ducatman, K. Hirschhorn, I. J. Selikoff, "Vinyl Chloride Exposure and Human Chromosome Aberrations," *Mutation Research* 31 (1975):163–8.

[15] J. N. Etteldorf, C. D. West, J. A. Pitcock, et al., "Gonadal Function, Testicular History and Meiosis Following Cyclophosphamide Therapy in Patients with Nephrotic Syndrome," *Journal of Pediatrics* 88 (1976):206–12.

[16] M. Kucerova, V. S. Zhurkor, Z. Polivkova, et al., "Mutagenic Effect of Epichlorohydrin. II: Analysis of Chromosomal Aberrations in Lymphocytes of Persons Occupationally Exposed to Epichlorohydrin," *Mutation Research* 48 (1977): 355–60; D. Picciano, "Cytogenetic Investigation of Occupational Exposure to Epichlorohydrin," *Mutation Research* 66 (1979):169–73.

[17] T. G. Silezneva and N. P. Korman, "Analysis of Chromsomes of Somatic Cells in

Of the 18 chemicals or chemical processes that could not be classified as to their carcinogenicity for humans, although data from animal studies indicate their potential carcinogenic activity, 4 were found to be positive for cytogenetic aberrations: chloroprene,[18] ethylene oxide,[19] lead and lead compounds,[20] and styrene.[21]

Each of these agents is suspected of causing cancer, and in every case where human cytogenetic studies have been reported, the results indicate significant chromosomal damage. In fact, almost every known mutagen or carcinogen in animals has been shown to produce chromosomal abnormalities.

The evaluation of an exposed and nonexposed group for chromosomal abnormalities is a comparatively simple procedure; it requires a trained cytogeneticist, a light electronmicroscope, and several milliliters of blood. Presently, two types of analysis can be performed. The traditional method is to determine chromosome aberrations in metaphase cells. In this study, the identification and classification of chromatid and chromosome abnormalities is carried out, and both numerical and structural variations are identified.

The second procedure is a new method that is referred to as "sister chromatid exchange" (SCE) analysis. SCE involves a crossing over of sister chromatids either at the time of (or after) DNA replication. With certain chemicals this procedure has been found to be more sensitive than the standard metaphase analysis. There are classes of active chemicals, however, that do not increase sister chromatid exchange, and the genetic basis for this test is not fully understood. Although metaphase analysis is the preferred method, analysis for SCE can be a valuable adjunct.

Another test that can be carried out concurrently with cytogenetic

Patients Treated with Antitumor Drugs," *Soviet Genetics* 9 (1973):1575–9.

18 L. D. Katosova, "Cytogenetic Analysis of Peripheral Blood of Workers Engaged in the Production of Chloroprene," *Gigiena truda i professional'nye Zabolevaniia* 10 (1973):30–2; I. V. Sanotskii, "Aspects of the Toxicology of Chloroprene: Immediate and Long-Term Effects," *Environmental Health Perspectives* 17 (1976):85–93.

19 V. F. Garry, J. Hozier, D. Jacobs, et al., "Ethylene Oxide: Evidence of Human Chromosomal Effects," *Environmental Mutagens* 1 (1979):375–82.

20 A. Forni and G. C. Secci, "Chromosome Changes in Preclinical and Clinical Lead Poisoning and Correlation with Biochemical Findings," in *Proceedings of the International Symposium on Environmental Health Aspects of Lead, Amsterdam, October 2–6, 1972* (Luxembourg: Commission of the European Communities, Directorate General for Dissemination of Knowledge), pp. 473–485; R. Garza-Chapa, C. H. Leal-Garza, and G. Molina-Ballesteros, "Analysis Cromosomico en personas profesionalmente expuestas a contaminacion con plomo." *Archivos de Investigacion Medical (Mexico)* 8 (1977):11–20.

21 T. Meretoja, H. Vainio, and M. Sorse, "Occupational Styrene Exposure and Chromosomal Aberrations," *Mutation Research* 56 (1977):193–7.

analysis is the analysis of body fluids for the presence of genetically active compounds. In this procedure, samples of blood or urine, or both, are either tested directly or are concentrated and then evaluated for the presence of mutagenic agents. *Salmonella typhimurium* is often used as the indicator organism to detect the presence of chemicals that can induce genetic mutations; this analysis is a valuable addition to the cytogenetic tests because it can identify compounds that cause genetic mutations rather than chromosomal aberrations. Agents that have been identified with the indicator organism *S. typhimurium* in the urine assay include cyclophosphamide,[22] epichlorohydrin,[23] ethyleneimine imine,[24] metronidazole,[25] and niridazole.[26]

Recently a unique test has been developed that allows detection of the increase in the frequency of Y chromosomes in sperm, indicating a disjunctional error of the Y chromosome. If mature spermatozoa are stained with quinacrine dihydrochloride, the Y chromosome is visible under conventional fluorescent microscopy, with the Y body appearing as a single bright spot within the sperm. When the Y chromosome has failed to separate normally at anaphase of spermatogenesis, there are two Y chromosomes and therefore two fluorescent bodies. This technique has already detected an increase in the frequency of sperm bearing two Y bodies in persons exposed to chemotherapeutic agents, radiation, and dibromochloropropane. This procedure not only detects compounds that could induce cancer, but it also shows a specific effect in germinal cells.[27]

Sperm analysis for abnormal morphology can also be used. Several known mutagens and carcinogens have been identified in both

[22] D. Sievert and U. Simon, "Cyclophosphamide Pilot Study of Genetically Active Metabolites in the Urine of a Treated Human Patient: Induction of Mitotic Gene Conversion in Yeast," *Mutation Research* 19 (1973):65–72.

[23] M. S. Legator, L. Truang, and T. H. Connor, "Analysis of Body Fluids Including Alkylation of Macromolecules for Detection of Mutagenic Agents," in A. Hollaender and F. J. de Serres, eds., *Chemical Mutagens: Principles and Methods for Their Detection*, Vol. 5 (New York: Plenum Press, 1979), pp. 1–22.

[24] T. H. Connor, unpublished data.

[25] L. Dobias, "Human Blood Mutagenicity for *Salmonella typhimurium* Tester Strains after Oral Application of Entizol," *Mutation Research* 77 (1980):357–60.

[26] M. S. Legator, T. G. Pullin, and T. H. Connor, "The Isolation and Detection of Mutagenic Substances in Body Fluids and Tissues of Animals and Body Fluids of Human Subjects," in B. J. Kilbey, M. Legator, W. Nichols, et al., eds., *Handbook of Mutagenicity Test Procedures* (Amsterdam: Elsevier, 1979), pp. 149–59.

[27] W. R. Kapp, Jr., M. C. Benge, D. Picciano, et al., "Monitoring Y Chromosomal Nondisjunction in Humans with the YFF Sperm Test," in P.F. Infante and M.S. Legator, eds., *Proceedings of a Workshop on Methodology for Assessing Reproductive Hazards in the Workplace*, PHHS (NIOSH) Publication No. 81–100 (Washington, D.C.: U.S. Government Printing Office 1978), pp. 307–26.

animals and man with this procedure. The genetic validity of this test, however, has yet to be established.

The procedures just described form a comprehensive battery of tests that, properly used, should identify most hazardous substances that can induce cancer and genetic abnormalities in the workplace. There can be little doubt that the success record of these tests to date has been rather remarkable, and to the best of my knowledge they have produced no false positive results. They can almost be considered diagnostic for hazardous materials in the workplace and can serve as advance warning signs with chemicals, much as a radiation badge can tell us that we are being exposed to harmful radiation. It should be emphasized that these effects occur well in advance of clinical symptoms. These tests are noninvasive and comparatively economical, and the answers are forthcoming shortly after the tests are performed. These studies can be conducted with comparatively small numbers of individuals, and there is no compelling reason why these tests should not be incorporated as an integral part of a medical surveillance program in industry.

Justification of short-term procedures in a medical screening program

A list of criteria that must be met to justify any medical screening program has been developed. It includes the following features:

1. The disease must have a major and significant effect on the quality or length of life.
2. The disease must have an asymptomatic period in which detection and treatment significantly reduce morbidity and mortality.
3. Treatment in the asymptomatic phase must result in a better prognosis than after symptoms appear.
4. Tests must be available to detect the disease.
5. Incidence of the disease must be great enough to justify screening for it.

In addition to the above five criteria, other considerations include the progression of the disease, sensitivity and reliability of tests, cost of detection, benefit from treatment, and risk factors.

If we look at the battery of tests that can be instituted to detect carcinogens in the workplace, it can quickly be seen that they meet the established criteria to justify a screening program.

1. Cancer and genetic abnormalities affect both the quality and length of life.
2. Induction of cancer includes a latency period ranging from five to forty years; induction of genetic effects spans several generations.

Indeed, a long asymptomatic period must pass between the chemical insult and the resultant syndrome.

3. If we can remove the individual or curtail his or her exposure to environmental carcinogens, there is little doubt this can greatly reduce the probability of this individual's having cancer or genetic abnormalities.
4. A series of tests is presently available that can detect these hazardous chemicals.
5. Certainly the incidence of cancer and mutagenic abnormalities will justify any screening.

Other criteria that should be considered, such as the rate of progression of the disease, the sensitivity and the reliability of the tests, the cost of detection and benefit from early treatment – all would further argue for using these procedures. In fact, it is hard to describe any commonly used screening test carried out in human subjects that is as ideally suited for detecting future adverse effects as are the tests available for detecting carcinogens and mutagens in the workplace.

Interpretation of test data

There is a key difference between the short-term procedures described above and other methods that are used to detect malignancies in high-risk groups. The short-term procedures, for the most part, cannot be interpreted on an individual basis. The individual who is exposed to a hazardous substance and is showing an increase in any one of the short-term end points will not necessarily exhibit a malignant tumor or experience an adverse outcome in giving birth; therefore, these procedures must be interpreted from a population standpoint. If we find, for instance, that the cytogenetic test reveals an increase in either stable or unstable configurations among an exposed group as compared to a control group, we can deduce from that study that the exposed group has been in the presence of a chemical that has the potential to cause genetic damage, including cancer.

These procedures, therefore, identify a situation involving exposure to a chemical or process in the industrial environment that, if it persists, could have serious health consequences. Once positive test results are obtained, remedial measures can be initiated. These remedial measures may take the form of the wearing of protective clothing or of the elimination of exposure of workers. As Beauchamp has shown (Chap. 1, this volume), once a harmful effect has been detected, it is the right and obligation of the employer to report the findings of the study to the employees, and indicate what remedial action has been taken. In certain instances, remedial action may not be feasible, and in these cases the obligation would be for the employer to notify the employee of potential danger. In my experience, workers readily accept such tests when they are informed of the purpose of the tests and of their noninvasive nature and are given the explanation that these are population studies to detect harmful chemicals in advance of any deleterious effects.

As with every other biological test, the information that can be derived from these techniques depends upon utilization of a proper experimental design, including statistical analysis. In cytogenetic studies a number of variables have to do with inter- and intralaboratory variation. Exposure of individuals to drugs, viruses, and radiation; success or failure in selecting a proper, nonexposed control group; seasonal variation; the number of cells that are counted; and a variety of other factors can influence the experimental outcome. It should be emphasized, however, that for almost any test conducted either in animals or in humans, a series of variables can be listed that are similar in many instances to those listed for the cytogenetic test. If the variables are understood and properly taken into account, and if the work is carried out by competent investigators, the determination, on the basis of the data, as to whether a group of individuals has been exposed to a deleterious agent will be valid. The major problem, as in most testing procedures, is the possibility that because of faulty design a toxic chemical will not be detected (a negative, or beta, error).

Our limited experience with cytogenetic analysis in high-risk populations attests to the power of this form of testing. When combined with the other tests previously discussed, it is my opinion that we have at our disposal the tools to substantially prevent the induction of cancer and genetic abnormalities in the occupational setting, and probably in the population as a whole. As discussed by Omenn in this volume (Chap. 19), our major goal is to reduce carcinogens in the work environment rather than to test for the consequence of an existing disease.

The successful experiment that failed

Cytogenetic testing at Dow Chemical

In 1964, the Texas Division of the Dow Chemical Company initiated a cytogenetic testing program. By 1974, 43,044 shorthand and conventional karyotype tests had been performed on 1,689 workers involved in the production of various chemicals. During the same period of time, 25,104 karyotype tests were performed on 1,302 individuals prior to their employment.[28] All prospective employees were given a preemployment physical examination that included a cytogenetic analysis. By incorporating a cytogenetic analysis and the preemployment examination, a reference point was obtained for future chemical investigation of the individual, thus allowing for the accumulation of a large pool of control data. This procedure also had the advantage of allowing the individual to serve as his or her own control for subsequent studies. Many of the problems one could anticipate from such an industrial screening program were successfully obviated in the Dow facility. Certain individuals were found to have chromosome abnormalities unassociated with their work exposure, and these people were referred to suitable medical experts for genetic counseling. Because the program was explained carefully to the employees, little resistance was encountered, and the program seemed to be well accepted.

During the course of Dow's extensive program of cytogenetic monitoring, many important improvements in technique was developed. The minimal number of cells to be analyzed were determined; questionnaires to be used prior to cytogenetic study were developed; and a computer assistance program was designed. The Dow Chemical Company management was pleased with its own program, and, in fact, during tours that were conducted of the facility, it was mentioned that Dow was protecting its employees by the latest state-of-the-art procedures.

For example, in 1976, Dr. D. J. Kilian, then the medical director of the Texas Division of Dow Chemical, made the following comments on the cytogenetic screening procedures:

> Cytogenetic monitoring becomes a reliable and objective method of evaluating change in the human genetic material when there is a sizable population under competent medical surveillance, the work force is relatively stable and healthy, and when there is known exposure to diverse chemical compounds. Many companies within the chemical industry can be considered as ideal sponsors for such studies. They employ stable populations and subpopulations of significant size that are involved – at least to some extent – in the production of

[28] D. J. Kilian and D. Picciano, "Cytogenetic Surveillance of Industrial Population," in A. Hollaender, ed., *Chemical Mutagens: Principles and Methods for Their Detection*, Vol. 4 (New York: Plenum, 1976), pp. 321–40.

chemicals, some of which are suspected of being genetic hazards. In many cases, precise records of continuously monitored exposure to specific compounds, alone or in combination, are available. The number of individuals is generally large enough for good study design and statistical analysis. Employee turnover rates tend to be low, there is adequate technology for accurate monitoring, and there are scientific capabilities for evaluating what changes may be found in the individual. A substantial number of individuals (clerical and administrative personnel, for example) have no contact with the substances to be studied, and data from these groups are suitable for matched control evaluation. Interdisciplinary study design is enhanced by the ready availability of findings from periodic medical examinations and records from comprehensive health insurance plans.

Study of the entire work force and its subgroups can be expected both to verify existing standards and to provide reliable information about changing toxicological safety factors well before an overt hazard is imminent. Findings of this nature, of course, serve the general population, as well as the employees, by ensuring that appropriate protective measures can be taken upon early identification of genetic risk.[29]

In 1977, the scientists at Dow Chemical Company in Texas evaluated workers exposed to epichlorohydrin and to benzene. Tables 20.3 and 20.4 show the results found with these chemicals.[30] Both studies showed an increased number of chromosomal breaks with exposure. The positive results detected with these chemicals are not surprising, since benzene is a known carcinogen and epichlorohydrin had been shown to induce chromosome abnormalities in a previous study.

Shortly after the finding was made that benzene and epichlorohydrin increased chromosomal aberrations in exposed workers, the extensive screening program instituted at the Dow Chemical Company in Texas was phased out. The effort that was initiated in 1964 with the laudatory goal of identifying carcinogenic and mutagenic agents in the workplace is no longer in operation. However, the achievements of the program should not be overlooked:

[29] Ibid.

[30] D. Picciano, "Monitoring Industrial Populations by Cytogenetic Procedures," in P. F. Infante and M. S. Legator, eds., *Proceedings of a Workshop on Methodology for Assessing Reproductive Hazards in the Workplace*, PHHS (NIOSH) Publication No. 81–100 (Washington, D.C.: U.S. Government Printing Office, 1978), pp. 307–26.

Table 20.3. *Distribution of chromosome breaks related to benzene exposure*

Group	No. in group	Individuals with No Chromosome Breaks (%)	Individuals with Chromosome Breaks (%)
Controls	44	59.1	40.9
Benzene workers	52	26.1	73.1

Note: $X^2 = 10.24$, $p = < .005$
Source: Data from Picciano, D., "Monitoring Industrial Population by Cytogenetic Procedures," in P. F. Infante and M. S. Legator, eds., *Proceedings of a Workshop on Methodology for Assessing Reproductive Hazards in the Workplace,* PHHS (NIOSH) Publication No. 81–100 (Washington, D.C.: U.S. Government Printing Office, 1978), pp. 307–26.

Table 20.4. *Distribution of chromosome breaks in epichlorohydrin workers and nonexposed individuals*

		Percentage of individuals with *x* chromosome breaks			
Group	No. in group	0	1–2	3–4	>4
Nonexposed	75	53.3	30.7	14.7	1.3
Exposed	93	26.9	50.5	10.8	11.8

Note: $X^2 = 18.43$, $p = < .005$
Source: Data from Picciano, D., "Monitoring Industrial Population by Cytogenetic Procedures," in P. F. Infante and M. S. Legator, eds., *Proceedings of a Workshop on Methodology for Assessing Reproductive Hazards in the Workplace,* PHHS (NIOSH) Publication No. 81–100 (Washington, D.C.: U.S. Government Printing Office, 1978), pp. 307–26.

The development of a comprehensive medical questionnaire to determine possible factors that could influence the cytogenetic outcome

The development of an educational program among the workers so that they understood the procedure and its significance

Referrals for genetic counseling of abnormal findings not associated with any chemical exposure

The development of a sophisticated statistical analysis program

The accumulation of a large data base, including data showing seasonal and temporal variation

Availability of factors concerning intralaboratory variation
The important finding of two chemicals that induced cytogenetic abnormalities in employees

The accomplishments of the Dow program were numerous, and indeed possible objections that could be raised about industrial monitoring were successfully overcome. From all available information, it would seem that the Dow model experiments on cytogenetic screening of workers was a highly successful endeavor. The 1977 results on benzene and epichlorohydrin indicated the successful operation of the program. During the years of the Dow program, a great deal of valuable information was accumulated, and we can hope that it will become generally available. Certainly the Texas Division of the Dow Chemical Company has played a leading role in developing the techniques of monitoring workers by short-term methods and has demonstrated a practical program that could successfully be integrated into a medical surveillance program.

Failure to reach closure in the dispute concerning industrial monitoring

The Dow experience notwithstanding, the broader question is, Given the availability of these procedures, why is industry not utilizing them to protect their employees? The reasons given by the medical director of Dow, that these methods are simply research tools and hence are not ready for general use, should be closely examined. The existing information generated by these procedures to date, even with limited use, indicates that they are far more definitive for detecting carcinogens and mutagens in the workplace than most other procedures that are used to monitor employees. Dow's own officials, prior to 1977, described the value of routine applications of these techniques. If the allegation is made that sufficient data are not available, then one would ask how much more information is needed, given the fact that Dow had used the procedures over a thirteen-year period and had screened well over 3,000 individuals.

Also, it would be hard to justify not doing these procedures because of the cost of the techniques.

Perhaps a better justification for not carrying on these procedures is the possibility of future litigation. If any employee participating in a study that detects a potential carcinogen develops a malignant tumor several years later, he or she could sue the manufacturer of the product. A chemical company, therefore, that routinely screens its employees and does detect potential carcinogens may be liable for future lawsuits. If this is the scenario, it is possible that some corporations would rather not do the procedure than be held accountable at a later date for having exposed employees to a carcinogen. This fear of litigation implies that the most responsible companies who use these techniques as a part of a medical surveillance program would be subject to future litigation. This would be an interesting case of the "good guys" being punished and the "bad guys" going free. There can be little doubt that the identification of hazardous chemicals in the workplace may have an immediate adverse economic effect upon the manufacturer of the product.

The advent of much of our regulatory legislation in the last decade is due in part to the failure of industry to take the measures with their products that are necessary to protect the general public or their employees. One can only hope that given the importance of the risk of cancer in the occupational setting and the availability of methods to identify these hazardous substances, some of our major corporations will soon take a leading role in instituting these methods in order to safeguard their employees. At the present time, these procedures are used in other countries, such as Finland.

It is difficult to contest the statement of the past Dow Medical director of the Texas Division, Dr. D. J. Kilian, who was intimately involved with the Dow program:

> Until more sensitive and appropriate tools are available, cytogenetic monitoring techniques give the most significant results in terms of human experience and should be regarded as our best instruments at the present time for the detection of genetic injury at a stage when corrective action can be taken for the benefit of the individuals involved. Cytogenetic monitoring of suitable industrial populations is both feasible and desirable for the protection of human beings exposed to possible genetic hazards.[31]

We have not only the cytogenetic techniques but also other short-term procedures that can extend the power of this approach. It may well be that the most significant contributions to cancer prevention in high-risk populations is within our grasp.

[31] D. J. Kilian and D. Picciano, "Cytogenetic Surveillance."

21

The power of efficiency: balancing benefits and costs in regulating occupational exposure to toxic substances

TAYLOR R. DURHAM

The costs imposed by federal regulatory activities serve as banner and litmus test for those seeking to curtail government involvement in the economy. Interest in deregulation encompasses both economic and environmental/public health regulation, although the purposes of each are quite different, as are the criteria appropriate to their assessment. The focus on costs has stimulated an effort to demonstrate the magnitude of benefits provided through regulation. This effort encounters serious methodological difficulties in the environmental/public health area, including the imputation of economic value to human life, and poorly defined health outcomes.

The Occupational Safety and Health Administration (OSHA) has resisted pressures to quantify the benefits expected from more stringent standards for occupational exposures, but recent and imminent Supreme Court decisions or congressional action may compel the agency to incorporate procedures for weighing benefits and costs to justify exposure standards.

This essay is an examination of some of the pros and cons of this possible development in OSHA's regulatory process. In particular, it argues that benefit–cost methodology is insufficiently developed for handling major elements of regulatory decisions in this area; that its use can only extend the range of controversy, and that the focus of the benefit–cost/efficiency paradigm ignores important aspects of social structure, discouraging efforts to enhance the health resource base of particular groups and neglecting the consequences of the form in which certain costs are imposed.

The past decade has been marked by legislation aimed at controlling human exposure to toxic substances. This legislation includes numerous acts that address various substances in diverse exposure settings. The guidelines and constraints imposed by these acts on the implementation process are neither unambiguous nor fully consistent, even within specific pieces of legislation.[1] Coupled with the fact that specific implementation procedures can substantially affect different interests, the resolution of these ambiguities within and across different legislative mandates has sparked a continuing and lively exchange between administrative agencies and the courts. The prognosis for the future scope of regulatory authority is uncertain, but there is substantial support within the academic and business communities for restriction of agency discretion.[2]

In congressional and judicial review of agency rule making, and within the executive branch, where concern focuses on inflation and productivity, an increasingly dominant theme is that agency activities should be constrained by economic considerations. The legislation on toxic substances at present runs the gamut of economic considerations, from ignoring costs of regulatory standards (clean air, water pollution) to urging some consideration of costs (safe drinking water), to requiring that a balance be established between the benefits and costs of standards (federal insecticide standards, toxic substances standards, and the "superfund" for hazardous waste cleanup).[3] Congress has explicitly denied an agency the opportunity even to consider benefits associated with certain substances, as in the U.S. Federal Drug Administration (FDA) treatment of carcinogens under the Delaney Amendment.[4] And the Supreme Court has held that even factory closings may not be considered too high a price to pay for compliance with Environmental Protection Agency (EPA) industry standards for water pollution control.[5]

Although consideration of compliance costs has been incorporated in to the regulatory process, even where the legislation does not explicitly call for it, as in the granting of variances by OSHA and EPA from industry standards for firms experiencing severe financial distress, the introduction of benefit estimates to be balanced against such costs has met with stiff resistance, principally from OSHA and despite efforts by EPA to develop benefit information for some controlled substances.[6] Court decisions and supporting judicial opin-

[1] Paul R. Portney, "Toxic Substance Policy and Human Health," in *Current Issues in U.S. Environmental Policy* (Baltimore: Johns Hopkins University Press, 1978).

[2] "OSHA, EPA: The Heyday Is Over," *New York Times*, January 4, 1981, p. E7; "Regulators Defend Their Turf," *Science*, November 11, 1978, p. 613.

[3] Portney, "Toxic Substance Policy," pp. 119–29.

[4] Richard A. Merrill, "Risk–Benefit Decisionmaking by the Food and Drug Administration," *George Washington Law Review* 45 (1977): 986–95.

[5] "EPA Upheld on Water Standards," *New York Times*, December 7, 1980, p. E2.

[6] "Dispute over Cancer Risk Quantification," *Science*, March 30, 1979, pp. 1324–5.

ions in the case of OSHA's proposed permanent standards for benzene exposure, and the Supreme Court's decision to hear a suit brought by the steel industry on OSHA's standards for coke oven emissions, indicate that the role of balancing benefits and costs in OSHA's rule making will become a prominent issue in the continuing debate over appropriate guidelines for occupational health regulation.[7]

From the standpoint of social policy, the move to place more reliance on striking some balance between the benefits and costs of occupational health standards raises several questions to be addressed here. The overarching question is, of course, whether the inclusion of requirements to balance benefits and costs in formulating rules improves the regulatory process, and, if so, according to what criteria and under what conditions. In addition to conceptual problems in obtaining estimates of effects in benefit and cost categories, there are difficulties in valuing and balancing these figures to arrive at regulatory decisions. These questions arise generally in the application of benefit–cost methodology to social decision making. There are, in addition, questions of whether this is a suitable approach to setting standards specifically for occupational exposures to toxic substances, and whether this approach in this particular context imparts a bias to decisions, in comparison with alternative criteria for establishing levels of worker protection. In addressing these latter questions below, I argue that the benefit – cost approach carries with it a conceptualization of the occupational health problem and a criterion for its appraisal (economic efficiency) that significantly bias the resulting decisions, and that this approach fails to consider the importance of the form in which costs of the economic system are imposed on particular groups in society, where differences in the types of cost borne by these groups may be relevant to social decision making.

An obvious, but perhaps necessary, distinction between types of regulation should be made in preface to the following discussion. There has been a tendency, particularly marked among proponents of deregulation, to lump economic regulation (antitrust regulations, and those of the Federal Trade Commission, Civil Aeronautics Board, and Interstate Commerce Commission) together with environmental and public health regulation, the province of EPA, OSHA, and FDA. Many criticisms of the latter category of regulatory legislation and its enforcement seek additional support by ignoring rather basic differences in the aim and rationale of the two categories. This tendency is

[7] R. Jeffrey Smith, "A Light Rein Falls on OSHA," *Science*, August 1, 1980, p. 567.

most noticeable when rising inflation and sluggish growth in productivity are cited as reasons for closer scrutiny of environmental and other health-related regulations. Economic regulation seeks to affect the prices and quantities of goods and services in ways deemed favorable to particular interests, usually consumer interests. Regulatory intervention in economic processes is urged, in these cases, on the grounds that special circumstances in production or in the organization of markets undermines the desired efficiency of competitive markets. The usual criticism is that owing to changing industry circumstances or to regulatory processes, the cure is worse than the disease, in terms of the expressed objectives of the regulatory intervention.

Environmental and occupational health regulation is directed explicitly to altering the mix of final social output, not to providing more of the same combination of goods at lower prices. OSHA, for example, is in the business of licensing production processes, and the criteria legislated for such licensing refer explicitly to possible changes in the form and distribution of certain real production costs – risks to workers' health – presumed not adequately taken account of prior to regulatory intervention. To some extent occupational and environmental regulation may be seen as part of an effort to reduce the total social costs involved in producing whatever society wishes to have produced, in the belief that in the long run it is cheaper to prevent the deterioration of specific productive resources than it will be to deal with the mess later. Such regulation, however, entails the consequence that some people will pay higher prices for commodities, whereas others will enjoy a standard of living that includes, specifically, greater health protection. Since this second type of regulatory activity changes relative prices through reallocating resources to the production of new goods, inflation and productivity concerns do not provide a sound basis for curtailing them.

OSHA and the benzene case

The recent Supreme Court decision to reject OSHA's proposed permanent standard for occupational exposure to benzene held that the agency had "exceeded its authority by reducing permissible exposure limits to benzene at industrial worksites without making a threshold determination that a significant risk was present at the original level."[8] That is, OSHA's standard was set aside because

[8] Gregory J. Mounts, "OSHA Standards: The Burden of Proof," *Monthly Labor Review*, 103 (1980): 53.

OSHA failed to present any evidence or estimates concerning the reduction in risks available through more stringent standards. The implications of this decision for OSHA's ability to propose standards are not clear, since the ruling turned on OSHA's failure to present any evidence, not on a failure to demonstrate a significant reduction in risk to health or to estimate the benefits, or to weigh benefits against compliance costs in a "reasonable" manner. The written opinions of the justices concurring in the Court's decision indicate considerable differences in their respective emphases on these matters. Four justices did not think OSHA's refusal to account for regulation benefits in even so minimal way was unreasonable. The Court will hear a case brought by the steel industry in opposition to OSHA's proposed standards for coke oven emissions, and this decision may confront more directly the issue of achieving "reasonable balance" of benefits with costs.

OSHA had originally established a standard of 10 parts per million (ppm) in the atmosphere of the work-place for benzene exposure in 1971, regarding benzene as a toxic substance affecting the central nervous system. Between 1971 and the 1977 proposal of a 1-ppm standard, studies had linked benzene exposure to excess incidence of leukemia. The agency considered the 1-ppm standard the lowest achievable, but not without risk.[9] Agency policy has been to prohibit exposure when substituties were available and to seek the "lowest feasible level" where this was not the case.[10] In order to expedite the setting of standards, OSHA has also sought to establish a streamlined plan to replace a case-by-case review system; under the new plan, a "confirmed" carcinogen would be regulated at the "lowest feasible level."[11] This new strategy arose from agency frustration with the slow progress it had made in setting standards during its first eight years: 23 permanent standards, compared with 2,200 suspected carcinogens in its list of 28,000 toxic chemicals.[12]

The Court's decision appears to undermine this strategy for dealing quickly with substances that, according to tests,[13] are classified as confirmed carcinogens. The decision does admit the use of the most conservative method for estimating risks at exposure levels below the

9 Luther J. Carter, "Dispute over Cancer Risk Quantification," *Science*, March 30, 1979, pp. 1324–5.

10 R. Jeffrey Smith, "Toxic Substances: EPA and OSHA Are Reluctant Regulators," *Science*, January 5, 1979, pp. 28–32.

11 Nancy K. Eskridge, "Conflicting Academy Reports Leave EPA Bewildered," *Bioscience*, May 1980, p. 309.

12 Smith, "Toxic Substances."

13 Eskridge, "Conflicting Academy Reports," p. 311.

prevailing standards – linear extrapolation from higher levels, rather than a demonstration of no safe threshold. However, OSHA has maintained that estimates of risks at lower levels of sufficient precision for setting standards are simply not supportable. This position has been endorsed by several reports from the National Academy of Sciences.[14]

Among agencies involved in regulating toxic substances, OSHA has been most resistant to quantifying the benefits, or expected risk reduction, expected from stricter regulatory standards. EPA has provided such estimates in support of its rules on benzene exposure.[15] OSHA's reluctance probably reflects either the professional background of the agency's staff or the inclination of a labor constituency to avoid making explicit trade-offs of workplace risks. The Act itself calls for a standard that "most adequately assures, to the extent feasible, that no employee will suffer material impairment of health or functional capacity even if the employee has regular exposure to the hazard dealt with by such standard for the period of this working life." The agency's reluctance is in keeping with the level of protection envisioned in this passage for each worker. The Court's decision emphasized the law's general requirement that all standards be "reasonably necessary or appropriate to remedy a significant risk of material health impairment."[16]

The agency's general lack of success in having its standards sustained by the courts during the 1970s doubtless underlies its opposition to imposing more stringent requirements on a case-by-case basis. Given the enormous number of toxic substances to which workers are exposed and the prospects for change in the number and identity of such substances, the testing requirements alone can drain available resources – resources that may be cut substantially.[17] But the delays of litigation and of possible congressional review of existing legislation would prove most costly in terms of the agency's objectives.

The regulatory problems confronting OSHA differ in important ways from those facing EPA. First, the population at risk for a given substance is likely to be relatively small; the exposure of millions of workers, as in the case of asbestos, is likely to be exceptional. Second, the exposures are contained, in the sense that certain segments of the population run little or no risk of exposure in occupational settings.

[14] Thomas H. Maugh, II, "Chemical Carcinogens: How Dangerous Are Low Doses?" *Science*, October 6, 1978, pp. 38–41.

[15] Luther J. Carter, *Science*, March 30, 1979, p. 1325.

[16] Mounts, "OSHA Standards," p. 55.

[17] *New York Times*, January 4, 1981.

Finally, there is some perception that workers voluntarily accept, at least in part, hazards of employment and may receive at least partial compensation for additional risks. Together, these factors tend to narrow the range of legitimate regulatory intervention, by restricting the magnitude of aggregate benefits, in comparison with other areas of toxic substance regulation, and by presenting such risks as voluntarily assumed and partially compensated.

Before appraising the significance of a regulatory shift that requires an explicit balancing of benefits and costs for setting standards, it should be noted that the present regulatory machinery does entail some consideration of benefits and costs of precisely the sort envisioned for a more formalized approach. This feature is not emphasized or even acknowledged by those who attack current procedures as economically capricious. First, of course, there is little in OSHA's record for the 1970s that suggests a headlong rush to impose meaningless but costly standards. The number of standards established by 1977 (4, covering 23 carcinogens and other substances) is small in relation to the number of suspected carcinogens alone, estimated at 2,200.[18] In addition, the agency has been hampered by industry reluctance to provide information on the ingredients in one-third of over 20,000 substances subject to regulation and by industry lawsuits contesting standards and testing procedures.[19]

OSHA's internal process of establishing priorities under the constraints of limited resources takes into account prospective benefits by focusing on potent substances coupled with large numbers of exposures. It is true that the information base used to establish these priorities is far from perfect, but this is not a particularly cogent objection to a process aimed in part at developing such information. On the cost side, OSHA has avoided promulgating strict standards where compliance costs were very large (e.g., in the case of noise levels) or concentrated in impact (e.g., in the case of arsenic trioxide). The "lowest feasible level" standard adopted for carcinogens has been interpreted to mean technically possible and "within the financial capability of the industry" or "not overly burdensome."[20] Stan dards may be stricter where firms are large and profitable or where costs can be passed on to a broad consumer base. These economic considerations may be inspired by political concerns, but they are nonetheless economic considerations. Focusing on court cases dealing with specific substances ignores the introduction of such consid-

[18] John Mendeloff, "Reducing Occupational Health Risks: Uncertain Effects and Unstated Benefits," *Technology Review*, 82 (1980): 67–78.

[19] Ibid., p. 68.

[20] For all of these points, ibid.

erations within the regulatory process, activated by limits on internal resources and anticipated political or judicial opposition.

The effect of requiring a formal and explicit benefit–cost justification, case by case on specific standards, is to shift a substantial burden upon the agency, in comparison with procedures that allow standards to be proposed early in a regulatory initiative, after preliminary screening through procedural considerations of benefits and costs, and then to be amended with variances where compliance costs are found to be unusually severe. The burden, of course, extends well beyond the agency itself. It is costly for industry to deal with the "prior restraint" of standards found later to be excessively stringent. Delay in setting standards carries its own costs, in a different form and for different people. In the former case, the burden of resolving uncertainty on the side of conserving workers' health is more likely to be widely shared and less likely to involve irreversible costs concentrated upon a particular group. When costs in the form of unemployment or limited availability of substitute commodities are severe and concentrated, they can be identified and dealt with through variances or through existing social mechanisms for handling similar costs.

The compliance costs for stricter standards are influenced by the technology and organization of production processes at the time when standards are set and enforced. With the prospect of future standards being imposed on the side of caution for workers' health, industry has a strong incentive to scrutinize and adapt production facilities in anticipation of future regulatory action. To the extent that the existing production technology is accepted as a basis for calculating costs of compliance, such incentives are reduced. Similar opportunities for anticipatory adjustment are not available to workers.

Benefit–cost principles and standards for occupational risks

Benefit cost analysis is touted as a method for establishing consistent regulatory priorities for risk reduction in a manner that preserves economically efficient use of scarce resources. Two issues are addressed in the following discussion:

1. Does the use of benefit–cost principles promise to reduce the uncertainties surrounding decisions in this area or to add its own fringe of uncertainty?
2. Aside from the possible increase in regulatory lags that benefit–cost analysis may introduce, does its application to occupational health risks introduce any systematic biases to regulatory decisions?

The starting point for advocating use of benefit–cost methods in regulatory decision making is the characterization of occupational standards as an investment in human resources. Government regulatory intervention may be supported by showing that such investment is too low, for whatever reasons, in the sense that investments in this areas are more productive of value than some alternative use of resources. This will leave open the question that bothers those who feel that regulation has gone too far: What level of investment in this area is socially optimal? Or, beyond what point will expenditures on this particular social objective become "too high"?

Benefit–cost analysis is proposed as a means of adjusting investments in human resources or health through occupational or environmental exposure standards to optimal levels, neither too high nor too low. This is to be done by avoiding the transfer of resources from one use to another when the benefits to be obtained are less than those provided prior to the change. Following benefit–cost guidelines promotes economic efficiency, in that it prevents resources from being assigned to uses less valuable than available alternatives. "Efficiency," in this context, refers to the greatest value of output attainable with given resources, where value is established by the pattern of people's wants. The pattern of wants to which the economy responds is related to the individual's willingness and ability to trade income for goods and services, and hence the value to be maximized under the efficiency criterion is sensitive to the prevailing income distribution, among other things.

Application of the method calls for a full accounting of all benefits and costs, to whomever they accrue, for each alternative to the status quo. Benefits and costs are to be evaluated according to a common metric, that is, to money values derived from actual or adjusted market prices or imputed through simulated market processes. Benefits and costs are separately aggregated, and these totals compared; where necessary, a discounting procedure is used to compare costs and benefits occurring in different time periods. The efficient allocation is guided by selecting among alternatives ranked according to net benefits.

Allocation by benefit–cost guidelines is intended to reconcile resource use in public sector decisions with those in the private market sector, essentially by simulating through benefit–cost rules the principles guiding allocation in the private sector. There is an extensive literature, marked by considerable controversy, on the conditions under which the benefit–cost algorithm mirrors market solutions[21]

[21] See Ezra J. Mishan, *Cost–Benefit Analysis* (New York: Praeger, 1971).

and on the conditions necessary for benefit–cost methods to conform with the underlying principles of welfare economics.[22] There is, in addition, widespread recognition of the failure of welfare economics to come to grips with major problems in social policy, particularly those involving equity, social justice, and income distribution.[23] The intent of the following discussion is not to review these and other matters relating to benefit–cost analysis in general but to consider certain elements of the method in the context of occupational health risk regulation.

Consistency and time

Changing alternatives

The efficiency gains contemplated by advocates of benefit–cost analysis are derived by applying consistent methods in appraising a set of alternatives for investing in protection of health and life. OSHA contends that available methods are insufficiently reliable and precise to provide estimates of exposure effects at levels relevant to regulatory standards. It is not likely that methodological advances will resolve this matter in the near future. What is far more likely is that new testing techniques will be developed and new substances will be introduced to the workplace. Test procedures will detect new sources of health risk and confirm or question old ones but will not reduce the range of uncertainty for effects estimates.

Useful application of benefit–cost analysis in establishing efficient regulatory priorities will require either successive iterations for evolving sets of substances or a wait for what appears to be a stable set. Without this, much of the benefit–cost rationale is undermined. Only after reaching a period when the goal of benefit–cost analysis is plausibly attainable can the regulatory process itself begin. This period of waiting for discovery, which the experience of the 1970s has shown can be influenced by industry strategy, may reasonably be expected to span three or four generations.

Discounting

In many public investment situations, if we accept a benefit–cost approach to setting occupational health standards, it is necessary to

[22] James Broadway, "The Welfare Foundations of Benefit–Cost," *Economic Journal*, 84 (1974):753–61.

[23] Eric Von Magnus, "Economic Theories of Justice," PhD diss., Syracuse University, 1978, a thorough and lucid treatment of equity formulations and application to policy.

compare benefits and costs that accrue at different times.[24] The discrepancy between expenditure and returns is quite marked for investments made in order to reduce risks that have long latency periods. One way to handle this problem is to express all benefits and costs in terms of a "present value." A discounting technique is used to arrive at some base year equivalent for benefits or costs estimated for future years. The higher the discount rate, the lower the present value of a given future magnitude. For a given discount rate, the more remote in time some benefit or cost, the lower its present value.

Although it is clear that individuals and financial institutions discount future returns, there is some disagreement even among economists over whether this is always rational. There is even more dispute over the appropriate rate of discount to be used for government programs – the "social" rate of discount.

In the context of occupational health standards, the benefits represent a reduced burden placed in the distant future upon members of the current generation of workers and upon members of future generations. These benefits, to use terminology congenial to benefit–cost analysis, are the values of preserving at different levels a stock of a certain asset held by members of current and future generations of workers – their health.

Using the present-value comparison of benefits and costs assigns a lower weight to distant returns; whether this is appropriate depends on how this particular rule compares with alternative procedures, according to selected critieria. An alternative to the present-value criterion is that of maximal sustainable yield. Although this criterion may be recommended for protecting a given resource base, it can be shown that under certain conditions the present-value criterion can actually be more conserving of future resources.

Several problems with the present-value criterion, of particular relevance to occupational health issues, may be noted. First, of course, the present-value criterion favors the present generation and its existing pattern of wants, knowledge, and productivities. Second, this criterion is not applicable to problems in the valuation of assets whose value depends upon beliefs about their future use. As discussed later in this chapter, in the section entitled "Compensation and the Value of Life and Health," whatever procedures are employed to assign values now to health and life based on data of wants or preferences cannot take account of how future generations will use or

[24] This section draws on the excellent discussion of present value and alternative criteria in Talbot Page, *Conservation and Economic Efficiency* (Baltimore: Johns Hopkins University Press, 1977), Chaps. 7–9.

value this asset. This difficulty is more severe when there exist no meaningful substitutes for the asset being valued.

Consistency and costs

Achieving efficiency through consistent comparisons requires that costs as well as benefits mean the same thing from case to case. If expenditures are made from a pool of tax revenues, the cost of any project, however difficult it may be to trace the actual incidence of the tax, is comparable to that for any other financed from the pool.

Compliance with regulatory standards requires expenditures by firms in the affected industries; different standards affect different industries, and firms within a given industry face different adjustment problems. A total cost figure for meeting a stricter standard can conceal opportunities to reduce risks at relatively low cost for a large number of workers by combining these with firms or industries where compliance costs are very high. This possibility is accommodated in a process that grants variances for cases of extreme economic hardship, but at the expense of leaving some workers exposed to greater risk.

There are short-term and long-term burdens of costs incurred by industry, some of which are indirect and quite difficult to estimate, such as costs of a change in market structure when smaller or older firms are forced out of business. One component of these and other costs is the value of output produced at lower cost with greater risk, and given up to comply with standards. This output can be quite diverse for a single standard, ranging, for benzene, from Tupperware to tires, and its valuation can be the object of considerable controversy.

Efficiency, benefit–cost analysis, and compensation

When resources are reallocated to reduce risks, there are gainers and losers – as there also are if no reallocation takes place. This is because the costs of selecting one alternative over another are the values produced in the other use, benefits for some other group. Rarely are the affected groups identical. The comparison required by the efficiency criterion involves values conferred upon different groups; this, in turn, requires comparison of values generated in part by individual choices contingent upon different income, wealth, and wants. Meeting the efficiency criterion of producing greater value with given resources entails accepting this comparison. This confronts a matter that has proved vexing to welfare economics, upon which the efficiency rationale of benefit–cost analysis is based: the basis and justification of interpersonal comparisons of well-being.

One way to resolve this dilemma, often invoked for benefit–cost analysis, is to assume that reallocation effects are small and randomly distributed over repeated applications and that hence the issue can be ignored. The suitability of this approach for occupational risks is considered later in this chapter, in the section entitled "Social Structure, Benefit–Cost Analysis, and Occupational Health Standards." Another proposal is to require that losers be compensated from the gains made possible through reallocation. This introduces several problems of its own, including the need to assess the adequacy of compensation, and the question of whether passing a benefit–cost test entails that such compensation is even possible (let alone whether it will be paid). This latter question is considered here; the meaning of adequate compensation is deferred to the next section.

Does the demonstration of positive net money benefits required by the benefit–cost test ensure that losers can be compensated from the increased value of output? The answer is no: Passing the test can, even with attempted compensation, lead to an overall deterioration in the satisfaction of wants for gainers and losers as a whole. One problem is that the transfer from gainers to losers of all money gains may not be sufficient to make the losers "whole," that is, to return them to the same position, according to their own estimation, that they held prior to the change.[25] This is because the adequacy of compensation depends on relative prices of the goods to be acquired by losers; these are affected by the reallocation and by the spending of income received in compensation. There is no assurance that prices will be such as to permit compensation; in this respect the benefit–cost operation does not receive blanket approval even of the efficiency criterion.

Compensation and the value of life and health

Proponents of applying benefit–cost techniques to occupational health standards recognize that a central and controversial aspect of this approach is the need to assign explicit values to human lives.[26] This is necessary for estimating major benefit components and for gauging the adequacy of compensation for bearing risks to life and

[25] Broadway, "Welfare Foundations."

[26] Richard Zeckhauser, "Procedures for Valuing Lives," *Public Policy* 23 (1975): 419–64.

limb. To some extent, advocates of the benefit–cost approach treat this feature as a virtue of the method, claiming that the analytic need will compel decision makers to confront and resolve in some manner issues that are customarily ignored or dealt with less openly.

There are several useful reviews of efforts to obtain empirical estimates of wage premiums paid for risky employment and related procedures for valuing lives,[27] and of the conceptual problems attending efforts of this kind to assign dollar values to lives or the protection of lives.[28] The following discussion touches on several issues selected for relevance to occupational risks to life and health and to the discussion of regulation and social structure in the next section.

A number of empirical studies have estimated wage premiums paid for accepting job-related risks of death or injury as a basis for inferring the values that workers implicitly place on loss of life or specific physical disability. These procedures, and related "willingness-to-pay" experiments to elicit such values, depend for their validity in economics on the nature of the choice presented to individuals and on the conditions in which individuals make the choices. These two elements affect the degree to which estimates can be taken as measures of individual valuations; the appropriate use of these estimates in social decision making is not determined by the conditions generating data used to obtain estimates.[29] The prescribed conditions are a basis for accepting the derived valuations of life or injury as commensurate, for efficiency purposes, with the valuations of resources that may be allocated to the reduction of risks.

The first difficulty is with the type of choice that is being made. One example of this is that insurance purchases are not a satisfactory basis for gauging the value that a person places on his or her own life, since the choice involved concerns a desire to provide for others under certain contingencies. Another relevant distinction is that between "ex ante" and "ex post" choices: What a person is willing to pay to reduce a risk (or willing to accept to bear a risk) can be quite different before and after the risk lottery has been run. To use a metaphor from another controversy, a person may accept some payment to enter a fair draft lottery for army service and then discover that this amount is inadequate once his number has come up. Which choice should be used to estimate adequate compensation is not apparent on a priori

[27] W. Kip Viscusi, "Labor Market Valuation of Life and Limb: Empirical Evidence and Policy Implications," *Public Policy* 26 (1978): 359–86.

[28] Zeckhauser, "Procedures for Valuing Lives."

[29] Ibid., pp. 35–6.

grounds; estimates that systematically exclude consideration of the "ex post" valuations may be viewed as biased downward.

One objection often made is that estimates based on willingness to pay are conditioned upon the existing distribution of income and wealth for their acceptability. Zeckhauser suggests that this objection implies that the procedure would be acceptable if the income/wealth distribution were equitable.[30] If this is the only objection, this response may be correct but not useful, as elaborate debates on equity issues have illustrated.

There remain several problems in specifying certain conditions of choice whose fulfillment is necessary for accepting individual valuations. These are that individuals must be fully informed and that their choices must be completely voluntary. The question of what might constitute "full information" is in part tied up with the distinction between "ex ante" and "ex post" conditions. Another difficulty is that the kinds of information involved may never be of a sort that individuals can handle meaningfully for individual choices.

Calibrating the degree of individual volition in accepting occupational or other risks is more troublesome than for choices of food and clothing. Here again the distribution of income and wealth is relevant, but not decisive. It is clearly inaccurate simply to lump all risky employments as "your money or your life" choices; it is also clear that voluntariness can be circumscribed by available alternatives and personal resources for mobility, as well as by income. In the case of occupational health risks, voluntariness may become increasingly constrained as the level of information improves, in the rather plausible case that alternatives and mobility are increasingly restricted over time at the individual level. Socially structured mobility may shift over long time periods because of demographical and economic conditions.

This method, based on the criterion of acceptable levels of information and voluntariness for individual choices, focuses on separate, isolated choices by independent individuals. Excluded from these choice situations is any consideration by individuals, or by groups, of the pattern of risk bearing across different social groups; of the processes by which these are imposed; or of the distribution of benefits accruing over time from these processes. The grounds for excluding such matters from the information required to make voluntary choices are not specified, although it is compatible with the inclination in economics to treat individual preferences, to which the

[30] Ibid., p. 65.

economic system is to respond efficiently, as exogenous rather than as determined within the system itself.

Social structure, benefit–cost analysis, and occupational health standards

The issue in setting occupational health standards, in shifting the odds of long-term burdens from particular employments, is the extent to which real but uncertain costs of production will enter the economic system accounts. This is equivalent to marking off some boundaries to workers' health as a valued output of the economic system. The instruments used to lay out these boundaries will determine the range of uncertainty beyond which the degree of protection may not be traded off through individual decisions within the context of employment.

The benefit–cost approach to setting up these boundaries assumes that this protection is provided to workers at some cost to society, possibly too great a cost. The social dilemma so portrayed is whether society is providing itself too much protection, or too much protection from certain sources of risk, by giving up too much of other things society wants.

This characterization ignores the existence of a social structure that ensures that whatever the balance struck between workers' health protection and other goods, the real costs will fall disproportionately on the same people. Workers and their families may bear the consequences of greater occupational exposure to risk and enjoy the accompanying advantages of better employment prospects and higher wages and the availability of some less expensive commodities; or they may receive greater protection but have poorer employment opportunities and less real income.

The benefit–cost approach to setting social priorities leaves out of consideration any consideration of how social structure mediates the distribution of available net benefits, concentrating costs of any allocation on those disadvantaged in terms of opportunity, personal resources, and social mechanisms. In addition, benefit–cost analysis fails to address a question of legitimate concern in regulation: the form in which net benefits accrue to different groups in society.

The preceding discussion of the conditions necessary for acceptable individual risk valuations points to several criteria for prescribing the form of net benefits afforded specific groups. The basic notion is that fully informed, voluntary, individual choice is the mainspring of optimal social adjustment. Taking personal objectives as given, these conditions permit individual adjustments to obtain the most from available personal resources.

Both the complexity of the information required for making adequate evaluations of occupational risk, and the results of empirical studies of information handling and decision making involving risk, suggest that individuals do not perform well with these choice situations. There is some evidence that workers regard exposure to toxic substances in the same way in which cigarette smokers view the risks of smoking, denying harmful effects if they are not immediate and obvious and discounting adverse effects over long latency periods.

There is also reason to believe that workers' organizations may not be willing or able to provide adequate protection at work. A small survey of Baltimore-area workers in hazardous occupations found that a major source of dissatisfaction among workers was their inability to get action on recognized or suspected health hazards through union grievance machinery. Unions may wish to avoid negotiating wages or employment against health and safety, in order to preserve internal cohesion or for other reasons. Job scarcity can affect union as well as individual behavior, as shown in a case (*EPA* v. *Reserve Mining Co.*) in which a United Steel Workers local opposed EPA standards controlling levels of asbestos in the community drinking water, a community inhabited by local members.[31]

These considerations suggest a possible objective for some regulatory policy, one easily overlooked in a benefit–cost approach. This is that regulations seek to affect the form of certain costs imposed on particular groups; in this case, to broaden the boundaries for health protection and to shift the costs onto wage and unemployment prospects. The thrust would be to enhance and preserve a particular type of asset held by present and future generations of workers, a form of wealth otherwise subject to greater risk of cumulative degradation. The costs of such a move are not negligible, and they are largely borne by the group being protected. But they are imposed in a form more manageable, on an individual and organizational basis, in terms of the desirable conditions concerning availability of information and voluntariness of choice.

The difficulties in ascertaining the value of such an asset for benefit–cost appraisals have been discussed earlier. The benefits to workers of a strictly protected health asset are not easily estimated in an unbiased manner in comparison with a markedly different basis

[31] Portney, "Toxic Substance Policy," p. 117.

for individual choices. In a sense, this problem is that of the influence of income and wealth distributions upon individual willingness to trade some goods for others. It is at least plausible that a guarantee of strict protection for a particular form of wealth can have a substantial impact upon individual valuations and upon individual behavior that affects the long-term value of that specific asset. This latter possibility may be worth weighing heavily, since many other health risks are concertrated upon the same people experiencing greater risks in the workplace.

The power of efficiency in the workplace

Benefit–cost analysis, when used to set priorities for regulating exposure to toxic substances at work, invokes efficiency as the criterion for determining standards for occupational risks: The benefits obtained through stricter exposure standards should not exceed the costs of achieving those standards. This approach calls for making consistent comparisons of benefits and costs across opportunities to reduce risks, using actual and simulated market valuations of benefit and cost elements.

Practically and conceptually, the requisite consistency is difficult to achieve, and in the context of occupational hazards the attempt to do so will require a considerable waiting period. Problems in the meaning and measurement of major elements necessary to sustain efficiency through benefit–cost guidelines – for example, elements such as compensation for risks to life and limb – restrict the promise of this approach to deliver on its own normative claims. By requiring comparisons of aggregate benefits and costs for proposed standards, benefit–cost analysis glosses over considerations of flexibility and equity, which can be accommodated more readily under existing regulatory procedures. Finally, benefit–cost analysis ignores the form in which net benefits are conferred upon particular social groups and accepts the boundaries for intangible assets, such as health, established by the interaction of market forces with social structure.

22

Closure in occupational safety and health: the benzene and cotton dust decisions

MARK MACCARTHY

Two court cases in the early 1980s have focused public attention on controversies in the regulation of occupational exposure to toxic substances. The benzene standard, promulgated by the Occupational Safety and Health Administration (OSHA) in 1978, was invalidated by the Supreme Court in 1980 on the grounds that the agency had failed to demonstrate a significant health risk from low-level exposure to benzene. In 1981, however, the court upheld OSHA's standard regulating exposures to cotton dust and ruled that the agency could not use cost-benefit analysis to set permissible exposure levels. I want to discuss some of the issues raised by these cases. What are the challenges created by the Court's assignment to OSHA of a duty to make a threshold determination of significant risk before regulating? Is the dismissal of cost–benefit analysis a movement away from rationality in the standard-setting process? What is the role of economic and scientific experts in resolving these conflicts? Are the courts the appropriate way to achieve closure in this area?

Two themes underlie my remarks. One is that efforts to control exposures to toxic substances are typically debated in a context of scientific uncertainty. The decision to act or not must usually be made before closure of scientific debate has been achieved. Normative and policy considerations therefore become crucial, not only in assessing what to do about known occupational risks but also in characterizing risk and developing a regulatory strategy in the face of scientific uncertainty.

The second theme is that public policy concerning occupational safety and health essentially involves the normative balancing of competing claims of rights, justice, and public values. If this is so, then, although better scientific and economic data may improve regulatory decisions, even perfect scientific and economic knowledge will not uniquely determine regulatory decisions.[1]

The benzene case

In February 1978, OSHA published a final standard reducing the permissible exposure level for workers in industry for benzene from 10 parts per million (ppm) to 1 ppm.[2] The agency's rationale was that benzene produced chromosome aberrations, blood disorders, and leukemia and that the 10-ppm exposure level, which was not designed to meet the carcinogenic threat of benzene, was insufficient to protect against these health effects. No estimate of the expected number of cases of leukemia to be avoided by reducing the exposure level was made, on the grounds that the scientific evidence available at the time was not sufficient to justify such an estimate and that in any case the lowest feasible level of exposure should be adopted to protect worker health. Costs of compliance were estimated at between $187 and 200 million in first-year operating costs, $36 million in recurring operating costs, and a $266 million initial investment for engineering controls. OSHA maintained that these costs were within the financial capability of the affected industries and were justified by the benefits that, although not quantifiable, were likely to be appreciable.

The American Petroleum Institute (API) challenged the standard before the U.S. Court of Appeals for the Fifth Circuit and, on preenforcement review, won a judgment declaring the standard invalid on the grounds that it was based on conclusions that could not be supported by the administrative record.[3] In particular, the agency had failed to show that the 1-ppm exposure limit was reasonably necessary to provide for safety in the workplace. OSHA appealed to the Supreme Court, the case was argued October 10, 1979, and on July 2, 1980, the court affirmed the judgment of the lower court by a margin of 5 to 4.[4]

[1] Regulatory decisions are often the product of political compromise as well as an evaluation of the relative urgency of competing social goals. Public officials may often accept a second-best regulatory option when the best choice is blocked by political opposition. This essay deals only with the considerations that go into the choice of the best regulatory option. The art of political compromise comes in a later stage, when officials face a choice between a second-best option and something even worse.

[2] *Occupational Exposure to Benzene*, 43 *Federal Register*, 5918–70, February 10, 1978.

[3] 581 F 2nd 493

[4] 100 S. Ct. 2844 (1980).

API's argument turned on the interpretation of two sections of the Occupational Safety and Health Act (OSH Act) of 1970. Section 3 (8) of the Act defined a standard as a measure that is "reasonably necessary or appropriate to provide safe or healthful employment." API argued that the words "reasonably necessary or appropriate" meant that the agency was required to demonstrate that the costs of a standard bore a reasonable relation to its benefits. Section 6 (b) (5) of the Act requires the agency to set standards regulating toxic substances that most adequately ensure, "to the extent feasible, on the basis of the best available evidence, that no employee will suffer material impairment of health or functional capacity." API argued that this constraint of feasibility implied that the costs of a standard must be reasonably related to the benefits. Finally, API argued that OSHA had presented no evidence that reducing exposures from 10 ppm to 1 ppm would result in health benefits commensurate with the costs imposed on industry. In sum, API's argument was that OSHA had failed a balancing requirement implied by language in its enabling legislation and hence had exceeded its standard-setting authority.

In its reply, OSHA argued that the "reasonably necessary or appropriate" clause of section 3 (8) of the Act was best interpreted as requiring that the standard actually be needed to carry out the purposes of the Act, namely, to provide for worker safety and health.[5] The intent of the section was to rule out standards that merely protected property or increased worker comfort, convenience, or efficiency. According to the agency, this clause meant to restrict standards to health and safety matters – not to impose a balancing requirement.

Moreover, OSHA contended that the feasibility constraint of section 6 (b) (5) was a practical matter and relied on purely technological and financial considerations.[6] It required that reducing exposures to the mandated levels must be possible with existing knowledge and technological capabilities. A permissible exposure level that could be achieved only by a major theoretical breakthrough, or even an extensive research and development program, was not technologically feasible.

According to the agency, section 6 (b) (5) further required that the standard be affordable for the affected industries. Affordability was defined so that it was consistent with the demise of some, but not all,

[5] *Industrial Union Department, A.F.L.-C.I.O.* v. *American Petroleum Institute,* "Brief for the Federal Parties," pp. 42–7.

[6] Ibid., pp. 55–60.

the firms in these industries.[7] A brief digression on the agency's notion of economic feasibility is important. This criterion is a purely financial one. When an industry is required to install control equipment, it expends funds to comply. It receives a return in the form of reduced workers' compensation costs, increased labor productivity, fewer and less costly liability suits, and so on. But the benefits that accrue to industry are typically less than the expenditures needed to install the control equipment. Hence, on balance the industry will lose more than it will gain. The following thought experiment may make clear the idea of economic feasibility. Imagine that industry suddenly misplaced an amount equal to the net compliance costs imposed by a standard. Would the industry still be all right? Would it be viable? Would a large number of workers have to be laid off? Would many firms go out of business? Could it pass the costs on to its customers? Would it have the funds to make needed capital investments? Would it be able to afford additional expenditures on other threats to workers' health?[8] As these questions suggest, a whole range of qualitative business judgments would be called into play. But the impreciseness of these judgments would not derive from a difficulty in comparing the economic consequences of a standard with

[7] It may seem odd that OSHA should accept the demise of some *firms* as an acceptable price for worker safety but reject the demise of an entire industry. The explanation is this. OSHA is not permitted to shut down an industry, because of a balancing judgment by Congress that the wholesale loss of jobs this would entail would not be justified by gains in worker health. Congress also felt that giving a government agency a life or death choice over the fate of an entire industry was an illegitimate extension of public control over the economy. (This is one case where a rigorous cost-benefit analysis might prove to be more stringent than the feasibility criterion under which OSHA operates. In a cost-benefit framework, if requiring an industry to pay the full social costs of production, including the costs of protecting worker health, causes prices to rise so high that consumers no longer will buy the industry's product, then the industry itself flunks a social cost-benefit test and should be allowed to go under.) If a few marginal firms in an industry have to go out of business, on the other hand, the loss of jobs would be less widespread and would be justified by the gain in worker health in the remaining parts of the industry. The relevant case is *Industrial Union Department, A.F.L.-C.I.O.* v. *Hodgson*, 499 F. 2nd 467 (D.C. Circuit, 1974). In practice, however, OSHA has a flexible variance procedure whereby firms that cannot meet explicit requirements are allowed to use less costly means of compliance when they can demonstrate that this will provide equivalent levels of protection.

[8] OSHA asserts that the phrase "to the extent feasible" in section 6 (b) (5) "is sufficiently flexible to allow the Secretary to take into account in setting a standard the need for the industry to meet future threats to employee health." "Reply Brief for the Federal Parties," pp. 18–19.

its health benefits. Benefits are out of the picture. The vagueness comes from the difficulty of determining when an industry is in trouble.[9]

OSHA's crucial point was that it was required to show only that the standard protected against a health threat and that it was feasible. No cost–benefit balancing was permitted.[10] OSHA contended that it had met these conditions in the case of benzene. It had found that no safe level of exposure to benzene had been established and had concluded from this that lowering the exposure limit from 10 ppm to 1 ppm was a measure needed to protect workers against blood disorders, cancer, and chromosome damage; although the precise number of cases to be avoided could not be quantified because of limitations on data and extrapolation techniques, the agency claimed that some adverse health effects that might otherwise occur could be avoided by lowering the exposure limit. Furthermore, lowering the exposure level to 1 ppm did not require the development of new methods of exposure control and was not exorbitantly expensive relative to the financial capability of the affected industries. On this basis, the agency found that the 1-ppm standard was feasible for the affected industries. Section 6 (b) (5), therefore, both permitted and required the agency to reduce the exposure level to 1 ppm.

OSHA's formal decision-making process can be pictured as follows. First, the question is asked, Is there any material impairment of health owing to exposure to a toxic substance? If the answer is no, do nothing. If yes, then ask whether to regulate now. This decision can be based broadly on questions of severity of health risk, ease of exposure control, availability of substitutes, and the like and would involve administrative decisions about how agency resources could most effectively be used. If it is decided to regulate, then the agency must determine the exposure level at which no material impairment takes place. For carcinogens, OSHA had developed a generic policy according to which this level was to be set at zero in the absence of

[9] The data for these decisions derive from studies financed by OSHA and performed by outside contractors. Typically, these contractors obtain information about processes, technology, current practices, likely methods of compliance, and needed operating and capital expenses from the affected industries. The cost estimation work itself, though, is the responsibility of the contractor, not the affected industry. For this reason, the affected industry usually prepares its own estimate of compliance cost, and it typically diverges from the OSHA estimate. OSHA's considerable experience in this area suggests that prospective cost estimates are normally substantially larger than actual compliance costs.

[10] N. 5 *Supra*, p. 70. See also *Occupational Exposure to Benzene*, N. 2 *Supra*, p. 5941.

proof to the contrary.[11] If this zero-level was not technologically feasible, then the agency would select the lowest exposure level that could be met with reasonably available technology. If this technologically feasible level was not affordable for the industries involved, then the agency would select the lowest economically feasible exposure level, where this meant affordability for the industry as a whole.

The Supreme Court invalidated the standard, but individual members of the Court differed in their reasons for doing so. Justice Stevens wrote the plurality opinion, in which Justices Burger, Stewart, and Powell concurred, holding that the standard was invalid because it was not supported by adequate findings.[12] He concluded that section 3 (8) imposes a duty on OSHA to make a threshold finding that a toxic substance poses a significant risk to workers' health before reducing exposures. In his view, OSHA had not even tried to meet this threshold duty and had instead shifted the burden of proof to industry.[13] Justice Powell, in contrast, argued that OSHA had attempted to satisfy its threshold duty of finding significant risk but had failed the substantial evidence test by not compiling a record that adequately supported this finding.[14]

Many have argued that the role of judicial review in cases like this is to oversee the procedures that the agency uses to reach a decision, rather than to review the agency's assessment of the evidence upon which the decision was based.[15] Was the Supreme Court moving beyond this procedural review in the benzene case? The clear answer is, yes. Both Justice Stevens and Justice Powell found the evidence in the record insufficient to warrant the agency's action, and neither found a procedural defect, such as failure to provide adequate notice or failure to respond to issues raised in the rule-making proceedings. The lack of clarity on this point, in Justice Stevens's opinion, reflects, I think, the Court's hesitation to become involved in substantive questions. On the other hand, the decision to invalidate the standard because of a lack of fit between evidence in the record and regulatory action signals the Court's willingness to intervene in a substantive way, when, in the Court's view, the agency has egregiously failed to carry its burden of basing its actions on substantial evidence.

[11] *Identification, Classification, and Regulation of Potential Occupational Carcinogens,* 45 Federal Register, 5002–296, January 22, 1980.
[12] N. 4 *Supra,* p. 2847.
[13] N. 4 *Supra,* pp. 2869–70.
[14] N. 4 *Supra,* p. 2876.
[15] David L. Bazelon, "The Judiciary: What Role in Health Improvement," *Science* (1981), 792–3.

The duty to make a finding of significant risk would still allow OSHA to regulate, wrote Justice Stevens, for this duty is not a mathematical straitjacket requiring a point estimate of the probability of harm. Furthermore, since questions of risk assessment are on the frontiers of scientific knowledge, he ruled that the evidence required to support the agency's regulatory action need not establish "scientific certainty." OSHA is free to use controversial, but reputable, conservative assumptions in estimating risks, as part of its mandate to use the "best available evidence." Moreover, the judgment that a risk is significant will inevitably and justifiably involve policy considerations. Justice Stevens addressed part of the policy question of how large a risk must be in order to be significant by setting upper and lower bounds. He asserted that a risk of one in a billion is not significant, whereas a reasonable person would consider a risk of one in a thousand significant.[16]

Justice Stevens, with Justices Burger and Stewart concurring, did not decide the cost–benefit question that had been raised so prominently in API's brief. Instead, he deferred judgment on whether exposure levels must be set by balancing benefits and costs. He argued that the agency must first determine that a significant risk exists before it reaches the question of whether to reduce this risk to the lowest feasible level or whether the exposure level should be set by comparing benefits against costs. Since the agency had failed in its threshold duty to establish significant risk, the balancing question did not arise.[17]

However, Justice Powell did rule on the balancing question. In his view, the Act required OSHA to determine that the costs of reducing exposures to a particular level bear a reasonable relation to the benefits; standards whose costs are "wholly disproportionate" to health benefits are not reasonably necessary and are not feasible. In Justice Powell's view, OSHA had not passed this balancing test. He wrote that the evidence in the record did not adequately support OSHA's conclusion that the costs of lowering the exposure level for benzene to 1 ppm were justified by the expected benefits.[18]

Thus, the plurality agreed that the standard was invalid because it was not based upon an appropriate finding of significant risk. They differed on whether the agency had attempted to meet this threshold responsibility and failed, or whether it had not even tried. Three of

[16] N. 4 *Supra*, p. 2871.
[17] N. 4 *Supra*, p. 2863.
[18] N. 4 *Supra*, p. 2878.

the justices in the plurality were silent on the balancing question, but Justice Powell agreed with API that OSHA had failed a balancing requirement implied by language in its enabling legislation.

Justice Rehnquist held the standard invalid on totally different grounds. He ruled that section 6(b)(5) of the Act constituted an invalid delegation of legislative authority to OSHA. The problem was that the constraint of feasibility was hopelessly vague, and so could provide no guidance to the agency on where to set standards. In the absence of such guidance, the agency had virtually unchecked and constitutionally illegitimate power over industry.[19]

Justice Marshall wrote the dissenting opinion. He argued that the plurality had simply substituted their judgment of sound policy for the agency's and by so doing had overridden the plain intent of the OSH Act.[20] Furthermore, the plurality's concept of sound policy in the area was disastrous on its face, for the requirement to demonstrate significant risk before regulating amounted to putting the burden of medical uncertainty on the backs of workers. He predicted that this unwarranted judicial interference in the decisions of regulatory agencies would not stand the test of time.[21]

Burden of proof: How much evidence is enough?

The impact of this ruling on OSHA was considerable, for it shifted the burden of proof onto the agency to show that a particular substance posed a serious health hazard at the levels being regulated. OSHA could no longer set lowest feasible exposure levels simply by noting that no safe level of exposure had been shown to exist. Plainly, the information that a substance was hazardous at high levels of exposure did not suffice, in the Court's judgment, to show significant risk at low levels of exposure. The quantity and quality of health effect information is available, regulators must decide whether flawed could impose mandatory standards.

It is important to suggest at this point that the question of how much scientific evidence is enough to warrant regulatory action is not a scientific matter. The question of how much evidence is sufficient to warrant a scientific conclusion is of course a scientific matter, and when the evidence is strong enough to establish scientific certainty it is good policy (not good science) to use the accepted scientific conclusion as a basis for regulatory decisions. But typically, regula-

[19] N. 4 *Supra*, p. 2887.
[20] N. 4 *Supra*, p. 2887.
[21] N. 4 *Supra*, p. 2905.

tors must act when the evidence is missing or mixed. Normally, the available scientific studies are flawed and point to different conclusions. In order to decide whether to postpone action until better information is available, regulators must decide whether flawed studies can be improved, whether it is likely that the resources to do further study will be available, and how long the new studies will take. Moreover, the officials must weigh the consequences of regulating when the health risks are really very small against the consequences of not regulating when the health risks are really very large. In the case of benzene the choice is especially difficult, since benzene does not produce cancer in animals, and the only way to establish with greater certainly whether low-level exposures are carcinogenic for humans would be to wait out the twenty to thirty-year latency period and see if those workers presently exposed develop cancer. This is not to argue that OSHA's decision to regulate in this instance was correct. It *is* to suggest that the choice of whether the available scientific evidence is sufficient to warrant regulatory action is a complex policy choice, requiring the evaluation of the consequences of not acting. In the abstract, there is no scientifically determinable threshold of evidence below which no regulatory action is justified.[22]

If this view is correct, then the Court's ruling that OSHA did not have sufficient evidence to act is not a scientific judgment. The Court did reweigh the evidence and did second-guess the agency's evaluation of the material in the record. However, the Court is best interpreted (despite some of its language) as saying that it is scientifically uncertain whether benzene is a human carcinogen at low levels of exposure, and, moreover, that the evidence in the record as a whole – including costs, exposed population, and so forth – does not, in this case, justify a prudential policy of regulating now. The court was acting as surrogate policymaker, but not as surrogate scientist.

A different burden-of-proof question was not settled. API had argued that the agency was required to demonstrate that the benefits of lowering an exposure level justified the compliance cost involved. Notice the form of the question: Do the benefits of regulating justify the costs? The agency's action was presumed to be unjustified until a sufficient level of benefit was demonstrated. But a different question could be asked: Does the benefit of using a toxic substance at present levels justify the extra health risk involved? On this view, the use of

[22] The implications of this point for the role of a "science court" are plain. The role of such a court should be to compile and evaluate scientific evidence. It should have no authority to force or prevent regulatory action.

the toxic substance is presumed to be unjustified until a sufficient level of offsetting gain can be demonstrated. The agency did not present such an alternative, since it assumes some sort of balancing framework, but the difference in the burden of proof brought out by these two ways of posing the question of balancing is enormous.

The use of conservative assumptions in risk assessment

Further questions arise concerning the meaning and extent of the threshold test of significant risk. A plausible view is that a significant risk is a risk large enough to warrant some further action to reduce it. Conversely, a risk is insignificant when it is tolerable or acceptable at its present levels – that is, when it is so small that nothing need be done about it. This characterization suggests the familiar division of the problem into one of assessing the size of a risk and one of assessing normatively whether a risk of that size is worth doing something about. Justice Stevens's remarks about a one-in-a-thousand risk being significant, but not a one-in-a-billion risk, suggest that it is the degree of the probability of harm that was his concern. He could be interpreted as suggesting that the agency determine some risk magnitude, say one in a hundred thousand, as the threshold level of risk, and that the agency regulate hazards posing a larger threat but not those that pose a smaller threat. However, evaluation on the basis of the size of probabilities can be misleading. In some cases, a low probability of a large harm is something of great concern. At the very least, the size of the population at risk must be known, so that the expected outcome, and not merely the size of the probability, can be evaluated. This problem can, of course, be avoided by thinking of risk not as the probability of an outcome but as some combination of probability and possible outcome.[23]

[23] The idea that an occupational risk is too low to worry about should not be confused with the idea that an event's probability is so low that it is virtually certain not to happen. An occupational risk is a frequency distribution stating how many cases will in fact arise in an exposed population. From an individual's point of view, an insignificant occupational risk may appear to be like the risk of being hit by a meteor next Thursday. From a social point of view, however, a certain number of cases will result from a public acceptance of insignificant occupational risks. Accepting occupational risks *is* accepting occupational injury and illness. This point reveals a little-noticed implication of the Court's rejection of a zero-risk policy: OSHA does not have the responsibility or authority to protect all workers. The Court's opinion suggests that some workers may suffer material impairment of health not because it costs too much to protect them but because their number in relation to the overall exposed population is too small.

There are, of course, enormous difficulties in assessing the size of a risk. As a result of the benzene decision, the agency will have to face these problems in a far more serious way. One problem here is extrapolating from high doses in animals or high exposure levels in humans to low exposure levels in humans. In many cases, there appear to be several dose-response functions that fit the observed data equally well. There is no agreed-upon theoretical basis for favoring one function over the others, and they differ enormously concerning the size of risks at low levels. The size of the risk involved at these levels of exposure, then, may be virtually unknown, even if the existence of a risk at low levels is a virtual certainty.[24]

In this situation, one suggested course of action is to rely on conservative assumptions at every step of the extrapolation process. Thus, the linear dose–response function is recommended over the probit-log function, since it implies larger health effects at low exposure levels. Upper confidence levels of risk are to be used, rather than the point estimates themselves. Lowering the level of confidence required for statistically distinguishing effects has been suggested. In extrapolating from animal studies, it has been suggested that we use the species or sex that has the largest response.[25] Could OSHA use these conservative assumptions in calculating risks? Surprisingly, the answer is not clear from the benzene case, despite assurances from Justice Stevens that reputable but conservative assumptions could be used to assess the significance of risk. OSHA had relied on the idea that there was a dose–response relationship of some kind in order to make the claim that the benefits of lowering the exposure level were likely to be appreciable. But Justice Stevens characterized this dose–response relationship as an "assumption" and a "theory," and concluded that there was "no empirical evidence" to support the conclusion that there was any cancer risk whatsoever at 10 ppm.[26] If the idea that there is a dose–response relationship is not sufficient to support a claim that there is some risk, how can the idea that this

[24] A further problem concerns the interaction of several different substances. It appears likely that the cancer risks from substances considered individually may be considerably less than their aggregate effects. A requirement to establish that each individual chemical hazard poses a significant risk over and above the risk posed in combination with other chemicals might not adequately protect workers from a significant risk posed by a group of chemical hazards. The relevant unit for the determination of the size of a risk may be a group of chemicals that react synergistically together.

[25] Jerome Cornfield, "Carcinogenic Risk Assessment," *Science* 198 (1977): 693–9.

[26] N. 4 *Supra.*, p. 2869. This is one of the places where, I think, Justice Stephens overstates his case.

dose-response relationship is linear be used to support a more precise estimate of the risk?

Regardless of whether this option is legally open to OSHA, the use of conservative assumptions is problematic. In some cases, the assumptions are not adopted for scientific reasons but for policy reasons. Some have objected to the idea of using conservative assumptions in this way. According to one critique, this distorts the resulting cost–benefit assessment by comparing extreme estimates of risks with more moderate estimates of benefits. Furthermore, according to this critique, using these extreme estimates of risk incorporates value judgments into the estimation of risk in a concealed and hence illegitimate way.[27]

In defense of conservative procedures, let me first note that not all of them are of equal value. The policy of altering confidence levels, for example, appears to have little to recommend it. But a blanket critique is unjustified. The policy of using a linear dose–response curve, for instance, is the scientific equal of using the competing probit-log dose–response curve. What then can we rely on to choose in this case, but nonscientific criteria? The generalized critique of all conservative assumptions misses the mark by assuming that in these cases we know what the facts are and what the judgments are. We do not know on scientific grounds alone what the risks are; that is, what it means to say that there are equally good dose–response relationships that provide widely different estimates of risk at low levels. So we do not know whether our estimates of risk are extreme or not. Yet we have to make a decision about what to do. How, then, do we proceed? Because of the consequences of not regulating when, in fact, the risks are high, it is probably good policy to use conservative assumptions in assessing the size of risks. This is a policy choice, but it is not therefore irrational. What would be irrational would be to substitute policy assumptions for accepted scientific facts. When there is no scientific consensus on the facts of the matter, however, value judgements must be made in order to characterize occupational risks at all.[28]

Another objection is to the use of conservative assumptions as a way of avoiding the disclosure of uncertainty.[29] A set of conservative

[27] Cornfield, n. 25, *Supra*.

[28] This point should be expressed with some care and certainly does not imply that policy judgments determine truth. But when scientific criteria fail to determine truth, policy dictates which assumptions it is best to use for regulatory purposes (not which assumptions are true). The dependence of risk assessments on policy judgments derives only from the inevitability of scientific uncertainty and need not rest upon more global theses about the lack of objectivity in science as such.

[29] N. 15 *Supra*, p. 793.

assumptions may still produce an extremely low estimate of risk. It is important to note that the experimental or epidemiological data from which these extrapolations are made are themselves subject to some uncertainty; they may be supplanted by new data indicating higher risk, as happened in the case of asbestos. This point is correct but suggests only that the use of conservative extrapolation procedures must be fully disclosed in rule-making proceedings. It in no way suggests that these procedures are illegitimate.

The significance of risk

These are all problems associated with assessing the size of a risk. It should be clear that typically this is by no means a completely scientific endeavor but is shot through and through with policy decisions about how to proceed with the characterization of risk in the face of scientific uncertainty. The assessment of the size of risk, therefore, is usually not simply a presentation of value-neutral facts. But the policy questions are enormously greater when it comes time to assess whether a risk of a given size is worth doing something about. The Supreme Court's decision to determine significance before regulating may be seriously misleading at this point. For whether a risk is worth doing something about is in part dependent on how easy it is to do something about it.[30] If a safe substitute is readily available for a known carcinogen, then why not use it? If control of exposures is absurdly inexpensive or easily affordable, then why not control them? In these cases, the need to assess the significance of the risk is less than obvious, but a possible interpretation of Justice Stevens's opinion is that the agency would be barred from regulating small risks that could be easily eliminated. The reasoning is this: According to Justice Stevens's opinion, the agency must first establish the significance of a health risk and then move on to the question of how strictly to regulate the risk. The assessment of significance, therefore, is done without any balancing of costs and benefits. A policy that allowed the agency to move quickly to reduce or eliminate risks that have no benefits or risks that are easily remediable might be more prudent. On the other hand, given the time and resources needed to move a proposed regulation through the review and rule-making process, it is probably good policy to aim at significant

[30] "The Supreme Court, 1979 Term," *Harvard Law Review*, 94 (1980): 242–51. The idea that risks with no benefits and easily avoidable risks should be eliminated regardless of their size is a staple of the risk assessment literature. See, for example, C. L. Comar, "Risk: A Pragmatic *De Minimis* Approach," *Science* 203 (1979): 319.

risks first. But it may be bad policy to limit agency action to the area of significant risks.

Principles to assess the importance of eliminating a risk of a certain size are obviously needed. A start can be made by noting that making this assessment solely on the basis of the size of the risk or expected harm is not always wise. Some kind of balancing framework is needed at this point. But so far the agency has only specified factors to consider in deciding which carcinogens to regulate first.[31] Principles, even rough and pragmatic ones, to structure the normative assessment of risks have not been made explicit.

The cotton dust case

The benzene decision was silent on the question of setting exposure levels by balancing costs and benefits. Hence, one option in the postbenzene climate of opinion was to retain the lowest feasible level of exposure policy but supplement it with a significant risk threshold test. The agency's proposed change in its cancer policy illustrates this possibility. In its response to the benzene decision, OSHA withdrew some language from its cancer policy and inserted new language. The initial policy called for setting exposure limits for carcinogens at the lowest feasible level. The agency withdrew this language and proposed a new requirement that exposure limits be set at the lowest feasible level that is reasonably necessary or appropriate to eliminate significant risk. Although this supplements the feasibility criterion with a significant risk criterion, it does not require the balancing of costs and benefits in setting levels. It allows the possibility of determining the significance of the health risks at one decision level and the affordability of the costs at another level. The principle that if it is worth regulating at all, it is worth regulating to the lowest feasible level, remains intact.[32]

That the structure of OSHA's decision making could remain intact after the benzene decision can also be seen by examining the agency's position on cotton dust. In June 1978 OSHA promulgated a final rule limiting exposures to cotton dust to 200 micrograms per cubic meter (mg/cum) in the yarn-manufacturing sector of the cotton textile industry, 750 mg/cum in slashing and weaving, and 500 mg/cum everywhere else.[33] The agency argued that this was necessary to protect workers against byssinosis, a permanently disabling respira-

[31] See the list of factors to be considered in prioritization in *Identification, Classification, and Regulation of Potential Occupational Carcinogens*, p. 5240.

[32] 46 Federal Register 7402, January 23, 1981.

[33] *Occupational Exposure to Cotton Dust*, 43 Federal Register 27350, June 23, 1978.

tory disease caused by inhalation of cotton dust. On the basis of epidemiological studies, it was estimated that reducing the current exposure levels to the newly mandated levels would reduce the prevalence of byssinosis among workers from 26 per cent to 13 per cent.[34] The costs of compliance were estimated at $206 million in annualized capital investment expenses and annual operating and maintenance costs.[35] Although some marginal firms were expected to go out of business as a result of the new regulation, the agency determined that the standard was within the financial capability of the industry considered as a whole.[36]

The American Textile Manufacturers Institute, Inc. (ATMI), sought to have the standard invalidated in preenforcement review before the U.S. Court of Appeals for the District of Columbia, but in October 1979 the court of appeals ruled that the cotton dust standard was valid. ATMI appealed to the Supreme Court. Oral arguments were held in January 1981, and in June 1981 the Supreme Court affirmed the judgment of the lower court.

ATMI argued that feasibility and reasonableness had not been established in setting the standard. In their view, OSHA's interpretation of feasibility as a purely technological and financial matter imposed no effective constraint on the agency. Their interpretation of the feasibility constraint implied that the agency must make responsible predictions of what its standards will cost and what impact they will have on output, employment, competition, and prices. But they argued that the agency must also present "intelligible criteria" for determining when changes in these variables constitute an unacceptable impact on the industries affected. In their view, OSHA had not done this and had, instead, adopted a policy of viewing compliance costs as irrelevant up to the point at which the destruction of an entire industry was threatened.[37]

On the question of reasonableness, ATMI argued that OSHA must show that expected benefits are significant in light of the costs imposed. Moreover, OSHA must show that expected benefits over and above those expected from less stringent alternatives (the mar-

[34] N. 33 *Supra*, p. 27379.

[35] N. 33 *Supra*, p. 27369.

[36] In fact, the cotton dust industry took the opportunity presented by the standard to extensively modernize its facilities. It purchased new Swiss equipment that is both more efficient and safer. Because the control technology was installed in the course of capital replacement, the prospective cost estimates, based upon the assumption of extensive retrofitting, are considerably higher than the real control expenditures.

[37] *ATMI* v. *Ray Marshall*, "Brief for Petitioners," pp. 31–6.

ginal benefits of increasingly stringent alternatives) are significant in light of the extra costs imposed. ATMI argued that OSHA had not performed such a relative risk assessment. In particular, the agency had not assessed the benefits of an ATMI alternative for respirator use, medical surveillance, and medical removal. Although ATMI did not argue that a formal cost–benefit analysis was required, it did argue that OSHA had failed to demonstrate, even in an informal way, that requiring the use of engineering controls to reduce exposure levels to cotton dust had significant benefits.[38]

In reply, OSHA argued that the cotton dust standard protected against a significant risk of material health impairment. Byssinosis is clearly a health effect and not a matter of worker inconvenience. The preexisting levels of exposure allowed for a 26 per cent prevalence of this disease among those working in the industry and probably a larger percentage of those retired from work. Based upon a dose–response curve constructed from epidemiological studies, OSHA estimated that the standard would substantially reduce the prevalence to 13 percent. Finally, the industry alternative was not equally effective, since its transfer provisions were unworkable, and respiratory protection was demonstrably not as effective as engineering controls in reducing exposure levels. OSHA concluded that the standard was reasonably necessary to reduce a significant risk as required in the benzene decision.[39]

The agency argued that it had also evaluated the effect of the standard upon the affected industries. The policy position adopted was that OSHA is required to consider the financial health and viability of an industry, but it is not required (and not empowered) to engage in an individualized and subjective balancing of costs and benefits in every case. As before, economic feasibility is defined as affordability; a standard is economically feasible when it is within the financial capability of the industry as a whole, but not necessarily for each individual firm within the industry. Furthermore, a cost–benefit test in each and every case is not needed, since OSHA uses several economic criteria that prevent the misallocation of resources. A balancing of the likely benefits of regulating a toxic substance to the lowest feasible levels against likely costs of this action is used in setting priorities for regulating. In addition, the agency evaluates each standard to make sure that it provides the required level of protection at the least possible cost. OSHA's economic criteria for evaluating priorities and standards, therefore, imposed a genuine

38 Ibid., pp. 37–9.

39 *ATMI* v. *Ray Marshall*, "Brief for the Federal Respondent," pp. 29–36.

restriction on its ability to direct the health and safety resources of the country. On the basis of information in the record, the agency had concluded that the standard was economically feasible in the sense that it met these stringent economic criteria.[40]

It should be clear that the model of decision making that emerges from OSHA's defense of the cotton dust standard is virtually the same as the prebenzene model. As before, the agency can be pictured as asking a series of questions. First, is there a *significant risk* of material impairment of worker health at current levels of exposure? If not, then do nothing. If it is decided, on broad cost–benefit grounds, to proceed with regulation at all, then the agency determines the level at which this risk is not significant. It is no longer possible to use a zero-risk policy at this point; each carcinogen or toxic substance must be assessed separately, for even if there is no absolutely safe level above zero exposure, the remaining risk may or may not be significant. But it is clear that the determination of significant risk is made on health grounds. Consideration of economic and technological matters does not arise until later in the process, and when these matters do arise, their treatment follows the same pattern as before the benzene decision.

As in the case brought to the court in defending the benzene standard, the agency asserts that no balancing of conflicting values is allowed. The agency claims the authority to pursue the goal of occupational safety and health single-mindedly. If a significant threat to worker health exists, and if it is possible to eliminate it without threatening the viability of the affected industry, OSHA claims the authority to allocate the resources to reduce the threat until it is no longer significant. On the question of balancing, it is asserted that Congress has already balanced this goal of workplace safety and health with conflicting goals in passing the OSH Act. It explicitly decided the balancing question by declaring that feasible standards that eliminate a significant risk to worker health are worth the costs. No further need for balancing exists.[41]

The Supreme Court upheld OSHA's position in this case, ruling that the agency is not required to employ cost–benefit analysis to set particular standards regulating exposure to toxic substances. Further, the Court declared that any such standard less protective than one based upon feasibility analysis is inconsistent with the OSH Act. The crucial question turned on the interpretation of section 6 (b) (5) of the Act. The Court held that

[40] N. 39 *Supra*, pp. 37–55.

[41] N. 39 *Supra*, p. 38.

> Congress itself defined the basic relationship between costs and benefits, by placing the "benefit" of worker health above all other considerations save those making attainment of this "benefit" unachievable. Any standard based on a balancing of costs and benefits by the Secretary that strikes a different balance than that struck by Congress would be inconsistent with the command set forth in Section 6 (b) (5). Thus, cost–benefit analysis by OSHA is not required by the statute because feasibility analysis is.[42]

In effect, then, the Court barred the agency from using cost–benefit criteria to set toxic substances standards, enjoining it instead to use feasibility analysis. Nor was there any doubt about what feasibility analysis was. The court defined "feasibility analysis" by quoting the "Brief for the Respondent Unions," in which OSHA's procedure is defined as asking a series of questions: "First, whether the 'place of employment is unsafe – in the sense that significant risks are present and can be eliminated or lessened by a change in practices.' ... Second, whether of the possible available correctives the Secretary had selected '*the* standard ... that is most protective.' ... Third, whether that standard is 'feasible!' "[43] The model of decision making outlined earlier was, therefore, upheld virtually intact.

Critique of cost–benefit analysis

The cotton dust case was a source of confusion to OSHA in the early days of the Reagan administration. Executive Order 12291, issued in February 1981, required all regulatory actions, including OSHA's, to meet an economic cost–benefit test.[44] As a result, OSHA went to the Supreme Court after oral arguments on the cotton dust case had been heard but before a decision had been handed down, asking the Court to refrain from further consideration of the case. The agency wanted to reconsider the cotton dust standard using cost–benefit analysis.[45] In its ruling the Court "declined to adopt" this suggestion and effectively prohibited the use of cost–benefit analysis in a reevaluation of the standard.

Did this rejection of cost–benefit analysis amount to taking a step

[42] *American Textile Manufacturers Institute, Inc. et al.* v. *Donovan, Secretary of Labor et al.* 452 U. S. (June 17, 1981). Slip op. p. 17.

[43] N. 42 *Supra*, Slip op. p. 15.

[44] Executive Order 12291, 46 Federal Register 13193, February 19. 1981.

[45] *ATMI* v. *Raymond Donovan*, Motion for Leave to File Supplemental Memorandum and Supplemental Memorandum for the Federal Respondent, October Term, 1980.

away from rationality in the standard-setting process? The policy of lowest feasible risk, even as constrained by a threshold test of significant risk, may appear extreme. It seems to allow the agency to override any combination of private interest and competing public values in achieving a maximal level of protection for workers. However, it is essential to recognize that the use of cost–benefit analysis will not help matters. The use of a cost–benefit decision rule is simply bad occupational safety and health policy, and its prohibition by the Court can only be seen as a step toward rationality.

Formal cost–benefit analysis calls for the equation of monetized costs and benefits at the margin. Standards should be tightened only when the dollar value of the health gains from a more stringent regulation exceed the costs of further control. The purpose of such a rule is to ensure that resources are allocated efficiently between control of occupational health hazards and other important goals. Despite this efficiency rationale, cost-benefit analysis is incomplete: first, because it is limited to items that have been traded on a market; second, because it ignores values other than efficiency. The only way cost–benefit analysis can incorporate these other considerations is by pricing them. But pricing nonmarket items is notoriously difficult and often involves dubious assumptions about implicit markets. For example, it is often alleged that workers reveal the price they are willing to pay for job safety by their acceptance of wage premiums for risky work. But it is well known that imbalances of power, information, and mobility prevent the full expression of workers' preferences for job safety. Cotton textile workers, for example, are not the fully mobile individuals of economic theory. Moreover, when health effects from exposure to toxic substances are delayed by long latency periods, as they may be for benzene, workers typically have little information concerning the health risks they face on the job.

Not only is pricing nonmarket health effects of workplace hazards extraordinarily difficult; it may also represent a lowering of the value we place on preventing these health effects. Often a "not for sale" sign distinguishes very highly valued items from normal commodities. Pricing an occupational fatality is removing a "not for sale" sign, and it may therefore suggest that human life should be removed from the class of specially valued things. The likely practical result of this symbolic act would be to reduce our national efforts to prevent occupational fatalities.

Safety and health on the job are also matters of rights, justice, and public values, as well as matters of efficiency. Very high levels of occupational risk are clearly violations of workers' rights. The creation of a class of expendable workers who bear high risks so that others may live more comfortably is a violation of elementary claims of justice. A large toll of death and disease on the job is a violation of the shared public value we place on treating workers as ends in themselves whose lives and interests are worthy of concern and respect. Since these considerations cannot be ignored in setting standards for occupational exposure to toxic substances, cost–benefit analysis must expand to include them if it is to be the sole or primary way of deciding these questions. The only way cost–benefit analysis can include these considerations is by pricing them. But pricing human rights, justice, and public values is even more problematical than pricing lives saved. Pricing values by determining what we as individuals are willing to pay for them reduces public disputes concerning the proper balancing of economic efficiency against other values to matters of personal preference. In a cost–benefit framework the decision would depend on how intensely people feel about the issue (measured by their willingness to pay), rather than on the merits of the arguments they put forward. Moreover, by reducing the question of what values we espouse as a community to the question of what we are willing to pay for as individuals, the use of cost–benefit analysis would, in the end, reduce our lives as citizens to our lives as consumers.

It is possible to reject econometric estimates of willingness to pay for safety on the job, and still attempt to impose a cost constraint on OSHA. A regulatory budget is one way to do this, for it would limit the amount OSHA can require industry to pay for workplace safety each year. But the appropriate size of such a budget is a matter of sheer speculation. An alternative would be to set a unit cost constraint. For example, it could be decided, perhaps by congressional vote, to spend no more than $1 million to save a life. Standards would be set so as not to spend more than this per life saved. The advantages of this approach are those of consistency and efficiency in the allocation of industrial safety and health resources. There are disadvantages, too. First, there is the disquieting symbolism of publicly pricing life. Second, there appears to be no basis for setting a unit cost constraint. How can it be decided how much to spend to save a life without knowing how much it takes to save a life, or how many lives will be saved, or what moral values will be sacrificed by not saving more? There is no guarantee that any given unit cost constraint will be sufficient to reduce unreasonable risks, to distribute risks more equally, and to satisfy our public values on the importance of life and bodily integrity. Third, this proposal assumes that economic cost should be of primary importance and that other considerations are less important. But other factors are equally important: the nature of the risk involved; how many people are at risk; how voluntarily it is assumed; who bears the cost; who gets the benefits; whether there is something especially dreadful about this risk; whether large amounts of individual freedom have to be sacrificed in order to eliminate it; exactly what economic goods might become more expensive; and so on. A multidimensional decision framework integrating irreducibly different considerations is required. Once all these other considerations are given their full due, it is likely that the implied cost per life saved for different programs will differ considerably. Enforcing a consistency along the cost dimension, then, is a way of discounting the importance of equally important factors.

It may at first seem that since the programs designed to save lives on the job accomplish their goal at a certain cost, any decision to proceed with such a program must be based upon cost considerations. There is, it might seem, an implicit value of life saving present in each decision to proceed. It may seem, then, that calling for the adoption of a unit cost containment figure simply makes explicit a previously hidden decision criterion and thus opens it up to public challenge. This position, however, confuses a consequence of a decision with a basis for it. A decision may imply the acceptance of a certain cost to save a life, but this figure may not have been the primary reason for making the decision. It may not even have formed any part of the justification of the decision. It is only when an explicit decision has been made first on how much to spend to save a life that a cost figure is a decision criterion. When a decision is based on a balance of other factors, the unit cost results from this weighing but does not determine it. Hence, it is not always correct to describe a cost-of-life figure as a hidden decision criterion.[46]

Let me repeat that a balancing framework of some kind is needed in regulating toxic substances and workplace hazards generally. But not all balancing need be based exclusively on cost–benefit comparisons. A distinction must be drawn between justifying a level of effort in an occupational safety and health program on the basis of a comparison of the monetary value of the associated costs and benefits and justifying such a program by weighing the reasons for and against it and deciding that, all things considered, the level of effort in the program is worthwhile. The first method is simply the cost-benefit approach and, in practice, treats efficiency as the only, or the most important, consideration. The second method considers efficiency and might sometimes give it pride of place but also considers

[46] Steven Kelman makes a similar point in "Cost–benefit Analysis: An Ethical Critique," *Regulation* 5 (1981):33–40.

individuals rights, justice, and competing public values as reasons for or against a level of effort in a program. In the first case, the basis of decision is already given, and the crucial questions are technical. In the second case, most of the technical questions remain, although some are less urgent (e.g., the monetary value of safety), but the bases for decision making are unclear. Much work needs to be done in formulating an alternative regulatory framework that integrates economic, health, and normative considerations without pricing them.

Closure

These cases suggest one way in which disputes about public control of occupational exposure to toxic substances can be resolved – namely, through the courts. The proper role of the courts, however, is complex. They must not only interpret what guidelines Congress provides for regulatory agencies; they must also decide whether the agencies have followed these guidelines. Thus, in the benzene case the Supreme Court made the significant risk criterion explicit and ruled that OSHA had not provided substantial evidence of significant risk. In the cotton dust case, the Court required feasibility analysis, not cost–benefit analysis, and ruled that OSHA had sufficient basis in the record to conclude that the standard was reasonably necessary and feasible. The courts must therefore become involved in the question of whether the scientific and economic evidence in a rule-making record is substantial enough to warrant regulatory action. This intervention, however, is for extreme cases only. The role of the court is preponderantly procedural: to see that all relevant issues are considered and to ensure that there is full disclosure of the normative and factual basis for regulatory decisions.

This suggests that the real issue of closure currently lies with the criteria used by the regulatory agencies. How *do* they resolve these disputes? I have tried to describe the significant risk and feasibility criteria that OSHA uses. Roughly, OSHA chooses regulatory priorities on broad cost–benefit grounds, and in particular rule-making proceedings it aims to reduce exposures to the lowest feasible level necessary to eliminate significant risk. Despite the formal denial that feasibility analysis implies balancing in particular rule-making proceedings, it is quite clear that OSHA ought to (and does) balance advantages against disadvantages in deciding how stringently to regulate. At the very least, the need to preserve some industry resources to protect workers against other health hazards requires a balancing framework. After the cotton dust decision, however, the technique of pricing espoused in formal cost–benefit analysis is not available to OSHA and would be a poor decision rule in any case. Nor are proposals to impose arbitrary cost constraints on the agency likely to improve matters.

The current situation with respect to closure, then, looks like this. Politically accountable officials weigh scientific, economic, and normative factors to arrive at regulatory decisions. In doing this, they are constrained by congressional instruction to give enormous weight to worker safety and health relative to other considerations. The role of scientific and economic experts is and should be subordinate to that of these officials. Those who decide whether or not to proceed with regulatory action in the face of scientific uncertainty must draw upon the expertise of scientists, but, ultimately, purely scientific considerations do not determine this choice. Those who decide whether or not the gains from regulatory action are worth the costs must rely upon scientific and economic experts to provide the best available assessment of anticipated health gains and economic consequences. Even here, however, purely scientific and economic considerations do not normally dictate the content of these apparently factual assessments. And even if they did, purely scientific and economic considerations cannot determine whether a regulatory action is in the public interest. Normative considerations inevitably play a dominant role in these decisions. Hence, some check on the power of regulatory officials is crucial. The courts can provide some of this through judicial review, although it is easier for the courts to prevent overly stringent regulations than it is to prevent regulatory neglect of health hazards. In the current scheme of things, then, the public's best assurance that regulatory officials reflect the public interest is the ballot box.

23

Nuclear fear: a history and an experiment

SPENCER R. WEART

Powerful emotions permeate the debate over nuclear energy. Anxiety and anger are evident in all the songs and shouts of antinuclear rallies, and at meetings where pronuclear engineers and executives gather, there too one may sense, beneath the controlled language, anxiety and anger. Perhaps this emotion is an uninteresting side effect such as may attend any debate that waxes strong. But what if the emotion is fundamental? What if there are feelings about nuclear energy that motivate the debate itself, so that the arguments over politics, values, and scientific claims are the side effects, the symptoms of a deeper conflict? Until we understand the origin and nature of the emotions aroused by nuclear energy, we cannot be confident that we know how to bring the nuclear debate to a satisfactory resolution.

Historical methods offer promise of uncovering the origin of these emotions, for as we go back in time, we can slough away transient phenomena that confuse the current scene. The reader must be warned, however, that the history of emotion is a field in its infancy. Not only in our subject matter but in our methodology, we enter uncharted waters.

When radioactivity was discovered in a few rare ores, the phenomenon was not at first recognized as separate from other newly famous forms of radiation such as radio and X-rays. Only around 1900 did radioactivity begin to arouse interest, when radium, an intensely radioactive element, became known. Among the interested scientists were Ernest Rutherford and Frederick Soddy, who late in 1901 made a revolutionary discovery: Radioactivity signals atoms that are changing from one element into another.

For use of facilities and support, I am grateful to the American Institute of Physics and its Center for History of Physics; to Johns Hopkins University; and to the Brookhaven National Laboratory.

"I was overwhelmed," Soddy recalled, "with something greater than joy – I cannot very well express it – a kind of exaltation." He blurted out, "Rutherford, this is transmutation!" "For Mike's sake, Soddy," his colleague replied, "don't call it *transmutation*. They'll have our heads off as alchemists."[1] Already at the moment the new science was born, it was flooded with emotion, with exaltation and anxiety.

What did the word *transmutation* signify to evoke these emotions? The question is important, for Rutherford and Soddy recognized that it is in the process of transmutation that nuclear energy is released. From 1901 until the discovery of uranium fission in 1939 (and to some extent since), ideas of nuclear energy, and indeed a large part of any work in nuclear physics, were based upon the paradigmatic concept of the transmutation of elements.[2] A comparison with alchemy was inescapable, and not only the popular press but many scientists, even Rutherford, spoke of nuclear research as a new alchemy.

Scholars have studied the history of the idea of transmutation back into the times of the alchemists, for transmutation has long been a symbol of high significance. These scholars inform us that the dominant tradition in alchemy was concerned as much with spiritual as with material change; the alchemist's apprentice was exhorted to see the transmutation of base metal into gold as secondary, an aid and symbol for the transmutation of the soul.[3] The idea of alchemy as a quest for self-discovery helps explain why alchemists and others showed anxiety about their work, concealing even their chemical formulas in arcane symbols, warning off the profane who might misuse their knowledge. For revelation is a fearful thing.

How can knowledge be fearful? Psychologists tell of clinical evidence that such anxieties are widespread. People confronted with powerful unknown forces, within themselves or outside, tend to

[1] Muriel Howorth, *Pioneer Research on the Atom: . . . The Life Story of Frederick Soddy* (London: New World, 1958), pp. 83–4. For this and the following discussion, see Lawrence Badash, "Radium, Radioactivity, and the Popularity of Scientific Discovery," *Proceedings of the American Philosophical Society* 122 (1978): 145–54, and *Radioactivity in America: Growth and Decay of a Science* (Baltimore: Johns Hopkins University Press, 1979); Thaddeus J. Trenn, *The Self-splitting Atom: The History of the Rutherford-Soddy Collaboration* (London: Taylor & Francis, 1977), pp. 42, 58–60, 111–17.

[2] Spencer Weart, "The Discovery of Fission and a Nuclear Physics Paradigm," *Otto Hahn and the Rise of Nuclear Physics*, ed. William Shea (Dordrecht: Reidel, 1983), pp. 91–133.

[3] See e.g., E. J. Holmyard, *Alchemy* (Harmondsworth, England: Penguin Books, 1957), p. 158; Carl G. Jung, *Collected Works*, trans. R. F. C. Hull (Princeton: Princeton University Press), Vols. 12–14.

recall childhood fears of arousing parental wrath by probing into forbidden areas.[4] Whether or not this explanation is accepted, one cannot deny that the fear of knowledge is ancient and pervasive in our culture. Adam and Eve, Pandora, Prometheus, Eurydice, Job's wife, Bluebeard's wife, the Sorcerer's Apprentice, Faust, and many others came to grief by seeking too much knowledge.[5]

The extent of the danger is particularly shown in the symbol of transmutation. In the crucible of the alchemist, substances were said to die and be reborn, undergoing a descent into corruption and putrefaction before they could be transmuted; this is a symbol of the agonizing descent into darkness that is necessary for psychological or spiritual transformations.[6] In short, there is abundant evidence that the symbol of transmutation calls to mind the great theme of death and resurrection.

Students of comparative mythology inform us that this theme has an old significance beyond the individual, corresponding with cosmic and social events and symbolized, for example, by the annual decline and return of the sun. The descent of matter into decay parallels a coming time of chaos, a time in which evil triumphs, humanity is afflicted by plagues and wars, and in the extreme the entire universe is destroyed. In our own culture, this universal theme has taken form as fear of divine wrath and Armageddon, a fear that haunted the Middle Ages and is felt by many people still.[7] In sum, there is reason to believe that the moment one brings up the idea of transmutation one must deal with ancient associations involving secret knowledge, personal and social transformation, and shattering dangers ranging right up to the end of the world.

The evidence from psychology, alchemical symbology, and comparative mythology may be unfamiliar to the reader, and I do not have space here for a more thorough exposition. Those who find the

[4] See Abraham H. Maslow, *Toward a Psychology of Being*, 2d ed. (New York: Van Nostrand Reinhold, 1968), pp. 60–3.

[5] Joseph Campbell, *The Masks of God*: Vol. 3, *Occidental Mythology* (New York: Viking Press, 1964), pp. 109–10.

[6] Mircea Eliade, *The Forge and the Crucible*, trans. Stephen Corrin (New York: Harper & Row, 1962), p. 169; Evelyn Underhill, *Mysticism: A Study in the Nature and Development of Man's Spiritual Consciousness*, 12th ed. (1930; New York: Dutton, 1961), pp. 140–8 and Chap. 9.

[7] Mircea Eliade, *The Myth of the Eternal Return, or, Cosmos and History*, trans. Willard R. Trask, 2d ed. (Princeton: Princeton University Press, 1965); Yi-Fu Tuan, *Landscapes of Fear* (New York: Pantheon Books, 1979), pp. 56–61; Campbell, *Masks of God*, 3: 269, and see also Vol. 1, *Primitive Mythology*, pp. 280–1.

argument far-fetched may just set it aside at this point, since I will now return to simple history and describe how nuclear fear has in fact been expressed.

The first public intimations of fear in connection with radioactivity occurred soon after Rutherford and Soddy announced their discovery and involved fear of dangers ranging up to the end of the world. Soddy said that the energy locked within the atom was so great that the earth must be regarded as a storehouse of explosives; a man who could unleash this energy "could destroy the earth if he chose."[8] Similarly, in 1904, a physicist declared in the *Quarterly Review*: "It is conceivable that some means may one day be found for inducing radio-active change in elements which are not normally subject to it. Professor Rutherford has playfully suggested to the writer the disquieting idea that, could a proper detonator be discovered, an explosive wave of atomic disintegration might be started through all matter which would transmute the whole mass of the globe into helium or similar gases."[9] Rutherford's unamusing joke that some fool in a laboratory might blow up the universe unawares was based on the possibility, seriously considered by scientists of the time, that radioactivity might be contagious from atom to atom, although the prospects for setting off such a catastrophe never seemed at all plausible.

The idea of a radioactive doomsday had become widespread by the 1920s. There were speculations that the vast stellar explosions occasionally seen in the sky might be the result of runaway experiments in induced radioactivity conducted by beings on other planets. The German physical chemist Walther Nernst warned that "we are living on an island of guncotton." By 1930 such warnings must have been familiar to most literate people, and the prominent American physicist Robert Millikan felt a need to deride the "advocates of a return to the 'glories' of a pre-scientific age," who, as he perceptively put it, "have pictured the diabolical scientist tinkering heedlessly, like a bad small boy, with these enormous stores of sub-atomic energy, and some sad day touching off the fuse and blowing our comfortable little globe to smithereens." Millikan was sure this was impossible, but five years later the French physicist Frédéric Joliot took the occasion of his

[8] Howorth, *Pioneer Research*, pp. 122–3; see pp. 56–8 for Soddy's interest in alchemy.

[9] W. C. D. Whetham, "Matter and Electricity," *Quarterly Review* 397 (1904): 100–26, quotation from p. 126; see p. 120 for mention of alchemy. Cf. Whetham to Rutherford, July 26, 1903, Rutherford Papers, Cambridge University Library, microfilm in Niels Bohr Library, American Institute of Physics, New York.

Nobel prize address to warn his colleagues not to be bad boys – to take care lest they blow up the earth.[10]

These fears were rational in the sense that they were not definitively contradicted by the facts of nuclear physics known at that time. And in general, the concerns I will be discussing in this paper cannot be dismissed as utterly baseless. However, talk of the destruction of the world cannot fail to call up irrational feelings. Who can contemplate such a thing without feeling a knot in the stomach, without recalling nightmares and childhood terrors? In this sense, all the concerns that I will discuss have important irrational resonances.

Anxiety about nuclear energy was also expressed from the beginning in a more plausible form: Radioactivity could be misused for wicked purposes. In 1904 Soddy told the Royal Corps of Engineers that nuclear energy might be used as a weapon, and the next year Pierre Curie noted in his Nobel prize address that "radium could become very dangerous in criminal hands." The popular press was already carrying speculations about radioactive death rays that could blow up battleships. By 1920 it was common knowledge that, as the British physicist Oliver Lodge put it, if the "wrong people" discovered how to unlock nuclear energy, "this planet would be unsafe." Probably the most influential writing on nuclear energy was H. G. Wells's novel, *The World Set Free* (published in serial form in 1913–14), which introduced the phrase "atomic bomb" to common use. Wells vividly described a cataclysmic world war of the 1950s in which cities were destroyed and left uninhabitable for generations by induced-radioactivity bombs the size of a handbag. It was after reading Wells's novel in 1932 that the Hungarian physicist Leo Szilard became the first scientist to attempt systematic experiments directed toward a wholesale liberation of nuclear energy.[11]

All this is the dark side of the early public response to radioactivity, but there was a dazzlingly bright side that outshone these anxieties. Again, we may find in Soddy some of the strongest initial statements. Explicitly and emphatically pointing out the connection

10 Nernst as quoted in Robert Jungk, *Brighter than a Thousand Suns: A Personal History of the Atomic Scientists*, trans. James Cleugh (New York: Harcourt Brace Jovanovich, 1958), p. 8; Robert Millikan, *Science and the New Civilization* (New York: Scribner, 1930), p. 95.

11 Howorth, *Pioneer Research*, p. 123, and "Shall We Ever Use This Power in the Industrial World?", *Current Opinion* 68 (1920): 838–9; H. G. Wells, *The World Set Free: A Story of Mankind* (New York: Dutton, 1914), first published in *English Review* 16–17 (1913–14); *Leo Szilard: His Version of the Facts*, ed. Spencer R. Weart and Gertrud Weiss Szilard (Cambridge, Mass.: MIT Press, 1978), Chap. 1.

with the ancient quest for transmutation, he insisted that if humanity succeeded in tapping nuclear energy, the alchemists' dreams would be fulfilled; "The future would bear as little relation to the past as the life of a dragon-fly does to that of its aquatic prototype." Soddy was concerned about the inevitable exhaustion of coal and other resources, a process that he and many others feared would eventually bring about the decay of civilization and regression to primitive conditions. (Soddy explicitly compared this with the old notion of a cosmic cycle of growth followed by destruction.) Nuclear energy might be humanity's salvation. "A race which could transmute matter would have little need to earn its bread by the sweat of its brow . . . such a race could transform a desert continent, thaw the frozen poles, and make the whole world one smiling Garden of Eden."[12] From about 1903 on, the idea that radioactivity could be a great boon for humanity spread swiftly and pervasively. Soddy's remark that a pint bottle of a radioactive element such as uranium had enough energy to drive an ocean liner around the world was repeated so often that by the end of the 1930s it had become a tired cliché. Scientists spoke of a possible "revolution of civilization undreamt of in the wildest speculations." Popular writers indulged in the wildest speculations – for example, that in the future "power will be broadcast free of charge for heating and lighting purposes from atomic energy plants."[13] Even the possibility of nuclear weapons had a sunny side, for such dreadful weapons might force humanity to put an end to war. In *The World Set Free*, Wells predicted that nuclear war would lead to the founding of a world government, with social justice, universal prosperity, and nuclear-powered garden cities in the arctic wastes. In short, from the beginning the hopes for nuclear energy were just as grandiose as the fears.

Optimism indeed dominated discussion of nuclear energy for the first few decades after Rutherford and Soddy's discovery. Expressions of fear were less common than hopeful visions. Presumably the initial response to radioactivity was so positive because optimism accorded with the belief, widespread through the nineteenth century and well into the twentieth, that scientific advance was leading toward an earthly paradise.

[12] Soddy, "The Energy of Radium," *Harper's Monthly*, December 1909, pp. 52–9, quotations from pp. 58, 56, and *The Interpretation of Radium*, 3d ed. (London: Murray, 1912), pp. 248–51.

[13] Arthur Haas, "The Atom as a Source of Energy," *Scientific Monthly*, February 1928, pp. 140–6, quotation from p. 140; Roger Babson as quoted in George Wise, "Predictions of the Future of Technology, 1890–1940," unpublished essay, 1976; see also Badash, *Radioactivity in America*.

It is not surprising that such ideas attended discussions of transmutation. In psychology and comparative mythology, the theme of transmutation is connected with hopes as much as with fears. The descent into corruption results in salvation, Armageddon precedes the millennium. Historically, millennial expectations in connection with science arose out of a milieu in which both chiliastic and alchemical ideas were widespread.[14]

The myth-ridden hopes and fears about nuclear energy, the excesses of the Sunday supplements, dismayed many scientists. By the 1930s, prominent physicists like Millikan and Rutherford made a point of telling the public that there was no process then known that could yield usable nuclear energy, although they admitted that it was not impossible that such a process might someday be discovered. The physicists' warnings against premature speculation about nuclear energy marked the first step in a divergence of attitudes. A gap was visible between some scientists' deliberate coolness, their show of rationality, and the tendency toward emotion and fantasy of the popular press.

There was one area where the benefits of nuclear energy could not be denied: medicine. Radioactive substances were of undoubted value in treating certain diseases, and radium therapy developed swiftly. This brought powerful influences to bear on popular ideas about radioactivity. For one thing, medical demand drove the price of the rare radioactive elements to great heights. Discussions of radium in the press often revolved around the metal's commercial value, and from Colorado to Siberia the discovery of uranium ore was seen in much the same light as the discovery of a lode of gold. This reinforced the popular perception that great value attached to radioactive matter.

More important was the link forged between radioactivity and health. Here again, hopes went beyond the bounds of fact. The press reported soberly that radium might well vanquish that paradigm of dread disease, cancer, and there were more speculative suggestions that radium might also halt tuberculosis, make the blind see, and perhaps someday raise the dead. Again we can turn to Soddy for an accurate perception of the mythic roots of these dreams: "The philosopher's stone was accredited the power not only of transmuting the metals, but of acting as the *elixir of life*." I suspect that there are deep psychological and cultural reasons for joining the concepts

[14] Franklin L. Baumer, *Modern European Thought: Continuity and Change in Ideas, 1600–1950* (New York: Macmillan, 1977), pp. 330–6, 369–70, 389–91; Robert Nisbet, *History of the Idea of Progress* (New York: Basic Books, 1980).

of transmutation and energy release with ideas about health, but since there has been relatively little scholarly investigation into the question, I will simply point out that the connections are evident in the history of science and made up part of the image that people held of radioactivity.[15]

The use of radium in medicine was not without danger, and this too was apparent almost from the start. Radiation from radioactive substances was found to cause skin sores and other damage, and Pierre Curie was quoted as saying that a lump of radium could kill a person. None of this was surprising, for such effects had already been noted with X-rays. For the first decade or so, the public perception of medical radioactivity simply followed in the footsteps of the public peception of X-rays.

The first reports of burns from X-ray radiation perturbed the public in 1897 and 1898, resulting in public outcry and lawsuits. Further concern was aroused when it was found that X-rays can produce sterility and cancer. Between 1903 and 1911, over fifty cases of X-ray–induced carcinoma were reported, and repeated amputations did not always stave off death. There were public demands for regulation, but since almost all of the injuries were sustained by physicians and technicians, criticism was muted. There is, after all, an ancient and honorable tradition that in treating a dangerous disease with a new therapy one accepts a certain proportion of deaths among the patients and a higher proportion among the physicians. The public did not see the hazards of radiation in a separate light nor find anything special about those killed by the new therapy (the names of 110 "martyred" professionals were inscribed on a monument erected in Germany in 1936).[16] From the 1920s on, the public, seeing medical organizations imposing regulations on the use of X-rays and other radiation, gladly accepted the benefits with no more fear of the risks than attends most medical procedures. Of course, since all medical procedures involve some anxiety, radiation was caught up in a pattern of images that was not altogether reassuring.

15 "Raise the dead": New London, Conn., *Day*, Aug. 27, 1903. This and other newspaper clippings are in the W. J. Hammer Collection, Division of Electricity, Museum of History and Technology, Smithsonian Institution, Washington, D.C.; Soddy, *Interpretation of Radium*, p. 250; see Eliade, *Forge and Crucible*; Weart, "Discovery of Fission."

16 Data from Daniel P. Serwer, "The Rise of Radiation Protection: Science, Medicine and Technology in Society, 1896–1935," Brookhaven National Laboratory Informal Report BNL 22276, 1976, p. 60; John M. Fowler, ed., *Fallout: A Study of Superbombs, Strontium 90 and Survival* (New York: Basic Books, 1960), p. 13. In addition, here and below I use Barton C. Hacker, "A History of Personnel Dosimetry and Radiation Safety in Nuclear Device Testing," Chap. 1, unpublished paper, 1979.

Radium, with fewer blemishes on its record than X-rays, was perceived as even more benign. There was widespread belief among physicians and the public that a little radioactivity might be a healthful stimulant. Spas were proud of the natural radioactivity in their water, and physics laboratories did a brisk business certifying the radioactive content of various mineral waters and tonic preparations. By the early 1930s, in Germany one could eat chocolate candies that had a small amount of radioactive material as an additive, then brush one's teeth with radioactive toothpaste.

A more cautious view had begun to spread after 1925, when radioactivity was identified as the cause of jaw necrosis, sometimes progressing to death, in women who painted dials with luminous paint at a factory in New Jersey. Tipping the points of their brushes with their lips, the women had ingested radium. The press took only a casual interest in the problem. As late as 1929 the *Literary Digest* (a predecessor to *Time* magazine), reporting the elaborate safety rules imposed by doctors on the use of radium, remarked that "many will be surprised to know the dangers of this healing substance."[17]

Perceptions shifted somewhat in 1931, when the death of Eben Byers sent, according to the *Literary Digest*, "a chill of apprehension" over the United States. Byers had suffered a gruesome death because he had regularly imbibed a radioactive tonic. Many thousands of people had used this patent medicine, and hundreds of thousands more patronized radioactive mineral springs and spas, so anxiety was widespread. Fears were expressed, for example, in a 1936 horror movie, *The Invisible Ray*, in which Boris Karloff, contaminated by a mysterious meteor, was cursed with contagious radioactivity. But the chief reaction in the press was of caution rather than fright; radium was seen as deadly only in the hands of quacks, remaining a powerful force for health in the hands of competent physicians. As late as 1939, scientists were still warning against the use of radioactive tonics. By the late 1930s, newspaper attention, if we may judge from the index to the *New York Times*, had drifted away from radium poisoning and back to commercial production of radium and the continuing demand for the substance by hospitals.[18]

In an attempt to probe the public mind more objectively, I have done a primitive analysis of the content of popular articles, using selected volumes of the *Readers' Guide to Periodical Literature*. I looked only at the titles of articles, suspecting that titles touch a less sophisticated level of thought than is found in the literary content of

[17] "Deadly Radium Gas," *Literary Digest*, June 15, 1929, p. 19.

[18] "Radium Poisoning,"*Literary Digest*, April 16, 1932, p. 13; Jeff Rovin, *A Pictorial History of Science Fiction Films* (Secaucus, N.J.: Citadel Press, 1975).

the articles themselves. I counted titles concerning radioactivity and atomic energy, noting titles that seemed by themselves likely to evoke in a naive reader either predominantly positive, optimistic feelings or negative and possibly fearful feelings. I also counted titles that implied both risks and benefits, and neutral titles. In samples through the period from 1900 to 1940, about a quarter of all the titles were polarized, falling into either the predominantly positive or the predominantly negative category. Of these, in selected periods from 1900 to 1921 I found no negative titles; what anxieties there were usually appeared in articles with an overall optimistic tone. In 1925–8, radium poisoning changed this, yet positive titles still outnumbered negative ones by nearly two to one. In 1932–5 the ratio rose to three to one, and in 1939–41, just after the discovery of nuclear fission, I again found no predominantly negative titles.

All this suggests that the public found nothing unusually troubling in radioactivity itself. We may hear claims that radioactivity, because it is invisible, malignant, and so on, inevitably arouses irrational fears. No doubt there is some truth to this, but the historical evidence suggests that radioactivity is inherently no more fear-provoking than the numerous other invisible, malignant substances connected with industry and medicine (germs and certain chemicals, for example). By 1939 the public was aware that radioactivity can cause debility, cancer, genetic damage, and sterility. Yet when the discovery of fission was announced, these hazards were almost entirely ignored in the burst of speculation about the coming nuclear age.

The discovery of fission revived the extreme hopes that scientists had tried to quiet in the 1930s. Almost as soon as the news of the discovery arrived, the *New York Times* editorialized that "romancers have a legitimate excuse for returning to Wellsian utopias where whole cities are illuminated by the energy in a little matter." Also near the surface was the current of extreme anxiety that had been present ever since the discovery of transmutation. "The tabloids love to write of blowing up the world with a gram of matter," *Scientific American* remarked, "and it's not such a sensational idea as one might think." The fear that nuclear energy might be misused, not accidentally but with evil intent, was also revitalized, and with one eye on Nazi Germany newspapers and magazines showed concern about the awful potential of uranium weapons. Nevertheless, the predominant tone, as the *Readers' Guide* analysis indicates, was optimistic.[19]

[19] *New York Times*, February 3, 1939, p. 14; Jean Harrington, "Two Elements for One," *Scientific American*, October 1939, pp. 214–16; see Harrington, "Don't Worry – It Can't Happen," *Scientific American*, p. 268.

In 1939, only a handful of nuclear physicists actively pursued the liberation of nuclear energy. Like the public, these scientists tended to be optimistic, and they devoted their main attention to developing controlled nuclear reactors for industrial use. But like the public, the scientists were also wary of the Germans and anxious about possible weapons. At that time physicists mistakenly supposed that a uranium bomb would be something like a reactor gone wild. Correctly perceiving that such a device would fizzle rather than make an enormous explosion, they considered uranium bombs unlikely.[20]

Naturally enough, in 1939–40 the press too was unclear about the difference between reactors and bombs. The old cliché of accidental catastrophe was therefore dragged in. *Newsweek*, after remarking that nuclear generators might run a car or light a home, warned that "the process ... might grow to uncontrollable proportions," giving an explosion that would make ordinary bombs "seem like firecrackers." More thorough and accurate was *New York Times* reporter Waldemar Kaempffert. Although he informed his readers that "the old possibility of destroying the earth is revived," he went on to discuss the genuine problems of nuclear reactors. They would have to contain an enormous quantity of dangerous radioactive materials, he noted, and if improperly controlled "would blow up like an overheated steam boiler." Kaempffert said he would prefer the safety of a conventional power plant.[21]

The theme was expressed in full detail in Robert Heinlein's story, "Blowups Happen," appearing in *Astounding Science Fiction* magazine in September 1940. At a future uranium power plant, an error is discovered in the calculations: If the plant gets out of control it can blow up, leaving the earth as barren as the moon. But the plant's board of directors, believing that nuclear energy is economically indispensable, refuse to acknowledge that any risk exists. Let us not dismiss this story by saying that it is a miraculously prophetic freak. It simply expresses themes that were automatically invoked by the concept of nuclear energy, concerns prior to and wholly independent of actual reactor engineering.[22]

Blowups may happen, and the scientists in the Manhattan Project who built the first nuclear reactors shared the public's concern. But in the secrecy masking the project, their views began to diverge from the public's, and by the end of the war they were rather more

[20] Spencer Weart, *Scientists in Power* (Cambridge, Mass.: Harvard University Press, 1979), Chaps. 4–10.

[21] *Newsweek*, March 27, 1939, p. 32; *New York Times*, March 5, 1939, p. II: 9.

[22] See also Lester Del Ray, "Nerves," September 1942. The Heinlein story is reprinted with some alterations in *The Man Who Sold the Moon* (New York: Signet, 1951).

confident of their ability to build reactors safely. Still, even the remote possibility of destroying the world seriously perturbed some project scientists, and Edward Teller was assigned the task of studying the question scientifically, just to be sure. On the nervous evening before the first bomb test, Enrico Fermi jokingly offered to take bets on whether the bomb would ignite the earth's atmosphere. The next morning, when the mighty fireball rose up, General Thomas Farrell reportedly turned in despair to a fellow officer, exclaiming, "The long-hairs have let it get away from them!"[23]

Surely the bomb test was awesome, but General Farrell was bringing something of his own to the sight. The roar of the explosion, he wrote immediately afterward, "warned of doomsday and made us feel that we puny things were blasphemous to dare tamper with the forces heretofore reserved to The Almighty." Many others who witnessed the "Trinity" test also recalled fears of doomsday and of forbidden knowledge. Others recalled visions of renewal. To the only reporter present, William Laurence, the roar was "the first cry of a newborn world," and caused some witnesses to break into a jig like "primitive man dancing at one of his fire festivals at the coming of spring."[24]

These old themes of millenarian hope, apocalyptic fear, and anxiety about the revelation of knowledge also characterized public opinion in the first few years after the news of Hiroshima and Nagasaki. On the hopeful side there were articles such as the one the editors of *Woman's Home Companion* rated among the most important they had ever published, in which William Laurence harked back to science fiction dreams of his childhood. He said that now nuclear power plants could transform the world's deserts into "blooming gardens" and turn swamps and jungles into "vast new lands flowing with milk and honey." Nuclear power, he declared, could "*make the dream of the earth as a Promised Land come true in time for many of us already born to see and enjoy it.*" He was not the only one to use the language of a new Eden. Of course, not everyone was so ecstatic, but public opinion polls showed that in the first few years after Hiroshi-

[23] I am grateful to C. Gruber for copies of Manhattan Project reactor hazards memoranda. See also Stanley A. Blumberg and Gwinn Owens, *Energy and Conflict: The Life and Times of Edward Teller* (New York: Putnam, 1976), pp. 117–19; Leslie R. Groves, *Now It Can Be Told: The Story of the Manhattan Project* (Boston: Little, Brown, 1967), p. 296; Lansing Lamont, *Day of Trinity* (New York: Atheneum, 1965), pp. 235–6.

[24] Farrell quoted in Martin J. Sherwin, *A World Destroyed: The Atomic Bomb and the Grand Alliance* (New York: Knopf, 1975), App. P, pp. 312–13; William L. Laurence, *Men and Atoms: The Discovery, the Uses and the Future of Atomic Energy* (New York: Simon & Schuster, 1959), p. 26.

ma most people expected a fairly swift application of nuclear energy for beneficial purposes and believed that more good than harm would flow from splitting the atom.[25]

But accompanying this optimism was a deep current of anxiety. The editors of *Woman's Home Companion* gave Laurence's utopian article the title "Paradise or Doomsday?" and illustrated it with a mushroom cloud. A typical summary was that of the *New York Times*, which on August 18, 1945, editorialized, "We now possess the means to blow ourselves and perhaps the planet itself to drifting dust or make this world of ours a paradise." A poll at that time found that amid their optimism, a quarter of the American public thought it not unlikely that nuclear experiments would someday cause an explosion that would destroy the world. The image of scientists as children tinkering with forbidden secrets naturally accompanied such views. The first significant political issue over control of nuclear energy revolved around restraints on scientists and imposition of security rules. The concern over secrecy seemed excessive to many at the time and in retrospect is clearly seen to have been partly irrational. But even the darkest thoughts about bombs could not overcome the strong currents of postwar hope, and many felt that the existence of nuclear weapons would at long last force humanity to abandon war; the time seemed ripe for the world government and universal prosperity that H. G. Wells had prophesied.[26]

We may summarize the views that the American public (and to a substantial extent the world public) held about nuclear energy shortly after Hiroshima by saying that these were the same as the views held about nuclear energy before Hiroshima, and ever since the turn of the century. This is confirmed by my analysis of the titles of articles related to nuclear energy in the *Readers' Guide*. Naturally there were many more such articles around 1948 than around 1940, and for the first time there were a number of articles primarily about war use. But of the articles about civilian use, the same proportion (about a quarter) had primarily positive or negative rather than balanced or neutral titles, and of these polarized articles the same proportion (nearly all) were positive.

Did Hiroshima and Nagasaki, then, have no effect on the public

[25] Laurence, "Paradise or Doomsday?" *Woman's Home Companion*, May 1948, pp. 32–3, 74. Early polls are reviewed by Hazel Gaudet Erskine, "The Polls: Atomic Weapons and Nuclear Energy," *Public Opinion Quarterly* 27 (1963): 155–90.

[26] Kenneth MacDonald Jones, "Science, Scientists, and Americans: Images of Science and the Formation of Federal Science Policy, 1945–1950," Ph.D. diss., Cornell University, 1975; University Microfilms, Ann Arbor, no. 75–24, 202, pp. 44–5, 50, 59, 228–34 and passim.

mind? I believe there was a profound effect, but one that tended to be unexpressed. Descriptions of the bomb's horrors, such as John Hersey's *Hiroshima*, were very widely read. It was commonly said at the time that the use of the bombs had made the public frightened about nuclear energy in general, and I take these statements seriously. Then why were pessimistic statements relatively uncommon? Some said that the public simply did not want to face dreadful things, and I doubt that napalm fire raids on Japan, the various death camps, and so on were discussed more than Hiroshima. And with some people there was a conviction, as the first head of the Atomic Energy Commission, David Lilienthal, admitted, "that somehow or other the discovery that had produced so terrible a weapon simply *had* to have an important peaceful use." So anxieties found indirect outlets. At the 1946 Bikini bomb tests, groups of sailors were swept with rumors about the terrible effects of radioactivity. Before the war, the death by cancer of hundreds of workers in the pitchblende mining and radium industries had scarcely been reported in the newspapers, but in 1947 there was a flurry of alarm over a single fatality at the Oak Ridge nuclear installation – not from radiation, as it turned out, but from hepatitis.[27] Images of Hiroshima were now ineradicably associated with radioactivity.

Fear of nuclear subjects became more widespread, or at least more open, in the 1950s. There was a wave of horror movies in which creatures were awakened or created by nuclear weapons: the "beast from twenty thousand fathoms" assaulted New York City in a 1953 movie of that name; monster ants in *Them!* (1954) invaded Los Angeles, and the greatest of all, *Godzilla* (1955), devastated Tokyo. Meanwhile, Seventh Day Adventists, convinced the Second Coming was at hand, declared that it would be preceded by Armageddon in the form of nuclear war.[28]

Evidently these fears were directly connected with military rather than civil use of nuclear energy. There was good reason for such fears to increase. In the first place, the Cold War was taking hold, and many people began to suspect that nuclear war was just around the

[27] Hersey, *Hiroshima* (New York: Knopf, 1946), first published in its entirety in the *New Yorker*, August 31, 1946; David E. Lilienthal, *Change, Hope, and the Bomb* (Princeton: Princeton University Press, 1963), p. 109; see Lilienthal, *Journals*: Vol. 2, *The Atomic Energy Years, 1945–1950* (New York: Harper & Row, 1964). On Bikini, see David Bradley, *No Place to Hide* (Boston: Little, Brown, 1948); "Radioactivity Scare," *Time*, September 1, 1947, pp. 56–7. For public concerns I have also benefited from "Public Information" files in the L. Haworth papers, Brookhaven National Laboratory, Upton, N.Y.

[28] Rovin, "Sect Anticipates Atomic Armageddon," *Life*, November 22, 1954, pp. 176–7.

corner. In the second place, there was a new perception of what such a war would be like. It is not easy to recall today that in the years 1945–54 people in most countries did not feel immediately threatened by nuclear weapons. After all, the destruction of Hiroshima was no worse than the obliteration of Dresden and Tokyo by incendiary bombs; nations could survive such blows and imagine themselves carrying on. All this began to change in the early 1950s with the advent of thermonuclear weapons. A thousand times more powerful than the Hiroshima device, hydrogen bombs were instantly seen to surpass the limits of known warfare. Now the prospects of destruction became immediate: the crossroads foreseen by the editorials of 1945 had been reached and the wrong fork taken. The end of the world was soberly, if inaccurately, prophesied in such works as the British film *The Day the Earth Caught Fire* (1961), in which bomb tests throw the earth into a collision course with the sun, and Nevil Shute's best-selling novel *On the Beach* (1957; Stanley Kramer film, 1959), in which nuclear war extinguishes humanity. Motion picture images of lifeless cities had a counterpart in the newspapers, which ran photographs of downtown areas emptied of people for civil defense drills. For example, in the 1956 Operation Alert, sirens sounded in seventy-five major American cities, calling people to file into sheltered places while radio and television went off the air. During these years many schoolchildren were taught to fall to the floor and cover their heads when their teacher cried, "Take cover!" The civil defense efforts of the 1950s not only demonstrated anxiety over nuclear war but also spread and perpetuated such anxiety. To make things worse, many people had little confidence that civil defense measures could save them.[29]

The realistic prospect of the sudden destruction of civilization was something new in human history, difficult or impossible to consider objectively. There were reports that most people preferred simply to ignore or suppress such awful thoughts. The alternative was to take action, and civil defense was not the only possibility, for one could directly oppose nuclear weapons. But this opposition often took an indirect form.

As early as 1953, public anxieties were heightened when it was discovered that radioactive substances from a bomb test in Nevada

[29] See, e.g., "Ducking for Cover," *Newsweek*, July 30, 1956, p. 28; Herbert Yahraes and Dixie Yahraes, "This School Is Ready for the H-Bomb," *Saturday Evening Post*, September 25, 1954, pp. 45 ff.; Ralph E. Lapp, "Fallout and Home Defense," pp. 303–10 in Morton Grodzins and Eugene Rabinowitch, eds., *The Atomic Age: Scientists in National and World Affairs* . . . (New York: Basic Books, 1963).

had been deposited by rain as local fallout as far away as New York State. "Anyone got a lead-lined umbrella?" asked a Long Island newspaper, skeptical of scientific assurances that plants and television sets would not be affected by the radioactive rain.[30] Anxiety turned to fear when it was revealed that the huge hydrogen bomb test of 1954 had deposited fallout on a Japanese fishing vessel, killing one of the crew. And the bomb had not only caused local fallout but had spread radioactive materials around the world. Scientists discussed the possibility that such tests would cause thousands or even millions of cancers and birth defects, as well as conceivably altering the weather or damaging all life on earth. In most countries the question of bomb tests now became the focus of public debate about nuclear energy. Until tests in the atmosphere were banned by the United States and the Soviet Union in 1963, these tests and their attendant fallout were heatedly denounced in official protests from Third World countries, in mass marches through Britain and elsewhere, and in political declarations such as Adlai Stevenson's 1956 presidential campaign speeches. The opposition to bomb tests greatly increased public anxiety about the dangers of radiation, to a point where some mothers worried about giving their children fresh milk that might be contaminated by fallout. There was a real political and ethical issue involved, for scientists could not guarantee that the tests would not cause at least some eventual deaths and birth defects at random among the world's peoples.

And yet, as an American scientist noted, "There are many other areas of life where we as a nation can and will affect more people's lives for better or for worse." He was not the only one to declare that "fallout is not precisely the major issue with regard to bomb tests. Behind this worry lies the much greater fear of nuclear war." A survey of attitudes toward fallout showed that the amount of anxiety that respondents felt was not related to the amount of knowledge they had about the topic: The source of anxiety was not in the facts of the controversy but was imported from elsewhere. Since more than half of the respondents believed that a nuclear war would destroy most of civilization on earth, the source of their anxiety can be guessed. Nor was it unreasonable to use bomb tests as a stalking-horse for nuclear war concerns. The opponents of testing generally believed that a moratorium on tests would be a first step, indeed the only step then feasible, toward slowing the nuclear arms race. A scholarly study of the bomb test controversy has concluded,[31]

30 *Newsday*, April 21, 1953, and see April 3, 1954. These and other clippings in files of Associated Universities, Inc., Brookhaven National Laboratory, Upton, N.Y.

31 Walter R. Guild, "Biological Effects of Radiation," in Fowler, *Fallout*, p. 91; see also Sidney Kraus, Reuben Mehling, and Elaine El-Assal, "Mass Media and the

> The horror of the hydrogen bomb was so great that most people could not face it directly. Yet try as they might, they could not ignore it. The testing issue gave them a way to deal with it obliquely. Instead of coping with the danger of all-out nuclear war, in which blast and heat and fallout might destroy most of the American population, they focused on the less drastic but still insidious threat of poisoned milk and contaminated air. The test ban became a kind of magic talisman, a way that the nation could confront a real and present danger without coming to grips with the true reality of the 1950s – the possibility of total destruction.

Along with the bomb test controversy came a political shift. Authorities in the countries that tested nuclear weapons insisted that the tests were harmless, and at times their arguments had as little objectivity as the arguments of their most extreme opponents. Many people, scientists as well as the public, began to lose confidence in the nuclear authorities. In the United States, the Atomic Energy Commission compounded its problems from 1954 on by insensitive handling of other issues. In the Dixon-Yates controversy over socialized versus capitalist electrical utilities, in the Robert Oppenheimer security hearings, and in a number of other disputes, many people concluded that the commission could not be trusted.

Civil nuclear energy was largely insulated from these concerns as work on "Atoms for Peace" went ahead rapidly. Applications of radioactivity to medicine, agriculture, and industry were hailed, and in the Soviet Union, the United States, and Britain, especially, it was felt that great progress was being made toward electricity production, ship propulsion, and desalination of water to make deserts bloom. If only humanity could avoid "the end foretold by the writer of the Apocalypse," a popular science book declared, "we can assuredly enter into the Promised Land of Plenty." An international survey of college students around 1959 found that despite some anxiety about radiation and other dangers, a large majority was optimistic about the future of nuclear energy.[32]

Given the increasingly politicized atmosphere after 1954, however,

Fallout Controversy," *Public Opinion Quarterly* 27 (1963): 191–205; Robert A. Divine, *Blowing on the Wind: The Nuclear Test Ban Debate, 1954–1960* (New York: Oxford University Press, 1978), p. 323. Cf. Carolyn Kopp, "The Origins of the American Scientific Debate over Fallout Hazards," *Social Studies of Science* 9 (1979): 403–22.

32 Lancelot Hogben, *Science for the Citizen*, 4th ed. (London: Allen & Unwin, 1956), p. 1122; Johan Galtung, *Atoms for Peace: A Comparative Study of Student Attitudes*, Institut for Samfunnsforskning, Report No. 5–1 (Oslo: Institut for Samfunnsforskning, 1960).

it was natural for the civil nuclear energy program to become a political issue, and not only in the Dixon-Yates sense. Scientists and engineers, as well as the public, had always had some concern about the safety of nuclear reactor programs. It seemed likely that given the undercurrent of public anxiety about radioactivity and explosions, any serious reactor accident would cause public outcry so intense that civil application of nuclear energy would be held back. For this and other reasons, in the 1950s it was the strong advocates of nuclear energy such as Edward Teller who laid the most emphasis on keeping reactors safe. But these advocates' warnings were taken up and used in opposition to nuclear reactors. The first such case (Lagoona Beach, Michigan, 1956) was explicitly political; the pronuclear United Auto Workers and others used official misgivings about the safety of a proposed reactor to belabor privately owned utilities and their supporters in the Atomic Energy Commission. In later cases the political character was less simple.

There were sound reasons behind concern over civil nuclear energy, and there were collateral developments that helped raise the level of concern, such as a rise in environmentalism and a boom in reactor construction. But here, I repeat, I am studying the irrational feelings that came along with the rational arguments. I am less interested in the association of nuclear energy with political disputes than with the feelings that made nuclear energy such a potent issue whenever it was drawn into politics. Where did anxieties about nuclear power plants originate?

The language in which opposition was expressed seems to have been taken over from the bomb test controversy and from talk about bombs in general. Reactors might "explode" and generate "fallout." When civil nuclear power programs were accused of spreading about low-level radioactivity, people resurrected all of the fallout-era disputes over the numbers of cancers and genetic defects that such radioactivity might cause. Surveys showed that the opposition to nuclear power was overwhelmingly based, not on economic or environmental problems specific to political issues, but on perceptions of danger; the words that were most often used were "accident," "explosion," "radioactivity," and "chain reaction." A quarter or more of the public believed, and believes to this day, that a nuclear power plant can explode like a nuclear bomb. Yet ignorance is not the source of opposition (nor of advocacy), for surveys have found scant correlation between attitudes for or against nuclear power and knowledge about it.[33] All of this suggests that anxiety about civil

[33] Barbara D. Melber, Stanley M. Nealey, Joy Hammersla, and William L. Rankin, *Nuclear Power and the Public: Analysis of Collected Survey Research* (Seattle: Battelle Memorial Institute, Human Affairs Research Centers, 1977), pp. 119, 124–7, 157–9.

nuclear energy draws much of its strength not from direct evidence but from suppressed fear of nuclear bombs.

An interesting check on this suggestion comes from analysis of the titles of nuclear-related articles listed in samples from the *Readers' Guide*. Up to the mid-1960s, there were few articles with mainly negative titles relating to civil use. The main change since 1946 was a rapid rise in articles concerning military use, including such subjects as bomb tests and civil defense. The number of such titles peaked in the early 1960s, then fell off rapidly. The decline of interest in civil defense was particularly abrupt; after a brief, intense period of controversy over fallout shelters in the early 1960s, the public and the United States government simply gave up any attempt to save the nation in the event of nuclear war. Although warheads multiplied and the danger grew, war was talked about less and less. In the movies, after *Fail Safe* (1964) and *Dr. Strangelove* (1964; subtitled, "How I Learned to Stop Worrying and Love the Bomb"), nuclear war films almost disappeared, and the bomb became a technological toy for James Bond to disarm. A fine, realistic television film on the effects of nuclear war, *The War Game* (1966), was the first documentary ever banned by the British Broadcasting Corporation.[34]

This decline of attention to the danger of nuclear war was accompanied by a decline of attention to Atoms for Peace as well. This was followed by a rise in attention to the negative features of civil nuclear energy. The ratio of *Readers' Guide* titles with predominantly positive to those with predominantly negative evocations abruptly reversed, from about 80 percent positive in 1964 to about 70 percent negative in 1978. Yet this shift did not represent a change in the general public's acceptance of nuclear power, for from the late 1940s until the late 1970s opinion surveys consistently found a 2 to 1 majority accepting civil nuclear power. What had changed was not so much the perception of danger, apparently, as the willingness to talk about it. All this implies that at least some of the increased expression of anxiety about civil nuclear energy was an outlet for the less and less often expressed fear of nuclear war. Indeed, a detailed survey of attitudes toward nuclear power showed that the largest difference between advocates and opponents lay in their view of the effects of nuclear industry on "a world at peace" and "family security," values obviously connected with anxiety about war.[35]

[34] John Brosnan, *Future Tense: The Cinema of Science Fiction* (New York: St. Martin's Press, 1978), p. 163; Jack G. Shaheen, ed., *Nuclear War Films* (Carbondale: Southern Illinois University Press, 1978), pp. xiii, 113–4.

[35] Melber et al., *Nuclear Power*, p. 118. See also William L. Rankin, Stanley M. Nealey, and Daniel E. Montano, *Analysis of Print Media Coverage of Nuclear Power Issues* (Seattle: Battelle Memorial Institute, Human Affairs Research Centers,

To summarize my conclusions, I can do no better than refer to the only detailed, international study of nuclear fear that I have found, a study conducted by the World Health Organization in 1957. The study committee believed that underneath the political concern and distrust of authority, which was only beginning to evidence itself in 1957, lay anxiety caused by many aspects of modern, secularized, technological society, but above all by the advent of nuclear weapons. Fear of nuclear war tended to confuse and block rational thought, contributing to pervasive anxiety. But beneath this anxiety lay still deeper problems. Fear of the consequences of discovering new knowledge was age-old, the committee noted, as shown in the legends of Prometheus, the Garden of Eden, the alchemists, and Faust. In clinical practice, childhood anxieties, aroused by feelings of impotence in the face of overwhelming power, were known to be related to destructive fantasies; the committee compared such fantasies to science fiction stories of the end of the world.[36] To this analysis I would only add that the committee, optimistic about civil nuclear energy, did not notice that the utopian promises of a nuclear age could equally well be compared with the legends of old and the fantasies of children confronted with uncomprehended forces.

To check how much these conclusions may have to do with our own thoughts today, I invite the reader's company in an experiment to be conducted within our heads.

This paper was written for a seminar in the peaceful village of Hastings-on-Hudson, New York. It happens that the much-criticized Indian Point nuclear power plant is nineteen miles north of our meeting place. Are we afraid of this station? Why? Let us investigate what we may fear.

I can imagine that on the day we meet, a complete meltdown, with emergency cooling failure, occurs at Indian Point. I can even imagine that the meltdown results in a failure of the containment vessel, so that a plume of radioactive material rises into the air, but not so high as to disperse overhead out to sea. A gentle breeze from the north

Footnote 35 (*cont.*)
1978): Rankin and Nealey, *The Relationship of Human Values and Energy Beliefs to Nuclear Power Attitudes* (Seattle: Battelle Memorial Institute, Human Affairs Research Centers, 1978), p. 46.

36 *Mental Health Aspects of the Peaceful Use of Atomic Energy: Report of a Study Group*, World Health Organization Technical Report Series No. 151 (Geneva: World Health Organization, 1958), pp. 16, 29–33. The report is summarized in "Mental Health and Atomic Energy," *Science* 127 (1958): 140–1. I reached my chief conclusions before finding this study.

blows the cloud directly toward New York City. We are in its path. I hasten to add that everyone realizes that the chance of such an accident is low – certainly less than the chance, for example, of a comparably dangerous chemical catastrophe (such as nearly happened farther down the Hudson the month before the meeting). Still I ask, what may we fear?

In this event we may suppose that those of us at the seminar, unable to be evacuated because of a massive traffic jam, are exposed to radiation far above natural levels. Many of our group vomit; most feel nauseous, fatigued, and upset. A few become seriously ill and require antibiotics. Subsequently we learn that we have been exposed to, say, 200 rem apiece. Some doctors say that our chances of contracting cancer have been increased by 2 percent – that is, whereas normally nine out of our group of fifty would eventually die of cancer, we may now lose a tenth person. But most experts assure us that the risk is not so great. Possible genetic effects among our descendents are even harder to assess, being too random and remote to demonstrate. We hear that to the north several hundred people died because of the accident; the loss is comparable to the effects of a flood, earthquake, or dam failure. More significant, perhaps, is the loss of industrial plant, particularly the rapid closing down of every nuclear reactor in the United States. We know that poverty is unhealthy, but the rise in infant mortality and other harm caused by economic loss is hard to evaluate, as is the cancer and genetic damage from increased reliance on fossil fuels.

How does this scenario make one feel? Speaking for myself, with this image before me I cannot love a nuclear reactor. Yet my mind files away the image in the same category as other and more common catastrophes – chemical dump explosions, dam failures, earthquakes, and floods.

Hastings-on-Hudson is near another high-risk site: New York City. There are missiles, tipped with nuclear devices designed for maximal rather than minimal danger, aimed at New York. The probability that these missiles will be launched is greater than the probability that the safeguards at Indian Point will fail altogether. What then may we fear?

We may conservatively suppose there might be a 1.5-megaton thermonuclear explosion over White Plains and a similar air burst, part of a New York City cluster, over the Bronx-Yonkers border. White Plains and the Bronx border are each about six miles from here, so the effects of the two bombs on us would be similar. Any of us who are in the open will be burned so severely that our chances of recovery will be slight, even if we survive the 130 mile per hour wind that comes shortly after the fireball goes up. Those of us who are indoors will be lucky if the building we hide in does not catch fire. The walls crack, and some interior partitions blow down, but perhaps many of us will be uninjured. (In a more likely case – a 1.5-megaton strike directly on Yonkers or a 20-megaton strike on White Plains – we are obliterated instantly.) Some hours later, fallout arrives from the Buffalo-Pittsburgh-Detroit triangle. Weather conditions and our luck in improvising a fallout shelter determine whether we live; we would be glad to escape with a radiation exposure of 200 rem apiece. The survivors, if any, eventually disperse in search of their families. Amid so much destruction, many will not live through next winter.

I leave it to you to evaluate this experiment. Do you, like myself, suspect that these images awaken deep fears that one would prefer to suppress? Do these fears center on nuclear bombs more than on reactors? If such is the case, it suggests why the conflict over civil nuclear energy is so difficult to resolve.

I believe that resolution can be reached only if the dread problem of nuclear weapons is kept firmly and continually in mind. If we do not suppress the fear, we will have to deal with it, and once that fear is dealt with we will be more free to consider civil nuclear energy objectively. Then we can speak without confusion from fear of unknown forces and secrets, images of apocalyptic disasters, and utopian dreams.

Of course, our fear can be extinguished only if the problem of nuclear war itself is successfully resolved, and this should encourage us the more to confront the true issue and responsibility of our generation.[37]

[37] This paper was written in early 1979, when public attention to weapons was disturbingly low and few signs of revival were visible. So far as I know, the idea that the civil nuclear power debate diverts energies more properly applied against nuclear war was first brought out by Victor Weisskopf.

24

Closure and controversy: Three Mile Island

CORA BAGLEY MARRETT

Closure through circumscription

The experiences of the President's Commission on the accident at Three Mile Island suggest that closure on a controversial matter might be reached by narrowing the terms of the debate.[1] I propose that the final report of the commission received rather wide support because it dealt with a limited and empirically based problem: What had taken place at Three Mile Island (TMI) Unit 2 and its environs beginning March 28, 1979. Had the document addressed the broader matter of the safety of nuclear energy, far fewer of its readers would have accepted it as decisive or significant. My purpose here is to trace the processes through which a limited definition of the problem emerged and to reflect on the reasons for and consequences of what I would term *closure through circumscription*.

Background to the formation of the commission

On April 11, 1979, President Jimmy Carter signed an executive order creating the President's Commission on the Accident at Three Mile Island. The order read as follows:

[1] The membership of the commission consisted of John C. Kemeny, Chairman, President, Dartmouth College; Bruce Babbitt, Governor of Arizona; Patrick E. Haggerty, Honorary Chairman and General Director, Texas Instruments Inc.; Carolyn Lewis, Associate Professor, Graduate School of Journalism, Columbia University; Lloyed McBride, President, United Steel Workers of America; Harry C. McPherson, Partner, Verner, Liipfert, Berhard, and McPherson; Paul A. Marks, Vice-President for Health Sciences and Frode Jensen Professor, Columbia University; Cora B. Marrett, Professor of Sociology and Afro-American Studies, University of Wisconsin-Madison; Russell W. Peterson, President, National Audubon Society; Thomas H. Pigford, Professor and Chairman, Department of Nuclear Engineering, University of California at Berkeley; Theodore B. Taylor, Visiting Lecturer, Department of Mechanical and Aerospace Engineering, Princeton University; Anne D. Trunk, resident of Middletown, Pa.

The Commission shall conduct a comprehensive study and investigation of the recent accident involving the nuclear power facility on Three Mile Island in Pennsylvania. The study and investigation shall include:

(a) a technical assessment of the events and their causes;
(b) an analysis of the role of the managing utility;
(c) an assessment of the emergency preparedness and response of the Nuclear Regulatory Commission and other federal, state and local authorities;
(d) an evaluation of the Nuclear Regulatory Commission's licensing, inspection, operation and enforcement procedures as applied to this facility; and
(e) an assessment of how the public's right to information concerning the events at Three Mile Island was served and of the steps which should be taken during similar emergencies to provide the public with accurate, comprehensible and timely information.[2]

The President required that a report be transmitted to his office and to the secretaries of the Department of Energy and the Department of Health, Education and Welfare not later than six months from the commission's first meeting. The charter for the commission further specified that the twelve-member body would meet about once a month.

The commission submitted its final report, *The Need for Change: The Legacy of TMI*, on October 30, 1979, six months after its inaugural meeting in April. Over that period the commission, supported by a staff of over eighty-five persons and large numbers of consultants, took extensive testimony, reviewed countless documents, and analyzed a wealth of technological and other types of data. Although the final report, and the special reports that accompanied it,[3] focused primarily on the accident, some commissioners, staff employees, and members of the wider public urged the commission to encompass more than the accident in the investigation. I want to suggest some reasons why this urging seemingly was not completely heeded.

[2] President Jimmy Carter, Executive Order No. 12130, signed April 11, 1979.

[3] The following, available from the Superintendent of Documents, comprise the published staff reports to the commission: *The Nuclear Regulatory Commission*, Report of the Office of Chief Counsel; *The Role of the Managing Utility and Its Suppliers*, Report of the Office of Chief Counsel; *Emergency Preparedness, Emergency Response*, Reports of the Office of Chief Counsel; *Reports of the Technical Assessment Task Force*, 4 vols.; *Reports of the Public Health and Safety Task Force*, *Report of the Emergency Preparedness and Response Task Force*; *Report of the Public's Right to Information Task Force*.

Setting the commission agenda

The commission did not begin with a clearly bounded problem; during our initial meetings we entertained diverse suggestions about the breadth and scope of our planned investigation. In fact, we spent the first two sets of public hearings – three of the fifteen days devoted to public hearings over the course of our work – eliciting advice on the topic we should probe. At the first set, held April 26 and 27, representatives from various federal agencies – the Environmental Protection Agency, the Nuclear Regulatory Commission (NRC), and the Department of Health, Education, and Welfare, among others – described the responses their agencies had made to developments at TMI and offered advice to us on the questions we should raise. Many organized their recommendations around events at TMI, but some raised more generic issues, such as the matter of educating the public about nuclear energy. We welcomed all counsel, for at that time we had no program to guide us. In fact, we had had no time to develop any such program, for the public hearing of April 26 represented our first assemblage as a commission. We took testimony before we had an opportunity to reflect on how we would use it.

The second set also produced advice on the investigation, although that was not what we had intended when we planned these hearings. We invited state and local officials, plant personnel, and private citizens from the Middletown, Pennsylvania, area – the site of the TMI facility – to appear before us between May 17 and 19 at a Middletown location and share their knowledge of and observations on the events that had transpired from March 28 on. Essentially this was to be a fact-finding mission and not a suggestion-gathering one.

Our plans went awry. On the first day, just before the hearing was to begin, we learned that our request for subpoena power had not been granted, although such power had been promised from the time the commission was established. We had assumed that the White House and the Justice Department had asked Congress to give us authority to subpoena witnesses and have them testify under oath. But the proposed legislation went to Congress later than we had expected, and when it arrived some members of the Senate Judiciary Committee questioned certain of its provisions. That was the situation when we assembled on the morning of May 17. After discussing the delays in the legislative approval and the importance of obtaining sworn testimony for our work, we voted to cancel all future hearings until we had the right to subpoena individuals.[4] With reference to the

[4] On the afternoon of May 17, the Senate Judiciary Committee passed the legislation; the House approved it at its next session.

scheduled appearances in Middletown, we agreed to invite statements from private citizens but not from the plant personnel, state officials, and all other individuals with some responsibility for the facility or for laws and policies governing its operations, individuals against whom legal challenges possibly could be brought.

Forty-five citizens appeared before us on May 19, the only day on which we held hearings. Few gave any technical details on the accident, but many importuned us to enter certain items on our ledger. A legislator said of his constituents, "They are seeking new ways of handling the dangers and the potential of nuclear power, and they are seeking a credible conviction that the lessons learned at Three Mile Island will mean the problems of Three Mile Island will never again be repeated . . . [The] Commission has an opportunity to provide the people of this area with those kinds of assurances."[5] A physician from the area proposed that the commission answer these questions: Is low-level ionizing radiation dangerous? How should radioactive waste be disposed of? How close was the accident to a full-scale meltdown? The commission report dealt with only the third of these questions. Some who testified contended that we should look not only at Unit 2, where the accident occurred, but also at Unit 1, which was not operating on March 28 because some routine work was being done there. Those who pointed to the undamaged unit suggested that it should not be allowed to restart, even though the routine work had been completed. Although much of the counsel came unsolicited, commissioners also actively encouraged the Middletown citizens to offer advice on the activities that the investigation should cover and the actions that should be taken.

The commission used yet another procedure to obtain public commentary about its agenda: It polled a variety of public interest and industrial groups. Before describing the poll, let me give the developments that preceded it. At the first executive session, commissioners considered any number of ways for involving a broad spectrum of the population in the proposed investigation. Some consensus emerged that an advisory group would be a useful mechanism for obtaining feedback. We first entertained the idea of a single committee but later abandoned that idea on the assumption that a committee made up of persons with conflicting views on nuclear energy would spend more time engaged in internal debate than in assisting the commission. Hence, the discussion turned to the creation of two separate commit-

[5] Robert S. Walker, testimony before the commission, May 19, 1979. The unpublished material cited here can be found in the commission records at the National Archives.

tees. The first would be composed primarily of representatives of the nuclear industry; such representatives would be asked to comment on a variety of technical matters that the commission was likely to pursue. The second would be a public advisory group, made up of individuals with strong interests and concerns about nuclear energy, including concerns about the safety of the technology. On June 28, 1979, we learned from a staff member that the senior vice-president of the Edison Electric Institute had agreed to establish an Industry Task Force. That task force would give advice and assistance to the staff on matters and issues requiring an integrated industrywide viewpoint. The task force met on July 16 and considered a list of issues that the staff prepared.

1. What do you consider to be the strong and weak features of the present regulatory process?
2. How can we best address the problem of maintaining a readily available source of engineering competence for use in times of emergency?
3. Can, and should, the utilities do more to develop public trust and confidence in nuclear power plants? In light of TMI experience, what should have been done?
4. Should a more advanced technology be introduced into the control rooms? If so, how much, and how most effectively incorporated?
5. Who should bear the costs of major accidents? How can financial pressures be best applied in the interests of safety?
6. What are the principal financial pressures on the industry today, and with what effects?
7. What do you visualize as the role of simulators in training, proficiency maintenance, and plant design?
8. What should the role of NRC be insofar as plant operations are concerned during an emergency?
9. What importance do you attach to the use of fault tree analyses in design and operations?

With the industry group operating, staff members moved to form the public advisory committee. The head of the Interfaith Coalition on Energy agreed to convene a group, and on July 24 he and eight others met with the staff to begin committee work. The committee did not materialize; its potential members viewed their obligations and needs quite differently from the commission staff. The convenor of the meeting described the problem in this way:

> Contrary to the group's clear understanding that we would be given access to the basic information and documents necessary for us to operate, we were informed at our first meeting this morning that the staff would provide us with nothing more than

> that which is already available to the public. We explained that, for us to give intelligent input and to make useful suggestions to the Commission, we would need to be permitted to talk with your staff and to see the basic information. Any panel unable to consult the basic source material would be little more than window dressing (letter to the commission from William Millerd, July 24, 1979).

One of the participants, the general counsel to the Union of Concerned Scientists (UCS), later reported:

> Although UCS had reservations about the group's ability to do a thorough job in the very short time available – our work had to be completed by the end of August – UCS agreed that I would serve. I was looking forward to the job, since it appeared that this Commission was sincerely seeking the viewpoints of those people who have for years predicted the occurrence of a major reactor accident . . . But after hours of attempting to convince the Commission staff of the obvious fact that we could perform a useful advisory function only in a candid and open atmosphere, we disbanded.[6]

The nine participants were especially disturbed by the set of questions that they were asked to address at the meeting of July 24. The issues, presented earlier to the industry panel, struck the participants as biased toward that panel and inappropriate for a citizen's group. Unable to reach any agreement with the staff, the group resigned.

The commission staff offered a different assessment of the situation. They believed that many of the group's plans would usurp commission responsibilities. Staff members pointed specifically to the fact that the group planned to "(1) critique the Commission staff's work plan; (2) critique staff's choice of consultants and the consultant's reports if necessary; (3) review lists of witnesses and lines of questioning; (4) critique staff inquiries to industrial interface committee; and (5) advise staff regarding adequacy and validity of testimony". (Memorandum from William Millerd, undated). The staff, in consultation with the commission chairman, issued a statement that the commission would welcome the advice of any group but would consider the active participation by an outside group as detrimental to a completely independent and objective investigation.

At the next meeting of the commission the controversy over the public advisory committee took center stage. It became apparent that we held varied views on the appropriate composition of any public

[6] Ellyn Weiss, "Kemeny Commission Fires Citizens Advisors," *Nucleus: A Report to the Union of Concerned Scientists Sponsors*, Vol. 2 (August 1979), p. 6.

group. Some thought that such a panel should consist only of individuals opposed to or at least strongly critical of nuclear power as a source of electrical generation. Other commissioners thought that the membership should represent no identifiable point of view; instead, it would consist of citizens at large. Quite clearly, we had no consensus on the nature of and reasons for a public advisory committee. But we did agree on one point: We had no authority to create any advisory committee. Having been established by an executive order, the commission itself constituted an advisory body. Given this circumstance, we abandoned the notion of a citizens advisory panel and dissolved the industry task force as well.

Still, we felt a need to involve more than the commissioners, staff, and consultants in the inquiry. We then decided to poll a variety of organizations for their ideas and suggestions about an appropriate agenda. Thus, on August 8 we wrote to over two hundred organizations and individuals asking them to identify the questions that they felt should be answered by the commission and any suggestions, leads, or materials they felt would help the investigation. Groups at extreme poles in the nuclear controversy were canvased, as were individuals and public interest groups with no stated position in that controversy. A host of organizations responded: Nuclear Fuel Services, Citizens against Nuclear Danger, the United Auto Workers, and the American Society of Hematology, for example. Although we did no systematic review of the kinds of organizations that responded, it is my impression that we received more and lengthier replies from large organizations, especially those with full-time staff, than from small membership-based ones. Unless the latter had already taken a position on Three Mile Island or related matters, they had no time to survey their members in order to meet our proposed response date of September 7.

The comments in the public hearings held in April and May, those from the short-lived advisory groups, and the suggestions from the organizations canvased in August did not point inevitably to a limited mandate for the commission. At the Middletown hearings a representative from TMI Alert, an organization that predated the accident but that was revitalized in its wake, ended with this statement: "This accident is attacking our property, our bodies, our minds, our children, and our children's children's children. For our sake, shut down Three Mile Island forever!"[7] In contrast, the leader of a business group asked that we helped dispel the notion that the area constituted a nuclear leper colony, forever doomed to economic and

[7] Kay Pickering, testimony for TMI Alert before the commission, May 19, 1979.

social decline. An officer of a nuclear power generation group bade us consider the benefit of nuclear power for future electrical growth and economic expansion; he wanted to broaden our vision beyond the accident itself. The fact that the commission limited its report to events surrounding the accident cannot be traced to the observations made by those whose advice we solicited as we set the agenda. It was not the consensus among these individuals that a narrowly focused investigation would be preferable to and more credible than a broader one.

Much of the advice we received came unsolicited; much of it was indirect, as well. Newspapers, and especially their editorials, served as one source of indirect, unsolicited advice. Several editors welcomed the formation of the commission, but they often had completely different reasons for doing so. The editor of the *Minneapolis Tribune* wrote on April 22:

> People want to know what happened at the Three Mile Island nuclear plant. People also want to know whether nuclear power is a safe way to generate electricity. *Really* safe.... These are serious questions – and they deserve serious answers. The President's Commission is not supposed to address such questions. Who then will – and in a way the public can understand?[8]

Other editors maintained that the commission should thoroughly investigate the accident at TMI but should not broach the question of nuclear safety in general.

Some of the unsolicited advice went directly to commissioners or staff members. Letters arrived from nuclear engineers, from health physicists, from environmentalists, and from "concerned citizens" describing the kind of inquiry the commission should pursue and often presenting the conclusions it should reach. A resident of Kansas wrote to the commissioners, "It is my opinion that all nuclear plants should be halted in construction.... There are many God given natural resources for the whole nation that are becoming available. Before we act too hastily let us keep in mind that we are representing the people of our nation for the equality of all men in their rights of life and freedom in the pursuit of happiness." (Letter from Mrs. R. L. Mathiasmeier to the commissioners, May 19, 1979.) From New Jersey a minister wrote, "I would like to make sure that you know there are concerned citizens, like myself, who seriously question the practicality of power produced in the face of so many grave risks.... Your

[8] Editorial, "Nuclear Questions Deserve Serious Answers," *Minneapolis Tribune*, April 22, 1979, p. A11.

commission must face these questions if it is to be honest and objective." (Letter from C. David McKirachan to the commissioners, undated.) In contrast, a Middletown citizen informed us, "I am definitely in support of the re-opening of Three Mile Island. Alternative energy sources like wind, water, geothermal and solar are not yet developed enough. Therefore, your support of the nuclear field is vital." (Letter from Mrs. James Floyd to Chairman Kemeny, May 17, 1979.) Another resident shared her views: "The nuclear industry and its complexities is not yet fully understood by the general public, but to alleviate our fuel shortage problem, this field is one that must be supported. I hope and pray that our Governor and all elected officials are presently not being swayed by the very well organized and vocal 'anti' groups." (Letter from Doris Whitman to the Commissioners, May 15, 1979).

Narrowing the focus

To explain the circumscribed terms of reference that guided the final report, one cannot turn to the comments communicated formally and informally, directly and indirectly to the commission. In other words, it was not the case that all who proposed lines of inquiry for the investigation restricted their proposals to TMI events. Perhaps the presentation of the comments made a difference: Opposing sides presented their arguments with such conviction that it was apparent that no report on nuclear energy in general would reconcile conflicting views. I suggest, however, that the circumscribed approach of the commission stemmed from the decision we made to organize our investigation around the executive order and the charter prepared from that order, not from either the content or the mode of the presentations others made to us.

Our request for suggestions from a range of groups was not a public relations strategy; we actually were searching for a framework, a set of themes to guide our work. But we did not rely only on public input; we, too, identified matters we considered significant. Commissioners individually presented and defended issues for the agenda and means for conducting the investigation. Often a suggestion elicited strong opposition; at other times it brought no response. The first few executive sessions tended to be unfocused, with persons advocating a myriad of issues, some of which clashed. We pursued no given line of questioning at the hearings held during April and May, for we had not decided what kinds of issues we needed to address. But the absence of a focus made both the commissioners and the staff members restive. We agreed that we could prepare no report by the deadline unless we structured the investigation in some way. Toward that end we canceled the hearings planned for June 13 through 15, an action we took for three reasons. First, another exploratory expedition such as we had conducted at earlier hearings was likely to represent an inefficient use of time. But unless we had a framework to use and considerable advance work around which to orient our questions, we could only follow the approaches we had used to that point. Thus, we had a second reason for rescheduling the public hearings: We asked the staff to spend their time preparing a master plan for the overall inquiry. A third matter also bore on our decision: The chief legal counsel had resigned on June 1, and as a result the legal staff had to be reorganized.

The staff did indeed offer a master plan at our meeting of June 13, a plan that generated considerable controversy among commissioners. Subsequently the commissioners took on the responsibility of preparing issues outlines to steer staff work and our own deliberations. Significantly, both the staff and the commissioners used the original charter as the organizing framework. Commissioners prepared outlines on issues to be raised with reference to technical problems at TMI, the structure and operation of the NRC, the response of federal, state, and local officials, the performance of the press, and the health effects of the accident – all topics that the charter mentioned specifically. The staff was organized into task forces, assigned to functions identified in the charter. The technical staff included a Technical Assessment Task Force, a Task Force, a Health Effects Task Force, and an Emergency Preparedness and Response Task Force, foı example. The legal staff, too, began to specialize in topics contained in the charter.

We chose the charter as our organizing instrument because it was the only document that laid out an agenda for the commission. But the choice of the charter as the point of departure undoubtedly influenced the course of the inquiry and the nature of the final document. The charter dealt almost exclusively with the accident; hence, in relying on it we were foregoing giving attention to some of the moral, ethical, and philosophical problems with which we had grappled. Additionally, once the charter became the focal point and our energies went into gathering data pertinent to it, we found it difficult to probe problems that lay outside its scope. But an interest in those matters did not disappear completely. Even at our final sessions some commissioners still insisted that our final report include some discussion of solar energy and of the problems associated with nuclear waste, for example. Such suggestions received little support at that time, however, for by then no information had been gathered to address these matters, and staff members were so busy compiling materials centered on the charter topics that they could not have turned to these other areas. Even some of the commissioners who initially would have preferred a more far-reaching report agreed at these last meetings that our own investigation had not provided us with any basis for writing such a document. Our reliance on the charter and the time pressure under which we operated were, in my view, factors that circumscribed the orientation of the final report. Contributing, too, was the position some commissioners took that to be credible our investigation needed to be intensive rather than extensive, deep rather than broad.

Perhaps, too, the nature of the data available when we began our work helped shape the kind of investigation we eventually undertook. By the time the commission was formed, several inquiries on TMI were under way. I have argued elsewhere that the tendency of those investigations to stress technical matters influenced the course of our own investigation.[9] The information most readily available dealt with what had happened from March 28 on, not with what might have happened or with the long-term political, economic, and social implications of the accident. We examined the technical and engineering data at length, and because we did we had little time to study the more sweeping issues. Admittedly, the final report went beyond the engineering questions to include various organizational, attitudinal, and human problems.[10] I am not proposing that all commissioners would have favored a broad-based investigation of nuclear energy, even if we had begun to undertake it from the beginning. But I am suggesting that time and some decisions we made about organizing served to constrain us.

The final report

The final report of the commission offered this general conclusion:

> "To prevent nuclear accidents as serious as Three Mile Island, fundamental changes will be necessary in the organization, procedures, and practices – and above all – in the attitudes of the Nuclear Regulatory Commission and, to the extent that the

[9] Cora Bagley Marrett, "The President's Commission: Its Analysis of the Human Equation," in *The Accident at Three Mile Island: The Human Dimensions*, eds. David L. Sills, C. P. Wolf, and Vivien Shelanski (Boulder: Westview, 1982), pp. 31–43.

[10] See C. B. Marrett, "The Accident at Three Mile Island and the Problem of Uncertainty," *Annals of the New York Academy of Sciences* 365 (April 1981): 280–91; "Organizational Issues in Accident Analysis," *Society* 18 (July/August 1981): 66–72.

> institutions we investigated are typical, of the nuclear industry."[11]

Most of the document described the ways we had reached that conclusion and the evidence supporting it. It noted, for example, that we took more than one hundred fifty formal depositions, heard testimony under oath from a host of witnesses, and collected enough material to fill about 300 feet of shelf space in a library. It pointed out as well that in our view too many analysts had believed in the inherent safety of nuclear plants and as a result had not encouraged enough vigilance over the operation of TMI. But we included certain disclaimers in the report, citing areas we had not addressed. With reference to the belief about safety, the report indicated that we had not examined the entire industry and thus could not determine if the "mind set" about the inherent safety of the engineering systems prevailed. We endorsed several other statements on the limits of our investigation.

> Our investigation centered on one accident at one nuclear power plant in the United States. While acting under the President's charge, we had to look at a large number of issues affecting many different organizations; there are vast related issues which were outside our charge, and which we could not possibly have examined in a 6-month investigation.
>
> We did not examine the entire nuclear industry. (Although, through our investigation of the Nuclear Regulatory Commission, we have at least some idea of the standards being applied to it across the board.) We have not looked at the military applications of nuclear energy. We did not consider nuclear weapons proliferation. We have not dealt with the question of the disposal of radioactive waste or the dangers of the accumulation of waste fuel within nuclear power plants adjacent to the containment buildings. We made no attempt to examine the entire fuel cycle, starting with the mining of uranium. And, of course, we made no examination of the many other sources of radiation, both natural and man-made, that affect all of us.[12]

Some commissioners argued that our report was too cautious and many of its statements too apologetic. Commissioner Babbitt, for example, contended that the commission should and could have addressed the issue of who should be allowed to run nuclear power

[11] President's Commission on the Accident at Three Mile Island, *The Need for Change: The Legacy of TMI* (final report), October 30, 1979.

[12] Ibid.

plants. In his opinion we had erred when we refused to recommend policies for the industry at large, for he believed that our record would sustain sweeping recommendations on the organization of all plants.[13] In addition, six commissioners maintained that our report should have recommended a moratorium on the granting of work authorization and construction permits for the construction of new nuclear plants.[14]

Several readers of the report concurred with the commissioners who complained about its limited scope. Yet the report received extremely favorable coverage from widely different segments of the population; both anti and pronuclear groups used it to support their positions; the President accepted it and began to implement many of its recommendations; and organizations within the nuclear industry offered complementary assessments of many of its findings and suggestions. Not all of the readers praised the results, of course. The editor of the Harrisburg *Patriot* complained that the report had little new to tell the people in the TMI area and "little more about which they might cheer."[15] An editorial in the *Philadelphia Inquirer* of October 31, 1979, protested that the analysis seemed to make the unacceptable sound routine, to render the Apocalyse reasonable. Yet, even with such scathing criticism, the editorial found favor with the "coldly, dispassionately descriptive" review of what had happened at TMI.

The report won bipartisan support in Congress; Senator Gary Hart, a Democrat, called it a candid and independent assessment and praised the commission for avoiding the gloss and platitudes of past government studies. Republican senator Peter Domenici considered the report an objective analysis by a group with no ax to grind.[16] Interestingly, congressmen with favorable comments on the report applauded two of its features: its descriptive nature, and its concentration on the accident itself. It would seem, then, that the fact that the commission chose to limit its presentation bore significantly on the response that the final report elicited. The analysis answered many questions about the accident – about the sequences of events that had taken place and about precursors to the accident throughout the industry, for example – that had previously been unresolved. Hence the report offered some closure although it did so on a

[13] Bruce Babbitt, "Supplemental View by Six Commissioners," in ibid.

[14] Bruce Babbitt, Carolyn Lewis, Paul A. Marks, et al., ibid.

[15] "TMI Report: Little New and Little to Cheer," Harrisburg, Pa., *Patriot*, November 1, 1979.

[16] Comments reported in "Nuclear Reports Call for Reforms Wins Bipartisan Praise in Congress," *New York Times* (November 1, 1979).

circumscribed matter rather than in relation to a broad, ethical problem. The commission dealt with a factual problem – what had happened at TMI – not the philosophical question of the extent to which the nation should rely on nuclear energy. Indeed, the commission agreed that the broad issues could only be handled in the public arena, not by any single task force. The factual question of what had happened was of course far less controversial and far more manageable than any discussion of nuclear power could have been.

To illustrate the problem of achieving closure on a nonfactual issue, let me describe our attempts to analyze and discuss what might have happened at TMI. Early on, some commissioners had argued that we needed to consider the courses the accident might have followed, not what actually took place. On the basis of this argument, we authorized an analysis on the results that would have obtained had different actions taken place or different conditions existed at Unit 2. The alternative sequences of events that the analysts considered ranged from some that differed only minimally from those that indeed happened to some in which the reactor core had melted completely through the containment vessel.[17]

Once the necessary simulations were completed and the data analyzed, we had substantial difficulty deciding on how the findings should be presented. The analysts reported that even if there had been a meltdown, it was unlikely that the release of radioactive substances would have been any greater than what in fact took place. This was attributable primarily to the fact that the unit rested on solid bedrock and that few pathways could be identified through which gases could have escaped from the containment building. But we feared that readers would read the assessment either too optimistically or too skeptically.

We devoted great attention to the wording of this section of the report.

> We explored each of several different scenarios representing a change in the sequence of events that actually took place. The greatest concern during the accident was that significant amounts of radioactive material (especially radioactive iodine) trapped within the plant might be released. Therefore, in each case, we asked whether the amount released would have been smaller or greater, and whether large amounts could have been released.
>
> Some of these scenarios lead to a more favorable outcome

[17] The report, "Alternative Event Sequences," appears in the *Reports of the Technical Assessment Task Force*, Vol. 2.

> than what actually happened. Several other scenarios lead to increases in the amount of radioactive iodine released, but still at levels that would not have presented a danger to public health. But we have also explored two or three scenarios whose precise consequences are much more difficult to calculate. They lead to more severe damage to the core, with additional melting of fuel in the hottest regions. These consequences are, surprisingly, independent of the age of the fuel.[18]

We went on to identify limitations of the assessments, including the fact that the models used did not take account of possible operator error; did not consider the possible vulnerability of the plant in case of various electrical and plumbing penetrations; and represented only a few of an infinite number of possible scenarios. Even with these precautions, just as we had feared, some readers exaggerated the results. Some accounts reported that we had found a meltdown to be a greatly overplayed event, if measured by the immediate and long-term health effects. Advocates of that view sometimes went on to criticize the commission for its failure to highlight this "encouraging" finding.[19] Other accounts found even more flaws in the assumptions and models on which the scenarios were based than had the commission. Supporters of this view often turned instead to a report that maintained that had there been significant releases to the environment, substantial loss of life would have resulted.[20]

Unlike the response to the events that had taken place, the response to the alternative events scenarios indicated that the question of what had happened remained unresolved. On this question the commission report did not offer any closure, any sense of finality.

Further observations on closure through circumscription

The commission investigation would not represent the only attempt to seek closure by limiting the debate. Consider the procedures followed in the licensing of a nuclear power plant. Once the NRC has received and reviewed the application, a public hearing must be scheduled and announced in the *Federal Register*. Those who wish to challenge the application – the intervenors – must indicate their

[18] Ibid., p. 20.

[19] See, for example, "The Secret Findings of the Kemeny Commission," in the *Newsletter of the Nuclear Legislative Advisory Service*, February 25, 1980.

[20] Jan Beyea, "Some Long-term Consequences of Hypothetical Major Releases of Radioactivity to the Atmosphere from Three Mile Island," Report to the President's Council on Environmental Quality (Princeton: Center for Energy and Environmental Studies, September 7, 1979).

interest in being heard and establish the bases for their challenge. The hearing boards have been directed to admit only some matters into discussion. No commentary for or against nuclear power as such or any such encompassing topic can be introduced; nor can presenters raise generic safety issues, issues that would apply to several different facilities. The hearing policies demand that presentations cover only the specific facility whose application is under review.

Based on the restrictions in the licensing hearings and the investigations of the commission, I suggest that those who must make decisions in controversial areas limit the scope of the debate in order to accelerate decision making. If they seek to reach a resolution as rapidly as possible, the reviewers will favor circumscribed debates over freewheeling ones. With reference to the license hearings, generic matters were removed from the individual plant hearings in order to streamline and accelerate the hearing process. I contend, too, that on controversial topics reviewers favor factual issues over hypothetical ones. The clearer the empirical nature of the problem, the greater the likelihood of consensus among the reviewers themselves, and the greater the probability that the adversaries will accept the resolution.

The circumscribing of an issue, then, facilitates the pace of decision making and enhances public acceptance of the results. But circumscription does not resolve the underlying question or questions. Perhaps, then, the expression "closure through circumscription" is a contradiction in terms, if even after the boundaries are narrowed the basic controversy remains. But closure is possible at several different levels. Moreover, for complicated controversies, which usually cover a bewildering array of subcontroversies, reaching closure on any single topic is no small accomplishment. Given the complexity and vagueness of many controversies, if reviewers can disentangle even one issue from the maze, if they can present a systematic analysis of that issue, then they and others who observe them will conclude that, at least to some extent, they have achieved closure.

25

Understanding the nuclear power controversy

DOUGLAS MACLEAN

I

The nuclear power controversy in the United States is not very visible these days, but this is not because it has been resolved. The industry is still there, along with its effective lobbying organizations, such as the Atomic Industrial Forum, and so are the vigilant antinuclear watchdog groups, such as the Union of Concerned Scientists. They still descend upon the Nuclear Regulatory Commission (NRC) every time a new safety standard is proposed and they still slug it out in the courts over each new operating license. But lately these battles have failed to arouse the public or attract the news media to the extent that they did just a few years ago. Nuclear power's proven potential – to electrify public controversy – is still there, to be sure, but the controversy is in abeyance and it may remain so for some time to come.

This waning of public concern over an issue that so exercised us only yesterday is a remarkable phenomenon, for the technological risks that focused the controversy in the past – the risk of accidents and careless regulatory oversight, the risk of proliferation of nuclear weapons, and the risks to future generations from the accumulation of radioactive waste – have not changed significantly. The accident at Three Mile Island resulted in some changes in operating and regulatory procedures aimed at improving safety, but any reassuring effects of these changes are offset by other events. The Republican administration of President Reagan is pledged to help the nuclear industry and especially to relax and speed up licensing procedures. It has replaced several skeptical commissioners on the NRC with men more sympathetic to its views. The concern over the proliferation of

My research has been supported by the National Science Foundation, grant number PRA-80-20019. For advice and comments, I am indebted to David Bodde, Claudia Mills, and Susan Wolf.

weapons-grade fissionable material is as much an issue as it always was, as government support for a breeder reactor program continues. And the quantity of radioactive nuclear waste, which would now more than fill the New Orleans Superdome, continues to grow, still without any plan for its permanent disposal.

We have to look elsewhere, however, to understand the changing dynamics of the controversy. Most obviously, the nuclear power industry today is economically moribund, despite the Reagan administration's stated desire to resuscitate it. The last commercial reactor ordered in the United States was in 1978, and cancellations of orders continue to plague the industry. Eighteen planned reactors were canceled in 1982 alone. Whether the industry is permanently dead in this country is a debatable issue, but the earlier plans for an atomic society are not to be taken seriously today.

A second reason is the watershed event of Three Mile Island (TMI), the major accident that resulted in negligible harm to the public. The only real catastrophe at TMI was financial, and the only thing the accident proved conclusively was that the costs of accidents are staggeringly high. But beyond that, this near disaster proved little. Each side has exploited it to confirm its own view of the risk of accidents. According to the nuclear industry, the system worked, and we also learned enough to make reactors even safer in the future; whereas the antinuclear groups claimed that we had lost control completely and were just plain lucky *this* time. Three Mile Island was a social catharsis, and as it recedes further into the past our level of concern falls off and we turn our attention elsewhere. But I want to suggest that to understand the nuclear power controversy, we must examine not only its risks and costs but also the particular cultural and historical aspects of this technology.

Both in Europe and in the United States, nuclear power has a history of provoking deep political divisions.[1] It is a big, centralized, complicated technology whose safe management demands both a heavy reliance on experts and a good measure of social cooperation. But the social consensus that such management presupposes is noticeably not forthcoming. Nuclear power has thus generated *the* technological debate of our time. Its critics oppose it for almost as many reasons as one can find to object to the impact of technological development on society. It should not surprise us, therefore, to find

[1] For discussions that compare the nuclear controversy in Europe and the United States, see Irwin C. Bupp and Jean-Claude Derian, *Light Water* (New York: Basic Books, 1978); Anna Gyorgy et al., *No Nukes: Everyone's Guide to Nuclear Power* (Boston: South End Press, 1979); and Dorothy Nelkin and Michael Pollak, *The Atom Besieged* (Cambridge, Mass.: The MIT Press, 1981).

that scientific studies, reports of new data, or economic forecasts have negligible effects on the division and are unlikely ever to lead to a consensus of popular or even expert opinion. According to surveys, the swings in public attitudes toward nuclear power are slight, and the scales never seem to tip very far.[2] The debate has a life of its own, which is independent of the scientific and technological issues that are its focus.

Were the opponents of nuclear power simply objecting to its risks, their position would be inexplicably paradoxical. The risks are real and complex, but it is hard to know why they provoke such fear. It cannot be because of our experience with nuclear accidents, for none of the accidents so far has seriously endangered human life. The actual record of the industry (even if only through sheer luck) has been pretty good. Is it because the particular risk involved – radiation – is extraordinarily dreaded? With little opposition or fear, the average American willingly exposes himself every year to more X-ray radiation than is needed or healthy in routine medical and dental checkups. Could it perhaps have something to do with risks being generated by large, centralized electric power plants? The hazards of burning coal are surely as great, but they do not inspire mass demonstrations. Try as we might, we will not find a basis for this opposition that divorces it from the technology per se. The strongest correlations between opposition to nuclear power and other public attitudes can probably be found in support for the Equal Rights Amendment or the Endangered Species Act.

Support for nuclear power, on the other hand, is no less paradoxical. There is no shortage of electricity in the United States and no immediate prospect of one, so we do not need nuclear power in the foreseeable future. Perhaps we have reason to fear oil shortages, but without major capital investments and infrastructure changes, nuclear energy cannot substitute for oil where shortages are likely to occur, that is, in gasoline and heating fuel. Nuclear power does not make economic sense, a fact that emerges more clearly every year and seems destined to resolve this terminally deadlocked political controversy.

The struggle over this technology is played out on the battlefield of risks and costs, but it is also a morality play. The positions are ideologically determined, as if the opponents are bound to see the risks as high, and the defenders bound to see the economics as

[2] Robert C. Mitchell, *Public Opinion on Environmental Issues* (Washington, D.C.: U.S. Government Printing Office for the U.S. Council on Environmental Quality, 1980).

favorable. The battle itself, then, is largely symbolic, serving as a surrogate for a referendum on some very basic social issues, which only occasionally emerge explicitly in the debates.

In the following sections I will further describe the risks and costs, and the balancing of these concerns, all of which make up the plot of this play. I will then offer an interpretation in terms of some deeper social issues and will comment on the problems for finding morally and politically justifiable solutions to technological controversies.

II

The issue that mobilizes opposition to nuclear power is the risk to health, primarily from possible catastrophic accidents, but also from normal operations and from mishaps we can expect. The widespread *Ur*-fear of nuclear power, a fear that was associated with atomic energy even before the discovery of nuclear fission, gives an incentive to nuclear opponents to discover and publicize these risks.[3]

Needless to say, both the effects and the probability of some dreaded accident, such as a core meltdown, are hotly debated. The record of nuclear power in this country so far is difficult to evaluate. On the one hand, it looks impressive, because no major accident resulting in a substantial release of radioactive material has occurred; but, on the other hand, there have been many small accidents, operator errors, construction flaws, and surprising events that experts assured us could not occur. As nuclear reactors get bigger, they become more unpredictable, and at Three Mile Island a large reactor was, for a short while, completely out of control.[4] The technology, moreover, is unforgiving, which is to say that a single major accident, which can only become more probable as reactors grow in number and size, might have catastrophic consequences. Such risks present difficult problems for rational decision making, especially since it might be impossible to assign meaningful probabilities to them.

A second kind of risk links nuclear power to nuclear weapons. The use and transport of fissionable material create a possible source of raw materials for atomic bombs. Some of this material could be

[3] See Spencer Weart, *Scientists in Power* (Cambridge, Mass.: Harvard University Press, 1979); see also Chap. 23, this volume.

[4] For details, see the "Kemeny Report," *Report of the President's Commission on the Accident at Three Mile Island*, John G. Kemeny, Chairman (Washington, D.C.: U.S. Government Printing Office, 1979) and the "Rogovin Report," U.S. Nuclear Regulatory Commission Special Inquiry Group, *Three Mile Island: A Report to the Commissioners and to the Public*, Vol. 1 (Washington, D.C.: U.S. Government Printing Office, 1980).

diverted by a government, or worse, by subnational terrorist groups. This possibility results from any process of enriching uranium, but it would be considerably more likely if we decide to reprocess spent fuel in conjunction, say, with a "breeder" reactor.[5] This risk is similar to the risk of a core meltdown in one respect: The consequences of a single dreaded event would be disastrous. They could be bad enough to make the entire nuclear power program a horribly bad decision. But whether or not a nuclear weapon is actually detonated, the mere proliferation of the possession of such weapons, which nuclear power makes more likely, is itself a risk. For if deterrence works at all like the theorists claim it does, one group can gain a powerful threat advantage that could itself be costly to its enemies, which might include us.

Further risks from nuclear power are created by low levels of radiation that are released during normal operations through the fuel cycle and by minor but likely accidents, which occur in uranium mining, in operating reactors, and in handling wastes. Except for the very difficult problems caused by the steady accumulation of mill tailings from mining, these risks may or may not be negligible. Alas, this too is a disputed issue.

We can add to these risks the further uncertainties of unsolved problems. Plants are licensed for construction before all the safety systems have been designed, and they have been licensed for operation before these systems have been demonstrated. Plants also remain in operation even after new problems have been discovered. But by far the most worrisome of these unsolved problems is the permanent disposal of radioactive waste. The mass of radioactive wastes we have already produced will remain radioactive for millennia. Yet, after more then thirty years of producing these wastes, we still do not have a program for isolating them permanently from the biosphere. Tremendous political opposition makes it difficult to find an acceptable solution, but not all the problems are people problems. Some of the thornier ones are technical and scientific.[6]

Amid so much uncertainty and so much division among experts, the prospects for achieving some consensus on the estimates of the risks of nuclear power are slight. Even if we could agree on the estimates, however, some deep problems remain in trying to determine whether the risks are acceptable. I will mention three of these problems.

[5] Amory Lewis, L. Hunter Lovins, and Leonard Ross, "Nuclear Power and Nuclear Bombs," *Foreign Affairs* 58 (1980):1137–77.

[6] Luther J. Carter, "The Radwaste Paradox," *Science* 219 (January 7, 1983): 33–36.

First, as with so many large-scale projects, the risks and benefits of nuclear power are distributed differently. The toxicity of radioactive wastes spreads risks over millennia, whereas the foreseeable benefits of using nuclear power are near-term by comparison. Thus, the question of moral obligations to future generations is raised in a dramatic form. Another distributive problem results from the policy of trying to localize the risky activities connected with nuclear power in sparsely populated areas. By reducing the magnitude of potential consequences in this way, the overall risks are reduced. Does this policy discriminate unfairly against people who live in rural areas? Those living in densely populated areas receive greater benefits from nuclear power, but our policy is to transfer the risks away from them. Not long ago, a decision was made to divert the transport of nuclear wastes from Brookhaven Laboratory on Long Island by sending them through Connecticut instead of New York City. The reason given was that fewer people would be exposed to risk. How can such a decision be justified to the citizens of Connecticut? The choice to impose the risks on them was not made randomly; it was made because there are fewer of them. It might be fairer (though surely more risky and costly) to dispose of nuclear wastes under Chicago, the city in which the benefits of nuclear power are supposed to be most heavily concentrated.

Another problem is the *qualitative* differences among the risks that must be compared and weighed. Risk is usually measured as the product of the value of some outcome and its probability. If we measure risk in this way, however, then a very low probability of a consequence with nearly infinite costs turns out to be indistinguishable from more garden-variety risks with higher probabilities and lower costs. Apart from the very real problem of assigning meaningful probabilities to the possible catastrophes, it is difficult to know how we should think about these risks. The Rasmussen Report, that now famous study of reactor safety published by the NRC in 1975, made a controversial decision to ignore some of the most serious possible accidents because they were too unlikely.[7] Some people think, on the other hand, that the low probability/high consequence risks should be treated with extra care, because their consequences would be so terrible. This reflects a view that one accident, killing 50,000 people, is worse than a series of 50,000 single-death accidents.[8]

[7] U.S. Nuclear Regulatory Commission, *Reactor Safety Study – An Assessment of Accident Risk in U.S. Commercial Nuclear Power Plants* WASH-1400, NUREG-75/014 (Washington, D.C.: U.S. Government Printing Office, 1975).

[8] See Barry Commoner, *The Poverty of Power* (New York: Knopf, 1976), pp. 96–7. For a suggestion for incorporating these differences into formal risk

I will give one illustration of the problems that can arise in comparing different kinds of risk. A 1980 study of safety goals for nuclear power plants by the NRC's Advisory Committee on Reactor Safeguards decided that genetic effects and latent cancers posed markedly similar risks. Therefore it did not seem necessary to treat genetic effects explicitly in the estimates used to set the goals, because controlling for latent cancers would be sufficient to control for genetic mutations.[9] An outside panel of experts that reviewed the study nevertheless rejected the substitution. It felt that, even if the risks were similar in ways the NRC study claimed, genetic effects should be treated separately and perhaps not even by the same methods of risk assessment.[10]

The fact that the risks of nuclear power include potentially catastrophic and irreversible accidents, cancers, and genetic mutations helps to explain why this particular technology is feared more than other equally probable causes of death. Some ways of dying are more dreaded than others, and several psychological studies have shown that the way people perceive the nature or quality of a risk determines their judgment of its acceptability.[11]

This leads us to the third problem, the lack of correlation between the best expert estimates of safety risks, on the one hand, and public perceptions and fear surrounding different risks, on the other.[12] The problem for policy makers is to determine whether the overriding goal in regulating nuclear power should be to protect the public or to satisfy the public (assuming, which is dangerous, that experts understand the risks better than the public does). Given the fear and

assessments, see Advisory Committee on Reactor Safeguards of the U.S. Nuclear Regulatory Commission, *An Approach to Quantitative Safety Goals for Nuclear Power Plants* NUREG-0739 (Washington, D.C.: U.S. Government Printing Office, 1980), pp. 89–90.

9 NRC *Quantiative Safety Goals*, p. 60.

10 U.S. Nuclear Regulatory Commission, *Workshop on Frameworks for Developing a Safety Goal* NUREG/CP-0018, BNL-NUREG-51419 (Washington, D.C.: U.S. Government Printing Office, 1981), p. 15.

11 Baruch Fischhoff et al., "How Safe Is Safe Enough? A Psychometric Study of Attitudes Towards Technological Risks and Benefits," *Policy Sciences* 8 (1978):127–52; also Daniel Kahneman and Amos Tversky, "The Psychology of Preferences," *Scientific American* 246 (January 1982):160–71.

12 Baruch Fischhoff, Paul Slovic, and Sarah Lichtenstein, "Lay Foibles and Expert Fables in Judgments About Risk," in *Progress in Resource Management and Environmental Planning*, Vol. 3, ed. T. O'Riordan and R. Turner (New York: Wiley, 1982); see also idem "'The Public' vs. 'The Experts': Perceived vs. Actual Disagreements About the Risks of Nuclear Power," unpublished, 1981. (Available from Decision Research, 1201 Oak Street, Eugene, OR 97401.)

concern, should nuclear power be regulated to levels of safety stricter than those of other technologies? Each incremental reduction of risk costs more, so we cannot treat the desired level of safety as an isolatable goal; it must be weighed against its costs and the other benefits forgone. The growing awareness of the need to address this very problem is primarily responsible for the development of the techniques of quantified risk assessment and risk–benefit analysis, which are coming to dominate the discussions of how to regulate technologies.

The gulf between expert and public perceptions of the risks of nuclear power is a complicated pheomenon. The acceptable-risk debate is sometimes a stage for expressing a variety of concerns, including distrust of experts, which may have little to do with judgments about their qualifications to evaluate the reliability of a feedwater pump in the emergency core cooling system. The nuclear power controversy is a political war, and the participants on both sides might pursue other goals on the battleground over health and safety standards, if this is where they think these goals can most easily be achieved. The perception of health and safety risks, moreover, is strongly colored by these other factors. Someone who thinks that nuclear technologies have bad political implications will tend to see greater health risks than others, and someone who is acutely aware of the economic costs of safety equipment in large pressurized water reactors might tend to see lower health risks than others.[13] Such issues cannot easily be separated out of the risk debate; nor can they easily be incorporated into it.

III

Any analysis of the nuclear power controvery must recognize the strategic role of economic considerations. Despite the fact that the first application of nuclear fission was the atomic bomb, there was at that time a euphoria about the potential of atomic energy that today can only strike us as incredible in its naïveté.[14] In 1945, David Dietz, science editor of the Scripps-Howard newspapers, wrote:

Instead of filling the gasoline tank of your automobile two or

[13] For an explanation of such psychological tendencies, see Amos Tversky and Daniel Kahneman, "Judgment Under Uncertainty: Heuristics and Biases," *Science* 185 (1974):1124–31.

[14] For an excellent account, see Daniel Ford, "The Cult of the Atom," Part 1, *The New Yorker*, October 25, 1982, pp. 107–59. The quotations in this and the following paragraphs from David Dietz, David Lilienthal, and Eugene Wigner are all taken from Ford, pp. 111–12, 133.

> three times a week, you will travel for a year on a pellet of atomic energy the size of a vitamin pill . . . . The day is gone when nations will fight for oil . . . . With the aid of atomic energy the scientist will be able to build a factory to manufacture gold . . . .
>
> No baseball game will be called off on account of rain in the Era of Atomic Energy. No airplane will bypass an airport because of fog. No city will experience a winter traffic jam because of heavy snow. Summer resorts will be able to guarantee the weather and artificial suns will make it as easy to grow corn and potatoes indoors as on the farm.

At about that same time, David Lilienthal, the first head of the Atomic Energy Commission, looked forward to "almost limitless beneficial applications of atomic energy." A decade and a half later, Lilienthal's tone had changed, and he conceded that the nuclear power program "has been quite a flop, and is not to be taken seriously."

The reason for this dramatic change in attitude was basically economics. In December 1953, when President Eisenhower was promoting the Atomic Energy Act in his famous "atomic power for peace" speech before the United Nations General Assembly, he made his promise of electricity so abundant that it would be "too cheap to meter." But those who were more intimately involved in the development of nuclear power than the president were to discover that the truth was quite different. By 1961, Eugene Wigner, who helped develop the reactor, said, "Our economic expectations concerning cheap nuclear power have, so far at any rate, failed us rather completely."

Despite massive subsidies from a government officially dedicated to promoting commercial nuclear power, the civilian program never appeared to be better than economically marginal. It is this fact that set the stage for the controversy, as Alvin Weinberg, who designed the pressurized reactor that has been used almost exclusively in the United States, stated:

> There was always a fundamental tug-of-war between "safe as possible" and "cheap as possible," . . . When the nuclear business got started, the price of electricity from coal was three or four mills per kilowatt-hour. Nuclear had to get into that ballpark, or there wouldn't be any nuclear power.[15]

The government was as encouraging as it could be: It subsidized the development of reactors and the price of fuel; it allowed the industry to ignore the costs at the "back end" of the fuel cycle, the

[15] Ford, "The Cult of the Atom," Part 1, p. 135.

decommissioning and decontamination of used plants after their thirty-year lives, and also the costs of nuclear waste disposal; and Congress passed a landmark piece of legislation in 1956, the Price–Anderson Act, that freed the industry from all but a nominal liability for accidents. But all this was this was not enough, and some further decisions had to be made.

The first, which was almost desperately reckless, was to increase drastically the size of plants, hoping that this would lead to economies of scale that would bring down the marginal costs. This happened in the mid- to late-1960s, the "bandwagon years" in the nuclear industry, when most of the reactors now in use or under construction were ordered.[16] The mean generating capacity of reactors ordered in 1968, for instance, was 926 megawatts (MW) electricity, yet the largest commercial reactor in operation at that time had a capacity of only 200 MW.[17] The *total* operating capacity in the United States at that time was only 5 percent of the ordered capacity. (This was in an industry where an investment-to-operating ratio of 2:1 was considered to be the outer bound of reasonable investment risk.)[18] So, from the beginning, commitment and hope far outstripped our experience with commercial nuclear power.

The result, not surprisingly, has been reactors mired in construction delays, cost overruns, and operating problems. The nuclear plants constructed during this period have operated at an average of 50 to 62 percent of their capacity, far below the 70–80 percent estimated minimum requirement for returning a profit.[19]

The need to cut costs was thus compounded; therefore, what were seen as redundant or excessive safety measures were ripe for compromise. Large plants were built closer to populated areas, for example, so that transmission costs would be reduced. The biggest safety problem, however, came with the increase in the size of reactors. This is why commercial reactors have risks that do not arise as issues, say, with research reactors or reactors that power submarines. During normal operation, nuclear wastes accumulate inside the core of a reactor that give off heat equivalent to 6–7 percent of its

[16] Bupp and Derian, *Light Water*, Chap. 8.

[17] David Bodde, "Regulation and Technical Evolution: A Study of the Nuclear Steam Supply System and the Commercial Jet Engine," Ph.D. dissertation, Harvard University, 1975. (See reference in Bupp and Derian, *Light Water*, p. 73.)

[18] David Bodde, "The Effect of AEC/NRC Regulations on the Development of Nuclear Steam Power," unpublished, 1981.

[19] Charles Komanoff, *Power Plant Cost Escalation* (New York: Komanoff Energy Associates, 1981). (See summary of Komanoff's Study, *Science* 212 (1981):652.)

full power. In smaller reactors, this is not a serious problem, for if the core ruptured, the containment vessel would prevent any further damage that might be caused by this "decay heat." In larger reactors, however, this decay heat is significant. It can threaten the core even after control rods have been inserted to stop the chain reaction, and it can rupture a containment vessel. The crucial line of defense for safety in large reactors thus shifts to the emergency cooling systems. Since 1966, the Atomic Energy Commission required all 1000 MW plants to have an emergency core cooling system (ECCS), but the ECCS was not even tested until 1971 – and then it failed.

Such technological problems have resulted in pressure for tighter and more costly regulations. Because the *margin* of safety in large plants had already been compromised for economic reasons, safety was now far more difficult to prove, and so the risks and costs have become the battleground for the actors in the struggle, whatever ideological commitments might actually be motivating them. Causing or reducing delays in licensing have become important strategies because there is no margin of profit and no margin of safety. Most of the economic problems in nuclear plants, I should add, are not caused by licensing delays or new regulations. Technological problems encountered during construction, along with increased labor and material costs during the 1970s, wiped out the anticipated economies of scale. The per unit capacity construction costs of larger plants actually increased. And, although the true construction costs are very difficult to determine, overruns for plants ordered in the mid- to late-1960s conservatively exceed 100 percent. Regulations can account for only 8 to 35 percent of these increases.[20]

How much does it cost to build and operate a 1000 MW nuclear plant. And how does this compare to the cost of generating electricity from coal? Answers to such questions are difficult to verify and highly disputed. It is generally conceded that nuclear plants are more expensive to build than the ones using coal, but they are perhaps cheaper to operate. These differences make comparisons more difficult, as do the massive subsidies and externalized costs in both nuclear and coal operations. Debates over costs still go on, but like debates over risks, they live apart from empirical facts. They are further manifestations of ideology and other values.

The remaining areas of economic controversy will concentrate on determining who pays the costs for the back end of the fuel cycle and for waste disposal, and who pays to clean up enormously expensive

[20] Bodde, "The Effect of AEC/NRC Regulations." See also Bupp and Derian, *Light Water*, Chap. 8, and Komanoff, *Power Plant Cost Escalation*.

accidents such as that at TMI. And then, of course, there is the breeder reactor. The breeder brings different risks and different uncertainties, and it is not at all unlikely that it will end up being as disappointing and controversial as today's conventional reactors, if it is ever developed and deployed. Its future depends entirely on public subsidies, for no private firm is now willing to take on the risks of developing it, and without the right incentives they may never want to buy them, even if they ever become commercially operational.

IV

Although the nuclear power controversy focuses on the putatively empirical issues of health risks and economic costs and benefits, it is fueled by something more fundamental and ideological. I will now attempt to describe what I take to be the sociodynamics of nuclear power.

Nuclear critics sometimes appear (at least to the other side) to be hopelessly misinformed about the technology and its risks. It seems that their fears are based on ignorance, partially as a result of being led astray by an almost equally ignorant and often irresponsible press. Should the management of nuclear technology, then, be left to the experts – the engineers, economists, and risk assessors – as many people claim?

The most ardent defenders of nuclear power and especially those with a professional or economic stake in its success are not without their own biases and axes to grind. Engineers and other experts have been shown to be overconfident in their predictions in general, and nuclear power has a history of extraordinary overconfidence in those most responsible for its development.[21] If these people are better informed about certain aspects of the technology, they might be more misinformed about some of the less technical bases of opposition.

But the entire notion of expertise is problematic with something as complex as nuclear technology. Any "expert" who wants to render some general judgment about nuclear power must render judgments beyond his or her area of special competence. The issues involved require a greater understanding of physics, engineering, medicine, epidemiology, geology, economics, systems analysis, psychology, management techniques, and so on, then any individual can muster. So, in a sense, there are no experts, no individuals who have special

[21] See the articles reprinted in *Judgment Under Uncertainty: Heuristics and Biases*, ed. Daniel Kahneman, Paul Slovic, and Amos Tversky (Cambridge: Cambridge University Press, 1982), Parts VI and IX.

insights into all the technical areas, let alone the nontechnical ones. There are people who know particular areas well, and they are often divided in their judgments. Then there are people whose business it is to promote nuclear power, people whose business it is to regulate it and protect people and the environment from its potential hazards, and the rest of us who will be deeply affected by the decisions that are made. The important decisions must be made by the society at large, and the proper role of different experts in influencing those decisions is one of the problems we must address.

Furthermore, rather than relying on experts to evaluate nuclear technology, people increasingly want a technology for which such blind reliance and, perhaps, the threat of curtailing civil liberties is unnecessary.[22] This is a prominent theme of those who would like to see the public subsidies for nuclear power go instead to developing forms of solar energy that can be decentralized in their deployment and managed locally. One technology might be preferred to another because it allows those served by it to make their own autonomous decisions about its risks and costs. This autonomy may even be worth some degree of economic inefficiency. If nuclear power requires a surrender to the authority of experts, this might itself constitute a reason to reject that technology.

Another feature of the sociodynamics can be found in the complaint that people fear nuclear power because they are extraordinarily risk-averse about nuclear power and basically pessimistic about new and sophisticated technologies.

The claim about technological optimism and pessimism can be dismissed as a red herring. Opponents of nuclear power are often quite optimistic about, say, the commercial feasibility of photovoltaic cells or sophisticated systems for converting biomass to fuel. Some nuclear engineers are paradigms of technological pessimism when they talk about solar power. Optimism and pessimism are technology-specific. In the nuclear debate we do find optimism about the technology and its economics and pessimism about its risks, both going beyond what the evidence supports. But these attitudes are not fundamental, as the claim suggests; they are what we need to explain.

The risk-aversion charge is the more serious. The public expresses fear about nuclear power far out of proportion to fear about other threats to its well-being. The risks of acid rainfall, for example, seem to inspire less general concern than risks of toxic radioactivity, even

[22] Russell Ayres, "Policing Plutonium: The Civil Liberties Fallout," *Harvard Civil Rights–Civil Liberties Law Review* 10 (1975):369–443.

though the probability and magnitude of the former may make them no less threatening. There is nothing necessarily irrational, however, about refusing a risk that is measurably lower than the accepted risks. This would be clearly irrational only if the *qualities* of different risks (as opposed to the measurable balance of expected harms and benefits) are irrelevant to a rational decision to accept or reject them. This is at least counterintuitive.

Finally, people may object to nuclear power for reasons that have little to do with risks to life and health. They may fear the kind of economic and political system – the kind of life – that they feel inescapably accompanies it. Following Amory Lovins, many people see our energy policy in the United States are embodying an exclusive choice between a nuclear future (the "hard path" of technological development) and a solar one (the "soft path").[23] Nuclear technologies are thought by these critics to be necessarily highly capital-intensive and highly centralized in their depolyment as a system for producing the socially basic good of electricity. Solar or soft technologies, on the other hand, are thought to be compatible with or even to encourage a decentralized superstructure of deployment and production. Nuclear technologies, then, are likely to concentrate capital and power in the utility and oil industries (who own most of the uranium) to an even greater extent than they are concentrated today.

Some people believe that this path of development, like some other trends toward concentration of wealth and power in fewer and fewer industries, threatens the ability of individuals or even of nations to control their own destinies. If the technology for producing energy requires centralized control and if political power depends largely on energy, then technological development in the area of energy production can determine the distribution of social and political power in our world.

Of course we might decide that Lovins is simply wrong, and his thesis about the stark choice of two incompatible paths is wrong or too simplistic. But most such rejections by defenders of nuclear energy have amounted to little more than hand waving.[24] This is just one more instance of empirical claims being accepted or rejected for basically nonscientific reasons. I am aware of only one serious attempt to study this issue, which was undertaken by the Swedish

[23] Amory Lovins, "Energy Strategy: The Road Not Taken?" *Foreign Affairs* 54 (1976):65–96. Reprinted and expanded in Lovins, *Soft Energy Paths* (New York: Harper Colophon Books, 1977).

[24] See essays in *The Energy Controversy*, ed. H. Nash (San Francisco: Friends of the Earth, 1979).

government to help guide their long-range energy planning. Its conclusion supports Lovins's thesis:

> Measures to create a solar industry therefore threaten the nuclear industry. And, conversely, measures to protect the nuclear industry and the demand for nuclear-generated electricity may very well create an insurmountable barrier to a nascent solar industry... [T]he challenge to policymakers posed by an open-ended energy policy can hardly be stressed enough.[25]

Whether or not these paths are exclusive or allow for a third alternative of a mixed strategy, the choice we make is not technologically determined but a basic political decision. The dangers posed by the concentration of wealth and power can fortunately be resisted, for people can govern themselves to block this concentration. They can define the individual rights and principles of distributive justice that they will act to defend and they can defend them by fighting economic and technological developments that threaten them. Far from being an irrational response motivated by fear of risk, this would seem to be one of the primary purposes of the government in a democratic society. We can make political decisions not only about the risks and benefits we choose to accept and that our children will inherit, but about the governmental and economic structures as well.

The sociodynamic reasons behind the nuclear power controversy call attention to the kind of society we will create by the economic and technological decisions we make. They force us to confront questions about the quality of life and also our moral requirements to future generations. They serve as reminders that economic and health considerations, though crucial, are not the only considerations. It is not only a level of risk or welfare that is at stake in the controversy, but a kind of life as well.

V

The nuclear power controvery is not basically about technical or scientific matters, and, therefore, we should not expect to find some technique or analytic procedure to resolve it. As helpful as decision analysis and risk–benefit analysis are for laying out our alternatives and clarifying our thinking about them, they cannot succeed in finding an acceptable solution to such deep problems. The kinds of values at stake simply cannot be assigned any meaningful quantity of

[25] M. Lönroth et al., "Sweden Beyond Oil: Nuclear Commitments and Solar Options," *Science* 208 (1980):563.

importance. We "weigh" them instead through argument, discussion, and debate.

Nor should this fact be taken to indicate a failure of rationality, for it is surely not an intellectual failing to understand that technologies and energy policies have the profoundest effect in shaping our society and determining the quality of our lives. It would be irrational not to be concerned about anything so important. This particular controversy may be playing itself out for reasons that have little to do with the arguments on each side, but others like it will surely arise to take its place.

I believe it is wise to reject analytic attempts to solve these controversies, because I suspect all such rational decision-making techniques distort our values if we expect them to settle issues upon which society is fundamentally and philosophically divided. We should regard it instead as a sign of health that such controversies exist, for they force us to think about the broader implications of our technological choices. We should bring such debates into the open and discuss the arguments at length, even if this means slowing somewhat the pace of technological progress. The price and supply of electricity, after all, are but one small component in determining the quality of life.

PART III. CONTROVERSY, CLOSURE, AND THE PUBLIC

26

The role of the mass media in scientific controversy

RAE GOODELL

Dartmouth College president John Kemeny, reflecting on his experiences as chairman of the President's Commission on the Accident at Three Mile Island, concluded that the two-hundred-year-old American political system is outdated, in that it is failing to cope adequately with modern scientific and technological issues. Many of the most serious weaknesses of the current system, according to Kemeny, are found in the mass media. If significant national debate and consensus are to be achieved, Kemeny argues, American democracy must be modernized, and with it the media (Kemeny, 1980).

As scientific controversies become an increasingly prominent part of the American political scene, many observers are sharing Kemeny's concerns. To address these concerns, it has become important to take a closer look at the role of the media in generating, shaping, and bringing to closure public issues with scientific and technological components. How do the media currently influence the course of scientific controversy in public issues? What assumptions and constraints underlie the media's role? What changes should be made and can realistically be expected in the media's handling of scientific controversy?

Types of media influence in scientific controversy

According to one estimate, most print and broadcast organizations receive ten times as much information each day as they can use (Sandman, Rubin, and Sachsman, 1976). And according to another estimate, better than 99 percent of scientific information never even reaches the media for consideration (Thistle, 1958). The need to select from this surplus of information provides the basis for much of the media's impact on scientific controversy.

As a result of the selectivity of the media, for example, the very appearance of issues in the media legitimizes the issues as candidates for political attention and increases their standing on the political agenda. It is assumed that topics in the media have met certain editorial priorities, somehow also reflecting public priorities. For instance, the fact that, as Allan Mazur has noted (1981), the media have given extensive coverage to the risks of nuclear power but not to those of wood-burning stoves can be expected to affect future energy regulation.

In general, the media amplify issues that have already attracted considerable interest and involvement from scientists or citizens. Occasionally, however, the media contribute significantly to the generation of public issues: The public sociobiology controversy, for example, arose largely through the efforts of a few scientists and a few national reporters, according to Mazur's study. Science journalists themselves have speculated that media emphasis on Laetrile contributed to pressuring the federal government to initiate research on the drug's efficacy ("Science, Technology," 1980).

Similarly, a diminution or cessation of media coverage decreases an issue's political urgency. In 1978 and 1979, for instance, the dramatic decrease in media coverage of the recombinant DNA issue contributed to scientists' ability to persuade the federal government that the issue of laboratory safety hazards had reached closure (see Goodell, 1980).

The criteria by which the media filter information are variable and sometimes even contradictory, a reflection of the media's dual function in American society. On the one hand, the media are private, competitive businesses whose survival and success depend on delivering readers, viewers, or listeners to advertisers. As such, the media must place highest priority on selecting information that attracts audiences. On the other hand, the media are considered to have a responsibility to meet the variously defined needs of society for information that will contribute to individual well-being and intelligent political decision making. The concept of social responsibility has become increasingly salient in media judgments and governmental regulation as economic and political conditions have led to a decrease and consolidation of media organizations.

Attempting to respond to public need in the context of private enterprise, the media have developed a number of rules of thumb for selecting issues, and for selecting information within issues, that further affect the conduct of scientific controversy.

First, the media view their role as that of informing, not educating, the public; as providing news, not strengthening general knowledge. Thus the amount of background information about science that reaches the public is limited, although it can be argued that such information is vital to forming the up-to-date concepts needed to evaluate newsworthy scientific events when they arise. Once his or her formal education has been completed, a member of the public generally depends almost entirely on the mass media for information about science (Dubas and Martel, 1975; Schramm and Wade, 1967; Krieghbaum, 1967). The situation is exacerbated by the fact that, at least until recently, editors customarily assumed that topics in science and technology (except for applied medicine) were remote and unappealing to most audiences. Editors also accepted the widely held notion that there was extensive public hostility toward science. Consequently, in spite of the growth of science in recent decades, the amount of science in the news has not increased, hovering, since the 1930s, around 1 percent of the total news in newspapers (Nunn, 1979; Krieghbaum, 1967) and about the same now on television.

There is some research suggesting that editors have underestimated reader interest in science (Nunn, 1979; Dubas and Martel, 1975; Curley, 1979; Krieghbaum, 1967). The recent explosion of popular science magazines – *Science 86, Omni, Discover* – could be viewed as a test of that hypothesis.

One kind of background information about science that has been not only sparse but distorted in the media is a cultural image of science and scientists. On the one hand, in informational material the scientist is often represented as all-knowing, the sole relevant source of wisdom in settling disputes ranging from what brand of aspirin is best to how much nuclear power is safe (for further discussion, see Pfund and Hofstadter, 1980). On the other hand, in entertainment material the scientist becomes a white-coated eccentric, hair flying, test tubes rattling, retorts bubbling – capable of unleashing untold power but incapable of coherent explanations of it (for further discussion, see Basalla, 1976). The image is omnipresent – in cartoons, in television entertainment programs, and, unfortunately, even in educational programs such as "Electric Company," an educational television series for grade school children.

A second manifestation of media customs in coverage of scientific issues is that although much of science receives little media coverage, controversy within science receives a great deal. There is an assumption, all too frequently validated by audience response, that conflict – violence, dispute, danger – sells. Accordingly there has been a marked increase in the coverage of conflict in science news in recent decades, at least in print media. A University of Minnesota master's thesis (Cole, 1974) found an increase in conflict stories from about 17 percent to 28 percent of science coverage between 1961 and 1971 in four major newspapers. The trend was nearly twice as common in science stories written by general reporters, who presumably are assigned their stories by editors, than in those by specialized science writers, who have more autonomy. The same thesis also found that editors give conflict stories more play: longer length and more prominent headlines, photographs, and positions in the paper. Another study (Babcock and Ostman, 1980) found that the press, reporting on the space program from John F. Kennedy's presidential news conferences, emphasized conflict – the race with the Russians – whereas Kennedy himself was emphasizing cooperation – joint programs with the Soviets.

Interestingly, the new popular science magazines and television programs have not generally focused on controversy in their coverage of science. On the contrary, controversy is frequently absent, even when such omission is seriously misleading, as when experts disagree on the feasibility of a new technology. It is striking that this same overemphasis on the positive aspects of science has apparently been present in American magazines at least since the beginning of the century. An analysis of magazine coverage of science from 1910 to 1955 (La Follette, 1979) gives a description of magazines then that could just as easily apply now: "Mass media discussion of science combined extensive reporting on the actual results of science, promises and predictions that science would cure any social problem, and images of scientists as omniscient, powerful, well-meaning, and heroic, to develop a climate of exaggerated expectations of what science could do for and to society" (p. v).

The fact that most media (although not the popular science magazines) give special attention to controversy in science takes on additional significance in the light of Allan Mazur's studies (1981), suggesting that the quantity of media coverage of scientific controversy may have more influence on public opinion than its quality. Mazur finds that an increase in media coverage of a technical controversy is accompanied by an increase in public opposition to that technology; decrease in coverage is accompanied by a decrease in public opposition – regardless of the extent to which the coverage emphasized risks or danger. It appears that the mere existence of discussion of risk or of disagreement among experts about risk has a marked influence on public opinion.

There may also be a similar disproportion between quantity of coverage and impact on the public in media treatment of sensational topics. Studies do not confirm intuitive observations that the media sensationalize scientific issues. Sensationalizing does occur, but infrequently, according to case studies of Love Canal (Swan, 1979), the swine flu immunization program (Rubin and Hendy, 1977), the Three Mile Island accident (Sandman and Paden, 1979; Rubin, Chaapel, Cunningham, et al., 1979), and the risks of marijuana (Shepherd, 1979; Shepherd and Goode, 1977). The problem may be that when sensationalism does occur, it makes a greater impression than balanced coverage, as in this example from the swine flu case: The *New York Post* of October 14, 1976, ran the headline "The Scene at the Pennsylvania Death Clinic," and the story described how seventy-five-year-old Julia Bucci "had winced at the sting of the hypodermic ... taken a few feeble steps, then dropped dead" (Rubin and Hendy, 1977, p. 772).

A third area of media influence concerns the boundaries of debate in scientific controversy. Frequently the media, intentionally or inadvertently, restrict their attention to certain kinds of information about an issue. Frequently, for example, epidemiological data and other kinds of statistical analysis are given short shrift in media coverage of technological risk, largely because statistics are poorly understood by reporters and media management and poorly explained to them by experts. This lack of statistical data could be observed, for instance, in most coverage of toxic shock syndrome. A more profound kind of limitation of relevant information occurred in the recombinant DNA debate, in which the press, following the pattern set by the scientific leaders of the discussion, largely omitted long-range social and ethical considerations from their coverage of genetic engineering (Goodell, 1980).

A special limitation on the dimensions of debate stems from the media's customary handling of "balance" in coverage of controversial issues. Issues are generally assumed to have two sides, of somewhat comparable merit, that should be reported in juxtaposition, one source routinely countered by an opposing source. The judgment of the journalist, if any, as to the soundness of each side is to be reflected only in the order in which views are presented.

These overall distortions in the media – omissions, artificial balancing, narrow definitions of the boundaries of debate – are probably much more serious flaws in media reporting than actual factual inaccuracy. A number of case studies (Pfund and Hofstadter, 1980; Shepherd, 1979; Rubin and Hendy, 1977; Sandman and Paden, 1979) have found factual accuracy in media handling of scientific issues to be very high. This is not to say that factual errors do not occur but rather that they seem more frequent than they are. Also, because mass media audiences tend to recall only general impressions of a news story, it is likely that small, specific factual errors have a relatively slight effect on public understanding of science.

A fourth area of media influence, arising partly from the emphasis on balance, is the altering of ground rules as controversy moves from internal scientific discussion to the public arena. In particular, critics whose views might hold little credibility among scientists sometimes receive proportionately greater attention in the media. Media are likely to place greater emphasis on such credentials as awards (especially the Nobel Prize) and institutional affiliation and less emphasis on the scientist's disciplinary area of expertise. Thus, some scientific celebrities, community leaders, and administrators are frequently asked to comment on a broad range of topics (Goodell, 1977).

The media's lack of adversarial role in scientific controversy

Perhaps the most striking aspect of media involvement in scientific controversy, however, is not the areas in which it has influence but the areas in which it does not. To some extent the media clearly assert their own values, such as in emphasizing human interest, prominent scientists, and two-sided balance of views. To a surprising extent, however, media coverage reflects the values of the scientific community, with relatively little of the compensation, skepticism, and investigation that customarily accompanies political reporting.

Given a choice of potential sources of information in scientific disputes, in other words, the press emphasizes the views of certain established scientific spokesmen rather than those of a broader range of experts and lay observers. The press relies heavily on these relatively few scientists for factual information and for news judgments, including judgments as to when an issue is appropriate for public attention and when it is ready to be brought to closure.

Such symbiosis between journalist and source is common in the American media but especially intense in science reporting. (It may have counterparts in other specialized areas of reporting such as business or the arts.) This close dependence requires that scientific sources be objective observers even in areas of conflict, and almost omniscient experts even in areas of broad social application. It fails to acknowledge the politicized role of scientists in modern society, the importance of the scientific community as an interest group, or as a number of interest groups, whose purpose must be to protect the progress of basic research, sometimes at the expense of other priorities, in issues that might threaten their own well-being.

The pattern of dependence on a relatively few scientific sources can be seen in a number of case studies of press coverage of scientific issues. According to a study of press coverage of the 1979 accident at

Three Mile Island, for example, journalists largely reported the comments of government and industry representatives at the scene, rarely seeking out the views of other experts around the country. When information handed out by the spokespersons was conflicting or confusing, the reporters grumbled and complained, rather than note that the confusion itself was revealing (Sandman and Paden, 1979). Another study showed that journalists similarly expressed helpless frustration about the inconsistent explanations offered by government representatives during the 1976–77 swine flu vaccination crisis (Rubin and Hendy, 1977). And, in coverage during the 1970s of the risks of marijuana use, a study found that journalists cited primarily administrators associated with government and private agencies, Nobel Prize winners, and other widely known scientists; they cited much less frequently the scientists actually conducting research on marijuana risks (Shepherd and Goode, 1977; Shepherd, 1979). In coverage of the scandal at the Sloan-Kettering Cancer Center, in which researcher William Summerlin was found guilty of painting coat "transplants" on mice, most reporters confined their interviews to Sloan-Kettering administrators and thus missed many of the facts behind the fraud (Goodfield, 1981).

Occasionally, such lack of initiative apparently leads to serious oversights in press coverage: According to one study, the California press overlooked a major asbestos occupational health problem in the state, because journalists were relying on government sources, rather than checking also with labor and consumer groups (Medsger, 1977). Another study suggests that reporters overlooked a number of indications that there were flaws in the Apollo space program. If the press had exposed the flaws, the study speculates, steps might have been taken to correct the problems resulting in the January 1967 fire that killed three astronauts (Skardon, 1967, 1967–8).

The scientific community has become so accustomed to the press's dependence, journalists find, that scientific leaders are sometimes offended when the press does seek out alternative points of view (Culliton, 1977; Greenberg, 1974). (Similar observations about the lack of initiative in science press coverage can be found in other reviews, e.g., Lambeth and Ferguson, 1980; Pfund and Hofstadter, 1980).

The problems posed by their symbiotic relationship have emerged most clearly, perhaps, in press coverage of the recombinant DNA controversy (see also Goodell, 1980 and 1981). Coverage of recombinant DNA research fluctuated dramatically during the course of the controversy, in response less to developments in risk assessment or regulation than to changes in the public stance of scientific leaders with regard to these developments. Briefly, recombinant DNA research was first reported in the media in 1974, largely only after scientists brought the safety issue to press attention as a side effect of bringing it to their colleagues' attention. The media accepted the DNA researchers' definition of the problem – an immediate question of health hazard – and their approach to a solution – self-regulation by the researchers. Politicians, scholars, and scientific critics, not the press, later occasionally broadened the discussion to include ethical and social ramifications, and even then coverage occurred primarily as the result of specific news events, such as conferences, speeches, and hearings. With the emergence of a group of scientist–critics, coverage settled into a proponent–opponent balance, with little representation of the range of less extreme alternative views. In 1977, when DNA proponents organized a lobby to oppose federal regulation of recombinant DNA research, the media reported the lobby's position, usually without reservation. The media's cooperation undoubtedly contributed significantly to the ability of the lobbyists to bring the safety issue to a close in about 1978. Once the safety issue was resolved, a major public relations campaign was launched within the new genetics industry, and this too was reported in enthusiastic terms, generally without critical comment.

In short, whatever was volunteered by scientific experts – whether they were the organizers of the initial concern about research safety, the established scientist-critics, the congressional lobbyists, or the industrial spokesmen – largely determined the extent, timing, and direction of press coverage of recombinant DNA.

There are a number of reasons why the relationship between scientific spokesmen and science journalists may be particularly close. First, there has evolved within the media a small, specialized group of science reporters. Partly because many editors are intimidated by and uninterested in science, these science reporters have unusual autonomy. For example, according to a study by Sharon Dunwoody (1980), many of these science writers initiate at least 80 percent of their own stories, whereas reporters on general assignment typically initiate very few. For the same reasons it can be expected that the reporting by such science writers is subjected to relatively little scrutiny and alteration by editorial superiors.

Science writers are further removed from some of the normal journalistic conventions by their existence as a closely knit group that shares interests and values, including a sense of second-class status relative to the rest of journalism. The result is sometimes a lack of competitiveness, competitiveness that would foster a wider range of sources.

This insularity extends to science writers' relationships with scientists, as well. One of the specialized science writer's most valued professional assets is a list of scientific sources on whom he or she can depend for guidance, story ideas, and information. Aware of scientists' sensitivity to public criticism, science writers are unlikely to question their sources' judgment lest they strain the trusting relationship they have worked hard to develop. Nor are these writers even interested in questioning their sources' authority, because they share scientists' enthusiasm for the process, values, and discoveries of science. Such writers tend not to be particularly interested in politics and controversy, and, like the scientists, they are concerned that political stories will tarnish the image of science and increase public hostility toward basic research ("Science, Technology," 1980). These writers even share scientists' doubts that lay people – including science writers themselves – are qualified to form intelligent judgments about technical issues.

Not all scientific controversy is covered by these specialized science writers, of course. In fact, if a scientific controversy becomes prominent enough, science writers are frequently relegated to sidebars, whereas general and political reporters are assigned to the main story, as *New York Daily News* science writer Edward Edelson, then president of the National Association of Science Writers, observed in 1979 (Edelson, 1979). General writers, however, are equally prone to depend uncritically on established scientific experts. For one thing, they lack confidence in their own judgment on the technical aspects of issues. For another, their editors, also lacking confidence, insist that sources cited in their stories have obvious credentials.

The writers most likely to exercise some independence and to conduct thorough investigation in reporting scientific controversy are those who free-lance for major magazines. Free-lance writers are notoriously poorly paid, however; those willing and able to take the time for in-depth reporting are necessarily rare.

Finally, writers who do seek out a greater diversity of sources learn that such sources are hard to find. The set of values within the scientific community that protects the integrity and progress of basic research discourages scientists from establishing relationships with the press. Reporters interact, then, with the scientists for whom pressures to keep silent are somehow countered by equally strong motivations – selfish or idealistic – to speak out.

The result is that scientists, with regard to their accessibility to the press, form a kind of pyramid. At the bottom are a broad base of scientists who are rarely heard from, generally because they are uninterested in publicity or because the media are uninterested in their work. Next are the many scientists, including most elite university scientists, whose research occasionally becomes newsworthy for a short time. Closer to the top are the regular sources, the inner group of scientists who comment on areas of research and technical dispute that are within their specialities. Still higher up the pyramid are the scientists who speak out frequently to promote or to criticize broad areas of science and science policy – the university administrators, Washington bureaucrats, consumer group leaders, and others who are involved in issues that they feel need public response – either public support or public intervention. Finally, at the top of the pyramid are the few scientists who are sufficiently motivated, quotable, colorful, credible, and accessible to become celebrities. Some of these celebrities are critics of science, and some are popularizers, that is, promoters of public education and support for science.

Thus the few scientists who are frequent news sources tend to hold strong views on science or science policy. They tend, for example, to be at the extremes in scientific controversy, exacerbating the press's tendency to create artificially two-sided representation of issues. Sometimes, too, the expression of alternative views is actively discouraged by scientific leaders, who fear that the research enterprise will be jeopardized by publicly exposed controversy. Such pressuring of critics occurred, for example, in the later phases of the recombinant DNA safety debate (Goodell, 1981).

Prospects for change

Realistically, one cannot expect major changes in the role of the media that would jeopardize their function as profit-making businesses. It would be inappropriate, for example, to demand large increases in background information about science, educational material that would probably have little audience appeal.

On the other hand, media can and should be asked to scrutinize some of its normal practices, such as the establishing of two-sided balances, and to extend some of its other customs, such as investigative political reporting. And, in fact, a recognition of the difficulties inherent in the symbiosis between a small group of science writers and a small group of scientists has been expressed occasionally by science journalists themselves (Stockton, 1981). There are also signs that the media are beginning to place a higher priority on science news, perhaps partly as a result of experiences like Three Mile Island and the increase in science controversy in general.

By the same token, one cannot expect major changes in the role of the scientific community that would jeopardize its primary function of protecting and fostering basic research. The scientific community will continue to reward primarily those activities that contribute to research, activities from which popularization sometimes distracts and with which it sometimes even conflicts.

On the other hand, efforts can and should be explored to mitigate the tension between research and popularization, and thus to increase the number and variety of scientific sources available to the media. A number of suggestions have been made, including the pragmatic proposal that universities increase the logistic support available to scientists who find themselves in demand from reporters – telephone service, secretarial service, counseling in media relations – to offset some of the more objectionable side effects of media attention (Eisendrath, 1979). Observers have also urged that universities and other research institutions hold more background briefings for reporters and editors on potentially newsworthy topics such as energy, genetic engineering, and arms development. The most important function of such briefings would undoubtedly be to expose journalists, particularly inexperienced ones, to a greater variety of experts upon whom they could call later when news stories developed.

And finally, there are a few possibilities for injecting into the reward system of science somewhat more support for activities to popularize science. Universities and foundations could, for example, offer fellowships to scientists for periods of popular writing; more prestigious awards for effective communication with the public; and special positions in which job security is based on popular rather than technical publication. (For example, the American Institute of Physics and the United States Steel Foundation offer an annual award for a scientist who publishes material for the general public in the area of physics and astronomy.) Similarly, the prestige and effectiveness of thorough, investigative science writing among journalists could be enhanced by offering better security, incentives, and rewards, including, as June Goodfield (1981) has suggested, positions within academic settings.

In summary, there are observable ways in which the acts and omissions of the media shape the portrayal of scientific controversy in the public arena. The influence of the media is subject to modest change by modifying the network of customs and assumptions that guide both media management and the scientific community.

What is less clearly understood, of course, is the extent to which the media's portrayal of issues actually affects public perceptions, and the extent to which such perceptions affect the conduct of debate and closure. It may well be, for example, that the media exert far less influence by altering public opinion on issues than by altering scientific and political leaders' assessments of prevailing public opinion on those issues. The reporting of science and science policy has only recently begun to be studied and analyzed critically. As the research in the area increases in quantity and sophistication, a number of further implications of the conduct of scientific controversy can be expected to emerge.

References

Babcock, William A., and Ronald E. Ostman. 1980. "U.S.-U.S.S.R. Space Exploration and Technology, John F. Kennedy, and the Press." Paper presented at the session on "When Experts Differ: The Role of the Mass Media in Scientific and Technological Controversy," Annual Meeting of the Association for Education in Journalism, Boston, August 10.

Basalla, George. 1976. "Pop Science: The Depiction of Science in Popular Culture." In G. Holton and W. Blanpied, eds., *Science and Its Public: The Changing Relationship*. Boston: D. Reidel.

Cole, Bruce Jon. 1974. "Science Conflict: A Content Analysis of Four Major Metropolitan Newspapers, 1951, 1961, 1971." Master's thesis, University of Minnesota.

Culliton, Barbara J. 1977. "Science, Society and the Press." *New England Journal of Medicine* 296: 1450–3.

Curley, Thomas. 1979. "Readers Want Latest News, Consistent and Complete Newspapers." *Gannetteer*, p. 9, March. (Gannett Co., Inc., Lincoln Tower, Rochester, N.Y. 14604.)

Dubas, Orest, and Lisa Martel. 1975. *Media Impact*: Vol. 2, *Science, Mass Media, and the Public*. Ottawa: Ministry of State for Science and Technology.

Dunwoody, Sharon. 1980. "The Science Writing Inner Club: A Communication Link between Science and the Lay Public." *Science, Technology and Human Values* 30: 14–22.

Edelson, Edward. 1979. "The President's Letter." Newsletter of the National Association of Science Writers, 28: 13.

Eisendrath, Charles R. 1979. "The Press as Guilty Bystander." In D. A. Jackson and S. P. Stich, eds., *The Recombinant DNA Debate*. Englewood Cliffs, N.J.: Prentice-Hall, pp. 279–99.

Goodell, Rae. 1977. *The Visible Scientists*. Boston: Little, Brown.

Goodell, Rae. 1980. "The Gene Craze." *Columbia Journalism Review* 19: 41–5.

Goodell, Rae. 1981. "Scientists and the Press: the Case of Recombinant DNA." Paper presented at the annual meeting of the American Association for the Advancement of Science, Toronto, January 8.

Goodfield, June, 1981. *Reflections on Science and the Media*. Washington, D.C.: American Association for the Advancement of Science.

Greenberg, Daniel S. 1974. "Let's Hear It for Science." *Columbia Journalism Review* 13: 16–23.

Kemeny, John G. 1980. "Saving American Democracy: The Lessons of Three Mile Island." *Technology Review* 83: 65–75.

Krieghbaum, Hillier. 1967. *Science and the Mass Media*. New York: New York University Press.

La Follette, Marcel E. Chotkowski. 1979. "Authority, Promise and Expectation: The Images of Science and Scientists in American Popular Magazines, 1910–1955." Ph.D. diss., Indiana University.

Lambeth, Edmund B., and Robert Ferguson. 1980. "The *Columbia Journalism Review* and Scientific Controversy: An Overview and Commentary." Address presented at the session on "When Experts Differ: The Role of the Mass Media in Scientific and Technological Controversy," Annual Meeting of the Association for Education in Journalism, Boston, August 10.

Mazur, Allan. 1981. "Media Coverage of Technical Controversies." *Journal of Communication* 31: 106–15.

Medsger, Betty. 1977. "Asbestos: The California Story." *Columbia Journalism Review* 16: 41–50.

Nunn, Clyde S. 1979. "Readership and Coverage of Science and Technology in Newspapers." *Journalism Quarterly* 56: 27–30.

Pfund, Nancy, and Laura Hofstadter. 1980. "Biomedical Innovation and the Press: Four Case Studies." Discussion draft, presented at the session on "When Experts Differ: The Role of the Mass Media in Scientific and Technological Controversy," Annual Meeting of the Association for Education in Journalism, Boston, August 10.

Rubin, David M., H. A. Chaapel, A. M. Cunningham, et al. 1979. *Report of the Public's Right to Information: Task Force to the President's Commission on the Accident at Three Mile Island.* Washington, D.C.

Rubin, David M., and Val Hendy. 1977. "Swine Influenza and the News Media." *Annals of Internal Medicine* 87: 769–74.

Sandman, Peter M., and Mary Paden. 1979. "At Three Mile Island." *Columbia Journalism Review* 18: 43–58.

Sandman, Peter M., David M. Rubin, and David B. Sachsman. 1976. *Media: An Introductory Analysis of American Mass Communication*, 2d ed. Englewood Cliffs, N.J.: Prentice-Hall.

Schramm, Wilbur, and Serena Wade. 1967. *Knowledge and the Public Mind.* Stanford: Institute for Communication Research, Stanford University.

"Science, Technology, and the Press: Must the 'Age of Innocence' End?" 1980. Panel discussion, *Technology Review* 82: 46–56.

Shepherd, R. Gordon. 1979. "Science News of Controversy: The Case of Marijuana." *Journalism Monographs*, No. 62.

Shepherd, R. Gordon, and Erich Goode. 1977. "Scientists in the Popular Press." *New Scientist* 76: 482–4.

Skardon, James A., 1967. "The Apollo Story: What the Watchdogs Missed." *Columbia Journalism Review* 6: 11–15.

Skardon, James A. 1967–8. "The Apollo Story (Second Part)." *Columbia Journalism Review* 6: 34–9.

Stockton, William. 1981. Remarks to the meeting of the New England Science Writers, Boston, March 31.

Swan, Jon. 1979. "Uncovering Love Canal." *Columbia Journalism Review* 17: 46–51.

Thistle, M. W. 1958. "Popularizing Science," *Science* 127: 951–5.

27

The National Commission on Human Experimentation: procedures and outcomes

STEPHEN E. TOULMIN

The National Commission for the Protection of Human Subjects of Biomedical and Behavioral Research (the "National Commission," for short) was set up in 1974 by the U.S. Congress as one of the provisions of the National Research Act, P.L. 93–348, and its members were sworn in during December of that year. The primary part of the commission's formal agenda comprised a series of questions about the ethical requirements to be insisted on as conditions for the federal funding of research projects involving human subjects who were recognized as belonging to certain especially vulnerable groups. (Three or four other more or less peripheral items were added to this primary agenda, but I shall not discuss them here.)

Because the public controversies that preceded the establishment of the National Commission, in the aftermath of the U.S. Supreme Court's abortion decision, had been excited by horrifying press reports about scientific experiments on human fetuses, that was the group of vulnerable research subjects whose protection the National Commission was instructed to consider first, and a four-month moratorium was imposed on federal funding of fetal experiments while the commission dealt with this first task. Subsequent groups of vulnerable subjects consisted of young children, prisoners, and "the institutionalized mentally infirm," to cite the idiosyncratic language of the Act. Incidentally, the National Commission was also instructed to report on the ethical requirements relevant to the performance of psychosurgery. The reasons for this addition were evidently political. It reflected in part the widespread disquiet generated by Ken Kesey's book, *One Flew over the Cuckoo's Nest*, but it represented, further, the Congress's first tentative intrusion, by way of a highly controversial procedure into the protected world of regular medical and surgical practice (unlike the medical fraternity, the research community was always fair game).

On completion of the National Commission's deliberations about any of its major topics, it was instructed to send a full report, including recommendations for action, to the secretary of the then Department of Health, Education and Welfare (HEW). It was then the secretary's responsibility to publish the recommendations, together with his own responses, and to promulgate corresponding regulations within a fixed period of time, after allowing opportunity for public representations and comments. Formally speaking, however, the National Commission was independent of HEW, although drawing its administrative support (and its funds) from the department. This separation of the National Commission from HEW was designed to insulate the commission from departmental pressures; and, despite some early ambiguities, that insulation was eventually achieved.

The National Commission met regularly for some three and a half years, normally in Bethesda, Maryland, for two or three days at a time, once each month; it also had a solid professional full-time staff, and it published a long series of reports and recommendations. Before the commission went out of existence in the fall of 1978, some, but not all, of its recommendations had been acted on by HEW. Where the secretary did act, his formal regulations were in general along lines close to those of the commission's recommendations, but in every case there were minor differences between the commission's recommendations and the specific regulations subsequently promulgated by the department. As things went along, it was never wholly clear how far those differences reflected a genuine perception of administrative needs overlooked by the commission, how far they were responses to changes in the current political situation, and how far they served (rather) to satisfy the demands of bureaucratic amour propre. At any rate, the unusually strong legislative mandate of the National Commission, which, in certain respects, came close to a "delegated power of legislation" over the matters concerned, made it politically risky for HEW to depart very far from the commission's recommendations. This was done very strikingly, in fact, only in the case of research involving prisoners as research subjects. There (as we shall see), the secretary's eventual regulations were more stringent and conservative than the commission's recommendations and conformed more closely both to current political sentiments in the electorate at large and to changed policies in the federal prison system.

In what follows, I shall try to do four things. (1) First, I shall summarize the procedures that were followed by the National Commission in addressing its central responsibilities. (2) I shall next describe how these procedures were exemplified in the commission's deliberations about the employment of prisoners as research subjects. (3) Third, I shall comment on the roles played by the various steps in the commission's deliberations and indicate their relative contributions to such measure of consensus (or closure) as the commission in fact achieved. (4) Finally, by way of postscript, I shall raise some tentative questions about the philosophical implications of the commission's work.

Procedures

The National Commission went through very much the same sequence of steps in addressing each of the chief questions in its mandate, questions dealing with human fetuses, with children, with prisoners, and with the institutionalized insane. (They also followed much the same sequence in their work on psychosurgery.) These steps were nine in number:

1. Initially, the commissioners held a preliminary discussion among themselves aimed at identifying and formulating the ethically significant issues relevant to the protection of a particular vulnerable group of subjects.
2. Next, they considered what further material they required to obtain from outside their own immediate resources and experience in preparation for a thorough and effective discussion of those issues and gave instructions for the preparation of papers, largely commissioned from outside scholars and scientists: These included both analytical papers dealing with those ethical issues from different points of view and also empirical papers expanding the body of available factual information relevant to the commission's discussions.
3. Fairly early in each set of deliberations, they organized public hearings, usually in Bethesda, Maryland, but sometimes also in other cities around the country, to receive presentations from organizations and individuals concerned about the issues currently under consideration.
4. Around the midpoint of each investigation, they held an interim debate about the central issues raised: This usually led them to develop a first working taxonomy for classifying the different subgroups of research subjects, experimental procedures, forms of consent, and so on involved in each case.
5. Around the same stage, they also made site visits to typical research centers to observe the actual conditions under which human experiments of the kind in question were in fact being performed.
6. In the light of these interim discussions or site visits, they

considered what additional papers (analytic and empirical) they required for the purposes of their more definitive deliberations and arranged to have these papers prepared.

7. Subsequently, with all the available material before them, they came to grips with the central issues in a series of discussions that continued for no less than two months and, in one case (children), for most of a year.
8. As an outcome of that debate, they formulated and voted on recommendations to the secretary of HEW.
9. Finally, after the practical recommendations were in, they assembled the supporting arguments to be presented as the "justifications" for those recommendations.

(*Notice:* All of the commission's substantive discussions and meetings were advertised in the *Federal Register* and were open to the public. Rarely, however, was the public audience larger than a couple of dozen. Only at one meeting in San Francisco, when a major topic for discussion was the ethics of psychosurgery, were the commission's proceedings crowded out and made the target of organized public protest. By being willing to discuss psychosurgery at all, even – or especially – *coolly*, the commission exposed itself to charges of "condoning psychiatric abuse.")

The case of research involving prisoners

To come to my chosen example: The National Commission took up the case of research involving prisoners as one of its most urgent concerns, immediately after the completion of its initial "crash program" on fetal experimentation. The timetable of its inquiries worked out as follows. The commissioners (1) had preliminary discussions of this item at its meetings in May and June 1975 and (2) commissioned papers from outside consultants for delivery in three months. They (3) organized public hearings in Bethesda in January 1976; (4) had a first general discussion on the basis of the material then available in October 1975, and (5) in November 1975 made a full-scale site visit to the Southern Michigan State Prison at Jackson, Michigan, where drug testing laboratories were being maintained by Parke Davis and Upjohn. A smaller group also visited the Washington State prison in Walla Walla to observe the "behavior modification" research being conducted there by the prison authorities. In the light of these site visits, the commissioners found it necessary to redirect the thrust of their central questions, and they (6) commissioned a small number of additional papers as a result. (I explain the nature of this change of emphasis, and the reasons for it, later in this chapter.) Their (7) definitive debate on this topic began in spring 1976 and continued for three months; formal recommendations (8) were finally voted in July 1976; and the supporting documents (9) were ratified in September 1976.

This timetable gives a good measure of the commission's typical mode of operation. The initial work on fetal research was done under pressure from the moratorium on funding and the associated four-month deadline. (Nobody questions that the results of this work would have been more satisfactory if it had been done in less of a hurry. Indeed, there are some who believe that, given time to refine the taxonomy one stage further, it would have been possible to find room for the views of the one dissenting commissioner within the general consensus.) The work on research involving children went on much longer, both because of the great complexity and variety of issues that it raised and because of the delicacy of the distinctions needing to be drawn, so, in that case, the definitive debate went on for ten months before formal recommendations could be voted. The deliberations on research involving prisoners were thus neither specially hurried nor unusually prolonged.

When the commissioners began to consider prison research, they had no special reason to challenge the presumptions that underlay current public disquiet about the subject. Because of the essentially coercive nature of any prison environment (this argument presumed), human experimentation involving prisoners as research subjects is inevitably "exploitative," one further mode of victimization imposed, for the administrative convenience of pharmacological or psychological researchers, on those who were already the victims of a coercive society and situation. Despite all of the commissioners' efforts to suspend judgment in the absence of reliable information, their early discussions accordingly tilted strongly toward the view that ethically acceptable conditions for such research could exist only within a quite unusually humane and well-run prison, that is, a prison of a kind that probably did not at present exist in the United States and might not even be capable of existing here under preexisting conditions. It was with this presumption in mind that the commission went through stages (1) to (4) and approached its main site visit to Jackson, Michigan.

A long busy day at Jackson gave the commissioners and staff some wholly unforeseen "empirical data," and this quite altered the commission's perception of the prison research issue. The very conception of an "essentially coercive situation" turned out to need a good deal of refinement and qualification. The normal situation within a large maximum security prison was (they concluded) at the same time much more wretched and deprived yet also much less destructive of scope for individual autonomy than – despite the best available written evidence, and the oral testimony of knowledgeable advocacy groups – any of them had foreseen. Institutionally, there is a whole complex society within such a prison; economically, some two or three vigorous "markets" operate, largely on a "free enterprise" basis, with cigarettes or drugs, tokens, and straight cash as their currencies, and those inmates who handle themselves successfully can often take advantage of these "social" and "economic" factors to work out quite a tolerable modus survivendi. (A personal note: My own boys' boarding school in England was no Dotheboys Hall, but once inside the Jackson Prison, I quickly picked up a "feel" for the not-so-distant family resemblances between the institutional operations of prisons and boarding schools). On the other hand, state legislatures are happy to vote only the minimum budgets for running their prisons. So, for instance, the only official medical facilities in the prison consisted of an "infirmary" in a superannuated cell block, with totally inadequate staffing and equipment. Besides this glaring shortcoming, relations between the largely white, rural guards and the largely black, urban inmates were marked by as much lack of mutual respect and understanding as one would suppose, and except for the actual machinery for maintaining the perimeter of confinement, everything else in the prison was unutterably drab, shabby, and dispiriting. (In being required to devise better means of regulating the research operations going on in this situation, the commissioners felt like watchmenders called in to repair the clock in a broken-down car.)

What did all this have to do with the ethics of prison research? To begin with, private conversations with inmates soon made it clear that there was strong competition for the chance of participating in the drug research projects. Far from these experiments being perceived by the participants themselves as a further humiliation of the already humiliated, imposed only on those who could not find some way of excusing themselves, it was the best-educated and most "streetwise" of the prisoners – quite often the whites, in a predominantly black inmate population – who ended by taking part. (Subsequently, indeed, a group of black prisoners in Indianapolis brought a class action suit alleging that they were unconstitutionally discriminated against by being denied the same opportunity as whites to participate in such experiments.) The attractions of the research were threefold. First, every experiment was preceded by a medical checkup: One prisoner on a twenty-year sentence for a second-degree family murder explained that since his admission nine years earlier, these had been the only medical checkups he had been able to obtain, and he had enough self-respect to value the reassurance this provided. Second, participation was rewarded by worthwhile monetary payments, ranging from approximately $10 to $150 per research project, and this was really the only "straight" way of accumulating a cash reserve while in prison. Third, the research laboratories were cheerfully decorated and pleasantly equipped, with superior living accommodations. They were (we quickly saw) the only part of the prison environment that could give inmates any continuing sense of what it was like to live "on the outside."

As for the hazards of the research: The great bulk of experiments were in satisfaction of routine U.S. Federal Drug Administration (FDA) requirements, measuring, for example, comparative rates of "take up" with different shapes of pills or capsules, and the like. As such, they were almost risk-free, and in any case our informants insisted that they were free to pick and choose among the research projects offered and had no hesitation in declining experiments about which they were not wholly content. (Furthermore, the commissioners were satisfied that the fact of any inmate's participation or nonparticipation in the research was never made available to a parole board.) To their considerable surprise, therefore, the members of the commission found themselves pleaded with by the inmates not to make recommendations that would lead to the closing down of the drug research facilities. It was bad enough being deprived of liberty for twenty years: Must they be deprived, also, of their last remaining access to the satisfactions of life "outside"? So, the whole emphasis of the commission's deliberations changed. The original legislation had been an attempt by Congress to protect prison inmates from injury and exploitation by unethical researchers. The "protection" the inmates themselves asked for was not protection against the scientists, but rather protection of their autonomous right to choose to work with those scientists.

It took two or three months for the commission to digest the lessons of this Jackson visit. During this time, the members found themselves debating several new topics, notably, the research subjects' own claims to individual autonomy and, in particular, the ethics of research subjects' "doing it for money." If the actual situation with prisons was not, after all, "essentially" coercive, what then of the inducements that drew inmates into the research? (The inmates answered, for their part, that these inducements are in themselves neither less nor more "coercive" within the prison than they are in the existing society or economy "outside.") After Jackson, to speak quite generally, the commissioners never again took entirely at its face value the paternalistic role that the Congress had designed for them of being there to "protect" vulnerable research subjects, whether they wanted to or not. Instead, the question became, for them, how they were to strike a good, right, and just balance between all the different moral claims arising within the context of research involving human subjects from one group or another. So, in the prison case, they tried their best to avoid ruling out in advance all possibility that prison inmates could choose to act as research subjects without being subject to unacceptable coercion or inducements.

To wind up this story: Before the National Commission had completed its deliberations, the U.S. Bureau of Prisons – which was, of course, administratively quite independent of HEW – changed its policies, so as to exclude effectively all research by drug companies and other outside agencies from federal prisons. Nonetheless, the commissioners went ahead and presented to the secretary of HEW a set of recommendations indicating the conditions and circumstances in which they would be happy to see such research continue. Because of the timing of this particular case, however, considerations of politics won out over those of ethics. The secretary of HEW found it administratively infeasible to follow up the commission's somewhat complex recommendations in a way that would have had him moving in a politically unpopular direction, quite opposite to the Bureau of Prisons' own policy change, and, in this prison case it was both too tricky and also too late in the day for the commission to challenge the secretary's rejection of their chief recommendations as they had earlier challenged his deviations from their fetal research recommendations.

Comment on these procedures

How did these different steps contribute to the commission's eventual recommendations about any topic? And to the extent that a consensus emerged in the course of the deliberations, what made this "closure" possible? The crucial step in the commission's consideration of any topic (I would argue) was the first one: that is, the initial identification and framing of the moral considerations that the commission took as relevant to the topic concerned. About the nature of those considerations, significantly, there was rarely much disagreement. In part, no doubt, that was because all of the chosen topics were matters of public controversy, in advance of the commission's establishment: To that extent, the commission was no doubt right to begin by showing itself responsive to preexisting public concerns. Yet, in all cases the commission's deliberations went a fair way beyond the scope of earlier public debates, even undercutting them in two instances, and the commissioners did not hesitate to take up whatever moral issues they themselves "felt on their pulses." Significantly, too, the commissioners no more disagreed about what kinds of considerations were relevant to those further issues than they did in the case of those issues that had already been the subjects of extensive public debate.

This first, issue-framing phase had one other notable feature. Unlike what Arthur Kantrowitz called for, in arguing his case for a so-called science court in 1976, the National Commission made no attempt to define issues of strict "fact" for determination *in isolation from* all ethical considerations or "value" questions. On the contrary, only by identifying the ethical issues relevant to any specific group of research subjects could the commissioners decide just what *kinds* of "factual" information they needed as grist for their debate. For instance, it was precisely the moral force attaching to Kesey's critique of frontal lobotomy in mental institutions that prompted the commissioners to obtain detailed reports on the history and current status of psychosurgical techniques; lacking this factual information, they could not address questions about the moral acceptability of present-day psychosurgical procedures in any kind of informed way.

Thus, although it was always open to outsiders to volunteer relevant information to the commission, the presentations that carried most weight with the commissioners were predominantly those that were prepared by authors specifically commissioned to provide explicit responses to questions raised early in the commission's own deliberations: that is, questions that sprang from the commissioners' own perceptions of the moral issues relevant to that particular type of human experimentation. By contrast, the "public hearings" played a disappointingly minor part in the development of the commission's thinking. Only in exceptional cases did representatives of outside organizations take the trouble to follow the commission's proceedings, before those public hearings, closely and perceptively enough to make a productive intervention in the discussion.

Typically, representatives of advocacy organizations turned up at the hearings and testified more or less "cold." As a result, they, too, often addressed the issues at a much less sophisticated and discriminating level of analysis than the commission itself had already reached. To give just one example: A young attorney from the American Civil Liberties Union used the commission's public hearings on experiments involving children to give a ringing and comprehensive denunciation of all such research as being inescapably "exploitive" of innocent infants. Since the commissioners had already taken the trouble to sort out, in its own taxonomic way, certain large classes of research involving children that are clearly free of morally objectionable features, they were not very receptive to such an undiscriminative challenge, one that, on the face of it, would have required them to find some artificial way of "decriminalizing" the work of Piaget.

Why were the advocacy organizations so little effective in the commission's work? They appear to have assumed, in advance, that the commission would operate in a political mode and that they should approach it in the same way that they would approach a congressional subcommittee or similar body, as frank partisans presenting one side of a contentious issue to the best of their ability, and with the most powerful rhetoric. If they had been better represented at earlier meetings, they might have realized that such tactics were self-defeating. The eleven commissioners responded in a very chilly way to partisan or one-sided arguments, and they particularly resented being lectured.

At no time, in fact, were the National Commission's proceedings conducted according to either a political (pressure group) or an accusatory (adversarial) mode. The commissioners did not see it as their business to "gauge public sentiment" and make their recommendations accordingly. In this respect, their methods of argument flatly rejected any view of "ethics" as concerned with labile public attitudes of approval and repugnance, such as the Supreme Court fell back on, for instance, when it "decentralized" the standards of obscenity. On the contrary, the members of the National Commission set themselves, throughout, the task of identifying clearly recognizable considerations that might serve as solid moral grounds for giving (or withdrawing) public approbation and funding to different kinds of human experimentation. Sensational novels and newspaper reports for or against different kinds of scientific research might heighten or inflame public sentiment, but that was none of the commissioners' concern. If the current public debate was based on faulty assumptions about the actual situation, they should not hesitate to set the record straight by explaining the sources of those misperceptions.

When the intended composition of the commission was announced, with eleven nominated members, of whom only five were to be biomedical or behavioral scientists, the *New England Journal of Medicine* quoted one worldly onlooker as remarking, "Now, no doubt, we shall see matters of eternal principle decided by a six to five vote." After working closely with the commission for more than three years, however, I cannot myself think of a single occasion on which ethical differences of opinion divided the scientists from the laymen. None of the commissioners, of course, undertook to speak unqualifiedly for or against human experimentation as a whole, on the contrary, scientists and laymen alike were concerned to put their fingers, as precisely as they could, on the specific features that would lay particular types of human experimentation open to solid moral objections.

In a couple of cases, indeed, the commissioners rethought their initial positions and ended by agreeing on recommendations substantially more liberal toward research than those generally advanced in the current public debate. One of these was the prison case, about which I wrote above, the other was the case of psychosurgery. In both cases what brought about this change in the commissioners' perceptions of the moral situation was the new and more accurate factual information obtained by the commission in the course of its investigations. The public debate about psychosurgery, which had continued for some years before the establishment of the commission, for instance, had rested on a widespread assumption that frontal lobotomies were currently being, or were liable to be, performed in state hospitals and similar institutions under conditions of dubious "informed consent" and in circumstances that suggested that the governing aim of the surgery was social or institutional control rather than psychiatric therapy. The National Commission therefore began its deliberations by arranging for a number of independent factual reports covering the circumstances, the anatomical characters, and the consent procedures adopted in connection with all current psychosurgical operations. With all the resources of skepticism at their command, the commissioners ended by having no doubt, on the basis of the resulting information, that these procedures were now taking place under strict and discriminating professional control and in quite different contexts from those popularly assumed. Whatever may or may not have been the case some twenty years earlier, by the mid-1970s frontal lobe procedures had been effectively abandoned in favor of much more exactly localized ("stereotaxic") amygdalotomies, in which small numbers of highly specific and well-identified neural pathways were severed, principally as a means of relieving intolerable pain. Furthermore, the subjects of current psychosurgical operations were almost exclusively fully consenting – not to say, eager – middle-or-professional-class educated whites, with long previous histories of unsuccessful electroshock and other drastic treatments. So, without endorsing these procedures as "safe and efficacious," the commissioners in fact recommended that they continue only under HEW supervision on a research basis, rather than being accepted too quickly as "routine and accepted practice"; they were at least convinced that the earlier public anxieties about psychosurgery were no longer appropriate to the new situation. (To repeat: I cite this example here, less for the sake of its substance than for its value as illustrating how the acquisition of unforeseen factual information played a powerful part in the commission's deliberative procedures.)

To sum up: "What, then, was the National Commission's basic methodology?" To answer that question in a nutshell: Its discussions very quickly fell into a taxonomic mode. All of the questions on the commission's central agenda could be construed as being of the form, "Just what kinds of experimental procedures, involving just what subgroups of human subjects, conducted in just what kinds of circumstances, with just what consent procedures, et cetera, are and are not open to solid objections on moral grounds?" The commissioners came to see their task as being to identify the factors relevant to questions of this form, and they accordingly set out to develop systematic classifications by distinguishing different sorts of experiments, different sorts of loci of research, different procedures of consent, assent, toleration or refusal, different immediate and long-term scientific goals, and so on. All the factual material, site visits, public presentations, and the rest, that provided the mass of the material for the commission's definitive debates were finally digested and brought to bear on the commission's mandate by considering what light they threw on those central taxonomic issues and how they could help the commissioners to draw more refined, perceptive, and discriminating distinctions between "morally acceptable" and "morally questionable" classes of research projects. It was this taxonomic methodology that enabled the commission to address the administrative requirements implicit in any regulatory process, and it was this same procedure that frequently made it possible for the commissioners to increase their agreement by adding further differentiae and so producing new and more refined clarifications.

Personal postscript

With the commission's basic methodology in mind, let me end by offering some personal generalizations about the commission's work. As I saw the commission's work, the central project of developing discriminating taxonomies for classifying different kinds of experiments involving human subjects by picking out features relevant to the moral acceptability or unacceptability of such research was extraordinarily successful. In advance of the commission's deliberations, many philosophers, particularly, those sympathetic with (e.g.) Alasdair MacIntyre's views about the current Babel ethical debate, would have been inclined to doubt whether any kind of consensus could possibly emerge on any of the central issues. In actual practice, by the time any particular issue had been thoroughly discussed, there was no substantive disagreement about the nature of the relevant moral considerations. That is not, of course, to imply that the votes on the recommendations were always, or often, unanimous. In several cases, notably that of research involving young children, some commissioners were significantly more conservative, others significantly more permissive, in their weighing of the morally relevant considerations. Yet, to underline the crucial point: The commissioners were never in any serious doubt *what it was that they were not-quite-unanimous about.*

By contrast, when the time came to write up for publication the arguments they relied on to "justify" those same recommendations, something much closer to genuine Babel began. Protestants, Catholics, Jews, and atheists; attorneys, theologians, physiologists, and radical behaviorists; utilitarians and deontologists, libertarians and paternalists, blacks and whites, men and women, academic scholars and public interest representatives – all of them were prone to "justify" their readiness to go along with the commission's general consensus in different terms and from different standpoints. Again and again, as a result, it proved far easier for members of the commission to say *what* they agreed about than it was for them to explain *why* they agreed about it.

Conversely, however, once the basic taxonomic work was done, this Babel on the level of theories and principles rarely prevented mutual understanding (and frequent near-unanimity) about practical issues. Presented with specific types of cases characterized in very particular concrete terms, the members of the National Commission apparently found little real difficulty in agreeing what substantive moral issues, factors, and considerations they should take into account and trade off against each other in arriving at their formal recommendations, and even their substantive disagreements were usually marginal ones. Failures of communication and understanding, like those that MacIntyre emphasizes, began at the moment, and only at the moment, at which the further question was raised of explaining by appeal to what "principles" the commission's recommendations were to be "justified." When it came to giving justifications, accordingly, it began to appear as though the "principles" of

Catholic, Jewish, and humanist ethics tell us much more about Catholicism, Judaism, and humanism than they do about ethics. For, in practice, they quite evidently served not to strengthen the commissioners' shared moral perceptions so much as to square these perceptions with each individual's different commitments in other directions.

Alasdair MacIntyre's own comment on the commission's proceedings, in conversation, was that this practical consensus was not surprising, given the commission's homogeneity, since it was comprised entirely of upper-middle-class suburban Americans, and also given the political imperatives under which it labored. This comment does not satisfy me. Not only were there, in fact, rather substantial differences among the backgrounds, interests, and persuasions of the different commissioners, but, in addition, the course taken by the commission's discussions left me confident that, given time, the same practical results could have been achieved, even if its membership had included (say) a Buddhist, a *bracero*, and a factory worker. As for the external "pressure to agree," some such pressure is probably a characteristic element in all urgent and substantive ethical issues, and as such it can do nothing to discredit the commissioners' particular consensus any more than it does any other ethical agreement.

Indeed, the whole experience of working with the commission has, in fact, had the effect of reinforcing in me a respect for the Aristotelean perception of ethical problems as matters of practical reasoning. If we return to this Aristotelean standpoint, a preoccupation with quasi-geometrical systematicity, with universal principles, with deductive justifications, and the like comes to appear something of an intellectual mirage. In the commission's deliberations, the "locus of certitude" (to coin a phrase) lay less in the general principles to which different commissioners were inclined to appeal in explaining their own individual views than it did in the particular types of case that the commission's taxonomies picked out as possessing (or lacking) the crucial features that marked them of as open to (or free from) significant moral objections.

The same kind of thing may be expected to hold good in other ethical cases equally. This redirection of our intellectual attentions explains why both Al Jonsen, who was one of the commissioners, and I myself are currently interested in reexamining the various historical traditions of "case reasoning" in morality, traditions to be found in (e.g.) the history of Catholic casuistics, in early Roman law, and in rabbinical judgment. For quite a time now, moral philosophers have been paying too little attention to Aristotle's claim that there are no essences in ethics and that philosophical ethics should be redirected away from a Platonizing preoccupation with universal principles and the like. Al Jonsen and I both believe that the resulting questions will be philosophically fruitful ones. Just where we shall finally arrive by pursuing these lines of thought in the context of the current philosophical debate, it is, however, still too soon to say.

28

The forms and norms of closure

RUTH MACKLIN

The essays in this volume all address in one way or another the subject of closure in science, technology, and public policy. In some of these papers, the primary focus is on a theoretical or conceptual question: What types of closure are there? In other papers, the emphasis is on a historical or sociological question: What forms of closure have been involved in the termination of historical and contemporary controversies involving science and technology? The aim of this concluding chapter is first, to comment on the conceptual and descriptive accounts offered by several of the authors whose essays appear in this volume. A second, related task is to offer a reply to the following normative question: What types of closure should occur or are appropriate for bringing an end to different types of controversies?

Before that normative question can be tackled properly, a systematic review of the papers would have to yield clear answers to this series of prior questions:

1. What are the different types of *controversy* for which closure has been sought or in which termination has occurred?
2. What forms of *closure* (resolution, termination) can be identified?
3. Which of those forms of closure have in fact occurred in *particular* historical and contemporary instances of controversy, as described in the answer to question 1?
4. What *pattern*, if any, defines closure reached in debates or disputes in each of the categories of controversy listed in answer to question 1?
5. Where such patterns can be identified, do they shed any light on what type of closure people have thought to be appropriate to different types of controversy?
6. Is there a clear and defensible notion of "appropriate closure" for particular forms of controversy? Put another way, are there "normative norms" of closure?

7. If there are normative norms of closure appropriate to different types of controversy in the various domains in which debates and disputes requiring closure erupt, on what are those norms based or how can they be justified?

It is fair to say that complete agreement was never reached among members of the Closure Project on the answers even to the descriptive questions in the above list, much less to the conceptual and normative ones. Let me then review briefly (though not exhaustively) areas of agreement and disagreement, to pave the way for the issue on which I will eventually focus, namely, the "norms of closure."

Forms of closure and forms of controversy

I begin with Tom L. Beauchamp's classification scheme, as presented in his essay entitled "Ethical Theory and the Problem of Closure" (Chap. 1, this volume). I believe that Beauchamp's forms of closure and his elucidation of them are quite fruitful and largely accurate, so I will use his terminology in my later remarks after discussing the taxonomies offered by others. To review, then, the Beauchamp forms of closure are as follows: (1) Sound argument; (2) consensus; (3) procedural closure; (4) negotiation closure; (5) natural death. This is a general typology, not designed for any particular domain in which controversy occurs. Beauchamp offers examples of areas in which each of these forms has occurred, with further observations about the frequency of their occurrence.

Ernan McMullin proposes a different taxonomy in his essay entitled "How Do Scientific Controversies End?" (Chap. 2, this volume). McMullin's typology is confined to controversies in science (for which there are more detailed typologies, to be reviewed shortly). His preferred terminology, however, employs the term *closure* for only one of these forms of termination, since he holds that the term itself carries a special meaning. McMullin's taxonomy (which he acknowledges is not exhaustive of the various ways in which controversies in science may terminate) includes the following forms: (1) resolution; (2) closure; (3) abandonment.

Resolution corresponds roughly to Beauchamp's sound argument closure. Its features are not only that agreement must be reached among the parties to the disputes and the scientific community of their time but also that agreement is reached on the *merits of the case*. "The factors involved in satisfactory resolution are necessarily *internal* ones," according to McMullin, and to use the concept that he employs elsewhere, "epistemic" considerations are central both to the controversy itself and to its resolution.

McMullin's concept of closure, on the other hand, "always involves external factors of one sort or another: the authority of the state, a controversialist's ambition or fear, the withdrawal of publication facilities from one or both sides." And further, "The factors that bring about closure do not adjudicate the merits of the case; they merely terminate the public expression of scientific disagreement that had constituted the controversy." These so-called external factors are, then, nonepistemic ones. The type of termination of controversy that McMullin calls "closure" does not correspond to any single one of Beauchamp's forms but seems instead to be a mixture of three of them: consensus, procedural closure, and negotiation closure. This is easy to understand, since McMullin explicitly states that this mode of ending controversy always involves external factors. Moreover, he holds that "the list of possibilities is virtually endless, since the variety of external factors that can bring about closure is boundless." McMullin makes further distinctions within this category, identifying "pure" and "mixed" closure (cases of the former, he claims, are relatively rare in science), and also "strong" and "weak" closure, depending on whether or not external factors play a very important or a minor role in terminating the controversy.

It is clear that for McMullin the important factors in resolving scientific controversies – more specifically, controversies within science – both are and ought to be internal or epistemic considerations. It is understandable, then, that he lumps together forms of termination involving external factors into the one mixed category that he terms "closure." This does not so much represent a disagreement with Beauchamp's conceptual categories as a focus chosen to elucidate controversies within science. The less the dispute is about intrascientific matters, the greater richness we need for our taxonomy of forms of termination (or closure, to use the general term of this project). McMullin's "abandonment" form of termination is virtually identical with Beauchamp's natural death closure and is the form of closure that is least interesting and about which there seems to be the least controversy.

Turning next to Everett Mendelsohn's candidates for forms of closure, as described in his essay "The Political Anatomy of Controversy in the Sciences" (Chap. 3, this volume), we find the terminological waters muddied still further. Mendelsohn, like McMullin, confines his discussion almost entirely to controversies in science, but aside from the penetrating scholarship and intriguing analysis that both authors display, the similarity ends there. Mendelsohn proffers two types of termination of controversy in science: resolution and closure. But his explication of these terms shows that neither is being used in the way in which McMullin has used those same terms and also reveals different suppositions about how closure has occurred historically in scientific controversies. Mendelsohn elucidates his forms of termination as follows: "Resolution of a controversy represents a coming together of the conflicting parties and the emergence of a consensus. Closure, on the other hand, might be used to mean a more formal structure for ending a controversy or debate that permits a partial resolution, but not necessarily a dissolution, of the disagreement." Mendelsohn's "resolution," according to his description of that mode of ending controversies, appears similar if not identical to Beauchamp's consensus closure, but it is not identical with Beauchamp's sound argument closure or, of course (by the law of transitivity), with McMullin's version of resolution. The key difference between McMullin's and Mendelsohn's positions turns on the role played by consensus. Unlike the form of resolution that McMullin claims to be appropriate for terminating controversies within science, the consensus method does not require full-scale agreement on the epistemic or internal factors among parties to a dispute.

Mendelsohn's closure is similar but not identical to McMullin's closure, since both writers identify a range of external or nonepistemic factors that may contribute to reaching closure. But Mendelsohn's notion is even broader than McMullin's, since it appears to encompass the latter's abandonment category: "Often the only real closure comes with the death of a participant in the debate. After all, if interests do inform judgment and action, we cannot expect a resolution or a consensus to emerge unless those interests themselves undergo change." Mendelsohn's own characterization of his concept of resolution remains somewhat ambiguous, however, since in his detailed and informative analysis of historical cases, he makes some remarks suggesting that perhaps resolution has more epistemic characteristics than his initial description allowed. In concluding his discussion of the spontaneous generation controversy, Mendelsohn says that "here, as in so many other debates, even though an element of closure was achieved, no clear, cognitive resolution was gained among the contestants themselves." Now a "cognitive resolution" is much closer to McMullin's notion of resolution than to the consensus closure that Mendelsohn's description of his own notion of resolution seemed to indicate. The concept of consensus embodies the idea of compromise, an agreement arising out of a recognized and acknowledged need to agree. It is, therefore, a politically useful method, one appropriate for terminating disputes that issue in policy-making. In the strictly cognitive domain, a method of resolving controversies that relies on the disputants reaching a consensus would still be politically useful, but according to McMullin's position, it would be an inappropriate method for ending controversies within science.

If there is one thing on which virtually all participants in the Closure Project seemed to agree, it is that the nature of the controversy or the domain in which it occurs is related to two factors regarding closure: First, types of controversy can be correlated with particular forms of closure that have taken place historically; and second, the nature of the controversy has implications for the type of closure that is possible or appropriate. Without reviewing in full detail the various types of controversy that different authors have enumerated, we should recall that these include at least the following: within science, controversies over fact, theory, methodology, and principle; related to science, social controversies over the content of science or over the process of science; controversies involving technology, which include both technological applications of scientific findings and social controversies involving technology; political and legal controversies; and moral controversies. Instances of these types were explored in a number of case studies during the Closure Project: Controversies within science are exemplified by the continental drift debate; controversies about science are illustrated by the efforts to show that homosexuality is or is not a disease, based on a scientific framework; controversies over science policy are well documented in the nuclear energy debate and in the disputes concerning occupational health and safety; and the role of scientific expertise comes into play in the Laetrile controversy and the Three Mile Island affair.

The questions that now deserve further probing are, first, Which features of the preceding types of controversies explain why one or another form of closure was employed in such cases? And second, Are some forms of closure more appropriate than others for resolving certain types of controversy? In other words, are there normative norms of closure, and if so, in what are they grounded?

Norms of closure

The type of controversy, the time and place in which it occurs, and the policies, practices, or rules governing the particular context not only contribute to determining how closure in fact comes about but also how it ought to take place. Actual controversies almost always involve a mixture of three irreducible elements – elements that may overlap or become indistinguishable from one another in practice. These elements are (1) science, including facts, theories, and methods; (2) values – primarily, but not exclusively, moral values; and (3) politics, whether within the conduct of science, involving science and technology, or involving moral principles, such as questions of justice.

Recognizing that the nature of the controversy places some constraints on the type of closure that is possible, as well as on that which is likely to occur, I offer the following normative principles regarding closure. Although not every author would concur with every one of these principles, I think substantial support can be found for each of them in other chapters in this volume.

1. There should be no negotiation closure in pure science.
2. There should be negotiation closure in politics and in matters of science policy.
3. Procedural closure is sometimes necessary or desirable in science policy or in politics but is almost always inferior to sound argument closure or negotiation closure in cases where the latter are applicable.

In cases of "pure" scientific controversy – that is, where the conflict is over empirical facts, the nature of evidence, scientific theory, or methodology – sound argument closure is the appropriate mode of ending controversies. This view is stated or implied by Ernan McMullin throughout his chapter.

Support for normative principles 1 and 2 is forthcoming from Robert F. Rich in his essay entitled "Politics, Public Policy-making and the Process of Reaching Closure" (Chap. 5, this volume). Rich writes:

> It is worth noting that in science one is not dealing with a "negotiated settlement." Instead, by using well-developed canons of research and evidence, a procedure is legitimately developed for settlement of disputes without having to rely on negotiation. Scientists characterize this process as a "rational" one.

Rich contrasts this mode of reaching closure with that which typically occurs in a polity. Among the "rules of the game" in the latter context, Rich includes the following three:

> Each stake-holder should have the opportunity to legitimately affect decision-making procedures.
>
> The best evidence in a system of negotiated settlement is often political influence and the ability to use political pressure . . .
>
> Settlement or closure is reached through negotiation or, alternatively, through the exercise of "raw power."

Although Rich gives an account that is largely descriptive of actual practice, his reference to the statements above as "rules of the game" reveals their normative aspect. These rules, in turn, derive from some key principles of American government, principles that Rich sets out earlier in his essay, along with several further assumptions inherent in our political system about the resolution of conflict. The norms for reaching closure in politics generally, and in the arena of science in particular, reflect the basic values embedded in the political process in the United States. These norms differ not only from those governing scientific inquiry and the resolution of disputes in science; they also vary markedly from those that govern the settlement of controversial issues in countries such as the Soviet Union that have a radically different political system.

This last point is nicely illustrated by Loren R. Graham in his essay entitled "How History and Politics Affect Closure in Biomedical Discussions: The Example of the Soviet Union" (Chap. 8, this volume). Graham contends that "recent discussions about biomedical ethics in the United States (think for a moment of abortion and of *in vitro* fertilization) are incomprehensible outside the context of American social, political, and religious history." Similarly, he holds, "Controversies over biomedical ethics in the Soviet Union display characteristics resulting from the specific culture and politics of that nation." Graham observes that some varieties of argument that are tolerated or even respected in the United States, on topics such as abortion and recombinant DNA research, are not considered legitimate in the Soviet Union because "they are based on officially unacceptable philosophical positions, those of religion, mysticism, and dualism." Thus, unlike in the United States, where the norms of closure in politics and science policy differ from those that reign within science, in the Soviet Union something akin to sound argument closure governs both realms. Graham writes, "Rather than seeking a compromise in biomedical ethics between viewpoints that differ radically in their basic assumptions, Soviet specialists in these areas are expected to find the 'correct' solution, using the world view of scientific Marxism-Leninism as their basic guide." In the Soviet Union, then, negotiation closure is as inappropriate a means for resolving controversies involving values and politics as it is for settling disputes that are strictly scientific or technical. If we in the United States are critical of that situation, our criticism is itself a reflection of the cultural and political system to which we adhere, along with the values embedded in that system.

Consistent with the acceptance of the first two normative principles, we can expect that in "mixed" conflicts involving both purely scientific and also political or social aspects, sound argument closure is the appropriate method for settling the scientific controversy, while negotiation closure is appropriate for achieving a resolution of the political aspects. An example would be the many different debates regarding occupational health and safety, such as what is an acceptable level of a carcinogen in the workplace once the carcinogenicity of a substance has been established. Although they do not state this point explicitly in their essays (Chaps. 1 and 19, this volume) both Gilbert S. Omenn and Tom L. Beauchamp acknowledged in discussions at meetings of the Closure Project that they favor negotiation closure as the method of choice for resolving the final policy issues in occupational safety and health. Beauchamp, in a paper entitled "Informed Consent to Risk in the Workplace" (Closure Project, May 1981), nonetheless expressed dismay at the confrontational character of typical approaches currently taken:

> At the present time "controversies" about hazards in the workplace are being handled in the United States in one of two ways: damage claims for injuries are being pressed in the courtroom, and federal and local regulations are being established that regulate exposures and demand various warnings about hazards. Both approaches are ultimately confrontational: they pit the interests of government or workers against the interests of industry. This is a situation much to be regretted, and it is to be hoped that some nonconfrontational informed consent model – growing as informed consent has out of fiduciary relationships – can eventually be substituted.

Gilbert Omenn, in his essay entitled "Values in the Debate over Workplace Safety and Health: The Rancorous Rhetoric about Regulation" (Chap. 19, this volume), presents a rich and authoritative account of the issues in occupational safety and health, focusing on the recent role of the Occupational Safety and Health Administration. Omenn, like Beauchamp, seems to lament the role that the courts have assumed in trying to settle these controversies. He quotes with approval Judge David Bazelon's statement that

> the judiciary ... is not the proper forum for developing relevant information or selecting values, at least in these issues ... What, then, is the Court's role? In brief, I believe that the judicial responsibility is to monitor and scrutinize the administrative process. Our task is to assure that the agency's decision-making is thorough and within the bounds of reason. The agency's decisional record must disclose the evidence heard, the policies considered, the agency's precise reasons for resolving conflicts in the evidence. This includes their basis for selecting one scientific view rather than another ... Only if decision-makers disclose assumptions, doubts, and moral and

> political tradeoffs can experts and citizens evaluate administrative action . . . And only then can Congress and the people understand and debate the value choices implicit in regulatory action.

Implicit, yet clear, in this statement is the view that the political process is the appropriate mechanism for settling disputes in the controversial area of occupational safety and health.

Conclusion

The norms of closure discussed in the preceding section address two of the three irreducible elements found in most of the controversies explored in the Closure Project, namely, the scientific and the political aspects. Conspicuously lacking is a discussion of the third element: values, in particular, moral values. The problem of reaching closure in moral controversies is a long-standing one that goes to the heart of debates in meta-ethics about justification in ethics. It is impossible to do justice to that topic here, since it would involve an excursion into the nether reaches of epistemology, but a few brief remarks are in order nonetheless.

Uncertainty about the form of closure appropriate for moral controversies, or for the moral aspects of mixed controversies, stems from uncertainty about the nature of justification in ethics. Three classical positions on this matter are as follows:

1. Justification in ethics is essentially the same as justification in science.
2. There can be no justification, strictly speaking, in ethics because ethical judgments are noncognitive, that is, they are not "genuine" propositions.
3. Justification in ethics is possible, but it is of an entirely different sort from that which takes place in science. Each of these positions concerning justification rests, in turn, on an assumption about the relationship between facts and values – either the assumption that values can be reduced to or derived from facts, or the assumption that values and facts constitute entirely separate realms. Three of the authors in this volume have referred explicitly to this issue in their discussions of closure.

Ronald Giere, in a paper presented to the Closure Project, wrote:

> It might be objected that the above account of the possibilities of formal closure in scientific, legal/political and moral controversies assumes a fundamental distinction between facts and values. I agree that such a distinction is being assumed, but this is not objectionable. On the contrary, it is essential if one is

> to understand the nature of controversies involving science. But this is no place to debate the viability of the distinction[1]

And Loren Graham, in his chapter in this volume, observes:

> In the West most discussions of biomedical ethics have been based on the assumption that ethics and biology belong to entirely different realms and are different in kind. Ethics deals with values, whereas biology deals with facts. To confuse these two realms is to make a "category mistake," to fail to notice, as A. J. Ayer put it, that ethical statements have no cognitive significance . . . In the Soviet Union . . . ethics and values are considered to be no less a subject of scientific study than is biology. As a Soviet author wrote . . . , Marxists believe that, in principle, values can be submitted to "strict scientific research." This assertion of what in the West is called the "naturalistic fallacy" places Soviet Marxists in a different position from most of their Western colleagues when they approach problems of biomedical ethics.

The topic of the relationship between facts and values, and the nature of justification in ethics, is treated at greatest length and in the most detail by Tom Beauchamp in chapter 1. My own view of the norms of closure in this domain is that we should strive to achieve sound argument closure. Unlike politics, where the interests of all affected parties need to be taken into account in arriving at an acceptable policy (at least in the American political system), matters involving ethical principles are not appropriately settled by negotiation. However, to the extent that some value conflicts appear to be unresolvable because of basic value disagreements, efforts to achieve sound argument closure will inevitably fail. This form of closure requires acceptance of the premises of any argument, and it is precisely that agreement that is lacking in intractable moral disputes. Preferable to negotiating a settlement, I believe, is to invoke procedural closure in such cases, since the controversy is bound to continue even after the matter is "officially closed." On the basis of my belief that moral principles should not be matters for negotiation and that people should stand firm on their ethical commitments, I conclude that in cases where a policy must be forthcoming, it is better to seek a form of closure in which the parties to the dispute can maintain their integrity by adhering to their most deeply held moral commitments.

[1] "Scientific Knowledge and Technological Decisions," paper presented at Hastings Center Closure Project, May 1980.

AUTHOR INDEX

Abel, I. W., 441
Acton, A. B., 476n10
Adams, John, 154
Agassiz, Jean Louis, 69
Alexander, H. G., 96n9
Allen, Garland E., 7, 173n2, 179n19, 180n23, 181n27
Ames, Matthew M., 365n5, 361n19, 372n74
Anderson, C., 270
Annas, G., 286n6
Aquinas, Thomas, 32
Aristotle, 43, 50, 74, 571
Ashford, N., 446, 453
Auchter, Thorne, 455
Austin, J. L., 307
Avery, R. J., 316
Ayer, A. J., 251, 624
Ayres, Russell, 579n22

Babbitt, Bruce, 562–3
Babcock, William A., 588
Babson, Roger, 534n13
Baer, Karl Ernst von, 82n39
Baker, Howard, 204
Barnes Barry, 84n44, 96, 97, 99n17, 100n19, 129n9
Bastian, Charlton, 106, 112
Bastian, H. C., 89
Baumer, Franklin L., 535n14
Bayer, Ronald, 11, 16, 413n8
Bazelon, David L., 265, 451–2, 510n15, 622–3
Beard, Howard W., 323n28
Beardsley, Edward H., 186n34
Beauchamp, Tom L., 5, 6, 8, 14, 27–48, 200, 244, 260–3, 376, 377, 460, 480–1, 616–18, 622, 624
Beckman, G., 476n11
Beckman, L., 476n11
Bell, A. P., 409n7
Benge, M. C., 478n27
Bentham, Jeremy, 32, 450
Berkeley, George, 73
Bernard, Claude, 87, 114–15
Berthollet, Pierre Eugène, 68
Beyer, Jan, 565n20
Bickis, I. J., 365n37
Bieber, Irving, 11, 384, 390, 393, 397, 408, 418–22, 424n6, 425n7, 428–9, 430n14, 435
Bieber, T., 422n5
Bilibin, A. F., 254–5
Bingham, Eula, 440, 450
Bird, Randy, 173n2, 179n19
Black, Joseph, 68
Blackett, P. M. S., 220–1, 224, 225, 240
Bloor, David, 85n47, 317–18, 326
Blume, Stuart, 317
Boakes, R. A., 298, 305
Boas, Franz, 186
Bodde, David, 576n17, n18, 577n20
Boffey, P., 265
Bohanan, Luther, 339, 367–9, 371–2
Bohm, D., 73
Boissier de Sauvagnes, Francois, 9
Boland, M., 466n2
Bonaparte, Louis, 109
Bondi, H., 71
Born, Max, 59
Boscovich, Ruggiero Guiseppe, 73
Bowie, William, 211
Brahe, Tycho, 60
Brandt, R. B., 45
Broadway, James, 496n22, 499n25
Brody, Baruch A., 4
Brooks, C. E. P., 213, 214, 215, 239
Brosnan, John, 547n34
Buck, Peter, 100n19
Bugliarello, G., 291n15

Bullard, Edward, 224, 225
Bupp, Irwin C., 576n16
Burdon-Sanderson, John, 100, 115
Burger, Warren, 510, 511
Burgess, W. A., 447
Burke, Dean, 327, 365
Bush, Vannevar, 201
Butlin, H. T., 456–7
Byar, David P., 322n25, 356n4, 361n19, 374n71
Byers, Eben, 537

Calkins, D. R., 445, 448
Callahan, Daniel, 3n3
Campbell, Joseph, 531n5, n7
Campbell, Norman, 49–50, 93
Carey, W. S., 226, 240
Carpenter, F., 297n1, 301n7
Carter, Jimmy, 440, 461
Carter, Luther J., 491n9, 492n15
Casper, B., 272
Castle, William E., 175, 178, 182, 187
Cavendish, Henry, 67
Celler, Emanuel, 195
Chaapel, H. A., 589
Chase, Allan, 179n20, 191, 200n48
Childress, James, 38
Chomsky, Noam, 296, 302, 308, 310
Christie, Richard, 418n2
Churchill, Frederick B., 123
Clarke, Samuel, 53, 59
Clegg, A. J., 221
Clemmensen, J., 456
Cobbe, Frances Power, 118
Cocks, Paul, 253
Cohen, Harold, 299–300
Cohen, Robert, 418n2
Cole, Bruce Jon, 587
Cole, P. H., 467n3
Coleman, A. P., 213, 215, 239, 240
Conklin, Edwin Grant, 175, 182, 187, 199
Connor, T. H., 478n23, n24, n26
Corn, Morton, 440
Cornfield, Jerome, 515n25, 516n27
Court, B., 476n12
Cox, Alan, 223–4, 233–4
Cox, Archibald, 159
Crick, Francis, 133
Crozier, Michael, 284n2
Culliton, Barbara J., 591
Cunningham, A. M., 589
Curie, Pierre, 533, 536
Curley, Thomas, 587
Cuvier, Georges, 108–9

Dain, H., 418n1
Dalrymple, Brent, 234
Darwin, Charles, 69–70, 82
Davenport, Charles B., 173, 175–9, 182–5, 187, 192–4, 197, 198, 199
Davis, Katherine B., 185
Derian, Jean-Claude, 576n16
Descartes, René, 42, 50, 85–6, 102–5, 308
Dickstein, Samuel, 195, 196
Dietz, David, 574–5
Dince, P., 418n1
Divine, Robert A., 545n31
Dixon, R. L., 448
Dobias, L., 478n25
Doell, Richard, 223–4, 233–4
Domenici, Peter, 563
Dostoyevski, Feodor, 302
Draper, Wycliffe, 185n32
Driesch, Hans Adolf Eduard, 82
Dubas, Orest, 587
Ducatman, A., 476n14
Dunn, L. C., 182
Durham, Taylor R., 18
du Toit, Alex, 214–15, 216, 218, 219, 240
Dyban, A., 263n23

East, Edward M., 181, 199
Easton, David, 157
Eddington, Arthur Stanley, 59
Edelson, Edward, 593, 607
Einstein, Albert, 73, 74, 121
Eisenhower, Dwight, 575
El-Assal, Elaine, 544n31
Eliade, Mircea, 531n6, n7
Ellis, Havelock, 385
Ellison, Neil M., 322n25, 328–9, 356n4, 361n19, 374n71
Ellsberg, Daniel, 29
Endicott, Jean, 405
Engelhardt, H. Tristram, Jr., 3n3, 19n12, 273
Erskine, Hazel Gaudet, 541n25
Esendrath, Charles R., 595
Eskridge, Nancy K., 491n11, n13
Etteldorf, J. N., 476n15
Evans, R. B., 418
Ewing, John, 229
Ewing, Maurice, 226, 228–9, 230
Eyerly, Robert, 316
Ezrahi, Yaron, 290

Fairbairn, W. R. D., 305–6
Fantel, J., 447
Farley, John, 87, 105–6, 108n37, 110n41, n42, n43, n44, n45, 111, 112n46, n47
Farnsworth, David, 194
Farrall, Lyndsay, 100n19, 122n58
Farrell, Thomas, 540
Feigl, Herbert, 47

Feinberg, Gerald, 131n11, 132n15, 150n34
Ferguson, Robert, 591
Fermi, Enrico, 540
Festinger, L., 295
Feyerabend, Paul, 74, 95
Finer, Herman, 165
Fischer, Eugen, 179, 198
Fischoff, Baruch, 285n4, 573n11, n12
Fisher, Irving, 185
Fisher, Ronald A., 222, 224
Fletcher, John C., 357n6
Fletcher, Joseph, 32n5
Flourens, Pierre, 87, 110
Floyd, Mrs. James, 559
Ford, Daniel, 575n15
Forman, Paul, 59n12, 60
Forni, A., 477n20
Fortas, Abe, 29, 359
Foster, Michael, 100, 115
Fowler, John M., 59, 536n16
Frankel, Henry, 7, 133n16, 220n26
Freedman, Alfred, 392
French, Richard D., 114n49, 116
French, W. N., 365n37
Freud, Sigmund, 305–8, 383–4, 424, 431n15
Friedman, Milton, 155
Friedman, R. D., 446
Friedrich, Carl J., 165
Funer-Cravioto, F., 476n14

Galileo, Galilei, 61, 74–5
Galton, Francis, 178, 181, 187
Galtung, Johan, 545n32
Gamson, William A., 318–20, 326–7, 329–31
Garry, V. F., 477n19
Garza-Chapa, R., 477n20
Geison, Gerald, 87, 105–6, 108n37, 110n41, n42, n43, n44, n45, 111
Gerber, C. R., 448
Gideo-Frank, Lotham, 392n12
Giere, Ronald N., 16, 18, 21, 22, 32, 34, 273, 623–4
Glick, J. C., 291n15
Gold, Ronald, 71, 432
Goldman, M. B., 467n3
Goode, Erich, 589
Goodell, Rae, 12, 586, 589, 590, 591, 594
Goodfield, June, 591, 595
Gordon, Linda, 200n48
Gould, Stephen, 81–2
Graham, Loren R., 7, 29, 30, 621, 624
Grant, Madison, 173, 174–5, 185, 187, 194, 195, 199
Green, Chad, 327
Green, Richard, 408, 432
Greenberg, Daniel S., 591
Greenwood, T., 450
Guild, Walter R., 544n31
Gurchot, Charles, 323

Haas, Arthur, 534n13
Habermas, Jürgen, 96–7, 98
Hacker, Barton, 536n16
Haeckel, Ernst Heinrich, 81–2
Haggerty, Patrick E., 551n1
Haldene, J. B. S., 197, 457
Hall, Prescott, 187
Haller, Albrecht von, 191–2
Halliday, M. S., 298, 305
Hamilton, Alexander, 152
Hammersla, Joy, 546n33, 547n35
Haraszti, Zoltan, 154
Harriman, Mrs. E. H., 182–3
Harrington, Jean, 538n19
Hart, Gary, 563
Hartley, H., 68n18
Harvey, Joy, 109n39
Harvey, William, 102
Hassencahl, Frances, 194n41, 195
Hatterer, Lawrence, 428–9
Heezen, B., 226, 240
Heinlein, Robert, 539
Heirtzler, J., 234, 235n51, 237
Heisenberg, Werner, 59
Hendy, Val, 589, 591
Herschel, William, 70
Hersey, John, 542
Hess, Harry, 226–8, 233, 240
Hill, Christopher, 105n27
Hirschfeld, Magnus, 385
Hirschhorn, K., 476n14
Hite, Charles, 393n13, 394n14
Hoffmann, Erik, 253
Hofstadter, Laura, 589, 590
Hogben, Lancelot, 197, 545n32
Holles, Everett, 325n38
Holmes, Arthur, 211, 212, 214, 215, 216, 224–5, 240
Holmes, Samuel J., 175, 187
Hooker, Evelyn, 385, 396, 428
Hospers, Jan, 222
Hoyle, F., 71–2
Hozier, J., 477n19
Hrdlicka, Arles, 195–6
Huitt, Ralph, 194
Hume, David, 32, 47
Huntington, Samuel, 284n2
Hutton, Richard, 118
Huxley, T. H., 106, 112, 119

Infante, P. F., 467n4
Ingelhart, Ronald, 284

Irving, E., 221
Isacks, B., 231n45
Ivaniushkin, A. Iu., 251n4, 255

Jacobs, D., 477n19
James, Henry, 302
Jeans, James Hopwood, 59
Jeffreys, Harold, 210, 211, 212, 214, 223, 240
Jennings, Herbert Spencer, 180, 182, 187, 193–4, 196, 197
Jensen, Arthur, 185n32
Johannsen, Wilhelm Ludwig, 181
Johnson, Albert, *see* Johnson Act
Johnson, Rosewell, 173
Joliot, Frédéric, 532–3
Jones, Donald F., 181
Jones, Kenneth MacDonald, 541n26
Jonsen, Al, 612–13
Joravsky, David, 81
Jordan, David Starr, 173, 187

Kaempffert, Waldemar, 539
Kahneman, Daniel, 278, 573n11
Kalogerakis, M., 425
Kameny, Franklin E., 402, 429–31
Kant, Immanuel, 32, 50, 73, 450
Kantrowitz, Arthur, 449, 609
Kaplan, Abraham, 318n14
Kapp, W. R., Jr., 478n27
Karloff, Boris, 537
Katosova, L. D., 477n18
Kay, Marshall, 230
Keith, Arthur, 189–90
Kellogg, Vernon L., 185n32, 187
Kelman, Steven, 525n46
Kemeny, John, 551n1, 585
Kennedy, Donald, 316n4, 327n41
Kesey, Ken, 599, 607
Khrushchev, Nikita, 80
Kilian, D. J., 482–3, 486
Kilpatrick, James J., 325
Kinsey, Alfred, 385, 396, 424
Kirwan, Richard, 68
Klein, D. F., 408
Klein, Melanie, 305–6
Kolb, Lawrence, 387
Komanoff, Charles, 576n19
Korman, N. P., 476n17
Kotin, Paul, 456
Kovach, John S., 356n5, 361n19, 372n74
Kramer, Stanley, 543
Kraus, Sidney, 544n31
Krebs, Ernst, Jr., 320, 323, 331
Krieghbaum, Hillier, 587
Krimsky, S., 274
Kucerova, M., 476n16
Kuhn, Thomas, 50–1, 83–4, 94, 96

Ladd, E., 270, 271
La Follette, Marcel E. C., 588
Lamarck, Jean Baptiste, 108, 109
Lambert, B., 476n14
Lambeth, Edmund B., 591
Landa, K., 476n13
Lankester, E. Ray, 116–17
Lasswell, Harold D., 318n14
Laster, W. R., 365n40, 366n46
Laughlin, Harry H., 173, 174, 179, 180, 182, 185, 187, 194, 195–7, 198, 199, 200
Laurance, William, 540–1
Lavoisier, Antoine, 67, 68
Lawrence, William L., 540n24, 541n25
Leal-Garga, C. H., 477n20
Lee, Robert, 418n2
Lee, W. R., 450
Legator, Marvin S., 18, 19, 478n23, n26
Lehmann, Heinz, 389
Leibniz, Gottfried Wilhelm, 73
Lenz, Theodor, 179
Le Pichon, X., 234, 235n51
Levi, Leo, 365n37
Lewis, Amory, 571n5
Lewis, Carolyn, 551n1, 563n14
Lewis, C. I., 307
Lichtenstein, Sarah, 285n4, 573n12
Lief, Harold I., 435–6
Lilienthal, David, 542, 575
Limoges, Camille, 108n38
Lindsten, J., 476n14
Lipset, S., 270, 271
Lipsett, Mortimer B., 357n6
Littré, Maximilien, 110
Llenado, R. A., 365n39
Locke, John, 344, 345
Lodge, Oliver, 533
Lönroth, M., 581n25
Lovins, L. Hunter, 571n5
Lowell, Percival, 65
Lowrence, W., 285n4
Lubrano, Linda L., 249n1
Ludmerer, Kenneth, 174, 180n22, 191–2, 195n44, 196n45, n46
Ludwig, Karl, 114–15
Lyell, Charles, 70
Lysenko, T. D., 80–1, 113n48

McBride, Lloyed, 551n1
MacCarthy, Mark, 18
McDevitt, Robert, 392n12, 433
Mach, Ernst, 73
MacIntyre, Alasdair, 7, 9n6, 147n30, 611–12

McKenzie, Dick, 231
MacKenzie, Donald, 84n44, 96n10, 97, 99n17, 100n19, n20, 187
McKirachan, C. David, 559
MacLean, Douglas, 21
Macklin, Ruth, 8
McMullin, Ernan, 5, 6, 8, 30, 31, 71n25, 72n28, 74n31, 127n5, 170, 273, 616–17, 618, 620
MacNaughton, A. R. L., 365, 374
McPearson, Harry C., 551n1
Magendie, François, 114
Manner, Harold W., 323n31
Mannheim, Karl, 63n14
Marcuse, Herbert, 98
Markle, Gerald E., 5, 317n8, 320n16, 323n32, 330n46
Marks, Paul A., 551n1, 563n14
Marmor, Judd, 385, 388, 390, 413, 429, 431, 432
Marrett, Cora Bagley, 21–2, 561n9
Marshall, John, 154, 512
Martel, Lisa, 587
Mathiasmeier, Mrs. R. L., 558
Matthew, W. D., 217
Matthews, Drummond, 131, 133, 227–8, 230–7, 241–2
Maugh, Thomas H., 492n14
Mayhew, Henry, 438
Mazur, Allan, 8, 13, 126n3, n4, 134n18, 138n23, 149n32, 265, 268, 270, 444, 450, 586, 588
Mazzocchi, Anthony, 438
Medsger, Betty, 591
Medvedev, Zhores, 81–2
Mehler, Barry A., 183n29
Mehling, Reuben, 544n31
Meinesz, Vening, 230
Melber, Barbara D., 546n33, 547n35
Menard, H. W., 226
Mendeloff, John, 454, 493n18
Mendelsohn, Everett, 7, 13, 83–7, 129n9, 317, 617
Meretoja, T., 477n21
Merriam, Charles, 290
Merrill, Richard A., 488n4
Merton, Robert K., 63, 75, 93
Meyer, John, 408
Mill, John Stuart, 43, 346, 450
Miller, D. C., 65–6
Miller, Robert, 253
Millerd, William, 556
Millikan, Robert, 290, 291, 532, 535
Misner, George, 270–1
Mitchell, Robert C., 569n2
Moertel, Charles G., 329, 356n5, 357n8, 361n19, 372n74, 373, 378
Molina-Ballestros, G., 477n20
Monson, R. R., 447
Moore, G. E., 38
Morgan, Jason, 231
Morgan, T. H., 181, 182, 193–4, 197
Morley, Lawrence, 227–8, 230–7, 241–2
Morrone, John A., 323, 366n42
Moss, Ralph, 328, 330
Mounts, Gregory J., 490n8, 492n16
Muller, H. J., 181, 182, 197
Murray, J., 270
Musgrave, Alan, 68n16

Nackerson, Barton L., 392n12
Nahrstadt, A., 366n41
Navarro, Manuel D., 366n41
Nealey, Stanley M., 546n33, 547n35
Nelkin, Dorothy, 8, 126n2, 130n10, 139n24, 145n27, 164, 274, 287n8
Nernst, Walther, 532
Newell, Guy R., 322n25, 328, 356n4, 361n19, 373, 374n71
Newton, Isaac, 73, 82
Nicholi, Armand M., 392n12
Nisbet, Robert, 535n14
Nixon, Richard, 439, 440
Nordenson, I., 476n11
Nowotny, Helga, 93n1, 96
Nozick, Robert, 346
Nunn, Clyde S., 587

O'Berg, M., 471–2
Oliver, J., 231n45
Omenn, Gilbert S., 18, 136n21, 149n33, 446, 447, 622
Opdyke, Neil, 235–6, 242
Oppenheimer, Robert, 545
Osborn, Frederick, 186, 187, 197–8
Osborn, Henry Fairfield, 186–7, 198, 199
Ostman, Ronald E., 588
Ovesey, Lionel, 428–9, 435
Owen, Richard, 108

Paden, Mary, 589, 591
Page, J. A., 454
Page, Talbot, 497n24
Paracelsus, Philippus, 334
Pascal, Blaise, 310
Passmore, John, 103n23
Pasteur, Louis, 86–7, 89–90, 98, 99, 105–13
Pearl, Raymond, 180–1, 187, 197–8, 199
Pearson, Karl, 100, 122–3, 187
Peel, J. D. Y., 123
Peirce, Charles S., 8
Pennetier, 110n41
Peters, J. M., 447

Petersen, James C., 317n8, 320n16, 323n32, 330n46
Peterson, John, 5
Peterson, Russell W., 551n1
Petrov, A. M., 255–6
Pfund, Nancy, 589, 590
Piaget, Jean, 608
Picciano, D., 476n12, n16, 478n27, 482n28, 483n30, 486n31
Pickens, Donald, 187n35
Pickering, Kay, 557n7
Pigford, Thomas H., 551n1
Pitcock, J. A., 476n15
Pitman, Walter, 235, 236, 237
Plato, 450
Playfair, Lyon, 119
Polivkova, Z., 476n16
Pollak, M., 274
Poponoe, Paul, 173, 198
Popper, Karl, 3, 50, 51, 304, 305, 308
Portney, Paul R., 488n1, n3, 503n31
Pott, P., 456
Pouchet, Félix, 86–7, 89–90, 105–13
Powell, Lewis Franklin, 510, 511–12
Powledge, T., 287n9
Priestley, Joseph, 68, 73
Privitera, James, 369n56
Provine, William, 173n3
Pullin, T. G., 478n26
Pupin, Michael, 290
Putnam, Hilary, 308, 309

Quine, Willard Van Orman, 50, 309

Ramsey, P., 287n9
Rankin, William L., 546n33, 547n35
Rawls, John, 3, 29, 146n29
Rehnquist, William Hubbs, 512
Reitnauer, P. G., 366n41
Relman, Arnold S., 372–3
Reuther, Walter, 441
Rich, Robert F., 8, 13, 16, 126n3, 137n22, 149n33, 327n40, 460, 620–1
Richardson, John, 324, 325
Robbins, Anthony, 450
Robins, Eli, 390n9
Roche, John P., 150
Rockefeller, John D., 183
Rockefeller, John D., Jr., 183–5, 192
Rockey, P. H., 447
Rogow, Arnold, 397n22
Roll-Hansen, Nils, 106n31
Ross, Leonard, 571n5
Ross, Nathaniel, 392n12, 395n15
Roux, Wilhelm, 82
Rovin, Jeff, 537n18, 542n28
Rowe, W. D., 285n4
Royer, Mme Clemence, 109, 110
Royko, Mike, 395
Rubin, David M., 585, 589, 591
Rudwick, Martin, 68–9, 70
Runcorn, S. K., 220–1, 222, 225, 240
Ruse, Michael, 16n10
Rutherford, Ernest, 529–30, 532, 535
Rutherford, Glen, 324, 361–2, 367–8, 371–2

Sabath, Adolf, 195, 196
Sachsman, David B., 585
Saghir, Marcel, 390n9
St. Hilaire, Geoffrey, 108
Salmon, Wesley, 47
Salomon, Jean-Jacques, 161, 164
Sandman, Peter M., 585, 589, 591
Sanotskii, I. V., 477n18
Schalul, F. M., Jr., 365n40, 366n46
Schiefelbein, S., 265
Schramm, Wilbur, 587
Schuchert, Charles, 214, 239
Schwartz, Robert L., 4
Secci, G. C., 477n20
Selikoff, I. J., 476n14
Serwer, Daniel P., 536n16
Shaffer, J., 308
Shaheen, Jack J., 547n37
Shapin, S., 100n19
Shepherd, R. Gordon, 589, 591
Sherwin, Martin J., 540n24
Shiriaeva, A. S., 255–6
Shockley, William, 185n32
Shute, Nevil, 543
Siegelman, M., 409n7
Sievert, D., 478n22
Silezneva, T. G., 476n17
Silverstein, Charles, 389
Simon, U., 478n22
Simpson, Sir George C., 215–16, 240
Simpson, George Gaylord, 217–19, 231–2, 239–40
Singer, Maxine, 257
Skardon, James A., 591
Skinner, B. F., 295–310
Sloan, Douglas, 251n3
Slovic, Paul, 285n4, 573n12
Smith, Adam, 155
Smith, P. G., 476n12
Smith, Richard D., 316n7, 317n17, n19
Smith, R. Jeffrey, 489n7, 491n10, n12
Smith, Robert S., 439, 453–4
Snegur, E. A., 253n6
Snortum, J. R., 418
Snow, C. P., 162
Socarides, Charles, 384, 387, 390, 393, 394, 396, 428–9, 431–5

Soddy, Frederick, 529–30, 532, 533–4
Solomon, Susan Gross, 249n1
Sorse, M., 477n21
Spencer, Herbert, 123–4
Speranza, Gino, 195
Spiegel, John, 394, 397
Spitzer, Robert L., 11, 388–91, 392, 394, 397, 405n3, 413n8, 432–3
Spivak, Jonathan, 364n35
Stalin, Joseph, 80
Stang, Alan, 325n39
Starr, C., 277
Steinbruner, John D., 157
Sternglass, Ernest, 267
Stevens, John Paul, 510, 511, 514, 515, 517
Stevenson, Adlai, 544
Stevenson, Charles, 37, 55
Stewart, Potter, 510, 511
Stich, H. F., 476n10
Stoddard, Lothrop, 173, 195
Stoller, Robert, 404, 433
Summa, H. M., 366n41
Summerlin, William, 605
Suppe, Frederick, 132n14
Swiniarski, Joseph K., 365n40, 366n46
Sydenham, Thomas, 9
Sykes, L., 231n45
Szasz, Thomas, 385
Szilard, Gertrud Weiss, 533n11
Szilard, Leo, 533

Talwani, M., 234, 237
Tasca, M., 366n41
Taylor, Frank, 204
Taylor, Theodore B., 551n1
Teller, Edward, 540, 546
Terkel, Studs, 438
Thistle, M. W., 585
Tomatis, L., 466n1
Tough, I. M., 476n12
Toulmin, Stephen E., 10, 51, 94
Tribe, Lawrence, 56n7, 292
Truang, L., 478n23
Truman, David, 157n4, 194
Trunk, Anna D., 551n1
Tuan, Yi-Fu, 531n7
Tversky, Amos, 278, 573n11
Tyndall, John, 89–90, 98, 99–100, 106, 109, 111–12

Vainio, H., 477n21
van der Gracht, W. van Watershoot, 210–11, 240
Vavilov, Nicolai, 80
Vine, Fred, 131, 133, 227–8, 230–7, 241–2
Viscusi, W. Kip, 500n27
Vissing, Yvonne M., 320n16, 323n30
Volkan, Vamik D., 392n12
Von Ardenne, M., 365, 366n41
Von Magnus, Eric, 496n23
Voth, Harold M., 392n12

Wade, Nicholas, 316n6, 321n18
Wade, Serena, 587
Wagoner, J. K., 467n4
Walker, Robert S., 554n5
Wallace, Alfred, 207
Ward, Robert De Courey, 195
Watanuki, Joseph, 284n2
Watson, James, 132–3
Weart, Spencer R., 20, 530n2, 533n11, 539n20
Weber, M., 75
Wegener, Alfred, 53, 133, 204–18, 239, 240
Wegenfeld, Morton O., 320n16
Weinberg, M. S., 409n7
Weisman, August, 123
Weisner, Jerome, 163
Weiss, Ellyn, 556n6
Welch, Robert, 325
Wells, H. G., 533, 534, 541
West, C. D., 476n15
Westwood, G., 418
Whetham, W. C. D., 532n9
Whewell, William, 70
White, J. H., 68n17
Whitman, Doris, 559
Whitney, Leon, 198
Wigner, Eugene, 575
Willis, Bailey, 211, 214, 239
Wilson, J. T., 242n54
Wilson, Paul, 389
Wilson, Tuzo, 227, 230–1, 233–4
Wodinsky, Isidore, 365n40, 366n46

Yalles, Stanley F., 428
Youns, James Harvey, 315n1
Young, Robert S. K., 4

Zeckhauser, Richard, 499n26, 500n28, n29, 501
Zhurkor, V. S., 476n16
Zudova, Z., 476n13

SUBJECT INDEX

abandonment closure, 6, 13–14, 81–2, 616, 617, 618
abiogenesis (*see also* Pasteur–Pouchet debate), 111, 112
abortion, 40–1, 260–2, 283, 287, 599
About Behaviorism, 301, 303, 309
Académie des Sciences, 86, 87, 107–9, 112
"acceptable risk" (*see also* risk), 275–7
acceptance (Gamson), 331, 332
accountability, 165
acrylonitrile, 472
acupuncture, 359, 360
Ad Hoc Committee against the Deletion of Homosexuality from *DSM-II*, 392–5, 431, 434
AFL-CIO v. *Ray Marshall*, 452–3
alchemy, 530, 531
American Cancer Society, 316
American Eugenics Society (AES), 174, 185, 198
American Industrial Health Council (AIHC), 448–9
American Medical Association (AMA), 316, 320, 356, 387
American National Standards Institute (ANSI), 440
American Opinion, 325
American Petroleum Institute (API), 506–7, 511, 513
American Psychiatric Association (APA), 381–2, 384, 387–97, 399, 417, 430–6
American Society of Hematology, 557
American Textile Manufacturers Institute, Inc. (ATMI), 519–20
amygdalin, 355, 365, 369, 375
amygdalotomics, stereotoxic, 609
amygdalotomy, 36
"Anatomy of a Coverup," 321
antipositivism, 129
Antivivisection and Medical Science in Victorian Society, 116
Applied Eugenics, 173
Armageddon, 531, 535, 542
Armed Forces Institute of Pathology (AFIP), 322
aromatic amines, 466–7
arousal cues, 422–4
arsenic, 443, 466, 469, 476, 493
asbestos, 442, 456, 461, 466, 467, 470, 492, 503, 591
assessment, risk (*see also* risk), 516n28, 517, 520, 524–5
Atomic Energy Commission, 545, 546
Atomic Industrial Commission, 567
auramine, 466, 468
authorities, 318

balancing
  in media, 589, 594
  of moral claims, 606
  requirement, 19, 507, 511–13, 518, 521, 525–6
bargaining (*see also* negotiation; negotiation closure), 21–2
benchmark model, 458–9
benefit–cost/efficiency paradigm (*see also* cost; cost–benefit paradigm), 487, 489, 494–504
benzene, 18, 19, 138, 359–60, 442, 451, 452, 460, 483, 485, 489–94, 498, 505–6, 509, 513, 515, 521, 523, 526
benzidine, 442, 467, 469, 476
beta-naphthylamine, 467, 469
biology, 251–3
Biology of Superiority, The, 180–1
biometrics, 100, 122–3
bis(chloromethyl) ether, 469, 476
N, N-bis(2-chloramethyl) 2-napthylamine, 466

Bitter Wages, 454
Black Lung Benefits Reform Act, 454, 455
blood, circulation of, 102–4
"Blowups Happen," 539
body fluids, analysis of, 477–8
British Medical Association, 117
Brown v. *the Board of Education of Topeka*, 159
Bureau of Social Hygiene (BSH), 183, 185
byssinosis, 518–20

cancer and carcinogens, 441, 456, 465–79, 481–6
Cancer and the Worker, 466
Cancer Control Society, 326
Cancer Foundation, 326
Carnegie Institute, 175, 180, 182, 183, 186, 193, 196–7
causal relationship, 297–300
cell, 474–5, 478
Center for Disease Control (CDC), 322
children and scientific research, 39–40, 599, 601, 607–8, 611
chiropractice, 10
chloromethyl methyl ether, 469, 476
chloroprene, 477
Choice, 325
chrosome-screening, 285–6
circumscription, closure through (Marrett), 551, 563–4, 566
Citizens against Nuclear Danger, 557
civil disobedience, 29
claims making, 322–4, 330–1
Clean Air Act, 446
"close-binding-intimate," 419
closure
  and abandonment, 6, 13–14, 81–2, 616, 617, 618
  biomedical, 260–4
  through circumscription, 551, 563–4, 566
  consensus: Beauchamp, 5, 30, 260–3; Engelhardt and Caplan, 14, 15, 16
  defined: Beauchamp, 27–8; Frankel, 203–4; McMullin, 6, 8, 78–81, 616, 617; Mazur, 268, 271; Mendelsohn, 101–2, 121; Schwartz, 355
  through force, 14
  and the judiciary, 355–62, 367–79, *see also* judicial responsibility and review
  through loss of interest, 13–14
  natural death, 5, 13–14, 31–3, 616, 617
  through negotiation: Beauchamp, 5, 15, 17, 22, 33–46, 260–3, 331, 378, 379, 460, 462; Gamson, 331; *see also* bargaining
  normative, 619–24
  and occupational health and safety, 460–2
  procedural: Beauchamp, 5, 15, 17, 30–1, 34, 46, 151, 153, 260–3, 377, 379; McMullin, 78–9
  and resolution: Giere, 127–8; McMullin, 6, 14, 77–8; Mendelsohn, 101, 121
  sound argument, 5, 6, 9, 12, 14–17, 22, 28–9, 30, 31, 33–4, 42, 130n11, 260–3, 376–9, 460, 606, 608, 620
Closure Project, 3–4, 125, 126n3, 127n7, 128n8, 131n11, n12, n13, 138n23, 170, 171, 243, 616, 623
"cloture," 31
Code of Federal Regulations, 326
coherence theories, 44
coke oven exposure, 441, 442, 446, 489, 491
Commissioner's Report on Laetrile, 336
Committee for Freedom of Choice in Cancer Therapy, 321, 325, 330
Committee of Concerned Psychiatrists, 434
community, 8–9, 11, 49, 52–3, 75–6, 84–7, 88–9, 129
compromise, 156
conceptual incommensurability, 307, 311
conflict, 153–6
consensus, 5, 14, 15, 16, 30, 64, 95, 101, 151, 153, 413, 415, 616, 617, 618
consensus closure
  Beauchamp, 5, 30
  Engelhardt and Caplan, 14, 15, 16
constraint (Gamson), 318, 330
Contagious Diseases Act, 118
continental drift, 132, 133, 203, 207–10, 219–20, 224–5, 227–8, 233, 241
contractionists, 211–12
controversy
  and the community, 49, 52–3, 75–6, 84–9
  defined: Engelhardt and Caplan, 2; Giere, 126–8; McMullin, 51
  epistemic considerations in, 2, 5–6, 170–1
  and ethics, 1, 54–6, 251–3
  and fact, 64–6, *see also* facts
  and individuation, 4–5, 9–10, 11–12, 22, 28
  and the law, 56–8
  mixed, 75–7, 139–40, 359–61, 621, 622
  modes for ending: Beauchamp, 5; Engelhardt and Caplan, 13–16; McMullin, 6
  of principle, 72–5, *see also* principle

controversy (*cont.*)
and public policy, 136–9, 158–61, 246–8
and rationality, 3
role of, 280–1
scientific, 1, 13, 51–4, 59–91, 96–8, 101, 127–8, 246–8, 317, 619
of theory, 13, 64, 66–72, 83, 619
cooptation (Gamson), 319
Copernican debate, 60–1
correctness, 16, 22, 28, 127, 128, 143, 144, 252
cost (*see also* benefit–cost/efficiency paradigm: cost–benefit paradigm), 19–20, 488, 493, 498, 506, 513, 519
cost–benefit paradigm (*see also* benefit–cost/efficiency paradigm; cost), 505, 508, 516, 521–7
cotton dust, 442, 452–3, 455, 505, 518–22, 526
Council on Research and Development, 413
creationism, *see* evolutionist–creationist controversy
Criminalistic Institute, 183–5
Critique, 32
Cruelty to Animals Act, 119–20
cyclophosphamide, 476, 478
cytogenetic analysis, 474–9, 481–6

Day the Earth Caught Fire, The, 543
decision making, 157
decision-theoretic model, 146–7
Delaney Amendment, 488
De motu cordis e sanguinis, 102
deviance, social, 405
Diagnostic and Statistical Manual, Mental Disorders
DSM-I, 384–5
DSM-II, 381, 384–5, 388, 390–4, 397, 399, 401, 413, 417, 426–7, 432–4
DSM-III, 401, 404–6, 410–15
Dialogues, 32
differential susceptability, 446
dilemma, 46
directional studies, 220, 221–3, 225
directives, 42–6
disability, 405
discounting, 496–8
Discours, 102–3
disorder, 382, 384, 389–90, 392, 402–10, 414–15
distress, 405, 406, 407, 433
Dixon-Yates controversy, 545, 546
Dr. Strangelove, 547
Doctrine of Phlogiston Established, The, 68
dose–response relationship, 138n14, 515–16
Double Helix, The, 133
Dow Chemical Company, 481–6
drifters, 206n4, 213, 214, 215, 224, 239–44, 246
Drug Amendments, 349
drugs
federal regulation of, 333–4, 336, 362–4, 368–9
generic, 163
"new," 362–4, 368–9, 371, 377
Du Pont and Company, 471–2

Earth, The, 223
effectiveness, social, 433
efficiency, 489, 495, 498–9, 504, 523
ego-dystonic homosexuality, 11, 16–17, 402, 407, 410–13
Eltanin–19 profile, 235, 242
Environmental Protection Agency (EPA), 488, 489, 492, 553
EPA v. *Reserve Mining Co.*, 503
epichlorohydrin, 476, 478, 483, 485
epistémé, 59
"epistemic," 2, 5–6, 59–61, 83, 88, 170–1
epistemological conflict, 317
equity, 287–8
Escherichia coli, K-12 strain of, 36
ethical advisory committees, 251, 257–8
ethics, 1, 54–6, 145–8, 251–3, 570–1, 604, 607, 608, 612, 623, 624
Ethics and Language, 55
ethyleneimine imine, 478
Eugenical News, 185
eugenics, (*see also* genetics; race hygiene and racial purity), 32, 100, 134, 169, 171, 245
American, 172–5, 181, 182, 188, 197, 201
Eugenics Record Office (ERO), 173, 175, 182–3, 185, 194, 201
Eugenics Research Association (ERA), 174, 185n32
Europe an an Emigrant-exporting Continent, 180
euthanasia, 37
evidence, 3, 6, 7–10, 12, 13, 17, 30, 42, 162, 292, 306, 336, 350, 444, 512–13
"substantial," 363
evolutionist–creationist controversy, 129–30, 287, 358–9
Executive Order
12044, 451
12130, 551–2
12291, 522
expanding universe cosmology, 70, 71
Experimential Study of Evolution, 173
experts and expertise (*see also* moral

experts; "science court"), 142, 160–202, 265–6, 269–71, 273–4, 278, 281, 283–4, 288–9, 291, 293, 340, 357, 363, 364, 396, 439, 444, 449, 461, 527, 578–9, 591

facts, 13, 55, 64–6, 266, 268–71, 273–4, 292, 460, 607, 619
Fail Safe, 547
fairness, 16, 22, 127, 128, 143, 144, 293
fallout, 543–4, 546
feasibility (*see also* lowest feasible level), 445–6, 491, 493, 507–10, 519, 520–2, 526
Federalist Papers, The, 152, 160
Federal Register, 565–6, 602
fertilization, *in vitro*, 283, 287
fetal research and fetal rights, 40–1, 260, 283, 287, 599, 601, 602, 603, 606
Fire Protection Association, 440
"Fitter Family" contests, 174, 185
fixists, 206n4, 207, 209–20, 223, 226, 229, 239–44, 246
Food, Drug, and Cosmetic Act, 334, 362, 363, 364, 367, 369, 370, 371, 372, 379
Food and Drug Administration (FDA), 4, 6, 11, 137, 164, 315, 322, 326–9, 334, 336, 339–40, 343, 349, 356, 364–5, 367–79, 444, 605
force closure, 14
foundationalism, 42, 50
Foundation for Alternative Cancer Therapies, 326
freedom of choice, 288
free market, 155

gay activism, 381–2, 385–9, 394, 397–8, 407, 417, 429–31, 435, 436
Gay Activist Alliance, 430
genetics (*see also* eugenics), 82, 122, 256
geological time scale, 205n3
geomagnetic polarity reversals, 228, 232–3, 235–7, 241
Glossopteris, 207, 217, 219, 239
Godzilla, 542
government, American, 152–3

Handbook of Physiological Experimentation, 116
harm principle, 43
Harvey–Descartes debate, 85–6, 102–5
Health, Education and Welfare, Department of (HEW), 552, 553, 600, 606–9
heart, 102–4
heredity, *see* eugenics; genetics
Heredity and Politics, 457
Heredity in Relation to Eugenics, 178–9
hermatite, 466
Heterogenie, 106
Hiroshima, 540–2, 543
homosexuality, 16, 17–19, 283, 286
  defined: *DSM-II*, 426–7; Marmor, 431; Spitzer, 407; ego-dystonic, 11, 16–17
  historical perspective on, 382–6
  parental relationships and, 419–22
  pathological, 11
Homosexuality: A Psychoanalytic Study of Male Homosexuals, 417–22, 428–9
hydrogen bomb, 543–5

Ideology and Utopia, 63n14
immanence, 63n14
immigration quotas, 174, 179, 180, 191, 194, 245
inducement, 318, 330
induction, 42, 47
infant formula, 163
inference, rules of, 6–9, 12, 13, 17
influence, 318, 322, 326, 329
inherent disadvantage, 407–8, 411
"injury in fact," 57
injury tax, 453–4
insane, institutionally, 599, 601
insulation, 319, 327–8
"intellectual technocracy," 289–90
intentional language, 309–10
Interagency Regulatory Liaison Group, 444
Interfaith Coalition on Energy, 555
International Agency for Research on Cancer (IARC), 474–6
International Association of Cancer Victims and Friends (IACVF), 325–6
Introduction to the Study of Experimental Medicine, 87
intuitionism, 450
investigational new drug license (IND), 363, 364, 374, 375
Invisible Ray, The, 537
in vitro fertilization, 283, 287

Jaramillo "reversal," 234, 235, 236
John Adams and the Prophets of Progress, 154
John Birch Society, 325, 330
Johnson Act, 174n4, 180, 191, 195–6
Juan de Fuca Ridge, 233, 234, 237
judicial responsibility and review (*see also* closure and the judiciary; science and litigation), 56–8, 154–5, 159, 163–4, 355–62, 367–79, 451–2, 510, 526
"justiciable" controversy, 57
justification, 41–2, 45, 47–8, 611–12, 623

knowledge, fear of, 530–1

Laetrile, 4–6, 9, 66, 136–8, 163–5, 245–6, 283, 288, 586, 619
  and closure, 315–16
  federal regulation of, 333–41, 348–53
  and holistic medicine, 323–4
  Indiana statute legalizing, 349–50
  and individuation, 4–5, 9–10, 11, 12, 324–5, 343
  and the judiciary, 316, 327, 352, 361–5, 367–75, 377–9
  political success of, 338–40
  toxicity of, 329, 337, 365–7, 370–1
"Laetrile, the Making of a Myth," 370
Laetrile Working Groups, 322
"Laetrilomania – Again," 372–3
Lamont group, 234–6, 242
land bridge hypothesis, 207, 217
latency period, 467, 472, 513, 523
lead, 446, 455
"legal standing," 57
Leibniz–Clarke controversy, 53, 59
Letter to the Grand Dutchess Christina, 61
libertarianism (*see also* quasi libertarianism), 344–7, 348, 350–3
life, valuing, 499–502, 523–5
logicism, 50
loss of interest closure, 13–14
Love Canal, 588
lowest feasible level (*see also* feasibility), 506, 511, 512, 518, 526
Lysenko affair, 31, 369n56

McNaughton Foundation, 323
Manhattan Project, 539–40
Marbury v. *Madison*, 154
marine geology, 220, 225–30
market metaphor, 21–2
Mattachine Society, 429
Mayo Clinic, 375
media, mass, 585–97
Medical Psychoanalysts, Society of, 418
medicine, 535–6
meltdown, 548, 564, 565
Memorial Sloan-Kettering Cancer Center, 316, 321, 328, 330, 591
Mendelian–biometrician controversy, 134
Mendelian genetics, 175–7, 187, 245
Mental Health Aspects of the Peaceful Use of Atomic Energy, 548
methadone, 37
methodological, 77
  defined, 73
metronidazole, 478
migratory routes, 218
misuse, fear of, 285–6
mobilists, *see* drifters
model, 445, 448
moral dilemma, 29
  moral experts, 41, 274, 277, 280
  moral implications, 286–7
  moral justification, 45
  moral principles, 45–6
  moral values, 55, 136, 246, 268–71, 273, 292, 415, 450, 607, 620, 622, 623

Nagasaki, 540, 541–2
β-naphthylamine, 467, 469
2-naphthylamine, 466
National Academy of Science (NAS), 265, 266, 280, 449, 492
National Cancer Institute (NCI), 316, 320–2, 328, 331, 356, 357, 361, 373–5, 378
National Commission for the Protection of Human Subjects of Biomedical and Behavioral Research, 35–6, 39, 599–611
National Gay Task Force, 394–5, 396, 434
National Health Federation, 326
National Institute of Mental Health, 428, 432
National Institute of Occupational Safety and Health (NIOSH), 450
National Institutes of Health (NIH), 449
National Research Act, 599
National Research Council, 445
National Toxicity Program, 449
National Training School, 299–300
natural death closure, 5, 13–14, 31–3, 616, 617
necrosis, jaw, 537
Need for Change, The: The Legacy of TMI, 552
negotiation (*see also* bargaining), 151–2, 156, 157, 163, 331, 381, 616, 617, 620, 621–2, 624
negotiation closure
  Beauchamp, 5, 15, 17, 22, 33–46, 378, 379, 460, 462
  Gamson, 331
neoplasms, 465, 471
new advantage, 331–2
nickel, 466, 469
niridazole, 478
Nomenclature and Statistics, APA Task Force on, 17, 388–9, 391, 401–2, 412, 413, 431, 432, 433
noncontroversy, 38
nonepistemic factors, 30, 60–1, 170, 244, 617
normalcy, 424–6, 428–9, 432

normative closure, 619–24
nuclear economics, 574–8
nuclear energy, 529, 530, 532–5, 538–9, 540–1
Nuclear Fuel Services, 557
nuclear plant licensing, 275–8, 565–6
nuclear power, 293, 569–70, 572–4
Nuclear Regulatory Commission (NRC), 552, 553, 554, 561, 562, 565, 567
nuclear sociodynamics, 578–81
nuclear weapons, 570–1

Occupational Health Committee, 447
Occupational Research Study Group, 447–8
Occupational Safety and Health Act, 359–60, 439–40, 445–6, 457–8
Occupational Safety and Health Administration (OSHA), 18, 19, 137, 437–8, 440, 445–6, 450, 452–3, 457–60, 489–93, 496, 505–7, 509–13, 515–16, 518–22, 524, 526, 622
oceanic ridges, 226, 227, 229–30
Office of Science and Technology Policy, 444, 449, 450
Office of Technology Assessment, 280, 449
olfaction, 45
One Flew Over the Cuckoo's Nest, 599
On the Beach, 543
ontogeny recapitulates phylogeny, 81–2
ontological, defined, 73
Operation Albert, 543
Origin of Species, 106, 109, 110, 304
Origins of Continents and Oceans, The, 133, 212, 216
Our Wandering Climates, 216
outcrossing, 181

paleogeography, 222
paleomagnetism, 220–5
"Pangea," 204–5, 217
"Paradise or Doomsday?" 541
paraphilias, 406
participation, 319
partisans, potential, 318
Passing of the Great Race, 173, 174–5
Past Climates, 216
Pasteur–Pouchet debate, 86–7, 89–90, 105–13
pedigree, family, 175–6, 183
permanency theory, 207, 217, 218–19
Permo-Carboniferous glaciation, 206, 208, 212–13, 215, 216, 217, 239
persuasion, 318, 319, 330
phlogiston–oxygen debate, 67
Pioneer Fund, 185n32
plate-tectonic model, 132, 217, 231–2
Plessy v. *Ferguson*, 159
pluralism, 153, 156
polar wandering, 205, 206, 208
pole-flight force, 208–9
"policy," 47
political process, 157–8, 620
polity, 152, 162–3
Population Council, 186
positivism, 110, 128–9, 132, 304–5
Power and Discontent, 318–20, 329
President's Commission on the Accident at Three Mile Island, 20–1, 551–66
Price–Anderson Act, 576
"Primacy of Economics over Eugenics, The," 181
Principia, 73
principle, 13, 64, 72–5, 77, 460, 611–12, 624
principles, moral, 45–6
Principles of Biomedical Ethics, 38–9
Principles of Geology, 216
Principles of Physical Geology, 225
prisoners, 46, 599, 600–1, 602–6, 609
probit-log function, 515, 516
procedural closure
  Beauchamp, 5, 15, 17, 30–1, 34, 46, 151, 153, 377, 379
  McMullin, 78–9
property rights, 345–7
prophylactic treatment of preadolescents, 431–2, 433
Psychiatric News, 392, 395, 431
psychopathology, 408, 409, 432, 433
Psychosexual Disorders, *DSM-III* Advisory Committee on, 411, 412, 413
psychosexual dysfunction, 406
psychosurgery, 36, 599, 601, 602, 607, 608–10
Public Scrutiny, 326

quantum theory of matter, 59, 73
quasi libertarianism (*see also* libertarianism), 345–7, 350–3

Race Betterment Foundation, 174, 185n32
race hygiene and racial purity, (*see also* eugenics), 179–81, 200–1
radioactivity, 138n23, 529, 532–6, 538, 542–4, 545
radium, 529, 533, 535–7
Rasmussen Report, 572
rationality, 3, 6–7, 45, 62, 143, 162
"Rationalising Mankind 'Big Business' Methods in Evolution Eugenic Reform," 189–90

Reader's Guide to Periodical Literature, The, 173, 537–8, 541, 547
reasoning, 300–3
recombinant DNA, 30, 36, 76, 140–1, 163, 249–51, 283, 286, 586, 589, 591–2, 594
referendum (APA), 381, 382–96, 434
"Reflections on Meaning and Structure," 303
regulation, 489–90, 506
reinforcement, 297
resolution variables, 331
"revealed preferences," 277
Rising Tide of Colour against White Supremacy, The, 173
risk, 19, 275, 277, 284–5, 287–8, 448–504
  "acceptable," 275–7
  assessment of, 516n28, 517, 520, 524–5
  nuclear, 569–70, 572–4
  significant, 505, 511–12, 514, 518, 520, 521, 526
  zero, 441, 509–10, 521
Roads of Glen Roy, 68–70
Rockefeller Foundation, 183, 186, 193, 198
"Role of the South American Basin in Biogeography and Evolution, The," 216
Royal Society for the Prevention of Cruelty of Animals (RSPCA), 115–19
Rutherford v. *United States,* 356, 361–2, 367, 372, 373, 374–8

Salmonella typhimurium, 478
sanction, 319
science, 49–50, 130–2, 619–20
  and litigation, 357–61, *see also* judicial responsibility and review
  media discussion of, 587–94
"Science court" (*see also* experts and expertise), 135n19, 149n32, 266, 291, 449–50, 607
scientific claims, 130–2
scientific community, 8, 9, 11
scientific controversy, 1–3, 13, 51–4, 59–91, 96–8, 101, 127–8, 162, 246–8, 317
scientific jury (*see also* "science court"), 290–1
scientific methods, 130
scientific racism, 191
scientific technical revolution (STR), 253–6
scientist, cultural perception of, 587, 590–4
"scientist lobby," 119
Scopes trial (*see also* evolutionist–creationist controversy), 359
screening program, medical, 474–86
sea floor spreading, 217, 226–7, 228, 230, 231, 233–4, 241–3
second opinion, 321, 328, 330
"self-management," 301
separation of powers, 56–7
Sherley Amendment, 349
sial, 204, 211, 212
significant risk (*see also* risk), 505, 511–12, 514–518, 520, 521, 526
"silent revolution," 284
sima, 204n2, 211, 212
sister chromatid exchange (SCE) analysis, 477
social control, 318–19, 330
Social Darwinism, 190
social policy costs, 18
social responsibility, 586
Société d'Anthropologie, 109, 110
soot, 456–7, 466, 468
sound argument closure, 5, 6, 9, 12, 14–17, 22, 28–9, 30, 31, 33–4, 42, 130n11, 376–9, 460, 606, 608, 620
Southern Michigan State Prison at Jackson, 602, 603–4, 605
sperm analysis, 478–9
spontaneous generation, *see* Pasteur–Pouchet debate
standard, 507
Station for Experimental Study of Evolution, 175, 182
steady state cosmology, 70–2
Stowe v. *United States,* 361
structural model, 143–5
styrene, 477
Summa Theologica, 32
Supreme Court, 56–8, 154–5, 159, 163–4, 316, 359–61, 490–2, 505, 506, 510–13, 517, 521–3, 599, 608
swine flu immunization program, 589, 591

tars, 466, 468
technological decision, 141–8
technology, 76, 125–50
thalassophilia, 176–7
Thalidomide, 363
Them!, 542
Three Essays on the Theory of Sexuality, 384–4
Three Mile Island, 567, 570, 589, 591, 594, 619
tidal force, 208–9
TMI Albert, 557
Toxic Substance Control Act, 471
toxic substance legislation, 487–8
Traité du monde, 102
Traité elémentaire de chimie, 67–8

transform fault hypothesis, 227, 230–1, 233
transmutation, 530, 531, 534, 535–6
tris(1-aziridinyl)phosphine sulfide (thio-TEPA), 476
trust, 319–22, 330

Union of Concerned Scientists, 556, 567
United Auto Workers, 557
universalism, 93, 146–7, 450
uranium fission, 530, 538
Ur-fear, 570

value ranking, 143–4
values, moral, 55, 136, 246, 268–71, 273, 292, 415, 450, 607, 620, 622, 623
Verbal Behavior, 296
Vine-Matthews hypothesis, 131–2, 133, 228
Vine-Matthews-Morley hypothesis, 227, 228, 230–1, 232–7, 241–2
vinyl chloride, 442, 466, 467, 469, 472, 476
vivisection, 100, 113–121

Walsh-Healy Act, 440
War Game, A, 547
Washington State Prison at Walla Walla, 602
wealth, distribution of, 345–6
weapons, nuclear, 570–1
welfare state, 344, 345
What Is Science? 49–50
Working, 438
World Set Free, The, 533, 534

X-ray radiation, 529, 536–7, 569

zero risk (*see also* risk), 441, 509–10, 521